At Home in Space
The Late Seventies into the Eighties

Ben Evans

At Home in Space

The Late Seventies into the Eighties

 Springer

Published in association with
Praxis Publishing
Chichester, UK

Ben Evans
Space Writer
Atherstone
Warwickshire
UK

SPRINGER–PRAXIS BOOKS IN SPACE EXPLORATION

ISBN 978-1-4419-8809-6   /   e-ISBN 978-1-4419-8810-2
DOI 10.1007/978-1-4419-8810-2
Springer New York Dordrecht Heidelberg London

Library of Congress Control Number: 2011937961

Cover design: Jim Wilkie
Project copy editor: David M. Harland
Typesetting: BookEns, Royston, Herts., UK

Printed on acid-free paper

Springer is part of Springer Science+Business Media (www.springer.com)

Contents

Illustrations

Author's preface

"Never judge a book by its cover", or so the saying goes, and a first glance at the cover of *this* book might leave the spaceflight enthusiast wondering *why* it is emblazoned with a picture from Skylab, taken in 1973, when the subtitle highlights 'The Late Seventies and Eighties'. I ask the reader to forgive me. When I set out to write a five-volume history of humanity's exploration of the heavens, it seemed a big project, though relatively straightforward. Starting with Yuri Gagarin's pioneering voyage in April 1961, the journey through five dramatic decades promised to be an exciting one, with specific breakpoints between the volumes: the resumption of manned lunar landings in the 1970s, the arrival of the Shuttle in the 1980s, the development of the International Space Station in the 1990s and the increased 'privatisation' of getting people into space in the opening years of the present century. My intention was for something a little more complex than a basic log of manned expeditions into space, but as time has rolled on, the project evolved into something much larger and more complex than I had envisaged. It has, therefore, been impossible to track an *entire* decade with each volume. The first volume, *Escaping the Bonds of Earth*, had to take into account some of the advancements of the 1950s as a prerequisite to focusing on 'its' decade, the 1960s. In a similar vein, the second volume, *Foothold in the Heavens*, needed the focus to fall in considerable depth on some of the most remarkable achievements of the Space Age – Apollo 11 being the obvious example – at the expense of covering an entire decade.

Furthermore, I quickly realised that spaceflight was not, and *is* not, a unique phenomenon, outside of public or political control. Rather, it has been an integral part of our social, economic and cultural fabric, and the lightning speed or snail's-pace slowness of its progress through the decades has been increasingly dictated by outside influences: the Bay of Pigs, a mythical 'missile gap' between the Soviet Union and the United States and the Cuban Crisis of October 1962 were all instrumental in determining the course of space policy. In the early 1970s, a progressive thaw in relations between the two superpowers similarly impacted their space programmes, allowing for the genesis of Apollo-Soyuz, but very quickly refroze within a few years, as disagreements over the Helsinki Accords, a resumption of American diplomatic ties with China and the Soviet invasion of Afghanistan drew them back into the icy waters of the Cold War once again. In a sense, the decade or so to be covered in this third volume, *At Home in Space*, from 1973 until 1982, presents a deeply depressing picture: one which began with so much promise for the future, offering not only genuine co-operation in space, but, hopefully, co-operation *on Earth*, as well, but which ended with hostile words of "evil empire" and equally hostile acts of Star Wars and Able Archer. I feel that it would be unconscionable to discuss our progress in space without paying due tribute to *why* we were doing so, the obstacles we had to overcome in order to get there and the opinions, attitudes and feelings of the political masters who controlled the purse-strings for such endeavours.

By the middle of the 1970s, the heady days of Apollo and the lunar landings had given way to an increasingly more frugal attack on the heavens. Astronauts, managers, scientists and even some politicians saw no reason why a manned expedition to Mars and a permanent lunar base should *not* be achieved before the end of the century. It might not be on the scale of Arthur C. Clarke's imaginings, but it was certainly more than just a dream. However, for an increasingly apathetic public in America and a largely disinterested Politburo in the Soviet Union, the costs were excessive. America's efforts switched from the Moon and Skylab to the development of what was advertised as a cheaper, more frequent and more reliable means of getting into space – the Shuttle – whilst Soviet Russia focused on gradually mastering the new frontier through the establishment of near-permanent orbital stations, the Salyuts.

My intention in writing this third volume has been to explore some of the reasons why the political, social, cultural and economic climate changed so markedly for both superpowers in the pivotal decade of the 1970s and the early years of the 1980s. More than three decades later, we continue to live with the consequences of those frugal times and the very *shape* and *size* of many components of today's International Space Station are dictated by the shape and size of a Space Shuttle, whose own shape and size was set, according to military requirements, all those years ago. Even the Russian segments of the station bear more than a passing resemblance to the design of the early Salyuts. However, 'frugal' or not, the period from 1973 until 1982 was a decade in which – far from 'stopping' or even 'stalling' – a new and exciting chapter in space exploration began ... and human beings truly found a new 'home' in space.

Ben Evans
Atherstone, England
March 2011

Acknowledgements

This book would not have been possible without the support of a number of individuals, to whom I am enormously indebted. I must firstly thank my wife, Michelle, for her constant love, support and encouragement throughout the time it has taken to plan, research and write this manuscript. As always, she has been uncomplaining during the weekends and holidays when I sat up late, typing on the laptop, or poring through piles of books, old newspaper cuttings, magazines, interview transcripts, press kits or websites. It is to her, with all my love, that I would like to dedicate this book. My thanks also go to Clive Horwood of Praxis for his enthusiastic support and to David M. Harland for reviewing the manuscript and offering a wealth of advice and guidance; I deeply appreciate not only their support, but also their patience in what has been an overdue project and one which has proven more difficult to write than I had imagined. Additional thanks go to Ed Hengeveld, who has been enormously gracious with his time in identifying suitable illustrations for this book, including many 'unfamiliar' ones which surely bolster the text. Others to whom I owe a debt of gratitude include Sandie Dearn and Malcolm and Helen Chawner. To those friends who have encouraged my fascination with all things 'space' over the years, many thanks: to Andy Salmon and Andy Rowlands and to Dave Evetts and Mike Bryce and to Rob and Jill Wood. Our two golden retrievers – the ever-hungry Rosie and the attention-seeking Milly – have provided a ready source of light relief and a regular opportunity for me to leave the laptop and either play with them or give them a biscuit.

Acknowledgments

1

Unlikely partners

THAW IN RELATIONS

In the bleakness of midwinter, Tyuratam is a desolate and windswept place, characterised by bitter snowstorms and hurricane-strength blizzards. It is a tiny railway junction in central-southern Kazakhstan, 200 km east of the Aral Sea, and in the local tongue its name is roughly translatable as the gravesite of Tyura, beloved son of the great Mongol conqueror Genghis Khan, whose medieval empire spanned much of Asia. According to some sources, it began as an ancient cattle-rearing settlement on the north bank of the Syr Darya River, although at least one Soviet-era journalist has given it a more modern origin, hinting at its foundation in 1901 as an outpost to replenish steam engines passing between Orenburg and Tashkent. Its importance over the last half a century, though, cannot be disputed. It was from this sparsely inhabited expanse of steppe, five decades ago, that the first strides of a journey far more audacious, much longer and considerably harder than any the Great Khan could have foreseen were taken.

It was from this place that Yuri Gagarin, the first man in space, began his historic flight in April 1961, changing our perception of the Universe forever. It was from here that Valentina Tereshkova became the first woman to travel into orbit and Alexei Leonov became the first person to walk in space and Georgi Dobrovolski, Vladislav Volkov and Viktor Patsayev became the first to live aboard a space station. Yet Tyuratam has suffered its fair share of disasters, too, including a series of catastrophic failures which prompted the cancellation of the N-1 Moon rocket. For all of its historic attributes, this remote corner of old Soviet Central Asia – an area swarming with scorpions, snakes and poisonous spiders, whose climate produces vicious dust storms, soaring summertime highs of 50°C and plunging wintertime lows of -25°C – was one of the most secretive, mysterious and closeted places in the world. In fact, even its *name* was kept strictly under wraps, as part of a deliberate effort to mislead and confuse prying Westerners. Today, it is still variously called 'Tyuratam', after the tiny railhead, or, more often, 'Baikonur', which covers a broader and different geographical location to the north-east.

A chink in the almost impregnable armour of this place finally opened in the early 1970s, when the first plans were laid for a joint US-Soviet manned space mission to cement a steady thawing of relations between two old foes. One of the greatest ironies of the last half a century has been that the human space programme owed its very existence to their distrust and fear of one another, yet the thaw led directly to a significant decline in space spending on both sides and drew both away from the nuclear trigger and closer around the negotiating table. Back then, as today, in the minds of the political leadership, space often represented little more than a pawn in a much larger game of international diplomacy. Ten years after Gagarin's flight, the need for a cheaper and more economical means of sending men into the heavens was acute and neither nation could afford to continue signing over limitless blank cheques to fund a series of politically motivated space spectaculars to outdo the other. Nor could they afford to sustain continued hostility: America desperately sought an exit strategy from the bloodbath of Vietnam and Soviet relations with China had reached a tense crossroads.

A series of US-Soviet meetings in Helsinki and Vienna, called the Strategic Arms Limitation Talks, hammered out the details of a chain of agreements to reduce ballistic missile stocks on both sides. One of the later deals, the Anti-Ballistic Missile Treaty, signed by US President Richard Nixon and Soviet General Secretary Leonid Brezhnev in May 1972, permitted each nation only two sites on which to base major defensive systems. A couple of years later, this was limited still further, and for three decades the treaty was regarded as a landmark act in the control of weapons of mass destruction.

By the summer of 1973, détente between the United States and the Soviet Union had reached its high-watermark with the Prevention of Nuclear War Agreement, in which both sides pledged to implement new policies to refrain from hostilities. Indeed, discussions of co-operation in the theatre of space had been ongoing for some time and the exchange of lunar soil specimens and the sharing of biomedical data was already well advanced. The idea of staging a joint *manned* space mission had steadily brightened and was set in stone when the Anti-Ballistic Missile Treaty was signed: it was agreed that a three-man Apollo spacecraft would rendezvous and dock in orbit with a two-man Soyuz, sometime in mid-1975. The ambitious venture, known as the Apollo-Soyuz Test Project (ASTP) in the West, would hopefully bring a degree of closure to the embittered competition which had dominated and despoiled the previous decade.

That sense of closure, though, did little at first to eliminate strong feelings of distrust on both sides. Even when Nixon and Premier Alexei Kosygin – the political head of the Soviet Union – came to sign the ASTP agreement on 24 May 1972, they did so under the equal pressure of brotherly optimism and cynical pessimism; optimism because they both *needed* this co-operative effort and pessimism because both still retained hundreds of ballistic missiles, tipped with nuclear warheads, aimed and primed at each other's major cities. At one stage, Alexei Leonov was taken to one side by Leonid Brezhnev and asked directly if he truly *believed* in the undertaking and whether he thought the two enemies could really work together. Even Nixon himself, speaking from the opulent surroundings of the Georgievsky

Hall in the Grand Kremlin Palace, acquiesced that the whole world would be watching the progress of the joint endeavour and would harbour its own hopes and reservations.

A year later, in the summer of 1973, the two crews for the mission appeared together at the Paris Air Show, where a special pavilion had been erected with a full-scale model of the linked Apollo-Soyuz spacecraft. Enormous photographs were hung of the Apollo team – Tom Stafford, Vance Brand and Deke Slayton – but those of the Soyuz crew, Leonov and Valeri Kubasov, were conspicuously missing. How could these two ideological opponents possibly co-operate, one journalist asked, without a hint of jest, if they could not even get *photographs* hung on time? It may have seemed a flippant question, but it underlined a growing sense that the communist East and the capitalist West were poles apart and simply *too different*, ideologically, culturally, technologically and politically, to work together. Some on each side had convinced themselves that the other side was bent on 'stealing' its 'secrets', but the general view was that each nation had much to learn from the other. For astronaut Dave Scott, who led a negotiating team to Russia in the summer of 1972, there was something else involved: curiosity, pure and simple. In Scott's mind, there would be no underhand spying on the Soviets – the United States was *already* in the lead, having achieved the world's first manned landing on the Moon – but ASTP would represent an exercise in learning more about how the two sides operated, how their respective crews prepared for missions, how they built their spacecraft and, indeed, what those spacecraft *looked* like. Moreover, Scott and others were intrigued by how little was known about this strange, closed society and their appetites were whetted by the opportunity to learn more.

Yet few people observed the mission through rose-tinted spectacles, for there remained a darker and more sinister side. Even in the first half of the 1970s, it was impossible to judge how long this new episode of US-Soviet détente would last ... and, indeed, it would not last for long and would worsen in the early part of the next decade. For ASTP, one of the greatest obstacles was the Soviet lack of transparency; they kept quiet about launch dates – their "tradition", they told the Americans – and this distrust eventually seeped into the ranks of the astronaut corps. When Tom Stafford asked his Soviet counterparts if he could *visit* the Tyuratam 'cosmodrome' and see the training installations and the Soyuz spacecraft with his own eyes, he was met with stiff and stubborn resistance. Even when permission was finally granted, it was obvious that the top-secret site had been specially prepared for them.

"We were loaded into a pair of very plush buses," Stafford wrote, "and driven 30 km north to the heart of the cosmodrome. The work crews ... had spruced up the place. They had even painted the sides of the train rails leading from the assembly building to the launch pad." He saw no uniforms, only a gaggle of crew-cutted personnel in ill-fitting civilian clothes, but it was obvious that Tyuratam was a military base. The Soviet system of running its manned programme as a component of the armed services was totally at odds with America's effort, which operated in full view of the world, under the auspices of a civilian organisation, the National Aeronautics and Space Administration (NASA). Over the years, many have seen the Soviets' aversion to transparency as a sort of 'reverse psychology', whose origins ran

In the same way that the Soviets rationalised their excessive secrecy as simply a tenet of their 'tradition', virtually every step taken by a cosmonaut crew at Tyuratam is heavily symbolic and respectful of those who have gone before. This was certainly true of the historic ASTP flight. Here, Alexei Leonov (right) salutes and asserts the full readiness of himself and Valeri Kubasov (left) to complete their space mission. There were other traditions, too. In the Soviet era, each crew signed their names on their hotel room doors, watched the patriotic film *White Sun of the Desert*, drank a little champagne ... and, wrote Leonov, even "pissed on the wheel of the bus". This was in respectful homage to their fellow countryman and friend, Yuri Gagarin, who had famously done the same before his pioneering launch in April 1961. Many of these traditions survive to this day.

back to the erroneous perception by the Americans of a 'gap' in missile-building technology in the late 1950s and early 1960s.

In essence, the Soviets *knew* that they held no major technological advantage over the United States. They knew that America's miniaturised computing technology was ahead of their own and they knew that infighting between their design bureaux and political manoeuvring from above made it a constant struggle to sustain an effective, rational space programme. The answer: to reveal *absolutely nothing* and to allow Western paranoia to do the job for them. When one considers the psychological impact of this cloak-and-dagger stance on the West, the absolute lack of reliable information allowed unfounded stories to grow, unanswered questions to fester and unsubstantiated rumours to linger. In the 1960s, the Soviets cleverly played this situation to their advantage; for all the mystery which surrounded, for example, their N-1 lunar rocket, it actually turned out to be a mere shadow of its American counterpart, the Saturn V.

Now, in the early 1970s, Tom Stafford found that 'becoming' more transparent after so many years of secrecy would be easier said than done. Months before his visit to Tyuratam, he had approached Konstantin Bushuyev, the deputy chief designer of the prestigious TsKBEM spacecraft bureau and head of the Soviet side of ASTP, with a few general questions about the Soviet programme.

"Where do you build your spacecraft?"

"In one of our factories." *That* was the entire answer.

Détente, it seemed, may have been formalised in the signatures of Nixon and Kosygin, but full openness was sadly lacking. Difficulties aside, it remains quite remarkable that many astronauts and cosmonauts – who had trained to fly fighter jets against one another a few years earlier – would actually cement genuine friendships, some of which have endured to the present day. Visits by Soviet officials and cosmonauts to the United States and vice versa were warm and cordial. Astronaut Gene Cernan once entertained cosmonauts Georgi Beregovoi and Konstantin Feoktistov, taking them to Sea World, to a San Diego Chargers football game, to Disneyland, to Hollywood and to the North American Rockwell plant in Downey, California, where the Apollo spacecraft was being built and tested. Hot dogs and pizza and meeting Clint Eastwood and Frank Sinatra underlined an immense cultural gulf between East and West, but by no means implied that friendships could not be cultivated. Cernan remembered Beregovoi, for instance, as a jolly bear of a man, who immersed himself into American culture, even donning a stetson and riding a bucking bronco on one occasion. On another, Alexei Leonov recounted visiting a liquor store to swap a bottle of highest-quality Stolichnaya vodka for a couple of crates of beer for his thirsty American 'comrades'.

In many cases, the cosmonauts' immersion in the lifestyle of a communist society did not detract from the reality that they had much in common with the astronauts – they were, after all, accomplished pilots and engineers, good-humoured and generally at ease with public speaking and, in Deke Slayton's words, "all the other political crap we had to go through". On the other hand, however, the Soviets were hardly ever without their KGB 'minders' and hardly ever left alone with the Americans. 'Hardly', that is, because ways were periodically engineered to sidestep this obstacle. On one occasion, in the United States, a private flight was arranged with 'just enough' seats for the astronauts and cosmonauts, but *not* for the minders, to join an antelope hunt in Wyoming. On another, in downtown Moscow, Slayton and Vance Brand managed to lose their 'tails' in the crowded throng of the GUM department store and sneak themselves a little free time. "It was like a big game," Slayton later wrote.

Notwithstanding the diplomatic irritations, many people within the Soviet Union were in favour of making overtures of peace and conciliation and engaging in the pursuit of co-operation with the United States. "We had suffered so greatly in the Second World War," explained Alexei Leonov. "Most were sick and tired of confrontation. People still remembered the time when American and Soviet troops had met at Torgau, on the River Elbe, in the spring of 1945. There was nostalgia for this time when our two countries stood together against the common enemy of Nazi Germany."

So it was that on the morning of 2 December 1974, when a pair of cosmonauts arrived at the base of the very same launch pad from which Gagarin and Tereshkova and Leonov and a dozen other 'star sailors' had departed, Tyuratam and the Soviet Union seemed just a little less mysterious and menacing than it had been for earlier missions. Anatoli Filipchenko and Nikolai Rukavishnikov had both travelled into orbit before, but on *this* mission, Soyuz 16, their six days aloft would be tracked and monitored extensively by their old adversary, the United States. It represented no dramatic breach of security, for Filipchenko and Rukavishnikov would be wringing out systems and procedures for ASTP, a mission which would have been unheard of a decade earlier. Twelve years before, the two superpowers had been literally on the brink of open nuclear conflict, their leaders exchanging words of threat and warning over international flashpoints like Cuba and Berlin. Now, though, relations were decidedly more cordial.

As already mentioned, when one considers this new reality, it is ironic that the very reason for the speed and tenacity with which both superpowers had pursued their space dreams in the 1960s was mutual fear and distrust. It represented a political engine which was used by President John Kennedy and Premier Nikita Khrushchev to score cynical points over the other. Kennedy, who had assumed the United States presidency in January 1961, had engineered a bold plan to land a man on the Moon, precisely because America was fearful of a gap between itself and the Soviets in missile technology ... a mythical gap whipped up by John Kennedy during his presidential election campaign. From Khrushchev's perspective, the need to maintain a pretence that the Soviets *were* ahead in the technological arena led him to advocate a continuous stream of space 'spectaculars': joint flights, remaining in orbit for days at a time, followed by a woman in space, then a three-man crew, then a spacewalk, each designed to contribute to an ideological advantage over the capitalist West. Thus it was five decades ago and it has been ever since, for short-sighted heads of state have continued to use space as a political tool, with each speaking honourable words of 'exploration' and 'peace' as their driving force, but each having one primary motivation: technological supremacy in the heavens. Even complaints from key politicians over the retirement of the Space Shuttle – and the 'gap' in manned launching capability which is expected to ensue – have mainly focused their argument on the perception that the United States will 'lose' its sense of primacy in low-Earth orbit. 'Primacy' in the 1960s and 1970s went hand-in-hand with the armed services and at least two Soviet space stations were devoted to exclusively military tasks, whilst America came close to creating its own. Moreover, the very *design* of the Space Shuttle – a vehicle whose presence has dominated the last three decades – was dictated by the need to satisfy one of its key customers: the military.

Moving from the later 1970s into the early 1980s, the distrust between the United States and the Soviet Union had intensified once more and much of the ground covered by ASTP was lost. President Ronald Reagan's tough words of an "evil empire" would spawn Star Wars and a renewal of plans to take systems of mutual assured destruction – the ironically-named 'MAD' – into the final frontier. More recent efforts of working together in space, through the Shuttle-Mir project in the

1990s and the International Space Station of today, have been deeply and distinctly rooted in politics and so too was ASTP entirely cemented by the political climate of its day. In the weeks before Anatoli Filipchenko and Nikolai Rukavishnikov launched aboard Soyuz 16, conversations between American and Soviet flight directors and senior managers were friendly, yet guarded. In *The Partnership*, a seminal study of ASTP, Edward Clinton Ezell and Linda Neuman Ezell noted that the Soviets told the Americans about Soyuz 16 several months before it flew and even offered to provide advance notice of its launch ... but on one condition: that NASA should reveal nothing to the press until the flight was underway.

In the West, this was utterly against the grain for the civilian space agency, which had long prided itself on its openness and transparency. In May 1961, America's first astronaut had been launched into space in the full glare of international publicity; a gutsy move which had won glowing praise and given the United States the moral high ground over the notoriously secretive Soviets. Then, in the late summer of 1974, the choice on the table was stark: the Americans would be given Soyuz 16's launch date and time, five full days before it was due to occur, together with its orbital parameters ... in return for absolute secrecy. This would represent a clear break with tradition, which NASA was not prepared to accept. In October, veteran flight director Glynn Lunney, the head of the American side of ASTP, telexed Konstantin Bushuyev and told him that NASA would prefer to know nothing about Soyuz 16 until it was in orbit, rather than breach its integrity with the press.

"We appreciate the Soviet desire to make their own announcement of ... [the] launch," Lunney told Bushuyev. "However, because of our own involvement in this activity, we would find ourselves in a difficult position if we could not report this information to our press. Therefore, we prefer to receive no information in this case, until you have released it or we can release it." As a result, at 6:35 am Central Standard Time on 2 December, Vladimir Timchenko, one of the senior Soviet managers, called NASA's Johnson Space Center in Houston, Texas, and asked for Lunney. The security guard who took the call told him that Lunney was not yet in his office and a second call was made a couple of hours later. Soyuz 16 had been launched at 12:40 pm Moscow Time (3:40 am in Houston) and a complex ballet of tracking of the spacecraft's orbital parameters began. "It was a surprise in some quarters," wrote space historian Phillip Clark, "that the launch took place so early ... because the joint flight [scheduled for mid-July 1975] called for a launch of the Soyuz at 12:20 [pm] GMT [or 3:20 pm in Moscow]." However, Clark continued, this could easily be explained by the Soyuz landing constraints, which were partly governed by lighting conditions in the prime recovery area. A summertime flight by the 'real' mission meant that local sunset would occur late in the evening and for a given mission duration the launch could come later in the day. In the case of the wintertime flight of Soyuz 16, on the other hand, sunset occurred much *earlier* in the day and in order to retain approximately the same landing conditions, an *earlier* launch was needed.

For the first time, the degree of information flowing from the Soviet Union was more than just a trickle and a clear picture emerged about the procedures leading up to a Soyuz launch and the activities of the cosmonauts in their first few hours in space. Clad in their pure white suits, Filipchenko and Rukavishnikov boarded Soyuz 16 about two and a half hours before liftoff and commenced their chores: the former, the commander, inspected the ship's spheroidal 'orbital module', whilst the latter, the flight engineer, checked the bell-shaped 'descent module'. At length, after verifying that all systems were functioning properly, Filipchenko joined Rukavishnikov in one of two couches in the descent module and waited for launch. The orbital and descent modules were the only habitable portions of a spacecraft which has currently operated for more than four decades and transported over a hundred crews of men and women into the heavens.

Soyuz was the brainchild of Sergei Korolev, the famous 'Chief Designer' of early Soviet spacecraft and rockets, with the original intention of undertaking both Earth-circling missions and lunar ventures to rival the United States' Apollo effort. As early as 1964, the design and definition of Soyuz was well underway and technical documentation and a mockup revealed it as a craft capable of lofting two or even three cosmonauts. Even its *name* was no accident: for the Soviet Union's official moniker – *Soyuz Sovietskikh Sotsialisticheskikh Respublik*, the Union of Soviet Socialist Republics – was often popularly known amongst its citizenry as 'Soyuz' (the Union). Therefore, the name of the spacecraft not only reflected its role in supporting rendezvous and orbital stations, but was also a highly symbolic and political statement. In *Challenge to Apollo*, a history of the early Soviet manned space programme, Asif Siddiqi noted that when Korolev first saw the mockup, he proudly declared that Soyuz was "the machine of the future".

With the exception of his vague title, it was not until long after his death on a hospital operating table that the world learned anything of substance about Korolev. Yet this man of outstanding engineering genius had masterminded some of the most remarkable triumphs in the exploration of space. It was he who had designed the R-7 intercontinental ballistic missile which launched Sputnik – the world's first artificial satellite – and later Yuri Gagarin. It was his Kaliningrad-based design bureau, OKB-1 (later renamed TsKBEM and, in the early 1970s, Energia), which assembled the first piloted spacecraft, Vostok, and it was he who oversaw the first three-man orbital mission, the first spacewalk and his nation's first (and ultimately fruitless) steps toward the Moon. His brilliance and unwavering devotion to a lifelong dream of exploring space was balanced by an all-or-nothing obstinacy which often manifested itself in a violent temper, capable of exploding without warning. Korolev lived a hard, thankless life of service to the Soviet state and it was this, ultimately, which consumed him.

Born in 1907 in the central Ukraine, his interest in aviation and rocketry emerged at a young age. Under Joseph Stalin's regime, with its ingrained fear of the power of the individual, there had been little opportunity for the *intelligentsia* to prosper and, as a highly regarded engineer, Korolev quickly found himself arrested and sentenced to ten years of hard labour in the notorious Kolyma gulag. The Nazi invasion of the Soviet Union in 1941 prompted his release to support the war effort. Later, he set to work developing an arsenal of rockets and missiles, which he hoped might someday ferry instruments into the high atmosphere and, finally, into space. His masterpiece, the R-7, though principally intended for the Soviet military as an intercontinental ballistic missile, would be used to put the first satellites and humans into orbit. By giving it this dual-purpose use, he displayed a trait of his canny character, keeping his military critics quiet by satisfying their needs in parallel with his own.

In the early 1960s, the regime of Nikita Khrushchev, who succeeded Stalin on the latter's death, was generally supportive of Korolev and his projects, which yielded a regular delivery of space 'firsts' that the feisty and erratic Soviet premier could use to enforce an ideological advantage over the United States. Khrushchev was far more interested in the glamour, political and military impact of spacegoing rocketry and, to an extent, this was fine with Korolev because it provided him with ready supplies of manpower and funding to pursue his space ambitions. It remains a pity, though, that he never received the recognition he deserved in life. After Gagarin's triumphant flight in April 1961, the Chief Designer was barred from publicly wearing his medals and even had to thumb a lift into Moscow when his car broke down. Efforts by the Nobel Prize Committee to establish an award for this unknown man similarly fell on deaf ears.

It is bitterly ironic, therefore, that his untimely death during a routine stomach operation in January 1966 should have finally uncovered something of the mysterious Chief Designer as a real person. Within the ranks of the cosmonaut corps, his death was immediately recognised for the calamity that it was: Yuri Gagarin, in a solemn eulogy, described Korolev as being "synonymous with one entire chapter in the history of mankind" and Khrushchev's successor, Leonid Brezhnev, was one of the pallbearers who carried his ashes for interment in the Kremlin Wall.

Many cosmonauts felt that the men who followed Korolev – his deputy, Vasili Mishin, together with Georgi Babakin, Vladimir Chelomei and the famous rocket engine designer Valentin Glushko – exhibited entirely different personalities which damaged the Soviet Union's chances of beating the Americans to the lunar surface. Had Korolev lived just a few years longer, wrote Alexei Leonov, "we *would* have been the first to circumnavigate the Moon". Korolev had always described Voskhod 2 – the mission during which Leonov became the first man to walk in space – as his life's last great work. However, remarkable as that achievement was, it was the development of Soyuz which has had the most long-lasting impact on the world. Since its first manned flight in April 1967, Soyuz and a modified version of the original R-7 continue to be used operationally today; a fitting legacy to an enduring talent.

Phillip Clark has traced its history back to a three-part 'Soyuz complex' – a manned craft, a 'dry' rocket block and a propellant-carrying tanker – which Korolev had envisaged in the early 1960s being assembled in low-Earth orbit to fly circumlunar missions. The first part, which Clark identified as 'Soyuz-A', but which the Soviets catalogued as 'Soyuz-7K', was closest in physical appearance to the spacecraft which actually flew and it was to the construction of this that Korolev committed OKB-1 in March 1963. Measuring 7.7 m long, Soyuz-7K had three components: a cylindrical 'orbital module', a bell-shaped 'descent module' for the crew and a drum-like 'instrument module' for manoeuvring equipment, propellant and electrical power. According to Korolev's earliest blueprints, it weighed around

6,450 kg, but, unlike the final design, was not equipped with solar panels, relying instead upon chemical batteries.

Supporting Soyuz-7K were the 'dry' Soyuz-B rocket block and the Soyuz-V propellant tanker, known to the Soviets by the designations of '9K' and '11K'. Clark hinted that a typical flight profile would have begun with the launch of a 9K, followed, at 24-hour intervals, by as many as four 11Ks, which would dock, transfer their propellant loads and then separate. When the 9K had been fully fuelled, a manned 7K would be despatched to dock with the rocket block. "Mastering rendezvous and docking operations in Earth orbit may have been one of the primary objectives of the Soyuz complex," wrote Asif Siddiqi, "but the incorporation of five consecutive dockings in Earth orbit to carry out a circumlunar mission was purely because of a lack of rocket-lifting power in the Soviet space programme." In fact, it was the sheer 'complexity' of the Soyuz complex which seems to have foreshadowed its restructuring sometime in 1964 and effected a delay of its maiden voyage until at least the spring of 1966.

By the end of the decade, seven manned Soyuz spacecraft would have rocketed into orbit. However, a key physical difference between these vehicles and the original 7K was that they employed a pair of large rectangular solar panels, mounted on the instrument module, to generate electrical power. The total surface area of these wing-like appendages was 14 m^2, with each wing measuring 3.6 m long and 1.9 m wide. The remainder of the craft's design was strikingly similar to the 7K: a spheroidal orbital module, 2.65 m long and 2.25 m wide, the bell-shaped descent module, itself 2.2 m long and 2.3 m wide at the base, and the instrument module, a cylinder 2.3 m long and 2.3 m wide.

This shape emerged at the end of almost a decade of planning, theoretical work and aerodynamic modelling. As early as 1958, Mikhail Tikhonravov and Konstantin Feoktistov, both engineers at Korolev's bureau, envisaged a multi-purpose craft capable of both Earth-orbiting and circumlunar missions. Space historians Rex Hall and Dave Shayler have noted that the shape of the descent module was decided at least partly by a desire to touch down on land, rather than in water, and several designs were sketched out. The first utilised aerodynamic surfaces, facilitating an aircraft-like return to a runway, whilst the second adopted a 'missile principle', entering space in a ballistic manner and descending beneath parachutes. By 1961, concerns about mass and the need for adequate thermal protection during re-entry had eliminated the winged design from consideration. The missile principle, though, needed further work to man-rate it: a ballistic descent would impose significant duress on the vehicle and its occupants and Tikhonravov and Feoktistov moved instead toward the concept of a 'glancing' re-entry to reduce stress. If the new craft was ever to undertake lunar flights, its return trajectory from the Moon would produce correspondingly higher re-entry speeds of perhaps 40,000 km/h, prompting the engineers to design a 'double-dip' profile, which, by reducing the velocity in stages, would lessen the G loads on the cosmonauts.

When consensus had been reached on the method of re-entry, OKB-1 engineers and researchers at the NII-1 and NII-88 aerodynamic institutes explored a trio of designs: one nicknamed the 'segmented sphere', another called the 'sphere with a needle' and a third dubbed the 'sliced sphere'. The segmented version emerged as the most promising design, with Vladimir Roshchin's group at OKB-1 promoting a descent module with a displaced centre of mass as a means of generating aerodynamic lift. By 1962, this had evolved into a shape approximating a car's headlamp, which aerodynamic simulations predicted would avoid the high deceleration and thermal loads of a ballistic descent and have sufficient lift to be able to steer towards a given landing site. A plethora of proposals also surrounded the means of landing, with helicopter-like rotors, fan-jet or liquid-propelled engines, controlled parachutes, ejection seats and shock-absorbing inflatable balloons all being considered. By 1963, however, Korolev had approved the design which remains in use today: a combination of braking parachutes and a soft-landing apparatus of solid-fuelled rockets.

Even as the descent module was taking shape, the appearance of the spacecraft remained somewhat fluid and early designs for a space station ferry and a lunar-going concept both utilised a descent module for the crew, attached to an instrument module for propulsion and power. Already, the design was expanding further to encompass a habitable orbital module and there was disagreement about where this should be located. In some initial drawings it appeared *between* the instrument module and the descent module and in others it was *above* the descent module. The idea of placing the orbital module below the descent module was soon rejected, since it would require cutting a hatch into the descent module's base, potentially compromising its heat shield. The final layout, with the descent module in the middle, was in place by the end of 1962. By this time, it had also received the name of 'Soyuz' ('Union').

In spite of Korolev's assertion that it was the machine of the future, Soyuz had been mired for some years in technical and bureaucratic problems, to such an extent that by 1964 its development was virtually paralysed by the Soviet drive for the Moon. Early plans called for it to carry one or two cosmonauts, but by December 1963 the basic design of the Earth-circling version, known as 'Soyuz-7K-OK' ('Orbitalny Korabl' or 'Orbital Ship'), had grown to accommodate a three-man crew. Its purpose was to support automated rendezvous and docking, spacewalking, manoeuvring and scientific research, thereby fulfilling the key requirements for a space station ferry.

During 1964, Korolev directed a small group under Boris Chertok, one of his deputies at OKB-1, to explore other uses for the basic 7K-OK craft. One proposal called for docking two Soyuz together in orbit to demonstrate their rendezvous capabilities and having a cosmonaut spacewalk from one ship to the other. Not only would this ambitious plan offer valuable engineering experience, but it also supported early ideas for a Soyuz-based Moon mission in which a cosmonaut would transfer from the command ship to the landing craft in lunar orbit by 'extravehicular activity' (EVA). In February 1965, Korolev presented this 'new' version of Soyuz, with an emphasis on near-Earth operations, to the Scientific-Technical Council of the State Committee for Defence Technology and was told to proceed.

Beginning at its base, the instrument module, also known as the 'service module', carried chemical batteries and two large solar panels to charge them, together with a thermo-regulation radiator and an integrated propulsion and attitude-control system. The latter, designated 'KTDU-35', comprised a pair of engines, one primary and one backup, sharing the same oxidiser and fuel supply. The primary engine had a thrust of 417 kg and was capable of a change in velocity of some 2,750 m/sec, equivalent to a specific impulse of around 280 seconds. On the basis of early reports, which speculated that this engine could boost Soyuz to an altitude of 1,300 km, Phillip Clark suggested that the spacecraft required a propellant capacity of 755 kg. Propellants took the form of unsymmetrical dimethyl hydrazine and an oxidiser of nitric acid, housed in spherical tanks within the instrument module. Attitude control came from 22 primary and eight backup hydrogen peroxide thrusters. Guidance, rendezvous, communications and environmental gear filled the remainder of the cylindrical compartment.

The descent module sat directly above the instrument module and housed the crew during ascent and re-entry. It had a habitable volume of some 2.5 m^3. The commander's seat was located in the centre, flanked by positions for a flight engineer and a research cosmonaut or 'test' engineer. Many of Soyuz' flight regimes were pre-programmed from the ground. Consequently, the main instrument panel presented the crew with readouts and visual displays of the performance of on-board systems, together with a monitor for the external television camera, an optical orientation viewfinder called 'Vzor' ('Visor') for attitude manoeuvres and the 'Globus' device to show the spacecraft's position above Earth. In the event of a failure of the automatic systems, and to facilitate rendezvous and docking, it was expected that the commander could assume manual control. As a result, two hand controllers (one for velocity, the other for attitude) were located directly underneath the instrument panel.

Rendezvous and docking were supported by the Vzor, together with a system of gyroscopes, attitude-control sensors and thrusters and the 'Igla' ('Needle') radar. The latter would automatically navigate the spacecraft to its target and draw to a halt at a range of 200-300 m, after which the crew would take charge and accomplish the final approach and docking. The systems to facilitate physical contact had undergone extensive development since 1962. At first, OKB-1 engineers Viktor Legostayev and Vladimir Syromiatnikov advocated a 'pin-cone' device to allow two vehicles to dock. At this stage, however, there was no provision for an internal transfer of cosmonauts from one craft to the other and, sometime in 1965, Korolev's

The R-7 rocket for Soyuz 19 – the Soviet half of the ASTP mission – undergoes final preparations for launch in July 1975. Note the closed 'petals' of the tulip-like support structure and the four tapering strap-on boosters. Originally designed by the canny and brilliant Sergei Korolev to satisfy both his military critics and his own dreams to someday send instruments beyond the atmosphere and into space, the R-7's descendants still endure to this day...and it is renowned as one of the world's most reliable launch vehicles. Only days before the 50th anniversary of Yuri Gagarin's historic flight, another of these rockets boosted cosmonauts Alexander Samokutyayev and Andrei Borisenko and their American crewmate Ron Garan towards the International Space Station.

proposal to change this was rejected by Feoktistov on the basis that a significant amount of work had already been done and additional revisions would put the development further behind schedule. The docking system featured a pin on the active spacecraft, which would be captured by a cone-like funnel on the passive one, essentially cancelling any remaining velocity or angular displacement.

The descent module would be the only component capable of surviving the intense heat of atmospheric re-entry and bringing the cosmonauts back to Earth. At the end of a mission, the instrument and orbital modules would be jettisoned and the descent module would employ half a dozen hydrogen peroxide engines, each producing a thrust of 10 kg, to provide roll, pitch and yaw controllability during the early stages of re-entry. To protect its occupants, it was coated with a heat-resistant ablator, together with a thermal shield at its base that would detach shortly before touchdown to expose the four solid-propellant landing rockets. A 14 m² drogue parachute would deploy 9.5 km above the ground in order to stabilise the craft, prior to deploying the main canopy. If a problem occurred, a secondary canopy could be deployed. Seconds before touchdown, an altimeter would command the landing rockets to fire to cushion the impact. Atop the descent module in space, the spheroidal orbital module held a bunk, a cupboard for food and water, life-support gear, controls for experiments, cameras and a variety of other equipment appropriate to each individual mission.

Difficulties aside, Soyuz promised to be one of the safest manned craft ever built, possessing as it did the Soviets' first 'true' launch escape system. This consisted of a tower atop the R-7's payload shroud and a multiple-nozzle, solid-fuelled rocket engine. In the event of an emergency during the period from 20 minutes before launch until about 160 seconds into the ascent, the shroud would split at the base of the descent module and the escape tower's engine would lift the descent and orbital modules to safety. At the top of the arc, the descent module would be released to parachute back to Earth, landing a couple of kilometres from the pad. Early predictions estimated that the crew could be exposed to acceleration loads as high as 10 G during such a scenario.

Launching Soyuz, which was considerably heavier and more complex than the earlier Vostok craft, demanded further improvement of Korolev's original R-7. The basic design of the missile, physically, remained the same: a two-stage behemoth, fed by liquid oxygen and a refined form of kerosene known as 'Rocket Propellant-1' (RP-1). Strapped around its lower stage were four tapering boosters, each 19.6 m long. The upper stage had an upgraded engine which enhanced its thrust from 27,210 kg to 27,573 kg. With the escape tower in place, the upgraded R-7 stood 49.3 m tall, and gave 411,650 kg of thrust at liftoff. This 3 percent increase over the earlier Vostok version enabled it to insert a 6,900 kg payload into a 200 x 450 km orbit. Like Vostok before it, the R-7 was rolled to the launch pad horizontally on a railcar; a method still used today. Four cradling arms, known as the 'tulip', supported the booster and a pair of towering gantries provided pre-launch access. In the small hours of 2 December 1974, the 15th team of cosmonauts to fly a Soyuz, boarded their craft through a side hatch in the orbital module and dropped into their seats in the descent module.

Originally designed for a crew of three, Soyuz had effectively become a two-person vehicle after the deaths of cosmonauts Georgi Dobrovolski, Vladislav Volkov and Viktor Patsayev in June 1971. During the spacecraft's development, Korolev felt that wearing a pressurised space suit would be just as uncomfortable and impractical as wearing a wetsuit inside a submarine and opted to do away with them. It was a fatal decision; a 'normalisation of deviance' which would return to haunt the Soviet manned space programme, for during the re-entry of Soyuz 11 a pressure valve inadvertently opened and Dobrovolski, Volkov and Patsayev died when their air leaked out of the cabin. In the wake of the tragedy, the valves were modified and it was decreed that, in future, space suits would be worn for *all phases* of a mission in which depressurisation was a possibility.

In response to this requirement, a 'Sokol-K' ('Space Falcon') suit would be tailored for each cosmonaut and was compatible with the seat liners aboard Soyuz. A prototype was completed within weeks of the disaster and by the spring of 1972 had been fully tested and signed off as flight-ready. Since the Soyuz 12 mission in September 1973, the suit and its descendents have been worn by every cosmonaut during launch, docking, undocking, re-entry and landing. "In the event of decompression," wrote Rex Hall and Dave Shayler, "the [Sokol] is automatically isolated from the cabin environment and supplied directly with either pure oxygen or an oxygen-rich mixture from a supply in the cabin or from self-contained systems." It included a soft helmet which could be pushed back over the head when not in use, a removable, white-topped 'skull-cap' for communications headgear and pressure-sealed gloves. The Sokol could also be used in the emergency transfer of cosmonauts from one spacecraft to another, with the aid of small hoses connected to the spacecraft's life-support system or through a portable backpack, although this has never been done. Testimony to its success is that, since 1971, no other cosmonaut has lost his or her life through the decompression of their spacecraft; indeed, the hardware has proven so reliable that there have been no other instances of depressurisation, *at all*, aboard a Soyuz. However, in order for the suits to be properly accommodated in the confines of the spacecraft, the third crew seat – that of the research cosmonaut or test engineer – was eliminated and its place taken by a system which could automatically pump air into the cabin in the event of decompression. Not until November 1980 and the arrival of an upgraded version of Soyuz would another three-man crew venture aloft.

Commanding the two-man Soyuz 16 was Colonel Anatoli Vasilyevich Filipchenko of the Soviet Air Force, one of a new generation of cosmonauts who combined advanced flying skills with correspondingly advanced engineering credentials. Born in the village of Davydovka, in the Voronezh region, close to the border with Ukraine, on 26 February 1928, he entered a specialised air force school after secondary education and received his first taste of flying. Within three years, in 1950, he completed the Kharkov Military Aviation School of Pilots and in 1961 graduated via correspondence from the Soviet Air Force Military Academy in Monino. By the time of his selection as a cosmonaut in January 1963, Filipchenko had built up an impressive résumé: deputy commander of a fighter squadron, senior flying instructor and accomplished parachutist. By the mid-1960s, he was part of a team preparing to fly the Soviet Union's winged orbital spacecraft, called 'Spiral'. He later transferred to the Soyuz training group and led a five-day rendezvous mission in October 1969.

Seated alongside Filipchenko, and also making his second spaceflight, was a sour-faced civilian engineer named Nikolai Nikolayevich Rukavishnikov; a man for whom technical matters were very much a way of life and who loved to repair old apparatus and build new machines. "His ambition," joked fellow cosmonaut Vladimir Shatalov, "is to convert a refrigerator into a vacuum cleaner!" Rukavishnikov had been born in the western Siberian city of Tomsk on 18 September 1932, the son of two railway surveyors, and grew to love geography, mathematics and physics from an early age. After graduating from school, he enrolled at the Moscow Institute of Engineering and Physics in 1951 and gained a diploma six years later, specialising in dielectrics and semiconductors.

His early work was at the Central Scientific Research Institute in Podlipkah, near Moscow, focusing on the development of the Ural computer and the testing of automatic control and protection systems for nuclear reactors. By the end of 1959, he was working for Sergei Korolev's OKB-1 bureau and received his first introduction to space by helping to design controls for unmanned interplanetary probes. Like several other civilian employees of the bureau, Rukavishnikov passed initial cosmonaut screening in the early summer of 1964 as part of efforts to select an engineer for the Voskhod 1 space mission. Two years later, a formal group of civilians was picked, but when four of their number failed the Soviet Air Force's medical exams, it was decided to nominate two more candidates, one of whom was Rukavishnikov.

The first few months of cosmonaut preparation were difficult. "I had to catch up on all the training that other cosmonauts had already passed," Rukavishnikov recalled later. "This included thousands of hours of intensive training, centrifuge, altitude chamber, simulated weightlessness flights and parachute training." His short, skinny stature surprised many of his peers, including Alexei Yeliseyev, who expected him to be dismissed on health grounds. Yeliseyev could not possibly have known that this serious technician would become his crewmate in a few years' time. Moreover, he could also not have foreseen that in 1979 Rukavishnikov would become the first civilian cosmonaut to actually *command* a Soyuz mission.

As a member of Russia's elite space-faring fraternity, Rukavishnikov quickly established a reputation for himself as hard-working and totally committed, routinely staying at OKB-1 both day and night until his tasks were completed. In addition to his technical knowledge and passion, he enjoyed motorbikes and travelling and during summer vacations he would venture off alone into the hills to explore. Very soon after completing initial training, he was assigned to the L1 circumlunar project. Then, in the spring of 1970, he received a short-lived assignment to a Soyuz flight that was to evaluate the 'Kontakt' ('Contact') rendezvous hardware for lunar landings. By midsummer, his career had taken a new turn and he was detailed as the 'research engineer' aboard the first flight to a space station. It is ironic, therefore, that despite flying three times into the heavens (and actually being assigned to a fourth mission in 1984), Rukavishnikov would *never* get to see the

The two trailblazers of ASTP: Nikolai Rukavishnikov (left) and Anatoli Filipchenko. Both are clad in their 'Sokol-K' ('Space Falcon') pressure suits, which were introduced in the wake of the Soyuz 11 disaster as a means of ensuring that cosmonauts could survive the decompression of their spacecraft. Note the soft helmets, pushed back over their heads, and the white-topped 'skull caps' for communications headgear.

inside of a space station. Nor, it seems, might he and Filipchenko have seen the inside of Soyuz 16 in orbit, had they had their way. In their history of Soyuz, Hall and Shayler noted – without elaboration – that the cosmonauts "had wanted their younger backups [Vladimir Dzhanibekov and Boris Andreyev] to fly...but they were overruled".

Fifteen minutes after launch, following a successful boost into orbit, Filipchenko and Rukavishnikov raised their visors, checked their craft's pressure integrity and finally removed their gloves. A little more than an hour later, they equalised the pressures between the orbital and descent modules and opened the hatch between them. Their initial orbit had a much higher apogee than was planned for ASTP itself and a subsequent correction highlighted that Soyuz was capable of making appropriate orbital adjustments if necessary. During their six-day mission, Filipchenko and Rukavishnikov tested the mechanism which would enable Apollo to dock with Soyuz, the new Androgynous Peripheral Docking System (APDS), by extending and retracting a simulated American docking ring. They also evaluated a modified environmental control system, improved solar panels and a new radar unit. As part of efforts to reduce the transfer and acclimatisation times between Apollo and Soyuz, which operated different gas mixtures at different pressures, the cosmonauts lowered and raised the spacecraft's pressure and increased the oxygen from 20 to 40 percent. Six and a half hours into the mission, they reduced the cabin pressure from 101 kPa – equivalent to terrestrial sea-level pressure – to just 72 kPa, and later 68 kPa, before increasing it to 110.7 kPa on 5 December. At length, five days after launch, the cabin was restored to terrestrial sea-level atmospheric pressure and the APDS, its job done, was explosively jettisoned.

When Filipchenko and Rukavishnikov touched down safely at 11:04 am Moscow Time on 8 December, two dozen kilometres north-east of the coal-mining city of Arkalyk in northern Kazakhstan, they completed a mission just shy of six full days ... and, as time would tell, matched the actual duration of the ASTP flight to within *ten minutes*. Their week in space had gone well and, to summarise Filipchenko's pre-launch comments to the Soviet news agency Tass, the cosmonauts had indeed checked "carefully all the ship's systems and ... its docking gear" and had fully acquired "100 percent confidence in the success of the first international space expedition". By the time that expedition took place in the "hot space summer" of 1975, it would be more than three years since Richard Nixon and Alexei Kosygin's agreement and would represent a quite remarkable example of achievement over adversity between two foes whose bitterness and mistrust of one another would, at least for a few days, be set aside.

POSITIVE BEGINNING

The roots of ASTP can be traced to the early summer of 1969, when newly-appointed NASA Administrator Tom Paine – a man whose expertise as an engineer was matched by a keen awareness that co-operation between nations was the only real way to mobilise the necessary funding and resources to explore the heavens –

made contact with Anatoli Blagonravov, chair of the Soviet Academy of Sciences' commission on space matters. At first, Paine suggested flying scientific experiments aboard each other's manned missions and, in *The Partnership*, Edward Clinton Ezell and Linda Neuman Ezell noted that in the months surrounding Neil Armstrong and Buzz Aldrin's Moon landing this proposal went totally against the grain. Previous administrations had built NASA and the space programme on a foundation of *competition* with the Soviets. Now, in Paine's mind, the time was ripe for the agency to stop waving the flag and "begin to justify our programmes on a more fundamental basis".

Throughout his time at NASA, Paine's commitment to the exploration of space never wavered. Appointed as Deputy Administrator in January 1968, it was he who had conditionally approved the audacious voyage of Apollo 8 around the Moon. When he took the agency's helm as its third Administrator in March 1969, he urged his political masters in Washington to boldly pursue a permanent lunar base, an advanced space station and a manned venture to Mars. He wholeheartedly endorsed John Kennedy's pledge to plant American footprints on the Moon ... but saw something else, too, in the late president's promise. "Paine believed that there was something implicit in Kennedy's challenge," wrote Andrew Chaikin in his landmark book, *A Man on the Moon*, "beyond its words; that it was a call for the United States to become a *space-faring* nation." When Richard Nixon's administration entered office in January 1969, however, it quickly demonstrated a clear disinterest in human space exploration and the vision of Paine, his deputy, George Low, and others was abruptly halted in its tracks. Paine would depart NASA in the summer of 1970, only weeks after Nixon had slashed NASA's budget and forced the agency to cancel two Apollo lunar landing missions.

Nevertheless, during his short spell as head of NASA, Paine made significant inroads in drawing the secretive Soviets to the negotiating table and extended the first real word and hand of friendship. He felt that 'competition' remained important, since it kept both sides on their toes, but considered it necessary to operate on a less aggressive level; that the 'peaceful war' of the space race should be presented to the American public as less of a scare tactic and more of an issue of national pride. Paine's correspondence with Blagonravov in April 1969 was met with interest and, although the Soviet academic declined an invitation to attend the launch of Apollo 11, noises from behind the Iron Curtain were clearly positive. When Armstrong and Aldrin landed on the Moon, for instance, Alexei Kosygin warmly congratulated former Vice-President Hubert Humphrey ... and expressed genuine interest in widening talks with the United States over future space matters. Real progress, though, came with a series of correspondence between Paine and Mstislav Keldysh, the influential head of the Soviet Academy of Sciences, with an early focus on joint planetary exploration.

Although these efforts came to nothing, a Space Task Group, established early in 1969 and run by Vice-President Ted Agnew, quickly identified greater international co-operation as an important next step after the Apollo landings. Surprisingly, for a government-sponsored document, the group's report admitted that the political desire to beat the Soviets *had* indeed been "one of the several strong motivations for

US space programme decisions over the previous decade". Now that the lunar landing had been achieved, the report continued, the imagination of the world had been captured and focused on a realisation that humanity now had at its *collective* disposal a new technology, whose applicability transcended national boundaries. Paine sent a copy of Agnew's report to Keldysh, together with a renewal of his wish that the two nations might someday "undertake major complementary tasks" and suggested a future face-to-face meeting between them.

On the political stage, Richard Nixon had already formed an inter-agency committee to review the impact of working with the Soviets and, with the exception of the Department of Defense, the response had been good. Initial focuses included working on a joint docking mechanism and a standardised rendezvous apparatus – perhaps to allow each nation to offer a 'space rescue' capability for the other – and in December 1969, when Blagonravov happened to be in New York, Paine raised this possibility in their first face-to-face discussion. A few months later, Neil Armstrong presented a paper in Leningrad, after which he and Buzz Aldrin were warmly welcomed at Tyuratam and George Low had significant private talks with several key Soviet officials, including Keldysh. Other meetings even drew inspiration from a recent Gregory Peck movie, *Marooned*, in which a hero cosmonaut helped to save an astronaut in a crippled American spacecraft. Early in July 1970, the new scientific attaché at the Soviet Embassy in Washington brought encouraging news: Keldysh and "appropriate" colleagues *were* prepared to discuss common docking mechanisms with the Americans.

By now, Paine was embroiled in the aftermath of the Apollo 13 near-disaster, the impending cancellation of several lunar landing missions and was only weeks away from departing NASA, but he was enthused by the response. He invited a pair of Soviet engineers to the Manned Spacecraft Center in Houston, Texas – renamed the Johnson Space Center in February 1973 – to discuss these docking concepts further, as preparation for more advanced joint talks. In a 31 July 1970 letter to Keldysh, Paine made reference to his imminent departure from the agency, but assured his counterpart that this would in no way alter NASA's position on co-operation. To reinforce this point, five weeks later, he even suggested the possibility of docking a Soyuz onto America's Skylab space station in 1973. Although Paine pointed out that it would be "feasible" to fit a compatible docking device onto Skylab, he doubted that the Soviets would accept the proposal, since they might consider themselves an unequal partner. Nonetheless, he was confident that it demonstrated his agency's sincerity.

The man who would next contact Keldysh was the new Acting Administrator, George Low, whose star had steadily risen within NASA since he joined its predecessor organisation, the National Advisory Committee on Aeronautics (NACA), in the early 1950s. Low had played a pivotal role in America's human space programme from the outset: as chief of manned spaceflight, he had been closely involved in the planning of Projects Mercury, Gemini and Apollo, before serving as deputy head of the Manned Spacecraft Center, then head of the Apollo office and ultimately deputy of the agency itself. His work with Paine also made him a key player in the genesis of Skylab, the Shuttle and ASTP. In a sense, therefore,

Low's quarter-century career with NASA involved him in almost every American manned space project; a remarkable involvement which would come full-circle in May 1984, only two months before his untimely death, when his son, David Low, was selected by the agency as a member of its astronaut corps.

When he took the reins of NASA as Acting Administrator in the autumn of 1970, Low quickly reasserted Paine's position and proposed sending a small delegation to Moscow in late October for inaugural discussions. At around the same time, however, any hope of someday using Soyuz as a potential rescue craft for Apollo fell on rocky ground, for the Advanced Manned Missions Planning Group of NASA's Office of Manned Spaceflight identified a number of fundamental obstacles. "Rescue possibilities appeared to be limited to an Apollo retrieving the crew of a crippled Soyuz," wrote Ezell and Ezell in *The Partnership*. "It would have been very difficult for the Soviets to accommodate all three Americans aboard *their* spacecraft, unless they attempted an automated rendezvous with Apollo, and Soyuz was essentially an Earth-orbital craft, while Apollo was designed for lunar missions." The differing orbital inclinations of both vehicles posed other problems and although Apollo had the capabilities to assist Soyuz, such an effort would require NASA to *know* the Soviet launch schedules and flight parameters in advance. Compounding the issue further was time: it would be possible, but complex, to create the androgynous docking hardware needed for Soyuz to dock with Skylab or Apollo.

Unperturbed by these worries, the four-man NASA delegation, led by Manned Spacecraft Center Director Bob Gilruth, arrived at Moscow's Sheremetyevo Airport on 24 October 1970 and over the next four days were shown the cosmonauts' training centre, Zvezdny Gorodok ('Star City'), on the forested outskirts of the capital, briefed on the intricacies of the Soyuz simulator and introduced to the docking apparatus and rendezvous equipment. One of the Americans, veteran flight director Glynn Lunney, later recalled how impressed he was by the simplicity of the Soyuz controls, the 'roomy' nature of the descent module cabin ... and the realisation that *this* was the mysterious machine that he had previously only *read* about in aerospace publications. Moreover, the men who were now cogently describing its workings to him were *real* cosmonauts who had actually *flown* it into orbit. During a series of lengthy presentations, Lunney talked the Soviets through American rendezvous, communications and propulsion capabilities, whilst fellow team member Caldwell Johnson described the Apollo docking apparatus and unveiled a proposal for an androgynous mechanism which might someday permit an orbital link-up. Known as a 'double ring and cone', it represented an early attempt to overcome the bulky and cumbersome probe and drogue system used to dock its Apollo command and lunar modules together.

The ring and cone mechanism had first arisen in 1963, during early definition studies for Apollo, but had been dropped from consideration. Johnson revived it a few years later as part of plans for NASA's orbital space station, later known as Skylab. "The cone was divided into 12 discrete fingers or guides," wrote Ezell and Ezell, "so that the cone of one gear would match the ring of the mating gear and vice versa. The fingers of one will exactly intermesh with the fingers of the mating gear." Its main advantage was that, unlike the Apollo probe and drogue, it did not risk blocking the connecting passageway between the two docked craft; additionally, it could accept a 'passive' partner and it was considered fail-safe. The Soviets responded with detailed descriptions of their Soyuz mechanism, which operated with a 'pin and cone' system. Unlike Apollo, which was limited to two prime and two backup retractions of the probe, the Soyuz apparatus could support repetitive dockings and undockings. In Lunney's mind, the Apollo concept was fine for lunar voyages, which nominally involved two dockings, but the heavier Soviet gear offered a much greater degree of flexibility.

Lunney and his colleagues were also introduced to the level of care needed during a Soyuz docking procedure, which demanded the precise alignment of electrical connectors in the face of the docking ring. When the 'head' of the docking probe was engaged in the drogue and basic alignment had been accomplished with a series of targets, further refinement was completed by running the guide pins of one craft into a series of 'sockets' on the other. At the time of the meeting, in October 1970, Soyuz had yet to transfer cosmonauts *internally*, in shirt-sleeves, between spacecraft, whilst Apollo had already long since demonstrated this capability. However, the Americans were shown a modified apparatus which *would* soon be used for dockings and internal transfers to an experimental space station. In this case, dockings would be done in a similar fashion to earlier missions, but when the craft were linked the probe and drogue would be unlocked and swung aside, opening up a 'tunnel' through which the cosmonauts could move. After four days in Russia, the NASA team returned to Houston with a renewed sense of optimism that the co-operative venture would bear fruit.

For Caldwell Johnson, the next few weeks involved sketching out a number of possible joint missions and a couple of obvious ideas came to mind: a Soyuz docking with Skylab – already proposed by Tom Paine in his September 1970 letter to Mstislav Keldysh – or an Apollo-Soyuz docking in which the former would act as a 'propulsive stage' to insert the latter into "a different orbital situation" or demonstrate the ability of an 'active' rendezvous partner. Other options existed beyond these possibilities, but would focus on *Marooned*-type rescue scenarios and as such would be considered as 'non-scheduled'.

George Low was excited by the positive comments from Gilruth's group and warmly accepted Keldysh's invitation to visit Moscow himself for face-to-face meetings. Low felt that, in addition to a manned venture, co-operation should also encompass four key areas which had been identified in the mid-1960s: working together to provide better worldwide weather forecasting, sharing scientific data more broadly, 'pooling' both nations' biomedical results and jointly exploring the oceans with Earth-circling satellites. By the beginning of December, Low and Keldysh had firmly scheduled a meeting in Moscow in mid-January 1971. At the same time, the notion of a 'real' manned mission, which offered something distinctly more tangible than "drawn-out discussions about abstract, hypothetical missions at some unspecified time in the future", was beginning to grow. In the weeks before the Moscow meeting, Low set to work drafting a series of agreements – and a press release – to highlight the outcomes that he hoped would emerge. However, the cordiality of the discussions between the head of NASA and the head of the Soviet

Academy of Sciences did little to improve the increasingly tense language being spoken by their respective political leaders. Shortly before leaving for Moscow, Low was briefed by Undersecretary of State Alexis Johnson about steadily rising diplomatic conflicts between Russia and the United States.

HIJACK!

Soviet attempts to portray the equality of the communist state as a better, fairer alternative to the inequality of the capitalist West had always been little more than skin-deep. In many ways, they were utterly hypocritical. Already, the unfair discrimination of Soviet Jews had reached the space programme, negatively influencing the career of Jewish cosmonaut Boris Volynov, who almost lost his seat in command of the Soyuz 5 mission purely on the basis of his religious convictions. Anti-Semitic feelings flared in the wake of Israel's victory over the Palestinians in the 1967 conflict: the Soviet Union had broken off diplomatic ties with the warlike Jewish state, religious observances were discouraged and even banned and, as a result, many thousands of Russian Jews applied for permission to return to their homeland. Such permissions were made deliberately difficult to obtain, for Leonid Brezhnev's regime had no desire to be seen as one from which thousands wanted to escape. Common excuses for the Soviet refusal to issue exit visas included claims that applicants had been made privy to classified national security information. The process was compounded further by the requirement for a 'request letter' from a family member living in the country to which the applicant intended to emigrate, plus documentation from schools and employers, evidence of financial status, approval from parents and even from divorced partners. Applications were slow, often taking six months or more to be processed. Moreover, in many cases, even *requesting* an exit visa was considered an act of social parasitism and an offence against the state. As a consequence, it is hardly surprising that only 4,000 exit visas were granted by the Soviet authorities during the 1960s.

The plight of the other would-be émigrés drew the international spotlight in May 1970, when a group of 'refuseniks' – people who had been denied permission to leave the Soviet bloc – tried to hijack an airliner and fly it to the West. Led by dissident Eduard Kuznetsov and including amongst their number a pilot with the rather unfortunate name of Mark Dymshits, the group bought up all the tickets for a flight from Leningrad to Priozyorsk. The attempt failed and Kuznetsov and Dymshits were both given the death penalty for high treason, but a tide of international protest succeeded in reducing their sentences to 15 years of incarceration. Both men were freed in 1979 as part of an exchange deal for a pair of Soviet intelligence officers who had been arrested in New Jersey.

Predictably, the failure of the Kuznetsov-Dymshits hijacking led to a harsh crackdown, including a 'diploma tax', whereby individuals who had received higher education in the Soviet Union had to pay many times their annual salary in a partially successful attempt to stem a 'brain drain' to the West. Notwithstanding these measures, continuing international pressure ultimately forced Brezhnev to increase the emigration quota and by the end of the 1970s more than a quarter of a million people had left for Israel and the West. When placed alongside the many *illegal* defections and numerous attempts by would-be escapees to scale the Berlin Wall from East to West (but rarely vice versa), this proved hugely embarrassing for the hard-line communist regimes of the Soviet bloc. The reality that hundreds of thousands of people *wanted* to escape the worker's paradise became even harder for the dictators to explain away.

Having said this, and despite an acute period of economic stagnation which forced increased attempts to foster détente with Western nations, the Soviet Union reached an unprecedented level of power and relative internal calm under Brezhnev. A Public Opinion Foundation poll in 2006 found a 61 percent approval rating for his brand of leadership. Many people would have preferred to live during the Brezhnev years than at any other period of Russian history in the 20th century. Still, the need to crack down on dissenters and the mass emigrations of the early 1970s would underline an equal reality that the communist life was not quite as idyllic as it was portrayed. The fall of the Soviet Union was still two decades away ... but in 1970, the first murmurings of discontent had already been heard and the first steps to bring about its collapse had already begun.

ROCKY ROAD TO PARTNERSHIP

The Kuznetsov-Dymshits affair was still playing out and the death penalties had only recently been declared when George Low prepared to visit Mstislav Keldysh in Moscow. The Jewish Defense League had already bombed several Soviet installations and intimidated Soviet personnel in New York and Washington and on 4 January 1971 the Soviet ambassador had delivered a stern note to the State Department, accusing the United States of "connivance" in these acts. More significantly for the impending NASA visit, the ambassador refused to guarantee the safety of American officials operating in the Soviet bloc. When Alexis Johnson briefed Low, he advised him to check with the embassy in Moscow before making a favourable press release ... just in case the diplomatic situation should worsen.

Shortly before departing the United States, Low met President Nixon's foreign policy advisor, Henry Kissinger, in California, and received assurances that the administration would support NASA's overtures of co-operation with the Soviets. He was granted virtually a free hand to negotiate in any area which was within NASA's remit. However, Kissinger cautioned that, no matter how well the talks went, Low should make *no* suggestion that a space deal might also solve *political* differences. Kissinger was acutely aware that past dealings between astronauts and cosmonauts had been friendly and the Soviets had shown themselves to be easy negotiators ... but political discussions were much more complex and the NASA team might compromise the work of diplomats on both sides. "As long as you stick to space," Kissinger told Low, "do anything you want to do. You are free to commit. In fact, I want you to tell your counterparts in Moscow that the President has sent you on this mission."

When Low arrived in the snowy Soviet capital on 16 January, he detected no hostility or even coolness on the part of his hosts; they spoke about NASA's upcoming Apollo 14 lunar mission, about Russia's recent Luna 16 flight and about the Lunokhod rovers, the first of which was at that very time traversing the surface of the Moon. Later that evening, during informal talks between Low's group and that of Keldysh, it was mutually agreed that while unmanned flights were important, a *manned* venture was essential "to lift the human spirit". Co-operation was important due to the sheer cost and the vastness of the technical problems facing both nations, but *competition* was also a vital tenet, in that it served to spur them on to greater things. After five days, Low and Keldysh concluded that several steps should be taken, with a primary focus on joint meteorological and Earth sciences research, the sharing of lunar and planetary exploration data and the regular exchange of biomedical results. Press releases were duly issued and the media response proved favourable. Behind closed doors, though, Low and Keldysh had a much more specific plan in mind.

On 20 January, whilst the negotiations were underway, the topic of rendezvous and docking on a manned mission was broached. Low took care to point out to Keldysh that the Americans did not yet wish to make a *formal* proposal, but rather wanted to present it as an option for the Soviets to consider. However, the discussion centred on a series of docking studies prepared by Manned Spacecraft Center staff in Houston in December 1970, which outlined a range of options, including an Apollo-Soyuz link-up by the use of some form of 'adaptor' between them, featuring either an external or internal transfer of crew members. A final, more ambitious proposal suggested docking an Apollo and a Soyuz onto an elaborate 'experiment module' – possibly some sort of temporary space station – for an extended programme of joint scientific research. This final option was quickly dismissed by Bob Gilruth and his deputy, Chris Kraft, on the basis that it was too complex and consequently it was not among the informal proposals tabled by Low that January.

In whatever form the joint mission came, the need for some form of androgynous docking apparatus was acutely necessary, not simply to overcome the differences in the Apollo and Soyuz mechanisms, but also to accommodate the differing cabin pressures; for the former employed pure oxygen at around 34 kPa (a third of sea-level pressure) and the latter a more 'terrestrial' mixture at 101.3 kPa. A new docking system would permit the installation of a drogue, akin to that used by the Apollo lunar module, to enter the conical receptacle atop the Soyuz, which normally served to guide the entrance of the probe. A more sophisticated alternative was to build an airlock-type adaptor – effectively a miniature spacecraft in its own right – to accompany Apollo into orbit, housed within the final stage of its Saturn IB launch vehicle. The Apollo would separate from the Saturn, perform a 180-degree rotation, dock and extract the airlock and subsequently perform a rendezvous and docking with the Soyuz. It was assumed that for crew transfers between the craft to take place, Apollo and Soyuz would operate at their standard atmospheres – pure oxygen and mixed-gas – accepting that this would require the men to undergo a lengthy process of 'pre-breathing' to avoid the bends.

The cabin pressure of Apollo could not be raised much above 55 kPa, due to structural limitations on the command module, and Soyuz could not drop significantly below that level without increasing the risk of fire as oxygen percentages increased in the total volume of the remaining gases. The obvious solution, wrote Ezell and Ezell, might have been to compromise on cabin pressure at around 55 kPa, but this would have required major modifications and defeated the object of making the fewest possible changes to the basic craft. In an effort to provide for the crew members' needs without adding supplemental oxygen to the Soyuz atmosphere, environmental control specialists at the Manned Spacecraft Center planned to develop a new closed-system portable mechanism to provide oxygen and recycle carbon dioxide for the American astronauts. In fact, work had already begun on just such a unit to help to eliminate the problems of 'oxygen enrichment' and increased fire risk during pre-breathing. Certainly, the development of an airlock would have solved some of these pressure problems, but if the crew members were obliged to pre-breathe in such a module it would demand a full life-support apparatus and, hence, a hefty weight and cost penalty. Even if pre-breathing took place aboard the Soyuz, a simpler device could be used, but the Americans would still need to transfer in their pressurised suits and *that* would necessitate cooling circuits inside the Soviet craft. Very soon it became obvious that unless a simpler means of tackling the problem was devised, life would become too complex for both crews.

Follow-on talks, planned for May 1971, were postponed due to Soviet efforts to launch their first space station, Salyut 1, and its inaugural manned flights. However, on 20 June, a delegation from Moscow arrived in Houston, midway through the long-duration Soyuz 11 mission, and were clearly ready to get down to business, crowing about the triumph of their cosmonauts. The Soviet group was shown the Apollo simulators and docking hardware by astronauts Fred Haise, Ken Mattingly and John Young. They also experienced, for the first time, a fundamentally different culture from Russia. One afternoon, they visited a Houston shopping mall, at which they purchased large quantities of children's clothes and a few 'unusual' items: a star drill that could bore into concrete, several pairs of tennis shoes with steel-arch supports and a saw with five interchangeable blades! They were astonished by the provision of free shopping bags, as well as by the American practice of opening and examining goods before buying them. At one point, a leading academician was met by a group of schoolchildren, who asked if they were Soviets. When he responded in English that they were, one of the girls broke into a spontaneous speech of welcome, telling him that she was happy they had come to the United States and hoped that their work was successful. The academician was moved to tears by the warmth of the greeting.

The work of the next few days was as successful as it was fast-paced, although the language barrier and a paucity of English-Russian translators meant that when a point of agreement had been reached, a document had to be prepared in both languages, verified as to meaning and technical content, then signed by both sides. This proved a slow and tedious process. At the end of the Houston meetings, Konstantin Bushuyev and Glynn Lunney were identified as the project directors, responsible for acting as focal points for future communications and technical exchanges between the two sides.

The relationship between Bushuyev and Lunney was an interesting one in itself, not least because of the wide gap in their ages; during the planning of ASTP, the former was a balding, bespectacled professor in his late sixties and the latter a young engineer in his mid-thirties. "He was kind of a sober-faced gentleman," recalled Lunney in an October 1999 oral history for NASA, "relatively unemotional, fairly grim of countenance, in a way, especially officially, but who became a good friend over time. There were times when, for example, the Soviets would be in town, we would invite them to our homes. I remember one occasion where he was at our house and our youngest son, Brian, was about seven or eight years old. He'd take Professor Bushuyev out in the woods behind our house and walk him around. Brian would show him everything that he played on, trees that he climbed, and jabber away in English and the professor didn't understand a word that he was saying! But they used to go off in the woods and they'd kind of wander around and Brian would take him up to the pasture and show him the horse and the fence and all this stuff. [Bushuyev] just used to get a real kick out of it. He openly was proud of his couple of grandchildren he had in Russia and I, likewise, had visited his apartment, had dinner with his wife and daughter and met some of his family that way. Over a period of time, I came to know him reasonably well. We never did speak each other's language ... we always worked through an interpreter. Everybody on the American side came to admire and respect him." In fact, one of Bushuyev's favourite anecdotes about how well the two project directors worked together was this: "In our joint work," he once said, "there has been just one complication. Dr Lunney drinks black coffee and I drink mine with cream!"

This cordiality did not detract from press and public cynicism that the talks were actually allowing the Soviets to 'steal' American technological expertise. Jim Maloney of the *Houston Post* suggested that NASA was "donating" its knowledge and Peter Mosely of Reuters asked Bob Gilruth if the work represented a 'true' exchange of technology or was actually one-sided. Gilruth responded that both nations had much to offer and that the run of talks had thus far focused merely on the possibility of working together, but the scepticism remained. Two days later, the Soyuz 11 crew – Georgi Dobrovolski, Vladislav Volkov and Viktor Patsayev – returned to Earth after a record-breaking three weeks aboard the Salyut 1 station. Their spacecraft made a perfect touchdown on Soviet soil ... but when rescuers opened the hatch, *all three* cosmonauts were dead. During re-entry, a pressure valve had inadvertently opened and all of the air had drained from the capsule, suffocating the men in a matter of seconds.

Throughout Russia, the tragedy brought about an unprecedented wave of mourning. People wept openly in the streets for the cosmonauts, who, for the better part of a month, had appeared nightly on their television screens – cosmonauts who were being presented as *human beings* and not cold, faceless supermen – and who had offered a clear Soviet response to the Apollo Moon landings. Now, instead of three heroes, bearing broad smiles and bedecked in medals and garlands of flowers, all Russia had was ... *three funerals*. The response from the United States was also one of shock, but for different reasons, and not just for its potential impact on the prospects for a joint mission. Since the summer of 1969, Tom Stafford had been chief of NASA's astronaut office and one of his duties had been to supervise the direction of the manned space effort after the Apollo lunar missions. In addition to the venture with the Soviets, America expected to launch a large space station of its own (Skylab) in the spring of 1973. It was planned that crews of three men would spend between one and two months aboard this outpost, performing scientific and biomedical studies in weightlessness. At the time of the Soyuz 11 disaster, Stafford was visiting Europe with his wife and daughters and was due to speak at the International Aeronautical Federation Conference in Belgrade.

"Before I reached Belgrade," he wrote, "I heard the news that the . . . cosmonauts had died on their return to Earth. My first worry was that the stress of a long-duration flight had killed them and wondered what that would mean to our Skylab crews. Clearly we needed to know more than what was in the news." Back in Houston, Deke Slayton – at the time head of the flight crew operations directorate and hence Stafford's boss – was of the same opinion: a year before, cosmonauts Andrian Nikolayev and Vitali Sevastyanov had hardly been able to *stand* after their 18-day mission and now Dobrovolski, Volkov and Patsayev had returned *dead* from orbit after 24 days. "Was there something about being weightless that long that could kill you?" Slayton wondered. *Flight International*, too, speculated on 8 July 1971 that "it is possible . . . that the degrading effect of weightlessness increased exponentially". The news from the Soviet Union, not surprisingly, offered a sketchy and none-too-helpful insight, but could not sidestep the inescapable truth that this was a tragedy of immense proportions. Yet even this truth was restricted to the periphery of the initial announcement, which emphasised the flight's strengths and tried to downplay the calamity which engulfed it. The report began in sombre fashion with "Tass reports the deaths of the crew of the spaceship Soyuz 11 . . . " and proceeded through a lengthy discussion of the extraordinary success of the mission and its re-entry, before ending abruptly with "upon opening the hatch . . . [the rescuers] found the crew of the spaceship in their couches without any signs of life. The causes of the crew's deaths are being investigated . . . "

Notions that weightlessness itself could have caused or contributed to the Soyuz 11 deaths was not entirely outlandish, since Gemini astronauts had shown 'signs' that the human heart grew lazy in the strange environment on flights lasting up to two weeks and a monkey named Bonny had returned from a nine-day biosatellite flight in July 1969 and succumbed to heart failure. However, both Soviet and American physicians soon independently concluded that decompression was probably to blame for the Soyuz 11 deaths and by the time of the cosmonauts' state funerals in Moscow on 2 July that had been accepted. President Nixon issued a statement of sorrow to his Soviet counterpart and appointed Tom Stafford as his official representative at the funerals. It was an interesting demonstration of how much the gap of distrust had closed in a just a few short years. In 1967, when cosmonaut Vladimir Komarov died during his return to Earth, America had been rebuffed when it asked to send a representative to the "private" funeral; *now*, however, Stafford's presence was welcomed. In fact, he acted as one of the pallbearers for the massive urns which carried the cosmonauts' ashes for interment in the Kremlin Wall. Many have seen Stafford's role in these events as not only

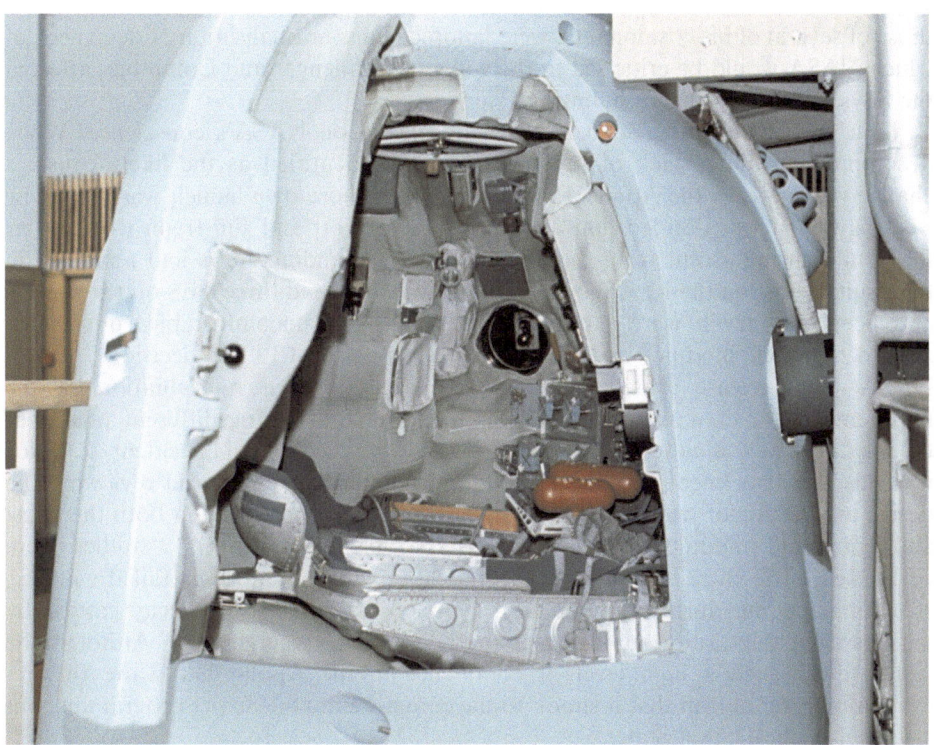

A cut-away view of the bell-shaped Soyuz descent module. Note the two custom-moulded couches for the cosmonauts and the wheel mechanism for opening the hatch into the orbital module above. Although this vehicle has been heavily and extensively modified over the years, and frequently criticised for its relatively primitive design, its basic form remains operational to this day and has outlasted other, more capable spacecraft, including the Shuttle. In fact, at the time of writing, Soyuz offers the only operational method for transporting astronauts and cosmonauts to and from the International Space Station for the first part of the next decade. In 2011, it evokes and lends weight to the words of the visionary Sergei Korolev: Soyuz, he said, was "the machine of the future".

remarkable, but also pivotal in his eventual selection to command the American side of the ASTP mission.

Despite earnest Soviet promises that the exploration of space would continue and that the 1970s would truly be a decade in which large orbital stations would be commonplace, it was apparent that if Dobrovolski, Volkov and Patsayev had been wearing space suits, they *would* have survived the depressurisation of their craft. Instead, they had been wearing only light woollen flight garments. Sergei Korolev's reasons for doing away with pressurised suits have already been discussed, but Nikolai Kamanin – commander of the military cosmonaut team until 1971 – objected strongly to the decision and, indeed, wrote to Nikita Khrushchev and Leonid Brezhnev several times to plead the case to keep them. His requests, and those of several of his cosmonauts, were ignored. This 'normalisation of deviance', of which NASA would be criticised in the wake of Challenger and Columbia, affected the Soviet space effort in a hauntingly similar way.

To a great extent, the Soyuz 11 tragedy played on NASA's conscience. When decompression and a lack of pressure suits was identified as the likely cause, a change was made to the Apollo 15 lunar mission, whose own launch was just a few weeks away. It was decided that astronauts Dave Scott and Jim Irwin would wear their suits during ascent from the Moon. (Previous landing crews had remained in their suits whilst on the Moon, but Apollo 15 was to spend three days on the surface and Scott and Irwin were expected to remove their uncomfortable suits whilst aboard the lunar module and, presumably, during ascent.) "The decision," read a NASA press release, dated 19 July, "was based on a re-evaluation of the requirements for crew members to wear pressure suits during different phases of the mission. The evaluation was conducted following the Soyuz 11 accident ... " Nor was this simply a knee-jerk reaction: the re-evaluation encompassed reviewing the design and testing of windows, hatches, valves, fittings and wiring in both the lunar and command modules. "In addition," the release continued, "studies were performed on re-entry effects on crew and cabin with a completely failed window, structural loading during lunar module jettison, cabin pressure decay caused by various-sized holes, suit-donning times and post-landing emergencies." Although the results established a high level of confidence in the Apollo hardware, this re-evaluation is notable in that it shook to the core not only the Soviet programme, but also that of the United States.

By the time of the Soyuz 11 disaster, NASA had a new man at its helm. Administrator Jim Fletcher, a physicist and veteran University of Utah president, had taken charge of the agency late in April 1971. In his first press conference, Fletcher declared that under his leadership, NASA would seek to find a "balance" between manned and unmanned exploration of the heavens ... but noted darkly that "exploration" in the vein of Apollo was henceforth out of the question. "It would be very exciting for man to go beyond the Moon," he said, clearly aware that two lunar landing missions had been scrapped the previous year to conserve funding in a period of ever-tighter space spending, "but that ... is just a little beyond the nation's budget right now." Fletcher proved a strong supporter of the joint mission with the Soviets, partly for the spirit of co-operation which it afforded, but also in pragmatic recognition of the reality that – after Skylab – there would be a hiatus of several years before America's next-generation orbital craft, the Space Shuttle, was due to enter service. Even optimists within NASA did not expect the vehicle to fly before 1979.

Fletcher and his head of manned spaceflight, Dale Myers, were keenly aware of this impending gap, although four Apollo command and service modules would be left over following Skylab. One of these *could* be set aside for a docking venture with the Soviets, whilst alternatives included a series of 'stand-alone' orbital missions, launched one per year from 1975, to undertake flights of 16-30 days for Earth sciences research. Other options included launching a second Skylab station (since a backup was being built), although *that* would prove costly and would unacceptably siphon funds from the Shuttle budget. Myers knew that there would be insufficient money for both the Shuttle *and* an interim Apollo effort. For the astronaut corps, the writing had long since been on the wall and the choice was stark: they could stay and perhaps wait a decade to fly again ... or they could leave. In the early 1970s, many picked the second option.

In the meantime, the airlock, or 'docking module', for the joint mission was steadily being defined. In August 1971, it was decided that all additional equipment should be housed within the docking module, thereby reducing the need for major changes to the basic Apollo spacecraft. Designers Jim Jaax and Gerald Mills felt that the new module should be able to draw on Apollo's power resources and that it should have its own environmental control system and not be dependent upon the life-support capabilities of the Soviet craft. By now, the emphasis had shifted away from docking an Apollo with a Soyuz and both nations were focusing instead upon a link-up with a Salyut orbital station. Irrespective of the target, the docking module had to be capable of withstanding atmospheric pressures aboard the Soviet craft *and* accommodating equipment to communicate on its frequencies.

A 110-page Docking Module Design Study, presented to representatives of the Manned Spacecraft Center and Apollo's prime contractor, North American Rockwell, on 29 September 1971, offered a full outline of the facilities to be provided by the new craft. Its primary purpose would be to serve as a structural joint between the docking mechanism of the command module and the new androgynous system; additionally, it would offer an atmospheric 'adaptor' between Apollo and Salyut and provide a habitable environment for the installation of communications gear and Earth resources experimentation. According to the study, the docking module would measure 2.5 m from its interface with the command module to the point at which the androgynous mechanism would be attached. Its interior diameter would be 1.4 m and it would provide a hatch diameter at the Apollo 'end' of 0.8 m and at the Salyut 'end' of 0.9 m. This compromise on hatch sizes was crucial if efforts to stage more than just one docking mission were ever to bear fruit. Admittedly, plans for the Shuttle and future space stations envisaged 1.5 m hatches, but Glynn Lunney considered it unreasonable to impose such demands on an Apollo-Salyut mission. Furthermore, he knew that a second-generation American orbital station had not even been defined yet and, in a best-case scenario, was at least a decade away.

As the next joint meeting with the Soviets drew nearer, the design of the docking module was further refined. North American Rockwell certainly felt that it did not pose them any special challenges – in fact, much of the technology had already been used during the Apollo lunar effort – and meeting a launch schedule of 1974 or 1975 seemed entirely possible. In the meantime, writing to Konstantin Bushuyev late in the summer of 1971, Lunney enclosed schematics of a possible joint docking mechanism to fit beneath Salyut's launch shroud. Bushuyev agreed that the mechanism would not present any major technical obstacles and the week-long meeting in Moscow was planned to begin on 29 November 1971.

Yet behind the positive negotiations, the thorny issue of 'technology transfer' remained. When Lunney briefed George Low, who remained in post as Deputy

The command module (lower) and docking module undergo electrical and mechanical checks in January 1975.

Administrator under Fletcher, a series of concerns were tabled. Would the United States build *both* halves of the docking mechanism, Low wondered, or would a common specification be agreed and each nation build its own part? The same question applied to the issue of radio communications equipment: would America *lend* the Soviets their own receivers or simply give them the technical specifications for the hardware and let *them* build them from scratch? Technology transfer fascinated the press, who were keen to know if the Soviet purpose was simply to get its hands on American technical goodies. In truth, the Apollo radios were obsolete and a new system was already planned for the Shuttle.

Regardless, the questions renewed the need for transparency on both sides. "Conduct of such a mission," NASA Headquarters was told, "warrants a measure of trust and the need to accept less than 100 percent knowledge and understanding of each other's equipment." This mention of 'trust' was underlined, too, by the initial reluctance of the Soviets to agree to weekly, unmonitored teleconferences and telex exchanges, which they described as "too expensive" and which would require them to "bring in their telephone people" before they could have direct conversations with the Americans. Eventually, Lunney and Bob Gilruth succeeded in persuading them that regular communication was a *necessity* for the mission to progress, but the cultural gulf remained wide and bridging it would not be straightforward. Subsequent teleconferences would be slowed not only by the language barrier, but also by technical deficiencies. Whereas NASA had long since employed a conference-style telephone set-up between its various facilities, the Soviets relied upon a more archaic system of using two telephone handsets, passed from person to person. Busy overseas circuits and lengthy delays to even *connect* a Houston call with the Soviet Academy of Sciences caused further problems. Still, Gilruth felt that such conferences were critical to permit an easy flow of information between the two sides and establish a system of mutual reassurance that progress *was* being made.

By the end of the November meeting, it was mutually agreed that "a test mission appears technically feasible and desirable" and, further, that it would be important to make an early decision about the practicality of scheduling it for the spring or summer of 1975. The emphasis, for now, remained on docking an Apollo with a Salyut and it was proposed that an American VHF ranging system and a series of 'docking targets' to provide visual reference aids for the command module's crew would be installed on the Soviet station. Elsewhere, the design of the androgynous mechanism was also moving ahead, with NASA mechanical engineer Bill Creasy and his team having built a mockup 'double ring and cone' system with four guide 'fingers' and attenuators on both rings, allowing either half to be active or passive during docking. A 16 mm movie of this system in action was presented at the November meeting and the Soviets refined it further by suggesting three guide fingers and, in lieu of hydraulic shock-absorbers, a series of electromechanical actuators. "In essence," wrote Ezell and Ezell, "the Soviets had accepted the idea of using a set of intermeshing fingers to guide the two halves of the docking gear from the point of initial contact to capture." The concept of using shock-absorbing attenuators on the active spacecraft's capture ring to buffer the impact of two spacecraft also proved perfectly acceptable to the NASA delegation.

Engineers on both sides planned to retract the active half of the docking gear using an electrically-powered winch; once retracted, structural or body latches would engage to lock the craft together. Three fundamental issues had to be resolved before a universal design could proceed further; the number of guides, the type of attenuators and the type of structural latches. The NASA engineers wanted four guides, arguing that it provided the best geometry when using hydraulic attenuators. Creasy added that the most probable failure situation using a hydraulic system of attenuators would be a leak which might cause one shock absorber to collapse on impact. The feeling in the NASA team was that four guides and eight shock absorbers was the optimum design. An electromechanical system might also be susceptible to the freezing-up or binding of one of its pairs of attenuators. Agreement between American and Soviet engineers on this point was harder to reach, with both sides preferring the capabilities of their own systems. At length, since NASA "had no significant engineering or hardware equity in its proposed design ... and since the USSR had considerable equity in its proposed design", the Soviet system was chosen as a baseline for the next period of study.

By the end of the November meeting, the two sides signed a series of minutes outlining the concept for a universal androgynous docking device. According to the formal agreement, "the design ... includes a ring equipped with guides and capture latches that were located on movable rods which serve as attenuators and retracting actuators and a docking ring on which are located peripheral mating capture latches with a docking seal". Basic data on the dimensions of the petal-shaped guides were also included, dictating that they would be *three* in number, as the Soviets had requested. On condition that the requirement for absorbing the docking forces was met, each side was now free to execute the actual attenuator design as it saw fit, with the Americans opting for hydraulic shock absorbers and the Soviets taking an electromechanical approach.

Plans to assign the command and service module for the Apollo 'side' of the mission were also well advanced. Already, North American Rockwell had advised NASA's Office of Manned Space Flight that its labour force was being reduced from October 1971. Only two more Apollo lunar landings remained on the flight manifest, followed by three Skylab missions, and a decision had to be made quickly on *which* craft should be set aside for the joint venture with the Soviets. Bob Gilruth favoured the newest ones out of the assembly shop – Spacecraft Nos. 115 and 115A – because both of their service modules were fitted with Scientific Instrument Module bays (SIMbays), in which a battery of Earth resources experimentation could be accommodated. Most of the other available spacecraft either did not have SIMbay provisions or were needed to support Skylab.

Using the newest command and service modules would cost in the range of $267 million; a figure which incorporated not only the craft themselves, but *three* docking modules – one for testing, one as a backup and another for the actual mission – and no fewer than *seven* androgynous adaptors for ground tests, flight operations and as spares. In February 1972, Dale Myers and other senior managers capped the cost of the venture at $250 million, which precluded the use of either Spacecraft No. 115 or 115A and focused attention on one of the older-specification vehicles instead.

Ultimately, Spacecraft No. 111 was set aside for the actual mission, with No. 119 pulling double duty as a potential Skylab rescue craft and later as a backup for ASTP. Since neither of these craft carried SIMbays, the Earth resources payload needed to be extensively simplified and a relatively small $10 million package of experiments was selected for installation in the command and docking modules.

The cost of this 'standalone' experiment package and its place on the mission is an important footnote in itself. From a political perspective, of course, NASA would be requesting authorisation from Congress to stage a joint rendezvous and docking mission ... but at the same time the agency had to be aware of the need to maximise the scientific return from that mission if, for whatever reason, the Soviets were unable to complete their side of the venture. One of the earliest criticisms from lawmakers was over the merits of spending $250 million to fly just $10 million of experiments. George Low was asked whether, in the event that the Soviets failed to step up to the plate, it would even be *worth* flying such a mission. "I think that answer would depend very much as to when this would happen," he replied. "Were it to happen *now*, when we have spent a substantial amount of money, which is still a small fraction of the $250 million, we might well decide and discuss ... the possibility of cancelling it altogether, because I am not sure whether it is worth remaining funds to be expended to go up there in 1975 for the $10 million-worth of experiments alone, without the rendezvous and docking. On the other hand, if the spacecraft were on the launch pad and ready to go and, for some reason, the Soviet portion of the mission [was] cancelled, then NASA would likely want to go ahead with the flight, but only after consulting with and obtaining the approval of the Congress and the executive branch." In Low's opinion, the best way forward in view of the sheer cost of the mission was to develop contingency plans, but to assume that the Soviets *did* intend to fly with them in 1975, since a joint effort was clearly the most desirable option. Yet this discussion of alternative strategies underlined the nagging doubts – political concerns, yes, but also technological ones in the wake of the Soyuz 11 disaster – which plagued many lawmakers. The reliability of the Soviets to stand by their pledges remained questionable in many minds.

In the meantime, efforts to finalise the design of the docking module continued and of pivotal importance in its development was a maxim to keep its systems and structures as simple as possible. At first, it seemed as if the Marshall Space Flight Center of Huntsville, Alabama – which had been responsible for constructing several key spacecraft systems, including the Apollo Telescope Mount for Skylab – might build the docking module, but the contract eventually went to North American Rockwell. In its final form, it would measure 3.15 m long and 1.4 m in diameter and would weigh a little over 2,000 kg. To avoid having to subject the crew to lengthy 'pre-breathing' to rid their bodies of nitrogen, it served as an airlock to raise or lower the atmospheric pressure between 34.4 kPa in the American craft and a slightly reduced 68.9 kPa in the Soviet craft with a series of equalisation valves. The docking module's pressure vessel was formed from a welded cylinder of aluminium, 1.58 cm thick, with a tapered bulkhead and tunnel at the American 'end' and a machined base and bulkhead at the Soviet 'end'. Controls and displays for pressure equalisation, VHF and FM transceivers, environmental subsystems, storage compartments, fire extinguishers, masks and a multi-purpose furnace for materials processing experiments filled the remainder of the module. On the exterior were four spherical gaseous tanks for 18.9 kg of nitrogen and 21.7 kg of oxygen and electrical power was routed from the Apollo command module. No provision was afforded for carbon dioxide removal; this function would be fulfilled by the existing environmental controls of each spacecraft.

The importance of having an 'alternate' mission, to which the American crew could move in the event of a cancellation of the Soviet launch, was also the highlight of an intriguing December 1971 study by North American Rockwell, which envisaged *two* missions in one, at *two* quite different orbital altitudes. It considered a command, service and docking module, launched atop a Saturn IB booster, initially into a 300 km orbit to rendezvous and link up with a Salyut station. Two days of joint activities would then be undertaken with the Soviet crew, after which Apollo – and its docking module – would separate and lower its orbit to around 285 km. The Americans would then spend no fewer than 11 days on a detailed Earth Resources Survey (ERS), utilising a large battery of cameras, instruments and other remote-sensing gear in the service module's SIMbay; for at the time of this study, it was still presumed to use an Apollo with a SIMbay. What is more, the North American Rockwell study even called for a *spacewalk*, five hours before re-entry, to recover film cassettes from the bay. In total, such a mission would run to just over 14 days. Although it never reached the official planning stages and represented little more than a feasibility study, it has been noted here since it clearly demonstrated the American need for 'alternate' mission scenarios to maximise scientific gain (and doubtless satisfy congressional critics), in addition to fulfilling the basic premise of rendezvousing and docking with a Soviet spacecraft.

As the joint mission steadily took shape, the political masters of Gilruth and Lunney and Bushuyev and Keldysh drew steadily closer to the negotiating table which would turn it from a relatively informal effort into a firm and 'real' international endeavour. NASA's senior managers were keenly aware that congressional authorisation and appropriations would have to be obtained before Spacecraft Nos. 111 and 119 could be modified or work begun to build the docking module. Politically, a bilateral agreement between the governments of the United States and the Soviet Union was needed *before* additional funds could even be requested and it was in the spring of 1972 that NASA Headquarters recommended to the White House that a formal agreement for a joint Apollo-Salyut mission should be included in a planned summit, due to be held in Moscow that May. Henry Kissinger wanted a firm recommendation from NASA by mid-April and a 'secret' visit to the Soviet Union was planned by Glynn Lunney, George Low and the agency's head of international affairs, Arnold Frutkin, to secure final agreement on technical proposals, organisational planning and a long-range schedule for the project.

The visit was unofficial and unpublicised – indeed, Low's secretary went so far as to purchase travel tickets from a commercial agent, rather than through NASA – and the three men arrived quietly in Moscow on 3 April 1972. Unfortunately, one of the key Soviet academicians, Boris Petrov, had been quoted by a *New York Times* journalist the day before that "meetings will take place early next week" between key officials. Luckily, no one picked up on Petrov's words and Low, Lunney and Frutkin were able to arrive in Moscow with no fanfare and got straight down to business with Petrov, Bushuyev and Vladimir Kotelnikov, newly appointed as acting head of the Soviet Academy of Sciences to replace a hospitalised Mstislav Keldysh. During their three days of meetings, Kotelnikov explained, almost matter-of-factly, that a joint mission with Salyut was no longer an option, since the station had only one docking port and the addition of a second port for a Soyuz would be both costly and technically difficult. In its stead, Kotelnikov advocated flying a joint mission using Soyuz, which was capable of accepting all of the modifications needed, including those associated with the docking module.

Surprise, wrote Ezell and Ezell in *The Partnership*, was too mild a word for the Americans' reaction, but Low agreed that – barring any technical difficulties – the switch from Salyut to Soyuz would be acceptable. Operationally, it would also prove much simpler, since it would necessitate only *two* co-ordinated launches (of Apollo and Soyuz), rather than *three* (with the space station), and would still permit the transfer of crew members between craft as the major public relations aim of the exercise. Having said this, it was a frustrating time, particularly for Low, who felt that the unresponsiveness of the Soviets and numerous quibbles over the wording of individual documents was hampering the talks; at one stage, he told Kotelnikov that it would be difficult for him to present the joint mission in a positive light to President Nixon. Openness, particularly with regard to the media, remained an overarching concern and Low stressed the absolute need for NASA to continue its time-honoured policy of disclosing *all* information available from its control centres and tracking stations. Despite the lengthy dialogue and the unexpected changes, Low's team returned to the United States, firmly committed to undertaking a joint flight, but with a difference: it would no longer be an 'Apollo-Salyut Test Mission', but an 'Apollo-Soyuz Test Project'.

THE MEN AND THE METTLE OF APOLLO-SOYUZ

The men who formally agreed the details for the joint mission and a ground-breaking series of arms-limitation deals in May 1972, ironically, had little interest in space. Leonid Brezhnev, General Secretary of the Communist Party, was *de facto* (though not *de jure*) master of the Soviet Union and by his hand the largest nation on Earth had undergone its most intensive phase of socioeconomic stagnation and political repression since the days of Joseph Stalin. At Brezhnev's direction, an attempt by Czechoslovakia's communist leadership to institute liberal reforms and improve human rights had been infamously crushed by 250,000 Warsaw Pact troops in August 1968 and his stance towards the space effort was far cooler than that of his predecessor, Nikita Khrushchev. In fact, wrote Asif Siddiqi in *Challenge to Apollo*, Brezhnev supported space "only if it brought political dividends".

Unlike the United States, in which the fundamental positions of power are vested in the President, the Soviet Union operated a confusing structure of 'collective leadership' after Khrushchev's 1964 downfall: the General Secretary was administrative head of the Communist Party, the Chairman of the Presidium was the *de jure* head of state and the Premier was effectively the head of government. For a time in the mid-1960s, many Western politicians and diplomats, including Lyndon Johnson and Henry Kissinger, viewed Premier Alexei Kosygin as the head of the Soviet Union. He had been sent on key missions to mediate between India and Pakistan in 1966, successfully persuading them to sign the Tashkent Agreement, and he fought vigorously to end an ideological 'split' between the Soviet Union and China. Kosygin's ultimate downfall, though, came from his desire to spearhead economic reforms, in which he sought to transform Russia from a strict (and failing) 'command economy' into something which drew closer parallels with the West, 'guided' rather than 'controlled' by the state. Many of his reforms were accepted at first, primarily because the effects of a harsh economic downturn were already being felt. Eventually, though, Kosygin triggered a strong backlash from hardliners within the Communist Party, few of whom wanted to relinquish their stranglehold on the economy or agree willingly to 'bourgeois' notions of private ownership and less rigorous oversight by the state. Consequently, by the early 1970s many of Kosygin's former supporters had flocked to Brezhnev's banner, allowing him to consolidate near-absolute power. Although it was Kosygin who signed the agreement for ASTP in Moscow, this would prove to be one of his last official acts on the international stage and he would scarcely be seen outside communist countries from then until his death in 1980.

The political status of ASTP as a symbol of friendship was of equal importance to Richard Nixon, whose name today is a notorious byword for the expression that absolute power corrupts absolutely. His five-year tenure at the helm of the United States began with an earnest attempt to slash military and space spending, to extricate hundreds of thousands of troops from an increasingly bloody campaign in Vietnam and to improve the situation at home, with rises in payments of government benefits, including Social Security and Medicare. Between 1969 and 1973, Nixon's Republican administration systematically cut NASA's funding from 2.1 percent of the federal budget to just 1.1 percent, whilst at the same time turning the Space Shuttle from a paper dream into the foundation of a real system which would guide America's ambitions in low-Earth orbit for more than three decades. On the down side, he had no interest in further manned exploration of the Moon or missions to Mars and this may have prompted the departure of the visionary (and Democrat) Tom Paine from NASA in the summer of 1970. Yet Nixon *was* fascinated by the possibility of ASTP, primarily from the standpoint of forging real and lasting co-operation between two bitter foes. The conflict in Vietnam had proven disastrous and the early 1970s marked a transitional time in which peace was acutely needed.

When George Low returned to Washington in April 1972, he brought positive news: despite the surprising, and seemingly last-minute, change from Apollo-Salyut to Apollo-Soyuz, he told Henry Kissinger that NASA saw the joint mission as a realistic goal and no further meetings would be required before the planned summit in Moscow. On the evening of 24 May, Richard Nixon and Alexei Kosygin duly signed the 'Agreement Concerning Co-operation in the Exploration and Use of

The commanders of ASTP – Alexei Leonov (left) and Tom Stafford – pose in a command module simulator in February 1975. Both crews carried quarter pieces of two commemorative plaques into orbit, which were connected together during the joint mission and returned to Earth aboard their respective spacecraft. These two men had played pivotal roles in their manned space programmes – Leonov as the first man to walk in space and Stafford as commander of the final, critical dress rehearsal before the first Moon landing – and had seen the competition and 'peaceful war' between the United States and the Soviet Union at its most intense. Now, with a nod of ironic humour, their final journeys into space would bring them together as *partners*.

Outer Space for Peaceful Purposes', a keystone pledge which would not simply encapsulate plans for Apollo-Soyuz, but would also set in stone hopes for a new era of sharing biomedical and scientific data and jointly tackling a plethora of pressing environmental and natural problems. That same evening, in Washington, Vice-President Ted Agnew invited Jim Fletcher to brief the press; the latter spoke of his enthusiasm for ASTP, which was now tied to a specific timeline to be completed by the summer of 1975 and therefore more likely to succeed. The words of conciliation went on late into the night, with Nixon telling the people of both nations that they would "sometimes be competitors", but "need never be enemies". Although ASTP was the start, the real deal came two days later, when he and Brezhnev signed the two arms-limitation treaties to lay the foundations of a new era of US-Soviet détente.

From NASA's perspective, ASTP was now 'real'. By the middle of June, its management had been transferred from the Office of Advanced Missions to the Apollo Program Office and one of the earliest moves was a $64 million arrangement with North American Rockwell to modify an existing Apollo spacecraft and to proceed with the construction of the docking module and androgynous mechanism. It was already known by this time that the Apollo to be flown for ASTP would be the last of its kind and was to be constructed by a workforce in steady decline; the Soyuz, on the other hand, was being assembled on a 'production-line' basis and would be used for many years to come.

The original intention, noted ASTP advisor Clarke Covington, was for Apollo to launch first, since it could support a longer mission of perhaps 14 days, compared to Soyuz, which "had this man-day limitation which limited them to about four days, plus a day margin". In effect, Apollo had the ability to launch first with enough consumables to wait in orbit for a few days if the first attempt to launch Soyuz was unsuccessful.

However, the Soviets were adamant that Soyuz should fly first. "We discussed this for several hours," Covington recalled in *The Partnership*, "not really knowing or understanding their reasons. They went through a lot of flight mechanics-type [explanations]. All of our ... people were ... trying to understand why they wanted to go first. It was really ... illogical. We were convinced – this was privately – that they had a fear of failure." Shortly thereafter, Covington received a draft copy of the Soviets' technical proposal for the project and tracked down a translator to help him read it. The Soviet rationale for flying ahead of Apollo, it seemed, was indeed rooted in the risk of failure: should something go wrong with the Soyuz, it would return to Earth and a *second* craft would be readied for immediate launch! Covington asked the translator to repeat the passage: surely the Soviets were not preparing *two* Soyuz in tandem to fly the same mission? Next morning, before meeting with the Soviet delegation, Covington cornered Glynn Lunney with the news. Lunney was equally amazed and when he asked the Soviet officials, they responded that, yes, there *would* be two Soyuz prepared for the flight. "That totally cratered our argument as to the need for Apollo to go first," remembered Covington. "If they were willing to launch two spacecraft, then there wasn't any reason why they should *not* go first!" Why the Soviets had not revealed this tiny nugget of crucial information earlier remained unknown and was passed off by the NASA team as just one of the numerous 'idiosyncrasies' around which they would be obliged to work.

By the middle of July 1972, the decision had been finalised: Soyuz *would* fly first, after which there would be as many as three launch opportunities for Apollo, beginning about seven and a half hours after the Soviet crew had reached orbit. The rendezvous between the two craft would require approximately one full day and produce the optimum conditions for station-keeping and a docking situation in which Apollo would take the 'active' role. After docking, Lunney told the assembled journalists, the two spacecraft would remain in a tight mechanical embrace for two days, during which an "agreed schedule" of exchanges of crew members would occur, together with the "performance of ... joint experiments". The spacecraft would then undock and complete their individual flight programmes before returning to Earth. For planning purposes, it was decided that the launch would take place on or around 15 July 1975, but that the timing and availability of a compatible trajectory plan would permit a mission during the ensuing three-month period, until October. The Soviets agreed to provide the Soyuz orbital parameters to NASA within six hours of launch, allowing the Americans to refine the Apollo trajectory accordingly.

NASA managers were initially perplexed as to why the Soviets insisted on Apollo being virtually leak-proof: in fact, they had implored the Americans from the outset to tighten their leakage rates, which were more than ten times greater than their own. It was not unusual for Apollo crews to lose as much as 40 g of oxygen per hour, for which NASA compensated by carrying additional oxygen replenishment stores. The Soviets, on the other hand, carried no additional reserves of pressurised gases aboard Soyuz. From their perspective, the Americans needed to better understand the Soyuz environmental controls before finalising the form of the docking module. Of critical interest to the Soviets was the safety of several valves which vented gases overboard and – haunted by the deaths of the Soyuz 11 crew the previous year – they insisted on the inclusion of a 'redundant' system as a precaution. By October 1972, it was decided that Apollo would continue to operate at its standard pressure, but that of Soyuz would be slightly reduced during joint operations in order to eliminate the need for lengthy pre-breathing. During the deliberations, external transfer (space-walking from one craft to the other) and the remote possibility of returning one crew member in a different craft in an emergency were also presented for discussion.

Years later, Clarke Covington recalled the Soviet awareness – *embarrassment*, even – that the capabilities of their craft were inferior to those of Apollo. When they discovered that they had an orbital 'ceiling' of some 225 km due to the weight of Soyuz when configured for ASTP, they promptly asked the Americans to lower the docking orbit from 232 km to 222 km, but made the request under the guise of a 'technical problem'. After much confusion, Covington eventually asked his Soviet counterpart whether the reason was because they had a weight problem and that their launch vehicle could not loft the spacecraft into a higher orbit. The response was in the affirmative. "It was no real problem to *us*," Covington remembered, "but we just couldn't understand why they wanted to do it. It *was* a big deal for them. They seemed to embarrass easily about the capability of their spacecraft, which they had no need to do. Their spacecraft was designed for a different thing than Apollo was. The Apollo ... is way over-designed for this mission: it was built to go to the Moon and back. We just had an inherent capability greater than theirs." The Soviet Union's desire to boast of its 'leadership' in space and its ingrained sense of national pride proved a difficult barrier to overcome.

As 1972 drew to a close, both sides agreed that an initial training session for the respective crews of the American and Soviet missions should take place no later than the summer of the following year in Houston. The next training event would be hosted in Moscow that autumn. It was agreed that six joint training sessions would be run in total, three in the United States and three in Russia. On the Soviet side, the plan centred on picking a pair of 'prime' crews and a pair of 'backup' crews. This would provide redundancy in the event of a Soyuz failure and the need to be ready to despatch a second craft. Both sets of crews would train alongside the Apollo team. When one considers the downsizing of crews in the aftermath of Soyuz 11, it was obvious from the outset that the Soviet ASTP team would comprise two cosmonauts. However, it was far from certain – at least until the late summer of 1972 – precisely how many men would fly aboard the American craft, even though it was nominally designed for a crew of three. When asked at a Houston press conference on the afternoon of 24 May 1972, immediately after the Nixon-Kosygin summit, Glynn Lunney was non-committal. "We have not yet made up our mind whether we will fly two or three," he explained. "The decision depends upon detailed timelining of the activities so that we can determine if there are any obvious advantages or disadvantages of having three or whether we would prefer two. Secondly, some better understanding for what experiments we might fly which would determine what kind of stowage … we would have in the command module, the space allowable being some factor in deciding how many people fly.

"The other factors," Lunney continued, "are precisely how the timeline will work and how satisfactory it will be. In other words, we will try to create a timeline with … two American astronauts and see how many people we have, where, for how long and how convenient the timeline is. We will do the same thing with three men and see whether there are any obvious advantages one way or the other, out of that kind of comparison. Likewise, we are going to try to consider what experiments we might have in the command module volume and the cockpit itself to see whether any space that those experiments might require would influence our decision on having the third couch in there." Certainly, the question of 'timelines' and 'schedules' arose at two joint ASTP meetings, one in Houston in July and another in Moscow in October, and it might be assumed that the eventual decision was made at this stage.

The names of the Apollo crew were announced on 30 January 1973. In command would be one of NASA's most seasoned astronauts, one of its most experienced pilots in terms of rendezvous and docking, a veteran of three previous spaceflights – including a mission to orbit the Moon – and the former chief of the astronaut office: Brigadier-General Thomas Patten Stafford Jr of the US Air Force. Joining him as command module pilot was civilian aviator Vance DeVoe Brand, an ex-Marine Corps fighter and test pilot who had previously backed up the Apollo 15 lunar flight and was at that time backing up two of the three missions to the Skylab orbital station. Just a few months after the ASTP announcement, but for a twist of fate, Brand might also have commanded the world's first 'space rescue', a topic to be discussed later in this volume. The final crew member was retired Air Force Major Donald Kent Slayton, who, on account of his initials, had gained the nickname of 'Deke' whilst a test pilot at Edwards Air Force Base in California. He would fill the role of docking module pilot. Although two members of the crew would be making their first flights, these men would bring an immense amount of experience to the table, none of them more so than Slayton himself.

Born near the small town of Sparta, Wisconsin, on 1 March 1924, he would make history on ASTP by becoming the oldest human ever rocketed into the heavens. Yet at the press conference, he responded to a question about his age with the typical blend of dry, gruff Slayton humour: "I guess I'd rather be a 50-year-old rookie than a 50-year-old has-been!" It should not have been that way, for Slayton was originally scheduled to fly in space in 1962, but was grounded for more than a decade by a persistent heart problem. During this time, he assumed the mantle of co-ordinator of astronaut activities – a precursor to the role of chief of the astronaut office – and later headed NASA's flight crew operations directorate, selecting and overseeing the training of each of America's space pioneers, including Neil Armstrong. In this position, he became a virtual father figure to many of them. He was a man labelled "the best" by Mike Collins and described as a dependable colleague who could be trusted "to get the job done, no matter what the job was" by John Glenn. In the words of a *Los Angeles Times* reviewer – writing on the flyleaf of Slayton's autobiography – it was indeed "one of the great fortunes of Apollo that [he] was grounded . . . for in 'Father Slayton' NASA had the man who could set the rotations, pick the teams and make the decisions the astronauts wouldn't have accepted from a bureaucrat or a scientist".

From earliest childhood until the very end of his life, Slayton was an adventurer; as a four-year-old child, whenever allowed outside into the front yard, he was tethered to a tree by his mother at playtimes – not as a punishment or a cruelty, but out of her sheer terror at the amount of traffic which hurtled past the family's dairy farm in small-town Wisconsin. "Eventually," Slayton wrote, "I convinced my mother that I wasn't going to go running into the road and I was set free. But I can make the case that ever since I was young I have wanted to explore . . . and people have tried to stop me."

In many ways, Slayton's upbringing on the family farm was the making of him. The work was hard, but rewarding, and it taught him the skills of business, entrepreneurship, book-keeping and even the fundamentals of veterinary science. It was, he said, a good place to be raised and years later he would write, only partly in jest, that "compulsory farm rearing" would be an excellent experiment in building character. One of the downsides came in 1929 when the five-year-old boy was helping his father to clear hay from a horse-drawn mower . . . and cleanly sheared off the ring finger of his left hand! "I was luckier than hell," Slayton wrote, "because I could have lost all of them. It didn't hurt, but of course, my dad was pretty upset about it." Physically, it hardly affected Slayton and when he took up boxing in his teens, he found that he could hit just as hard with his left fist as with his right. However, for many years, he was convinced that his missing finger was the first thing that anybody noticed about him.

Another interest to which Slayton was drawn in his teens was aviation, although by his own admission he had always feared warfare as a child. Nevertheless, four months after the Japanese attack on Pearl Harbour, in April 1942, he signed up with the US Army Air Corps and underwent initial instruction in San Antonio, Texas, before moving to Vernon for flight training. His missing ring finger almost eliminated him from selection by the military, "but they checked the regulations and discovered that the ring finger on your left hand was the only finger you could have

Curled inside the docking module, which he had nicknamed "the world's fastest, highest-flying sewer pipe", Deke Slayton familiarises himself with the systems of the tunnel-like adaptor in February 1975. As well as taking responsibility for its development, testing and pre-flight processing, Slayton – one of the original seven Mercury astronauts, but barred from flying for more than a decade, due to a heart murmur – would also enjoy his first night's sleep in orbit within the cramped confines of the docking module.

missing on either hand and be qualified as a pilot. Your ring finger, they decided, is the most useless finger on your hand!'' Doubtless, many divorcees would be inclined to agree ...

Soaring through the skies in the Stearman biplane and the single-winged Fairchild PT-19 and North American Aviation's AT-6 Texan, Slayton found a natural affinity for flying. Within a short time, he was assigned to advanced multi-engine training and in late 1943 was designated a combat-ready pilot for the B-25 Mitchell bomber. It was at this stage that he saw action for the first time, attached to the 340th Bombardment Group, initially in North Africa and later in the bombed-out Italian city of Naples. "There's a lot of just pure horseshit luck in flying," Slayton later wrote in his typically straightforward manner and his experience flying bombers over Italy and the Balkans was certainly demonstrative of this. On more occasions than he could count, he nursed B-25s back to base with blown tyres and missing hydraulics or riddled with hundreds of holes of anti-aircraft gunfire. By the end of his time in Europe, Slayton had flown 56 combat missions and after a short spell back in the United States he was assigned in April 1945 to the 319th Bombardment Group on the island of Okinawa. His seven combat missions over Japan were, in his own words, quite different to those over Italy and the Balkans and by early August he was flying the A-26 Invader bomber, strafing coastal towns in readiness for an amphibious invasion by US troops. When Slayton flew a combat mission on the 12th, he did not know until after his return to base that the Second World War was over. A few days earlier, two "big bombs" had been dropped on Hiroshima and Nagasaki ... and Slayton's combat career was over.

After a year as a B-25 instructor, he left military service in November 1946 and enrolled at the University of Minnesota to study aeronautical engineering. Whilst there, he maintained his proficiency with the Air Force Reserve – the Army Air Corps having become a separate branch of the military services in 1947 – and later joined the Minnesota Air National Guard, flying P-51 Mustang fighters. Two years later, in 1949, Slayton graduated in aeronautical engineering and was hired by Boeing to work on its new B-52 Stratofortress bomber. When the Korean War broke out in the summer of 1950, however, he was drawn back to the cockpit and active service. A rude awakening came when he tried to secure readmission into the Air Force: since he had not applied for inactive reserve status within the required six months of leaving the Minnesota Air National Guard, he was officially now a civilian ... and had to resume his career from scratch. Luckily, a former squadron commander managed to secure him a place as the Guard's maintenance officer. Two incidents of note occurred whilst there. The first was an immense storm which damaged perhaps 90 percent of the aircraft in his care ... and the second was his sighting of an unidentified flying object whilst wringing out a newly-repaired P-51.

It looked like a kite at first, but as Slayton drew closer he became convinced it was more likely a high-altitude weather balloon ... until, that is, he approached it from a slightly different angle and it took the form of a disk on edge. It seemed to climb at a 45-degree angle away from him and then disappear. Slayton did not report the incident for a couple of days, then happened to mention it in conversation – only to be hauled before an Intelligence panel to make a formal statement. The official position was that a local company had been testing a weather balloon and that, in addition to Slayton's sighting, it had been tracked by a light aircraft and two ground-based observers with a theodolite. The only point of confusion was that its velocity was clocked at more than *six thousand kilometres per hour*! "I don't automatically presume that it came from Alpha Centauri," Slayton wrote, "just because I can't identify it. It's still an open question to me." Years later, though, he would wonder if his sighting ended up in the pages of the Air Force's official UFO investigation, Project Blue Book.

Slayton's desire to participate in the Korean conflict ultimately came to nothing and he served three years in Germany as a maintenance inspector and squadron maintenance officer, securing his first experience in jet fighters. Then, in June 1955, he entered test pilot school at Edwards Air Force Base in California – from which Chuck Yeager had broken the sound barrier in the Bell X-1 – and qualified six months later. By his own admission, test flying in the mid-1950s was risky and the pilots did not know the performance of their aircraft until they got back on the ground and examined the data. "Today they do everything real-time with telemetry and computers," he wrote. "You do a test point and you can do analysis on it and say, obviously we're going to the next one, and proceed. We had to lay out a flight plan and say, okay, I *think* I can go this far this time. If you guessed right, you were okay. If you didn't ... "

Graduation in December 1955 led to Slayton's assignment to fighter testing. The presence of another Donald – Don Sorlie – caused confusion in the air-to-ground radio chatter. For a while, Slayton was called 'D.K.' and finally 'Deke', a nickname which would remain with him for the rest of his life. Whilst at Edwards, he tested several supersonic Air Force fighters, including the F-101 Voodoo, F-102 Delta Dagger, F-105 Thunderchief – for which he was responsible for determining spin-stall characteristics – and the F-106 Delta Dart. Shortly afterwards, the Soviets launched Sputnik and the first mutterings of putting a man in space reached the ears of the test-flying community. One day, late in January 1959, Slayton received mysterious orders to attend a classified briefing in Washington.

Over the course of the next few weeks, he and more than two dozen other pilots were poked and prodded at Dr Randy Lovelace's aerospace medicine clinic in Albuquerque, New Mexico, to evaluate their suitability for Project Mercury, the goal of which was America's first manned space mission. Every spot on their bodies was sampled and measured, with scarcely a muscle or gland left untouched. Throats were scraped, stool and semen samples were taken, electricity was zapped into hands and intensely uncomfortable 'steel eels' were inserted into rectums. Still more tests followed at the Aeromedical Laboratory of Wright-Patterson Air Force Base in Dayton, Ohio, where the pilots withstood cold water being pumped into their ears, sat for hours in overheated saunas, endured soundproofed and darkened isolation chambers, blew up balloons until they were out of breath, walked on treadmills until their heart rates soared to 180 beats per minute and were photographed from every conceivable angle and into every conceivable orifice. Many perceived the whole thing as excessive and a waste of time. "I'd flown combat missions and done operational test flying for 17 years by that point," Slayton wrote. "The fact that I'd *survived* should have told them everything they needed to know about stress."

Together with six other top-notch pilots – Scott Carpenter, Gordo Cooper, John Glenn, Gus Grissom, Wally Schirra and Al Shepard – Slayton was selected by NASA in April 1959 as a member of the United States' first team of astronauts. Training for Project Mercury was fierce and unforgiving, but after three years Slayton seemed to be in pole position for a flight into space: assigned to his first mission, he was scheduled to launch in May 1962 and would have spent five hours in orbit, completing no fewer than three circuits of Earth. He picked a name for his spacecraft – 'Delta 7' – which was "a nice engineering term that described the change in velocity". Unfortunately for Slayton, his own velocity, both in spacecraft and high-performance jets, would decline markedly, thanks to a minor, yet persistent, heart condition, known as 'idiopathic atrial fibrillation'. This took the form of occasional irregularities of a muscle at the top of his heart, caused by unknown factors and extremely rare in highly-fit adults. It had first arisen during a training run in the centrifuge in August 1959, prompting NASA flight surgeon Bill Douglas to obtain a clinical electrocardiogram at the Philadelphia Naval Hospital. This concluded that Slayton had a 'flutter' in his heartbeat, but further tests verified that the condition should not impair his work as an astronaut.

The problem resurfaced a couple of years later, with speculation that one of the astronauts, perhaps John Glenn, had a heart problem. Apparently, Slayton wrote, the call to Bill Douglas came from Air Force physician George Knauf, attached to NASA Headquarters, and had originated from "a source higher than the Department of Defense". Douglas denied that Glenn had a problem. Knauf asked next if Glenn's backup, Scott Carpenter, had a heart murmur; again, the response was negative. Then, to reinforce the point that the matter was of little relevance, Douglas explained that Slayton had long been known to have a minor heart condition. He expected this to be the end of the matter, but he unwittingly opened a can of worms.

Back in 1959, flight surgeon Larry Lamb had routinely examined Slayton and became convinced that heart fibrillation should disqualify him from the selection process. "He hadn't said so in 1959," wrote Slayton, "but he said so now. I don't think it was anything personal – this was just his medical opinion." Although Lamb's judgement was very much a voice in the wilderness, he also happened to be Vice-President Lyndon Johnson's cardiologist and in the spring of 1962 began to question the astronaut's suitability to fly. (Matters were not helped when the publicity-seeking Johnson was refused access to John Glenn's home during the latter's historic spaceflight on 20 February.) In early March, NASA Administrator Jim Webb reopened Slayton's medical file and the astronaut and Douglas were summoned to the office of the Air Force surgeon-general in Washington. A panel of military physicians signed Slayton off as fit to fly and their decision was endorsed by Air Force Chief of Staff Curtis LeMay.

For Webb, this was not enough. The Secretary of the Air Force, Eugene Zuckert, insisted that a panel of civilian physicians should also examine Slayton at NASA Headquarters. On 15 March 1962, only weeks before his scheduled launch, Slayton had his heart monitored by Proctor Harvey of Georgetown University, Thomas Mattingley of the Washington Hospital Center and Eugene Braunwell of the National Institutes of Health. As Slayton waited afterwards, NASA Deputy Administrator Hugh Dryden entered the room and told him, point-blank, that he could not fly. None of the physicians had found a specific medical reason to keep him off Delta 7, but their consensus was that if NASA had pilots available who did not have his condition, one of them should fly instead. Slayton, not surprisingly, was devastated.

Years later, he was convinced that the decision to ground him was a political one. "NASA knew it would have to publicly disclose my heart condition prior to my flight," he wrote. "There would be medical monitors at tracking stations all over the world who wouldn't know how to react otherwise. Everybody expected this to be a big deal. NASA would be opening itself up to a lot of medical second-guessing." Bill Douglas felt that problems could arise if the astronaut were to start fibrillating whilst on the pad – "do you scrub the launch or go ahead?" – but he was confident that Slayton was the best person to fly. However, Jim Webb's fear that it could trigger adverse headlines for NASA drew a line in the sand. "It didn't matter that a whole lot of doctors thought I didn't have a problem," Slayton wrote of Webb's actions. "He was only going to listen to the few who did." A launch abort could subject the astronaut to acceleration loads as high as 21 G and Webb feared the very real possibility that Slayton, dehydrated and perhaps fibrillating, could die during descent. The impact on the agency would be profound. Less than a year earlier, in May 1961, the newly-elected President John Kennedy had boldly committed the nation to landing a man on the Moon before the end of the decade, in order to score a major technological and political triumph over the Soviets. Any incident in which an astronaut died could stop this plan in its tracks.

The next day, 16 March 1962, the grounded and furious Slayton was forced to sit through a lengthy press conference, in which the minutiae of his case were examined. Hugh Dryden remarked that, despite the decision, Slayton might remain eligible for future flights. When a journalist asked if the problem had been caused by stress, Slayton responded no, and further that he did not even *know* about it until being hooked up to the electrocardiogram in 1959. Bill Douglas' own departure from NASA within days of the announcement was leapt upon by some journalists as 'evidence' of his bitterness over Slayton's treatment, but in truth he was already at the end of a three-year detachment to the space agency and his return to the Air Force had been in the works since mid-1961.

Despite his grounding, Slayton did not give up on flying. "I made some changes in my lifestyle," he wrote, "gave up drinking, started working out more regularly – quit doing everything that was fun, I guess!" Douglas also secured him an examination by Dwight Eisenhower's cardiologist, Paul Dudley White, in June 1962. White advised him that two-thirds of people with his condition would die early, whilst the remainder would probably never know they had it and might never be affected. The verdict: "Young man, you're going to live a long time." However, White's report, which highlighted that Slayton did not appear to have a problem, concurred that if astronauts were available without the condition, it would be preferable to assign one of them to the mission. When it became clear that he would not draw assignment to any of the remaining Mercury missions, Slayton turned his attention to the two-man Project Gemini, at that time scheduled to begin flying in 1964, only to be told that his ailment would make him a "hard sell" to senior management. Shortly thereafter, the Air Force decreed that Slayton no longer met the qualifications for a Class I pilot's licence – he could no longer fly solo – and, at the end of November 1963, he resigned from the service.

Although he would eventually get his ride into space, it would not be until ASTP in 1975. A lesser man might have thrown in the towel and departed NASA for pastures new, but not Slayton. With no guarantee of a mission, he decided to stay and in the summer of 1962, as the agency prepared to expand its corps by picking nine new pilots, he was appointed as co-ordinator of astronaut activities. NASA's initial plan to bring in a manager from the outside to oversee the corps was quashed by the astronauts themselves. "What we wanted the least," wrote Wally Schirra, "was somebody who would outrank us and issue orders in a military way. We wanted someone who knew us, who trained with us. Deke was the one and only choice." As America pushed for the Moon, Slayton was all-powerful within the astronaut office, deciding the career paths of the men who would someday walk on the lunar surface ... and those who would not.

Even in the early 1970s, his chances of ever regaining a *pilot's licence*, let alone securing the opportunity to ride a rocket, seemed infinitesimally small. Somewhere within that infinitesimal smallness, though, was a glimmer of hope. One day in the summer of 1970, during an antelope hunt in Wyoming, Slayton experienced his first heart fibrillation in several months; he had not suffered one ever since flight surgeon Chuck Berry loaded him up on vitamins following a rotten cold. As a personal experiment, he started taking the vitamins again ... and the fibrillations went away. "It wasn't the kind of thing you could use as hard medical evidence," Slayton wrote, "but it got me thinking that I might still have a more realistic chance." More than a year later, Berry happened to attend a medical conference in Istanbul and broached Slayton's case with Hal Mankin, a specialist who agreed to run tests. In December 1971, Slayton flew to Rochester, Minnesota, and checked into the Mayo Clinic under the false name of 'Dick K. King'. Whilst there, Mankin put him through a battery of tests, hanging him upside down on a treadmill, poking holes in him, pumping dye into his system and examining "parts of my body I didn't even know I had". At last, an angiogram yielded the final judgement. "Your man," Mankin told Berry, "is as good as gold!" Slayton had passed with flying colours, requalified for a Class I pilot's licence and in March 1972 – now aged 48 and a full decade since being grounded – he was restored to active astronaut status.

His problem now was getting a flight ... and the list of options was short. Two more lunar landing missions were scheduled for April and December, both of whose crews had already been immersed in training for several months. So too had the crews for three flights to the Skylab space station in 1973. For Slayton, the Apollo-Soyuz venture with the Soviets was his last-chance saloon. As head of flight crew operations, it was nominally Slayton who oversaw the astronaut selection process, but since he now considered *himself* a candidate, he asked Chris Kraft (who, following the retirement of Bob Gilruth, had taken charge of the Manned Spacecraft Center in January 1972) to handle the assignment on his behalf. A few weeks after Nixon and Kosygin signed the ASTP agreement and Kraft duly asked Slayton for his recommendation for the American half of the mission. Despite having never flown in space, Slayton was convinced that his seniority would be more than enough to carry him through ... and recommended himself for the position of command. For his crewmates, he identified a pair of astronauts whom he held in extreme high regard: Jack Swigert, a veteran of the ill-fated Apollo 13 mission, and Vance Brand, who was at the time training for the backup command of two Skylab flights.

Unfortunately, Swigert's name disappeared from the ASTP list almost as soon as it had arrived, in the wake of a particularly ugly affair which tarnished the reputation of NASA in the early 1970s. During the Apollo 15 mission, astronauts Dave Scott, Al Worden and Jim Irwin had carried several hundred first-day covers to the Moon and back, part of an arrangement with a German stamp dealer named Walter Eiermann. The intention was to sell them after the flight and split the profits to set up trust funds for their children. However, the covers – four hundred in total, a hundred for each of the three astronauts and the remainder for Eiermann – had *not* been authorised by NASA. The deal called for the covers to be sold exclusively to collectors, privately and with no publicity, but the German had other ideas and began to sell them within weeks of Apollo 15's return to Earth. The furious astronauts cancelled their agreement, but it was too late: the story had leaked into the European press and by the late autumn of 1971 some of the covers were fetching several thousand dollars apiece.

Although none of the Apollo 15 crew accepted any money, NASA quickly found itself in the midst of a full-blown public scandal ... and as the agency's inspector-general dug deeper into the astronauts' finances he found that others had signed first-day covers for Eiermann. One of them was Tom Stafford, who quickly proved that *his* payment went directly to charity and never touched his hands. Jack Swigert, though, revealed more money in *his* bank account records than he theoretically should. His response was one of defiance and self-defence: "It's *none* of your damn business!" He took legal advice and challenged NASA to say which law he had broken, but none could be found. Chris Kraft was in a quandary: some colleagues told him to drop the case, whilst others were aware that the covers were still in circulation and had actually aroused a degree of notoriety. Ultimately, the remaining covers were confiscated by the inspector-general and turned over to the Department of Justice. "Our own NASA lawyers didn't expect indictments," Kraft wrote, "but those Justice hotshots knew the law better than anyone else. If there was a crime here, they'd find it. They didn't." Congress demanded an investigation into what it perceived as "improper conduct" and many astronauts were mixed in their feelings; some were convinced that court-martials were in order, others that it was simply a dumb mistake. Eventually, all three Apollo 15 astronauts and Swigert were formally suspended from flight status. Jack Swigert took it very badly, Kraft recounted, and left NASA shortly afterwards. Ironically, he ended up becoming a politician.

Veteran astronaut Tom Stafford knew that he had a real shot at commanding the American half of ASTP, having served as chief of the corps and represented

President Nixon at the state funerals of the Soyuz 11 cosmonauts in the summer of 1971. He had also participated in several ASTP working parties in Moscow, developing good relationships with officials and cosmonauts alike. However, he knew that with Slayton back in the game, he would have stiff competition. The removal of Swigert also meant that there was fierce competition for the command module pilot's seat ... on what would be the *last* American mission until at least the end of the decade. Rookie astronauts Don Lind, Vance Brand and Bruce McCandless had all been with NASA since 1966 and all stopped by at Stafford's office to express their interest. Eventually, Stafford went with Slayton's judgement and recommended Brand to Chris Kraft: not only had he served in two backup positions, but he was a gifted pilot and had even paid for private Saturday lessons in Russian.

So it was that in January 1973, shortly after the final Apollo lunar mission, Deke Slayton found himself in Chris Kraft's Houston office, along with Stafford and Brand, being greeted with the immortal words: "*You* guys are the crew for Apollo-Soyuz." Slayton's euphoria at finally receiving a flight assignment after such a long wait was balanced by the disappointing news that Stafford – ostensibly his junior, yet by virtue of three previous missions far more experienced – would be in command. Brand would serve as command module pilot and Slayton would take responsibility for the docking module. Slayton remembered in his autobiography that the bittersweet news was "a little deflating", but understood the reasons behind having a veteran astronaut (and, in particular, someone whom the Russians had grown to trust) in command and acknowledged that "Tom was one of the best". For his part, Stafford remembered that Slayton was eternally gracious in his new role and there was "never a moment's tension" over the issue of command. "I told him I knew we had a good crew," Stafford wrote, "and a good mission to fly and that was as close as we got to talking about the issue."

Yet the fact that, on Earth, at least, Slayton remained Stafford's boss prompted a journalist at the 1 February press conference in Houston to ask the obvious question: How did he feel about taking orders from a deputy? "I see absolutely no problem with that at all," Slayton replied. "I think there's a lot of precedence in the country on this particular subject. Anytime you have a guy flying an airplane in the military service, he's a second lieutenant and you got the highest general in the Air Force. The commander is the commander, and there's no doubt in this flight who's going to be the commander. It's Tom Stafford. Now, when we're on the ground, working the *other* programmes and the other problems, then obviously we've got a normal and working relationship. But I see absolutely no problems at all and *I'm* responsible to Tom to be ready to fly this flight and *he's* responsible to me to see the crew's ready to go!" At this humorous irony, Slayton's eyes twinkled and the auditorium broke into laughter. It is also rather ironic that with such a wealth of experience now under his belt, Stafford had been considered "too green" for a real chance in the Project Mercury selection in April 1959. At that time, as America's first team of astronauts were introduced to the world, he was 28 years old and about to graduate from test pilot school. Moreover, his height would have rendered him ineligible and, had it not been for NASA's decision to increase the height limit for its

During a water survival training exercise in March 1975, Deke Slayton prepares to disembark from a mockup command module, watched closely by Vance Brand and Tom Stafford in the life raft. The relationship between Stafford and Slayton was a paradoxical one and inevitably generated its own share of humour, for Slayton was Stafford's boss on the ground, yet Stafford was Slayton's commander in the spacecraft. Slayton summed up the situation perfectly at their first press conference: "*I'm* responsible to Tom to be ready to fly this flight," he said with a grin, "and *he's* responsible to me to see the crew's ready to go!"

larger Gemini spacecraft, he would not have been picked at all. "The most important change for me," Stafford wrote, "was that [when I reapplied] they had raised the height limit from five feet 11 inches to six feet even."

By the time of his assignment to command the American half of ASTP, Stafford had already flown three times, in each case performing an intricate rendezvous and docking and in one case – Apollo 10 in May 1969 – a full dress rehearsal for the first Moon landing. After finishing his third mission, Stafford was promoted to chief of the astronaut office, which he held until June 1971. He subsequently rose to become Deke Slayton's deputy in the flight crew operations directorate, a position he held at the time of the ASTP selection. His promotion to brigadier-general, ratified by the Senate in March 1972 and confirmed by NASA the following December, made him the youngest flag officer in US military history. He would make further history by becoming the first person of general officer rank to fly into space.

Stafford was born in Weatherford, Oklahoma, on 17 September 1930, the son of a dentist father and a teacher mother. An avid reader and an enthusiastic watcher of the silvery Douglas DC-3 airliners which soared above his childhood home, he took a paper round to buy parts for his own balsa wood model aeroplanes. His dream was to someday become a fighter pilot and help win the war. Although he would not engage in conflict during the Second World War, his dream to fly would become a reality. In high school, he excelled in football, becoming team captain, although by his own admission he was far from perfect: shooting out streetlights with a BB gun, throwing a firecracker into the police station and attempting to disrupt English lessons with a cleverly orchestrated symphony of coughing. "The neighbours could always tell when I had been caught," he wrote. "I would be out front, painting the fence as a punishment, like Tom Sawyer."

Stafford's football abilities eventually drew the attention of the University of Oklahoma's coach, although he had applied for, and would receive, a full scholarship from the Navy to study there. He had already undertaken military training in 1947 as part of the Oklahoma National Guard and was called to temporary duty when the small town of Leedey was hit by a tornado. He also worked on manoeuvres to plot howitzer targets and his calculations contributed to his battery receiving an award for being the most outstanding artillery unit. The following year, 1948, brought both success and tragedy: acceptance into the Naval Academy and the death of his father from cancer. During four years at Annapolis, he was assigned to the battleship USS *Missouri*, where he met another midshipman named John Young. "We would have laughed," he wrote, "at the suggestion that someday we would become astronauts flying in space and circling the Moon together." After graduation in 1952 with a bachelor's degree, his eagerness to fly the F-86 Sabre – "the hottest thing in the sky" – led him to join the Air Force, rather than the Navy. He achieved his coveted silver wings from Connally Air Force Base in Waco, Texas, late the following year. By now married to Faye Shoemaker, he received advanced training in the F-86 and the T-33 Shooting Star. He was then assigned initially to an interceptor squadron, based in South Dakota, and later to Hahn Air Base in Germany as a flight leader and maintenance officer for the Sabre.

A paucity of opportunities for promotion almost prompted Stafford to resign from the Air Force in 1957 and he even drafted application letters to numerous airlines … before deciding to continue in the service when he first saw the F-100 Super Sabre and the forthcoming F-104 Starfighter. "If I went to an airline," he wrote, "I'd be flying the equivalent of cargo planes and could say goodbye to high-performance fighters." He was promoted to captain the next year, then selected for test pilot school at Edwards Air Force Base in California, where he found himself working harder than ever before. "Each morning's flight generated a pile of data from handwritten notes, recording cameras, oscilloscopes and other instruments. We had to reduce this data to a terse report that we submitted to the instructors and we had a test every Friday." From such schools, pilot astronauts were, are and continue to be drawn.

On graduating first in his class in May 1959, Stafford stayed at Edwards as an instructor and co-authored a pair of flight test manuals: the *Pilot's Handbook for Performance Flight Testing* and the *Aerodynamics Handbook for Performance Flight Testing*. By the spring of 1962, he was due for a permanent change of station and confidently expected to study for an advanced master's degree in a technical field, but was picked to attend Harvard Business School. This suited him fine, as a business administration credential would benefit both his military career and any subsequent plans he had. Nevertheless, when he learned in April that NASA was recruiting its second class of astronauts, he sent in his application. By July, after passing the Air Force's initial screening, Stafford was summoned to Brooks Air Force Base in Texas, where "we had the expected blood tests and EKG stuff, but no centrifuge testing ... we were all pulling Gs on a regular basis in high-performance aircraft. You didn't need to be some kind of physical superman to fly in space."

Some tests, though, were pointless. Stafford recalled looking into an ocular device for long enough to see a sudden flash of light, part of an evaluation of how an astronaut's eyes might respond to a thermonuclear explosion. "It wasn't enough to damage your eye – at least, I don't think it was – but you couldn't see for several minutes after the test," he wrote. The Brooks tests were followed by hour-long technical interviews in Houston in August and, although Stafford felt confident he had a good chance of being selected, his attention was primarily focused on the impending start of classes at Harvard. In fact, his interview in Houston came partway through his family's move eastwards from California. Arriving in Boston, Massachusetts, early in September, they unpacked enough belongings to live on temporarily and Stafford worked through three days of inaugural classes. Returning home on 14 September, he was greeted by his next-door neighbour and the news that Deke Slayton had called. The gruff Slayton told Stafford that, if he was still interested in becoming an astronaut, he had been officially selected. Three days later, on his 32nd birthday, Stafford sat alongside eight fellow pilots – Neil Armstrong, Frank Borman, Pete Conrad, Jim Lovell, Jim McDivitt, Elliot See, Ed White and John Young – for a press conference at Ellington Air Force Base, near Houston. Little could he have imagined how closely entwined their professional lives would become during the next few years, as successes and disasters and other shifts and changes in fortune would earn each of them a place in history.

Now, more than a decade later, on this January day in 1973, sitting next to two other astronauts for one of the most politically ambitious space missions ever attempted, Stafford surely reflected on how well fortune had smiled upon himself and his crew. Slayton was well aware that simply *being* grounded from flying might have saved his life: fellow astronauts Ted Freeman and Clifton 'C.C.' Williams had been killed in aircraft accidents and the Apollo 1 crew of Gus Grissom, Ed White and Roger Chaffee had died in a fire during a launch pad test. "I guess I felt that I was as competent as anybody else to command every flight that has flown," Slayton told a journalist, "and I missed them all, so that's been a thorn, obviously, but ... on balance, I think I've been pretty fortunate and I don't believe in looking back at the past too much, except in relation to what benefit it will give me in the future." Slayton's fortune at missing death and returning to active flight status in the nick of time to secure a seat on ASTP is clear, but good timing and good fortune had also honoured the third crew member, Vance Brand. Since his selection as an astronaut in April 1966, he had earned a reputation as one of NASA's most gifted fliers – "a hard-nosed aeronautical engineer and an experienced test pilot," to quote Slayton –

and yet, despite serving as a backup crew member for no fewer than *three* missions, he had still to savour his first taste of weightlessness.

Born in Longmont, Colorado, on 9 May 1931, Brand was an active Boy Scout during his formative years and achieved the title of 'Life Scout'. After high school, he entered the University of Colorado at Boulder, receiving his business degree in 1953 and accepting a commission into the US Marine Corps. For four years, he served as a naval aviator, which included assignments as a fighter pilot in Japan, then resigned from active military duty to return to his *alma mater* for a degree in aeronautical engineering. A master's credential in business administration from the University of California at Los Angeles followed in 1964, by which time Brand was working for Lockheed. In this role, he acted as a flight test engineer for the US Navy's P-3 Orion aircraft. His ties with the military remained intact, however: he continued to serve in the Marine Forces Reserve and Air National Guard jet squadrons until 1966 and graduated from the Naval Test Pilot School in 1963, working for a time at West Germany's F-104G Flight Test Centre in Istres, France, as an experimental test flier.

After being chosen by NASA as a member of its fifth group of astronauts in April 1966, Brand initially focused on thermal vacuum chamber tests of the Apollo spacecraft and was then a member of the support crews of the Apollo 8 and 13 lunar missions. During the latter, in April 1970, he served as a 'capsule communicator' (or 'capcom') in Mission Control, providing a direct voice link with the astronauts during one of the most harrowing space missions ever attempted. At around the same time, Brand was named as the backup command module pilot for Apollo 15 and, mindful of a three-flight 'rotation' system, surely anticipated assignment to Apollo 18 and his own journey to the Moon. When that flight was cancelled in September 1970, Brand continued with his Apollo 15 duties and later transferred to Skylab, serving as backup commander for its second and third missions. At one stage, in August 1973, he and crewmate Don Lind would come within days of being called upon to stage a rescue mission to Skylab. Since Brand was occupied with Skylab for much of that year, it was not until the early part of 1974 that he could devote all of his energies to ASTP.

Despite being a test pilot and an astronaut, Brand was not immune to the reality that the mission with the Soviets was far more than just a rendezvous and docking. It was true that one of the goals was to test a new androgynous docking system, which might someday be used by either nation to 'rescue' crew members from the other or to engage in further co-operative ventures, but on a more fundamental level it was an experiment in international relations. Astronauts and cosmonauts had met several times at the Paris Air Show and had gotten along famously, despite all being members of the military and – ostensibly – fierce opponents. Together with Slayton and Stafford, part of Brand's job was therefore to set an example of friendliness and co-operation. "I think we were helping all of the engineers on both sides feel free to be more friendly," he told a NASA oral historian in July 2000. "One thing that's hard to realise now ... is what the *climate* was like back then. In a way, after the Stalin years and the Khrushchev years ... why, Soviets were very foreign to us. After some of the things that happened, we thought they were pretty aggressive people and – I won't say 'monsters' – but they probably thought *we* were monsters!"

In time, the joint endeavour reassured both sides that neither possessed horns or forked tails and Brand found that, aside from technology, the biggest difference was culture, history and tradition. "They were probably a more secretive society," he recalled, "partly because their population had been overrun by invasions for a thousand years. You know, in this country we don't have that experience or that burden, so we tend to be much more open as a people." The closed, distrustful nature of the Soviet Union was evident: from security monitors, bugged hotel rooms and a stubborn, dogged reluctance to reveal more information to the Americans than was absolutely necessary. Tom Stafford's attempt to gain access to Tyuratam, simply to see the launch facilities, has already been mentioned, and Brand added that the openness of the Americans in offering their Soviet counterparts flights aboard their high-performance T-38 Talon jets was *not* mirrored by a similar offer of rides aboard MiG-21s. Even so, Brand quickly endeared himself to his Russian counterparts, who recognised the amount of personal effort he had applied to learn their language. In fact, when asked about this at the 1 February 1973 press conference in Houston, he told journalist Abby Brett that he had taken 30 hours of private tuition, "as a hobby", and went on to point out that he possessed a smattering of German and his wife, Beverly, had a good command of French.

Of course, having worked with Stafford and Slayton for several years as a colleague and, in fact, a subordinate, Brand felt that he knew and understood his two new crewmates and could operate effectively alongside them ... but until the Soviets *selected* their own team for ASTP, it was impossible to begin to train as a joint unit. At first, the Soviet position was that its Soyuz crew would not be named until January 1975 – a mere *six months* before launch – which Stafford told them at the outset was unacceptable. "We had too many challenges," he wrote, "such as simply being able to *communicate*; learning to be fluent in Russian required at least two years of full-time language training. That was how long you studied at the Defense Language Institute in Monterey, California, for example. We weren't required to be conversational or fluent, just functional. But even that would require two years, given the other demands of training and travel." By announcing their crew in January 1973, therefore, NASA forced the Soviets' hand into selecting their own.

At first, it seemed likely that either Vladimir Shatalov and Alexei Yeliseyev or Andrian Nikolayev and Vitali Sevastyanov would receive the Soyuz mission. All four were highly experienced cosmonauts and, in the case of Shatalov and Yeliseyev, were masters of rendezvous and docking and held senior managerial positions within the Soviet space hierarchy. Indeed, since the early summer of 1971, Shatalov had been the commander of the military cosmonaut corps. The Western media seemed convinced that Shatalov and Yeliseyev were the leading contenders, too, with one journalist posing the question to Stafford at the 1 February press conference. "I would hate to take a second guess," Stafford replied, "what's the selection choice of the Soviet Union." The speculation was heightened when Shatalov and Yeliseyev sat alongside Stafford, Brand, Slayton, Glynn Lunney and Konstantin Bushuyev at a second conference in Houston on 19 March. It therefore came as something of a surprise when, in May 1973, veteran cosmonauts Alexei Leonov and Valeri Kubasov were announced instead. At the time, no one in the West knew that both men had recently been training to fly a two-month mission to a Salyut orbital station. For Leonov, the decision to transfer him from Salyut to ASTP was disappointing at first. "I still really wanted to continue my work with the Salyut space station," he wrote. "Eventually, I said I would agree to command the Soyuz craft for the joint mission – on one condition: that Valeri Kubasov ... would be the engineer. Unknown to me, he had also been approached independently to join the crew and had agreed – on condition that I was his commander!"

All this was unknown to Tom Stafford when he arrived at Le Bourget Airport for the Paris Air Show on 24 May 1973, anticipating a public announcement of the Soviet crew that day. On prominent display in a special pavilion at the show was a full-size mockup of an Apollo spacecraft docked with a Soyuz spacecraft and the week-long event would attract more than 400,000 visitors, over a dozen times more than expected. When he spotted Yeliseyev disembarking at Le Bourget, Stafford felt sure that he would be aboard the Soyuz, probably with Shatalov in command. Also aboard the aircraft were two cosmonauts that Stafford did not know well – Leonov and Kubasov – and it was only later that evening, over a small reception 'feast' of black bread, vodka, cognac, caviar and crab, that he learned that *they* would fly instead. The crew of the second Soyuz, standing by in the event of failure, was to be composed of veteran cosmonauts Anatoli Filipchenko and Nikolai Rukavishnikov. Backing up both teams were four 'rookies': commanders Vladimir Dzhanibekov and Yuri Romanenko and flight engineers Boris Andreyev and Alexander Ivanchenkov; it was the first time that the Soviets had ever identified unflown cosmonauts ahead of a flight. Yeliseyev's presence at Le Bourget was reflective of the fact that he would serve as the Soviets' ASTP director.

Unlike the Soyuz team, the backups for the American crew would all be veterans: Al Bean had walked on the Moon, Ron Evans had recently returned from the Apollo 17 mission and Jack Lousma was scheduled for a two-month Skylab expedition with Bean in the late summer of 1973. For this reason, like Vance Brand, they would be unavailable to begin direct training for ASTP until the spring of the following year. Ordinarily, a three-man 'support crew' was also assigned to each Apollo mission, but ASTP was granted four – rookie astronauts Karol 'Bo' Bobko, Bob Crippen, Bob Overmyer and Dick Truly – since one (Overmyer) would be needed in Moscow during the course of the joint flight. In his autobiography, Stafford noted that whenever any cosmonauts visited the United States, they were always accompanied by ever-watchful KGB 'minders'. "We assumed that the cosmonauts did their own recon," he wrote, but added "so did we!" In fact, all four members of the support crew had been assigned to the US Air Force's classified Manned Orbiting Laboratory project before this was cancelled and they were transferred to NASA, so their Top Secret security clearances were even higher than Stafford's own. Similarly, the managers and flight directors who undertook trips to the Soviet Union – including Glynn Lunney and astronaut Gene Cernan – went directly to CIA Headquarters in Langley, Virginia, upon their return home to be debriefed. Getting himself entangled in the murky world of CIA affairs was something that Stafford wanted to avoid. "Given my prominence," he explained, "I knew any contact between me and the CIA would eventually become public ... so I made it clear to

The ten astronauts charged with prime, backup and support duties for ASTP are pictured in the Mission Control Center a week before launch. From the left are support crew members Bob Crippen, Bob Overmyer, Dick Truly and Karol Bobko; prime crew members Deke Slayton, Tom Stafford and Vance Brand; and backup crew members Jack Lousma, Ron Evans and Al Bean. This assemblage would represent the last US crew assignment before the Shuttle. Unlike previous practice, in which support crews comprised three men, ASTP received an extra astronaut (Overmyer), who would be based in Moscow during the course of the flight.

NASA management that I wanted no contact with the agency and didn't want to know of any. If ever asked by a reporter, I had plausible deniability."

Stafford's counterpart in command of the Soyuz half of the mission revealed little in his own autobiography about whether or not he regarded his trips to the United States as an 'intelligence-gathering' exercise. In fact, a sense of childlike wonder, unbridled enthusiasm and an eagerness to experience every facet of Western culture could easily sum up Colonel Alexei Arkhipovich Leonov of the Soviet Air Force. This might seem surprising, when one considers the brutality of his childhood and the challenges he overcame to secure a place as one of the world's most celebrated spacefarers. Born on 30 May 1934 in the tiny Siberian village of Listvyanka, near the point where the Angara River leaves Lake Baikal, Leonov grew up at the height of Stalin's purges and his young eyes beheld terrible events. One of the most traumatic was in January 1938, when neighbours arrived in the bitterness of midwinter to strip bare the family's belongings, their clothes, their meagre furniture and even their food. The boy, one of a dozen Leonov children, was even told to remove his trousers. It was the punishment meted out by the harsh Soviet regime on Leonov's father, a staunch Bolshevik falsely accused of being an 'enemy of the people'. He had, it was said, deliberately allowed seeds for the next year's harvest to dry out.

"My father was thrown into jail without trial," Leonov wrote. "As we were then regarded as the family of an 'enemy of the people', we were branded subversives. Our neighbours were encouraged to come and take from us whatever they wanted." His elder siblings were removed from school and the family was forced to leave Listvyanka. Ultimately, his father was absolved from blame (thanks to glowing testimony from a former commanding officer), received compensation and was offered the headship of his local collective farm. He declined and chose to work instead at a power plant in Kemerovo, on the Tom River, to the north-east of Novosibirsk, with his sister and brother-in-law. It was here that the man who would one day become the first to walk in space experienced his life's two passions: art and aviation.

In his autobiography, Leonov remembered drawing pictures on the whitewashed stoves of his neighbours' rooms, earning extra bread and receiving pencils and paints for his efforts. "I loved to draw," he wrote, and his interest was indulged by his parents, who stretched bed sheets over wooden frames to provide rough canvasses for his work in oils. His ambition to become a professional artist was eclipsed one day in 1940 by the desire to fly. "It happened the first time I set eyes on a Soviet pilot," he explained, "who had come to stay with one of our neighbours. I remember how dashing he looked in his dark-blue uniform with a snow-white shirt, navy tie and crossed leather belts spanning his broad chest. I was so impressed. I used to follow him everywhere, admiring him from a distance."

The pilot noticed the young boy's interest and demanded to know why he was being followed. When Leonov told him that he, too, wanted to be a pilot, the aviator smiled and explained that he would need to grow physically strong and study hard at school. Equally importantly, Leonov would have to wash his face and hands each morning with soap. "Like most little boys," wrote Leonov, "I was not too keen on soap and water," but he followed the pilot's advice, running to him every morning to proudly display his clean face and hands.

As German forces rolled into the Soviet Union in a sweeping advance from the north to the south, Leonov saw truckloads of wounded Russian soldiers arriving in Kemerovo and the mass construction of chemical plants, two of which were blown up by Nazi sympathisers. In the autumn of 1943, when Leonov started school, the Red Army had repulsed the Germans at Stalingrad, although times remained grim. Years later, he would remember chanting thanks to Stalin in school for his 'happy childhood' and, indeed, would grow up believing the despotic Soviet system to be the best in the world. Not until his mid-teens would he begin reading and learning of other, happier worlds beyond the borders of Russia. When Nikita Khrushchev came to power, Leonov took the black armband he had worn in mourning of Stalin's passing and burned it.

His ambition to become a professional artist culminated, in early 1953, with a journey in the back of an open lorry to Latvia to apply for a college place in Riga. This, sadly, came to nothing. Despite being accepted by the principal, Leonov's realisation that the cost of living in the Latvian capital would be simply unaffordable pushed him in another direction, toward his second love: aviation. In the autumn, he was offered a place at the Kremenchug Pilots' College in Ukraine and for two years learned to fly propeller-driven aircraft, then moved on to the higher military academy in Chuguyev to train on MiG-15 jets. Shortly afterwards, the 1956 Hungarian Uprising placed Leonov and other young fighter pilots on full combat alert. Graduation from Chuguyev coincided with the launch of Sputnik, at which time he was flying MiG-15s, specially modified to take off and land on soil airstrips at night and during the daytime.

One harrowing incident brought him to the attention of a mysterious recruiting team ... and paved the road to the cosmonaut corps. Whilst flying in heavy cloud, a pipe in his jet's hydraulic system snapped. Alerted by his on-board instruments to a fire, Leonov shut off the fuel supply to the engine and performed an emergency landing. He was too low, he wrote, to parachute to safety. His actions prompted the recruiters to ask "if I would be willing to join a school of test pilots". In October 1959, two years after Sputnik, he was one of 40 semi-finalists selected from a pool of thousands of highly-qualified MiG-15 and MiG-17 fighter pilots. Interestingly, the same questions had arisen in Russia as the United States over what kind of individuals were best suited to space travel: and *pilots* were deemed best, owing to their ability to work under extreme conditions, react with lightning speed and demonstrate a range of complex engineering skills.

For a month, Leonov and the other candidates underwent gruelling physical and mental evaluations. "We were put in a silent chamber," he wrote in his autobiography, "and set a series of complex tasks while blinking lights, music and noise were played to distract us. We were given mathematical problems to solve while a voice was piped into the chamber giving us the wrong answers. We were put in a pressure chamber with very little oxygen in extreme temperatures to see how long we could withstand it." By now, it was becoming clear to the 25-year-old pilot that this was for something more serious than test flying and rumour abounded that missions into space were on the agenda. At length, the candidates returned to their air bases to await further orders. In the meantime, Leonov was posted to East

Germany, barely 20 km from the Inter-German border and, before leaving, he married his girlfriend, Svetlana. In March 1960, he was recalled to Moscow for cosmonaut training. It was then that he first met Sergei Korolev, the man whom he would credit with masterminding the early Soviet space effort. "Our training was intensive," he wrote, "a punishing regime which pushed us beyond what we thought we were physically capable of. Every day started with a 5 km run, followed by a swim, before we even began our individual programmes. Every aspect of our daily routine was carefully monitored by a team of doctors and nutritionists." The trainees were also enrolled in the Zhukovsky Higher Military Academy for engineering accreditation.

Five of Leonov's colleagues rocketed into orbit between 1961 and 1963, together with a hastily recruited woman, and it was at around the time of Valentina Tereshkova's flight that he gained his first introduction to a new type of spacecraft: the 'Voskhod'. During a visit to Korolev's OKB-1 bureau, he was captivated by the "more interesting design" of one capsule in particular. It had, he wrote, "a transparent airlock attached, with a movie camera installed". Korolev explained that all sailors were required to swim and, by extension, all cosmonauts should learn how to 'swim' in open space. The airlock, which extended like a large blister from the Voskhod, would be used for just such an exercise. Leonov was told to don a training suit and evaluate the airlock. At that moment, he recalled years later, Yuri Gagarin, the first man in space, clapped him on the back and whispered that Korolev had just chosen Leonov for the assignment.

Although that assignment, which reached fruition on 18 March 1965, made Leonov the first man to walk in space, his pressurised suit ballooned and almost prevented him from re-entering the spacecraft. Worse was to come. Leonov and his commander, Pavel Belyayev, returned to Earth the following day, but landed hundreds of kilometres off-course, in the wild Siberian taiga, and narrowly escaped with their lives. Despite the celebrity that subsequently came his way, by the early 1970s, Leonov's fortunes seemed to have taken a downward turn: originally scheduled to command the Soviet Union's first manned circumlunar mission – and perhaps also its first landing on the Moon – he and his crewmate Oleg Makarov were permanently stood down shortly after Neil Armstrong took his historic steps on the Sea of Tranquillity. He was then assigned, with Valeri Kubasov, to fly a two-month mission to a Salyut space station, which also came to nothing when not one, but *two* Salyuts failed shortly after launch. Now, more than ten years since his Voskhod flight, he was looking forward to returning to space. "I was proud I had been chosen as commander," he wrote, "but I was ready. I had been ready for a *long* time."

Leonov's flight engineer, Valeri Nikolayevich Kubasov, had also lived through his own fair share of hairy moments during a cosmonaut career which had seen him perform the first in-space welding experiment ... an apparatus which almost burned a hole in his spacecraft's outer wall! Born in the town of Vyazniki, in Vladimir Oblast, some 200 km east of Moscow, on 7 January 1935, Kubasov seemed destined to become a cosmonaut from his earliest days as an engineer at the OKB-1 spacecraft design bureau. In May 1964, working for Sergei Korolev, he was among a handful of civilians who survived preliminary medical screening for a seat on a Voskhod mission, but did not fly. Two years later, after some 'relaxation' of existing medical rules, he was accepted into the newly-established civilian cosmonaut team. During his first flight, aboard Soyuz 6 in October 1969, one of his tasks was to operate the 'Vulkan' ('Volcano') furnace, provided by the Institute of Electrical Welding in Kiev. This required the internal hatch to be sealed and the orbital module depressurised, with the entire procedure running automatically and Kubasov monitoring the instruments from within the descent module. Shortly before returning to Earth, Kubasov duly closed the hatch between the orbital and descent modules and flipped switches to begin the experiment. Naturally, the Soviet propaganda machine lauded the 'success' of the experiment, but it was not until 1990 that it became clear that the low-pressure compressed arc had inadvertently aimed a beam at the wall of the orbital module. When Kubasov and his commander, Georgi Shonin, entered the orbital module to retrieve the welding samples, they were shocked to discover the damage and, fearful of an imminent depressurisation, retreated back to the descent module. Kubasov would later admit that the Vulkan was a strictly experimental device – not an operational one – and said that some of its elements could someday be used to perform emergency repairs! It is ironic today that the only 'emergency' on Soyuz 6 was very nearly caused by the Vulkan. Moreover, if any 'repairs' *were* called for, their very need would have arisen from damage caused by the furnace itself . . .

In early July 1973, with two years remaining to the scheduled launch of ASTP, the five men of the first US-Soviet joint mission met for their first familiarisation session. In his autobiography, Leonov recalled touching down in New York for a brief stopover on the journey to Houston. Culturally, it was utterly alien to him: standing at the foot of the Empire State Building, he found it hard to imagine how such vast skyscrapers could be constructed. The sheer number of luxury cars on the streets and the *noise* – the wailing of sirens, the blaring of horns, the constant buzz of traffic – astounded him. The trappings of capitalism abounded. "In the Soviet Union," Leonov wrote, "it was difficult to get *food*, let alone a beautiful shirt, but in America you could buy whatever you wanted at any time of day." In Houston, the Soviet delegation, which also included Kubasov, Alexei Yeliseyev and Vladimir Shatalov, toured the Apollo simulators at the recently renamed Johnson Space Center and later visited North American Rockwell's plant in California to observe Spacecraft No. 111 (the prime ASTP command module) under test and inspect a high-fidelity mockup of the docking module. Interestingly, the three-week visit was cut short by a few days. Privately, Tom Stafford and Deke Slayton felt that the early return home was engineered by the Soviets to avoid being invited to attend the launch of the second Skylab crew on 28 July. "If *they* accepted *our* invitation," Stafford wrote, "it would obligate *them* to invite *us* to a launch at Baikonur."

In Glynn Lunney's mind, the Soviets generally seemed to have an ingrained perception of the Americans as the sons of capitalists or members of the ruling elite; a perception which their communist masters continued to perpetuate. "Tom and I talked to them about how his grandparents were in the Oklahoma land rush and they all went out there in wagons and got a place to live," Lunney recalled in his NASA oral history. "I talked about how my dad had worked in the coal mines back in Pennsylvania. It was funny. You could tell that we were continually mismatching

Language instructor Nina Horner works with Deke Slayton (left) and Tom Stafford in June 1974. Although Slayton and Vance Brand had pursued their own studies of the Russian language and Stafford had been a familiar face to the Soviets for several years, the Americans considered themselves woefully ill-prepared to master the complex language. Consequently, from January 1974, a team of four instructors, including Horner, shadowed them during every waking and working hour to ensure their proficiency.

what we were and how we lived with what their expectations or what all of their teachings had led them to believe about us. You could see them continually shift ground as one sort of assumption after another would kind of fall apart on them. To their credit, they were willing to observe and to arrive at their own conclusions about things and you could tell that inside their own heads, they were re-evaluating us and our way of life."

Deke Slayton recalled that, at this point, none of the astronauts and cosmonauts were very conversant in the other's language, but when the next joint training event took place in Moscow the following October, significant progress had been made on the part of Leonov and Kubasov. However, even Slayton, with no fewer than 245 hours of instruction, did not consider himself to be proficient enough in the Russian language. The Soviet cosmonauts, it turned out, had been assigned their own individual English teachers, who were tutoring them for six to eight hours per day.

Not wishing himself or his crew to look bad, Stafford asked JSC Director Chris Kraft for a team of four language instructors to be with them at any time of the day or night, including weekends, to ensure that they were sufficiently prepared. In January 1974, Kraft brought in former US Army linguist Jim Flannery, two recent Soviet émigrés named Nina Horner and Vasil Kostun and a Belorussian-born language professor from the University of California at Riverside, Anatole Forostenko. Horner provided their classroom instruction, whilst Forostenko, Kostun and Flannery accompanied them on trips and worked out with them in the gym, encouraging them to 'think Russian', even whilst playing handball or lifting weights. This made life easier, but the difficulties remained. As Stafford later wrote, "Russian has more consonants than English and my tendency to mumble them, combined with my regional [Oklahoma] accent, caused Alexei Leonov to claim that I was speaking not *Anglisky* (English) or *Russky*, but *Oklahomsky!*"

Entering the mysterious, closed society of the Soviet Union on their three training visits came as a surprise for the astronauts, particularly the cosmonauts' training centre, Zvezdny Gorodok ('Star City'), on the forested outskirts of Moscow. By the time that the 50-strong American delegation arrived on 1 October 1973, the town was home to several thousand permanent residents, including cosmonauts, their families, training officials, instructors and support staff. "We were never alone [at Star City]," wrote Slayton. "For one thing, we learned pretty quickly that there were quite a few air force cosmonauts, about 45 of them at the time." Several of the Americans, including Bob Overmyer, had already visited the Soviet Union, and so brought with them ready supplies of peanut butter and savouries to ease the culinary rite of passage as much as possible. "They eat *caviar* for breakfast," Overmyer told Slayton incredulously at one point. Fellow astronaut Dave Scott, who participated in one of the earlier negotiating trips to Russia, had sampled worse: raw sturgeon and vodka in paper cups . . . for *breakfast*. "It was a far cry from scrambled eggs and bacon," Scott wrote, "and tough on the stomach!"

Hotel accommodation for the Americans was better, at least from Tom Stafford's point of view, than on his earlier visits, where he had stayed in a cavernous, monolithic place called the 'Rossia'. As a senior astronaut, he had been given the plush top-floor suite, but had been unimpressed by the standard of Soviet construction: some of the buildings were practically falling apart, with netting erected to prevent bricks falling onto pedestrians. For the October 1973 session, the Americans were housed in the Hotel Intourist in downtown Moscow and bussed to Star City each morning with police escorts. In the hotel's basement, Deke Slayton recalled, was a bar which was very popular with Finnish and Swedish businessmen, but the Americans were keenly aware that they were being watched at all times: by the notorious 'key ladies' on each floor, by the KGB 'tails' who followed them everywhere – through the GUM department store, on their daily jogs along the streets of Moscow and even staggering back from late-night drinking sessions – and by the unseen presence of bugging devices into their rooms.

At the same time, there were differences in appreciation of what constituted hospitality. In his autobiography, Alexei Leonov remembered his frustration that the Americans would fail to appear for breakfast or dinner. His words illustrate that he and others took personal offence from this discourtesy: "It seemed the Americans did not at first understand the trouble we had taken in attending to every detail of their stay." In failing to attend a meal, food was wasted and "as head of all matters concerning training at Star City, this meant I had *personally* to reimburse our canteen for the meals missed." From Glynn Lunney's perspective, the thoughtfulness and kind-heartedness of the Soviets with regard to the giving of gifts was both remarkable ... and unexpected. "Frankly, we were buffoons when it came to that," he told the NASA oral historian. "It's very important to them and they do it very well. They had these suitcases and they just seem to [have] a perfectly appropriate gift for a circumstance; a book or a statue or whatever. We were a bit cloddish, at first. We didn't think of gifts. We were all focused on this job. We came to realise that there really was a lot of redeeming value in the way they went about that and the way they made that kind of contact."

Certainly, NASA Deputy Administrator George Low, who made several trips to the Soviet Union between January 1971 and May 1975, noticed the growing warmth of the Russian people. "My first impression ... was that the country was drab and cold," he wrote in a series of diary entries, "that the people appeared to be afraid and subdued, that there was very little laughter and very little colour." Meetings with Soviet officials, he added, were similarly cold, excessively formal and standoffish. "The contrast now is quite remarkable. There are more people in the streets, they seem to be moving at a quicker pace, they are more colourful, you hear a great deal of laughter, there are many more cars, more lights. My first meeting with [Mstislav] Keldysh ... was extremely formal, the Soviet position was completely prepared and it was quite clear that the Soviets were unable to deviate even one inch from the prepared position. This time, we had a much freer discussion and the Soviet side seemed to be willing to respond on the spot." To conclude, Low noted that his own outlook had changed in those four years and that this very outlook had, at first, unfairly coloured his perception of the Russians.

Training continued at a feverish pace throughout 1974 and into the spring of 1975. Alexei Leonov and his delegation returned to Houston in April 1974 for their second visit, during which several simulations of rendezvous and docking were conducted. Two months later, the Americans were back in Moscow. This time, they were housed in a newer hotel, the Hotel Kosmonaut, which was roomy and typically Russian: there were no curtains on the showers, insufficient towels and bars of soap, no plugs for the sinks or bathtubs and doorknobs were frequently missing. On the other hand, more furniture than necessary was crammed into each room. A humorous incident was related by George Low in his diary of a visit to the Soviet Union in May 1975; his suite at the Hotel Kosmonaut had two bedrooms, both of whose beds collapsed during the night! "There I was," Low wrote, "at three in the morning ... completely taking apart both beds, cannibalising pieces to get enough good pieces for one bed, rebuilding that one and going to bed for the rest of the night."

The rooms, naturally, were still bugged, although the Americans quickly learned to turn this situation to their advantage. At length, Tom Stafford found that talking to the walls could be more effective than a call to room service. "To test the

Despite outward overtures of détente, the early dealings between the Americans and the Soviets remained fraught with uncertainty and mistrust ... to such an extent that the Westerners were frequently 'tailed' by the KGB and monitored by bugging devices. Yet astronauts and managers *were* granted the opportunity to tour Moscow and gradually the relationship grew warmer and friendships began to develop. Here, Tom Stafford (centre, in black hat), tours Red Square in November 1973 with a delegation of astronauts, cosmonauts, managers and officials. The spires of St Basil's Cathedral and Vladimir Lenin's imposing mausoleum form an impressive backdrop. To Stafford's left is fellow astronaut Gene Cernan, part of the ASTP negotiating team, whilst at the far right of the image is cosmonaut Valeri Bykovsky.

surveillance system," he wrote, "I started complaining loudly: 'The Russians are very wonderful and hospitable people, but it's too bad they decided to be so cheap about this hotel. There's not even a *fly swatter*!'" A few hours later, returning from a training session, the Americans found that *every room* now had its own fly swatter

... and, furthermore, *every fly* had been killed and dumped in the *unflushed* toilet. On another occasion, Slayton bemoaned the lack of a pool table ... and one appeared, as if by magic, shortly thereafter. "It was sort of archaic," he wrote, "the kind of table you'd find in a museum back in the States, but it was still a pool table." Complaints about inept training staff had only to be directed towards the nearest lampshade or the closest vase of flowers and, hey presto, the astronauts would never see that person again. Nor was the sense of paranoia restricted to the hotel. One day, when Bob Overmyer tried to move his chair at a meeting in the Kaliningrad mission control centre, he found that it would not budge. After a few yanks, he finally managed to move it ... and a bunch of wires from a listening device popped out of the base. "Everybody just sort of looked the other way," wrote Slayton.

However, there were instances of surprising openness on the part of the Soviets, not least of which had been an invitation for a group of American aerospace writers, including two from *Aviation Week* and two from *Time*, to visit Star City. The journalists were impressed by a space complex which seemed to be growing rapidly and provided a clear indication of the Soviet Union's "continued dedication to the exploration of space". A considerable amount of scepticism had revolved around exactly how 'open' the Soviets would be to journalists in the weeks and months surrounding ASTP and questions about the free movement of the Western press behind the Iron Curtain and the accessibility of Star City and Tyuratam had been raised on several occasions. The surprising sense of transparency, though, reached its zenith late in June 1974, when none other than Richard Nixon himself – in the midst of making a world tour – arrived in Moscow. It would be one of his last official acts as America's head of state, for Nixon was at that very time in the midst of the ugliest and most destructive affair of his presidency; an affair which would ruin his reputation, crush his administration and bring him within a whisker of impeachment on charges of corruption and abuse of power.

WATERGATE

When Nixon reached Moscow on 27 June 1974, one of the places he intended to visit was Star City, along with Leonid Brezhnev, although plans changed and the ASTP crews were invited instead to a grand reception in the magnificent surroundings of the Georgievsky Hall in the Grand Kremlin Palace. To the accompaniment of 'Ruffles and Flourishes', Nixon and Brezhnev entered the hall, followed by Henry Kissinger, Andrei Gromyko and Anatoli Dobrynin, where they were greeted by rows of admirals, generals and senior politicians. During the course of the evening, Nixon turned to one of the keynote guests, Tom Stafford, and asked him pointedly about progress in arms control and the future Apollo-Soyuz effort. "It was obvious to me," Stafford wrote, "having arrived only a couple of days before from the States, where all hell was breaking loose ... that Mr Nixon wasn't facing reality."

In fact, Nixon was only weeks away from the end of his political life.

The circumstances leading to the ignominious self-destruction of his presidency began two years earlier, on the night of 17 June 1972, when five men were arrested for breaking into the Democratic National Committee headquarters on the sixth floor of the Watergate complex in Washington, DC. Even today, the precise purpose of the burglary has not been conclusively identified, although the target was most likely the office of the committee's chairman, Larry O'Brien, who may have acquired evidence of illegal contributions – including some from aviation magnate Howard Hughes – to Nixon's re-election campaign. Indeed, an FBI investigation soon linked payments to the burglars with a fund used by the 1972 re-election committee and the first seeds of suspicion of a high-level conspiracy were sown.

All five men were tried and convicted on 30 January of the following year and the suspicions of the judge, John J. Sirica, were piqued when one of the burglars wrote to him, claiming to have been under 'political pressure' to plead guilty. In his letter, James McCord explained to Sirica that several high-ranking government officials, including former Attorney-General John Mitchell, had been involved in planning and executing the Watergate burglary. When questioned, Mitchell – who was then head of Nixon's re-election campaign – denied involvement or knowledge of any of the burglars – but the evidence of some kind of conspiracy was clear. As early as six weeks after the incident, the FBI had traced a $25,000 cashiers' cheque in the bank account of one of the burglars, earmarked for the re-election campaign. Subsequent inquiries would find that many more thousands of dollars had also passed through their accounts from the same source.

Within two months, the plot had begun to thicken. Whilst serving as attorney-general, it was found that Mitchell had controlled a secret Republican fund to finance intelligence-gathering against the Democrats and on 10 October the FBI revealed that the Watergate burglary had been just one part of a massive political sabotage campaign, orchestrated by the Republican officials in charge of Nixon's re-election. These revelations, for now, did little to harm the president himself, who was re-elected to a second term in a landslide victory on 7 November. However, the media was beginning to smell a rat and their interest kept the case in the public eye. The *Washington Post*, for instance, uncovered evidence that the Justice Department, the FBI, the CIA and even the White House were aware of the incident and had all played a part in covering it up. In April 1973, embarrassingly, Nixon was obliged to request the resignation of two of his most influential aides, Bob Haldeman and John Ehrlichman, both of whom were later indicted and imprisoned, as indeed was White House Counsel John Dean.

As the summer wore on, Nixon's position steadily worsened, with former officials revealing that 'recording devices' in the White House – and particularly in the Oval Office – had taped several damning conversations in the days after the burglary which implicated the president himself in the affair. Nixon initially refused a subpoena to release the so-called 'Watergate tapes', claiming 'executive privilege' and insisting that their contents were essential to the national security. In October 1973, as the noose began to close around him, Nixon demanded the resignations of recently-appointed Attorney-General Elliot Richardson and his deputy William Ruckelshaus and attempted to fire Special Prosecutor Archibald Cox. A few weeks later, when Nixon finally managed to get rid of Cox, in a particularly underhand manner, he exposed himself to a barrage of allegations of wrongdoing and famously stated "I'm *not* a crook!" to a crowd of several hundred Associated Press editors.

The situation, though, was moving beyond his control. Despite successfully firing Cox, Nixon was still obliged to permit the appointment of a new special prosecutor, Leon Jaworski, who pressed on with the investigation. Shortly thereafter, 'transcripts' of several of the Watergate tapes were released, but in early December controversy flared when it was found that an almost 19-minute-long portion of one recording had been erased. Nixon's personal secretary, Rose Mary Woods, admitted partial responsibility, claiming that she had inadvertently pushed the foot pedal on her tape player whilst answering the telephone. Three months later, in March 1974, seven former presidential aides were indicted for conspiring to hinder the investigation and late in July, not long after returning from Moscow, the Supreme Court overturned Nixon's claims of executive privilege and demanded that he turn the tapes over to Jaworski.

Circumstances now moved rapidly. The House of Representatives began a formal investigation into possible impeachment of the president and the House Judiciary Committee voted to recommend the passage of its first, second and third articles: obstruction of justice, abuse of power and contempt of Congress. On 30 July, the very day that Nixon handed over the subpoenaed tapes, all three articles were unanimously passed. Four days later, a previously unknown tape was released, in which Nixon and Haldeman discussed the Watergate burglary and formulated a plan, involving both the CIA and FBI, to block its investigation. The contents of this tape, which has since come to be known as the 'Smoking Gun', clearly showed that Nixon was complicit in the scandal and had lied to the nation, to his aides and even to his lawyers for over two years. His position as head of the government was now untenable and key Republican senators told him that enough votes already existed to remove him from power.

In a nationally televised address from the Oval Office – the very room from which many of the conversations which destroyed him had been made – on the evening of 8 August 1974, Richard Nixon, 37th President of the United States, became the first to tender his resignation. He admitted no responsibility or guilt for the crimes levelled against him, telling his audience that giving up power before his term ended was "abhorrent" to him, but that "America needs a full-time President and a full-time Congress" and stressed the need of the nation to focus on "great issues of peace abroad and prosperity without inflation at home". Thus he cleverly attempted to turn the situation to his advantage, claiming that to waste time fighting for his vindication could only harm the United States. "Therefore," he concluded, "I shall resign the presidency effective at noon tomorrow. Vice-President Ford will be sworn in as President at that hour, in this office ... "

A month later, Gerald Ford – who had replaced the corrupt Ted Agnew as Vice-President in October 1973 and so became the first person to have been appointed to the United States' two most senior political positions, without being directly elected to either of them – extended a full and official pardon to Nixon for his involvement in Watergate. He expressed a staunch wish that the matter should be laid to rest, but sparked outrage that he had enabled the former president to evade justice. Rumour would abound for years that some sort of shady 'deal' had been struck between the two men, in which Nixon agreed to resign in favour of Ford, on condition that the latter would pardon him. Whatever the truth of the matter, Nixon went to his grave in 1994, still professing his innocence, and Ford lost his bid for re-election in 1976 to Democrat Jimmy Carter. In many minds, both Nixon and Ford had struck a devil's bargain and both had ultimately paid the price. Indeed, in the days after Ford's defeat, the *New York Times* would write that his pardoning of Nixon had been "profoundly unwise, divisive and unjust" and that, in making the decision, the new president had destroyed his own credibility in a stroke.

FINAL DOUBTS

Trust was in distinctly short supply in the dying months of 1974, and not just in the wake of Watergate and the Nixon pardon. Several prominent senators, including William Proxmire of Wisconsin, one of NASA's fiercest critics, leapt upon each and every fault and failing in the Soviet manned programme ... and as ASTP drew nearer, there was no shortage of problems from which to choose. In late August, the Soyuz 15 spacecraft returned to Earth after failing to dock with the Salyut 3 orbital station. The Soviets tried to explain this away by declaring that their cosmonauts had merely been 'testing' an automated docking system and had come home early to evaluate landings under different weather and lighting conditions. This explanation stretched credibility too far. Shortly thereafter, *Time* told its readers that "the landing, made at night and in bad weather, seemed to underline the urgency of the return. What had gone wrong?"

With ASTP only months away, NASA wanted to know what had happened. "It was ridiculous," wrote Tom Stafford, "to believe that the Soviets had sent a crew to fly around and inspect a station they had previously occupied." A few days later, Stafford met Vladimir Shatalov in Washington and insisted that the Soviets *must* come clean about whatever difficulties Soyuz 15 had faced. Shatalov continued to claim that the flight had gone according to plan. "Look, Washington *isn't* Moscow," Stafford explained bluntly. "*Everything* leaks to the press. What's secret today winds up on the front page of the *Washington Post* tomorrow; top secret will be in the *New York Times* in a week. If you say you didn't have a problem and somebody from an intelligence agency knows differently and Congress leaks that you really did have a problem ... ASTP is *dead*!" Shatalov understood. Both men knew that ASTP was an important political symbol to foster better relations between the United States and the Soviet Union. Digesting Stafford's words, Shatalov went to make some calls. Three days later, on 11 September, the two generals sat side-by-side at a Houston press conference and laid it all bare for the world's media.

It was a small detail, perhaps, since the actual rendezvous and docking system for ASTP was of a different design, but it highlighted the need for more openness and honesty on this first co-operative venture in space. To be fair, Shatalov's admission illustrated how difficult the Soviets found it to make even minor admissions of fault in a space effort which they saw as a bastion of their national pride. For William

Alexei Leonov (centre) and Valeri Kubasov (right) prepare for their duties during one of several training visits to the United States in September 1974. On the left is Vladimir Shatalov, commander of the Soviet cosmonaut team, who had earlier been forced to explain to a Houston press conference the truth behind the failure of Soyuz 15 to dock with the Salyut 3 space station. The incident underlined an urgent need for more openness on the part of a closed society which, to be fair, had suffered greatly under Nazi Germany and was distrustful of the motives of the West, but whose technological accomplishments were far inferior to those of the United States.

Proxmire, though, this was not enough. A long-time opponent of NASA, his very *name* has been used as a verb for someone who opposes scientific progress for political gain and two science-fiction novellas, penned by Arthur C. Clarke and Larry Niven, have included him as anti-hero, an obstruction to the exploration of space. As far as ASTP was concerned, Proxmire felt that the project was taking chances with the lives of American astronauts "for the sake of some intangible diplomatic benefit". Of course, the astronauts saw things differently; Deke Slayton poured scorn on the senator for having "done everything he could to keep [NASA] from getting any money" and Tom Stafford wrote him off as someone who simply disliked space exploration in general. However, both men were keenly aware that Proxmire had the ears of other senior politicians, who remained very much on the fence over ASTP and working with Russia.

The situation worsened in the spring of 1975. In February, the Soviets returned to the United States for their final training session, visiting Spacecraft No. 111 – which was by now ensconced in the cavernous Vehicle Assembly Building at the Kennedy

Space Center in Florida, undergoing final preparations for launch – and undergoing two weeks of rigorous rendezvous and docking simulations in Houston. A detour to Disney World allowed the exuberant Leonov and Kubasov to enjoy several rides, including Space Mountain, after which only one more trip remained to Moscow in mid-April. Tom Stafford had long pressed the Soviets to allow him to see Tyuratam and, particularly, the inside of a Soyuz spacecraft, but had been rebuffed numerous times on account that it was a top-secret military base. Eventually, he told them that if he was not allowed to see a Soyuz before launch, he would refuse to fly ASTP. The ruse worked, particularly when Stafford's backup, Al Bean, also refused to fly. "Look," Stafford explained, "I've *seen* many detailed satellite photos. I *know* what the place looks like. Fly us in and out at night, if that's all you're worried about." On 28 April 1975, a little over ten weeks before ASTP was due to fly, the American delegation finally visited Tyuratam.

Surprisingly, they were also able to tour Tashkent, Bukhara and Samarqand in today's Uzbekistan. "The Soviets had been reluctant to let us loose in Asia," wrote Stafford, "because we would be exposed to the poverty and would see just how little impact the Communist Party had in these places. Yes, we saw the Hammer and Sickle on certain buildings, and there was always a facility named after Lenin, but once you got away from the airport and the main square, you were in a culture that ... seemed to have changed little in hundreds of years." Indeed, the Uzbeks have seen conquerors come and go over the centuries – from Alexander the Great to the Mongols and later Tamerlane – and the Soviets represented just the latest in a long line of occupations. Today, the Soviet era is confined to the pages of history; another age that came and went. After crossing the Syr Darya River and seeing Tyuratam, the American delegation was invited to experience the way of life of the Kazakh herders of the steppe. Seated cross-legged in traditional, leather-walled yurts and dressed in bright Kazakh robes and tall felt hats, they joined their hosts in toasts of friendship ... and sampled the local delicacies. Unluckily, what Deke Slayton and Anatole Forestenko mistook for a pastry actually turned out to be congealed sheep's fat! "Our hosts were also passing around a drink that was liquid sheep fat," wrote Stafford. "Fortunately, I managed to avoid it." As commander of the American half of the mission, though, he could *not* avoid the next horror. It was traditional, Alexei Leonov told him, that the main host and the guest should each eat one eyeball from a boiled, skinned ram's head which sat in pride of place on a platter. Perhaps fortuitously, Stafford was intoxicated on vodka by this point and, perhaps recalling the 'right stuff' with which all astronauts were meant to be imbued, promptly dug a fork into the head, plucked out the eyeball and began chewing. Support crew member Karol 'Bo' Bobko, seated at an adjacent table, threw up and even Slayton and Forestenko got up and left shortly afterwards.

By now, something else had given William Proxmire additional ammunition to level against NASA and the Soviets. Early in April, cosmonauts Vasili Lazarev and Oleg Makarov had launched aboard Soyuz 18 for a scheduled two-month expedition to the Salyut 4 orbital station. Their ascent proceeded normally until, at an altitude of 145 km, things went badly wrong. Ordinarily, two sets of pyrotechnics should have fired to shear the central core of the booster away from its upper stage and six latches should have blown. It did not happen. "An excessive degree of vibration caused the relay in half of the upper sequencer to close down and to signal three of the six latches to fire prematurely," explained Rex Hall and Dave Shayler, "with the lower core and upper stages still firmly attached. This was activated only seconds prior to the planned separation, but with the latches armed. The connection *that* triggered was in the same location as the electrical link between the upper and lower segments of the structure. Therefore, when the electrical contacts were severed due to the premature explosions, so [too] were *all* links to the *lower* latches, causing an uneven linkage between the core and upper stage as the vehicle continued to climb." The upper stage ignited on time ... but the central core *was still attached.* Although the thrust of the upper stage engine broke the remaining locks and separated the two stages, within seconds the unanticipated strain threw the booster off its intended trajectory. The on-board gyroscopes detected a deviation beyond the mandated safety limit and automatically commanded an abort. Two hundred and ninety-five seconds into Lazarev and Makarov's mission, the escape tower fired to pull the orbital and descent modules away from the booster.

"Inside Soyuz," wrote Hall and Shayler, "Lazarev felt the vehicle pitch and roll and reported a heavier pitch than on Soyuz 12. At that moment, the Sun suddenly disappeared from sight and a loud siren sounded. On the instrument panel, the red 'Booster Failure' light flickered on. For a few moments, the crew wondered what was happening as the sound of the booster stopped and for a second or two they became weightless as the forward velocity faltered." Now on the very fringes of space, the cosmonauts could only grit their teeth as the abort system automatically released the bell-shaped descent module to plunge back to Earth. Normally, they could have expected an acceleration of perhaps 15 G in such an emergency, but since Soyuz 18 was already pointing *directly downward,* its rate increased and the two men were subjected to 21.3 G of deceleration. "We began to experience a creeping and unpleasant pull of gravity," Lazarev recounted later. "It increased rapidly and its rate was much greater than I had expected ... Some invisible force pressed me into my seat and filled my eyelids with lead ... Breathing was becoming increasingly more difficult ... " The two men could barely communicate with one another, scarcely able to utter more than a few guttural grunts, wheezes and puffs under the immense loads.

The descent module's parachutes deployed as intended and it came down in the snow-covered Altai Mountains, about 830 km north of the Chinese border, landed on a slope and began to roll towards a sheer precipice. Thankfully, the parachute snagged on some vegetation. In temperatures of -7°C, the cosmonauts donned their cold-weather clothing and clambered outside. Having radioed a request for tracking information during the descent, but met with silence on the airwaves, Lazarev was afraid that they had landed in China and one story relates that he burned a pile of classified papers for a military experiment which he would have performed aboard Salyut 4. In fact, their flight path passed to the north-west of Xinjiang, which had entered the headlines in March of the previous year when a Soviet helicopter landed there in error and its pilots were captured by a Chinese patrol. At length, the two cosmonauts encountered friendly locals who spoke Russian and a helicopter rescue team quickly made radio contact.

"Oleg calculated the landing place almost precisely," Lazarev said later of the flight engineer's navigational measurements during the descent. "We landed a bit to the side of the place he had indicated. It was painfully disappointing and somewhat unpleasant. We had prepared ourselves for the mission for so long, only to wind up like this ... The very fact of failure was quite discouraging." They had touched down close to the town of Aleysk, about 200 km south of Novosibirsk, in southern Siberia – well within Soviet territory. However, it was the *next day* before a rescue team, battling the terrain, chest-deep snow drifts and high altitude, reached them. Plans to drop a team of physicians onto the mountain were called off by Lazarev, a veteran parachutist himself, after deciding that it was too dangerous. Next morning, it was hoped to extend a ladder from a helicopter, but the instability of this option made it impractical; another effort led to a group of would-be rescuers getting stuck in an avalanche. Finally, a civilian helicopter dropped a forest guide next to the two cosmonauts to render assistance and a military helicopter arrived and extracted all three of them.

Tass revealed the shocking truth the following day, when it announced that the booster had "deviated from the pre-set values" and "an automatic device produced the command to discontinue the flight". Vladimir Shatalov had learned little from his 'tell-the-truth' conversation with Tom Stafford a few months earlier, for he quickly made a statement asserting that both men were in good shape and ready to fly another mission. This was far from accurate. "Makarov says that under the G forces which they experienced, they could easily have first lost vision and then consciousness," wrote Hall and Shayler. "Although this did not happen, they experienced black-and-white vision and then tunnel vision." Makarov, it is true, *would* make a full recovery and undertake two further missions, but Lazarev suffered undisclosed internal injuries and never flew again. In the days after the flight, a rumour circulated that Lazarev and Makarov had died in the accident; this was scotched by ordering them to play *football* with some Americans to show that they were still alive ...

The cosmonauts' flight was effectively suborbital, reaching an apogee of 119 km and lasting barely 21 minutes. With the joint mission looming in July, Proxmire and other leading senators became even more vocal in their questioning of the Soviets' safety record, prompting Konstantin Bushuyev to tell Glynn Lunney that Lazarev and Makarov had been launched atop an *older* version of the Soyuz booster, which would not be used for ASTP. After this rationale was repeated to George Low at the Flight Readiness Review in June, Administrator Jim Fletcher conveyed his satisfaction to both President Ford and, by letter, to Senator Proxmire. Fletcher knew that Ford was supportive of ASTP and cordially invited him to attend the Apollo launch, if his schedule permitted, or at least to speak to the crews in orbit. In his letter to Proxmire, dated 7 April, Fletcher pointed out that the fault was with the upper stage of the rocket, rather than with the Soyuz craft. "As you may recall," he wrote, "the ASTP mission plan calls for a US Apollo launch *only* after a fully satisfactory Soviet Soyuz launch; therefore the US safety concerns focus on spacecraft, rather than launch vehicle questions." To close, Fletcher assured the senator that all pertinent data had been examined by NASA and "intelligence facilities" and the incident would not place Apollo or its crew in danger.

From Proxmire's perspective, such explanations missed the point. Whether it was a rocket failure or a spacecraft failure, the reality was inescapable. The Soviets had now launched 17 manned Soyuz spacecraft, of which two had ended in disaster – Vladimir Komarov's ill-fated flight in April 1967 and the tragic re-entry of Dobrovolski, Volkov and Patsayev – and now Lazarev and Makarov had cheated death by a hair's breadth. Of the other 14 flights, two others – Soyuz 10 and 15 – had experienced docking problems and could hardly be considered successes. Effectively, Proxmire concluded, that meant five manned missions (more than a quarter of the total) had either totally failed to achieve their objectives or had ended with a loss of human life. By early July, only two weeks before launch, he made public the testimony of the CIA's deputy director of science and technology, Carl Duckett, who had raised questions in June about the ability of the Soviets to control two manned missions simultaneously. At the time, they had a crew of two cosmonauts aboard the Salyut 4 station, whose two-month flight was expected to overlap that of ASTP.

"This warning from the nation's top scientific intelligence expert," Proxmire said, "should not be taken lightly." In his testimony, Duckett expressed concern that Soviet communications capabilities and central management facilities were grossly inferior to those of the United States and operating *two* missions in tandem offered a disturbing indication that ASTP might not receive the full support necessary for it to succeed. Proxmire urged NASA to postpone the launch until after the Salyut crew had returned to Earth. Responding to Proxmire's letter on 3 July, Fletcher noted that Konstantin Bushuyev had assured Glynn Lunney that the Soviets would employ *two* control centres – one at Yevpatoria in the Crimea for the Salyut flight and a newer complex in Kaliningrad, just outside Moscow, for ASTP. Bushuyev also explained that ASTP would be given priority if the two Soviet missions passed simultaneously within the same zone of coverage of a tracking station, which itself was unlikely, owing to their differing orbital paths. Reinforcing Bushuyev's words, NASA tracking specialists made their own calculations and confirmed that the Salyut and ASTP crews would be in communication with the same ground station only *twice* during the whole mission ... and even then for barely a minute apiece. "The [Salyut] mission ... does not constitute a hazard to ASTP," Fletcher told Proxmire, "and there is no reason to delay the launch of ASTP if the Salyut mission is still in operation."

The senator remained unconvinced. On 11 July, a few days before launch, he entered anti-ASTP articles in the *Congressional Record*, and on the 14th cited CIA data which seemed to indicate that the Soviets had encountered "severe problems" in space and that their technology was fundamentally inferior to that of the United States. Specifically, he cited the fallibility of their docking apparatus, their general lack of preparedness for the ASTP mission, the poor standard of training facilities and a history of failings in both their Soyuz and lunar programmes. He concluded with a list of damning national security concerns: there had been a great deal of 'technology transfer' since 1972 and the Soviets had learned US management and operational techniques. The United States, he explained, had a clear technological lead over the Soviet Union in the areas of communications, medicine, project management, quality control, handling of emergencies, co-ordination of launches,

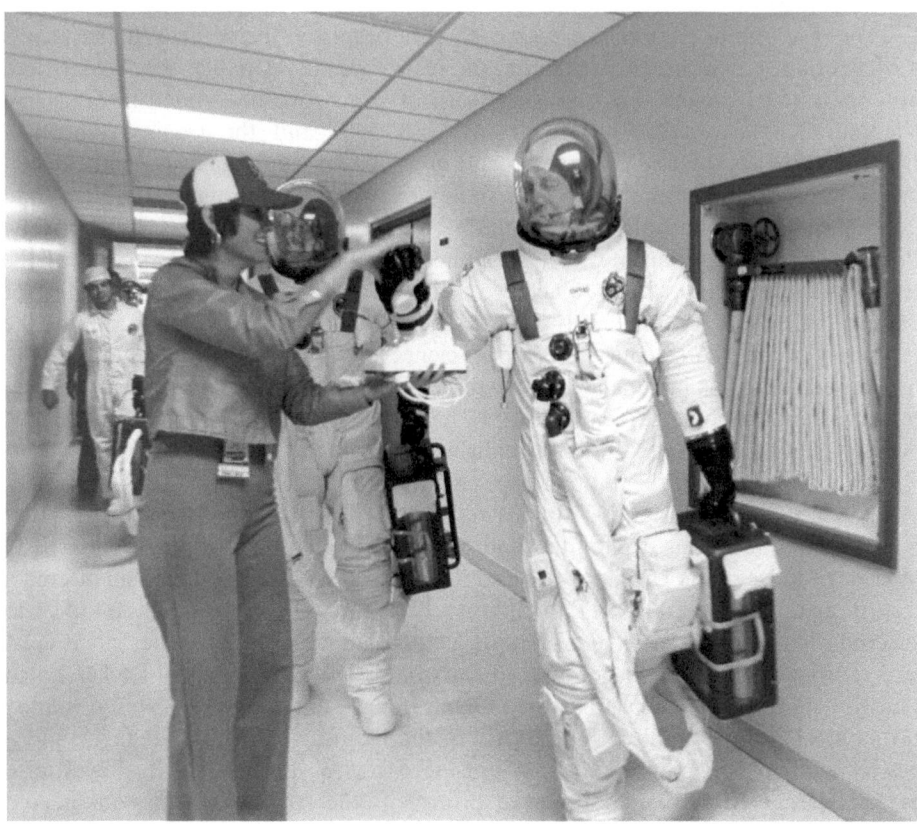

In the high-risk business of spaceflight, it was essential to keep a sense of humour, fun and perspective ... and to remember the many support personnel on the ground who made missions possible. Here, as Tom Stafford leads his crew out to the pad on launch day, he is presented with a mock telephone call by his secretary, Charlotte Ober.

procedures, computers, the capability for making in-flight changes and the training of crews. The United States had gained very little in return. By this point, such criticism was irrelevant, because the mission *would* go ahead, but, interestingly, despite Tom Stafford's admonitions that Proxmire simply disliked space exploration generally, the senator's concerns about inferior technology and particularly about poor safety and quality control oversight would arise again two decades later during the Shuttle-Mir effort and during the genesis of the International Space Station.

THE "HOT SPACE SUMMER"

Although the beginning of ASTP – first the Soyuz spacecraft and then, seven and a half hours later, the Apollo – had long since been scheduled for 15 July 1975, it was essential to map out alternative dates, just in case problems arose. Mission planners

Ken Young and Oleg Sytin were obliged to consider constraining factors, such as the lighting at both Tyuratam and Cape Canaveral *and* at their respective landing points. Experiments which were 'keyed' to the position of the Sun or specific astronomical targets had to be worked into the schedule, so that they would be performed at the most optimum time, *whenever* ASTP took place. Flight plans and trajectory plots for each alternate launch date had to be individually verified and then signed off both in Russian and English, before they could be formally approved. Day-to-day operations would be run by flight directors in Kaliningrad and Houston, with each team having basic responsibility for 'its' respective share of the mission. It would be these flight directors who would jointly orchestrate contingency plans in the event of an emergency. Having said this, in situations which directly affected crew safety, the nation whose men were in danger would be granted the unilateral right to take action – calling for an immediate undocking, for example – which would then be considered by their flight director.

As launch drew nearer, everyone within NASA was keenly aware that ASTP would mark the final flight of Apollo and, indeed, of the workhorse launch vehicle; the Space Shuttle was on the horizon, but was not anticipated to begin operations until the end of the decade. The mood was undoubtedly comparable to that which prevails within the space agency today, as 2010 heads into 2011 and the Shuttle itself enters the twilight of its life; back then, as now, a pervasive sense of gloom was juxtaposed with an overwhelming desire to make this last roll of the dice the very best that NASA could offer. One of the key problems at the Flight Readiness Review in June surrounded the Saturn IB booster, codenamed SA-210, which had been built in 1967. Ellery May, the Saturn manager at the Marshall Space Flight Center, noted that eight years in storage meant that many of the components were subject to deterioration, although periodic inspections and a lengthy refurbishment effort between October 1972 and early 1974 had monitored its condition. During one such inspection, May told the review, several hairline cracks had been found in two of the mounting points for the large tail fins on the first stage. The cracks had been caused by stress corrosion. NASA Headquarters decided to replace all eight fins. Installed on the launch pad late in March 1975, the final Saturn was ready to fly.

Perhaps the most fitting description of the two-stage Saturn IB was offered by the two men who should have been among the first to ride it into space. In late 1966, astronauts Gus Grissom and Wally Schirra served as prime and backup commanders, respectively, for Apollo 1, a mission which should have tested both the booster and the Apollo spacecraft in a manned capacity for the first time. They light-heartedly called the Saturn IB "the big maumoo" and, indeed, at the time it was one of the largest rockets ever used to carry humans, standing 68 m from the base of its first stage to the tip of its escape tower. Grissom, of course, would never fly it – he was tragically killed, along with his crew, in a flash fire on the launch pad in January 1967 – but Schirra and his team of Donn Eisele and Walt Cunningham would ride Apollo 7 in October 1968. From the point of view of spectators on the ground, the 590,000 kg rocket ascended slowly, which was perhaps not surprising when one considers that its first-stage thrust was just 725,750 kg, and for Schirra, Eisele and Cunningham this ponderous, stately rise from Earth was a very calm, very gentle, almost elevator-like climb to the heavens. The G loads on the Saturn IB were less oppressive than Mercury-Atlas or Gemini-Titan launches, both of which Schirra had experienced, and he told ground controllers that the ascent passed "like a dream".

Built under the direction of the famed German rocket engineer Wernher von Braun, who headed the Marshall Space Flight Center throughout the 1960s, the Saturn IB was originally intended as an 'interim' vehicle to wring out the Apollo spacecraft in low-Earth orbit, before its much larger cousin, the Saturn V, entered service for expeditions to lunar distance. Although the Saturn IB did not have the impetus to transport men to the Moon, it was an awesome monster, nonetheless. Its first stage, designated 'S-IB' and built by Chrysler, was 25.5 m tall and under the combined thrust of its eight H-1 engines lifted the Apollo spacecraft to an altitude of 68 km. Fed by Rocket Propellant-1 (RP-1) and liquid oxygen, the S-IB contained a 'cluster' of nine fuel and oxidiser tanks (held together by a 'spider beam') and a barrel-like tail unit on which the engines were mounted. Eight fins around the base provided the rocket with stability on the launch pad and then served to improve its aerodynamics during ascent. Meanwhile, the second stage, known as the 'S-IVB' and made by McDonnell Douglas, measured 17.8 m tall and was equipped with a single J-2 engine. It was connected to the S-IB by means of a skirt-like 'inter-stage' and was designed to pick up the thrust approximately two and a half minutes after launch and burn until orbital velocity had been achieved, some ten minutes into the mission. Its single propellant tank was divided by a common bulkhead into liquid hydrogen and liquid oxygen chambers and the thrust of the J-2 engine amounted to 102,000 kg. The S-IVB was used as a component on both the Saturn IB and the Saturn V and also provided a kind of 'garage' to carry additional payloads into orbit, such as the lunar module or, for the ASTP mission, the docking module.

The casual observer could be forgiven for thinking that the appearance of this rocket on Pad 39B – one of two manned launch complexes at the Kennedy Space Center – was a little unusual, like a child in adult's clothing. The facility was built for the Saturn V, a booster some 43 m taller than the Saturn IB, and the smaller rocket had to be mounted on a pedestal in order to elevate it to match the levels of the crew, equipment and propellant servicing arms of the 115 m umbilical tower. This pedestal was colloquially known as 'the milk stool' and had originally been proposed by Boeing in February 1969 as a solution to the height problem. Its design officially began in July 1970, following a lengthy series of pull tests, wind tunnel simulations and dynamic and stress analyses, and despite its wimpy-sounding name the milk stool was an impressive and sturdy structure. It stood 39 m tall and tapered from a 14.5 m square at its base to a 6.7 m square at its apex. It could withstand engine exhaust temperatures of 2,700°C, could endure hurricane-strength forces and could support a launch at maximum wind speeds of 59 km/h. Weighing a quarter of a million kilograms, it could more than prove its mettle under duress.

The milk stool was introduced to launch crews to the Skylab orbital station. The early Saturn IB launches had been staged from Pads 34 or 37, part of the US Air Force's Eastern Test Range at Cape Canaveral Air Force Station, and initial plans called for subsequent manned flights to do the same. Unfortunately, after Apollo 7

The Saturn IB rolls out of the Vehicle Assembly Building (VAB) in March 1975, bound for Pad 39B. The reader will recall that the launch complex was originally built for the much larger Saturn V lunar rocket and a supporting 'milk stool' was duly constructed to elevate the smaller Saturn IB and match the levels of the crew, equipment and propellant servicing arms of the umbilical tower. The VAB is an impressive structure in itself – to this day, in fact, it remains the world's largest single-storey scientific building – and in its early years, before the installation of an air conditioning system, its vastness even caused *clouds* to form. At the lower left of this image is the Launch Control Center.

in October 1968, it was realised that there would be a hiatus of at least two years before another Saturn IB was scheduled to fly. In the meantime, the launch complexes themselves steadily deteriorated; in particular, Pad 34 was old, under-sized, underfunded and suffering from air and salt corrosion. Pad 37 was designed for Saturn IBs, but had not yet been configured to support a manned launch and *that* change was expected to require almost two years. In May 1970, therefore, after lengthy negotiations and approval from Administrator Tom Paine himself, Skylab Director Bill Schneider announced the intention to employ the "more modern" facilities of Pad 39B for all future Saturn IB launches and design work on the milk stool got underway. Many benefits were outlined. Employing the milk stool meant that much of the Saturn V equipment at the pad could be 'reused' for the Saturn IB, thereby saving the agency an estimated $13.5 million. Moreover, the presence of a large assembly building in relatively close proximity to the newer launch complex offered better environmental protection during processing than the exposed Pad 34.

The 39B complex lies 2.6 km north of its sister, 39A, on a vast, three-hundred-plus-square-kilometre tract of land on Merritt Island, just off Cape Canaveral in Florida. The 'Canaveral' name derives from the Spanish *Cañaveral* or *Cañareal* – meaning 'Cape Canebrake', reflective of its dense thickets of cane vegetation – and is one of the three oldest European place names in the United States. It was home to farming and fishing communities until the coming of the Space Age and was chosen in preference to such candidates as Christmas Island and White Sands as a primary launch site. By virtue of being close to the equator, it enabled rockets to take better advantage of Earth's rotation and its population was sufficiently low to lessen the impact of accidents. For a time, after the assassination of the eponymous president who committed his nation to reaching the Moon, it was renamed 'Cape Kennedy', but the locals had it changed back to 'Cape Canaveral' in 1973.

Designed and constructed mainly in the early to mid 1960s as the primary foundation of America's drive for the Moon, the main 'spaceport' has retained the name 'Kennedy Space Center' to this day. One of its most iconic structures is the Vehicle (originally 'Vertical') Assembly Building (VAB), in which the Saturn lunar rockets were assembled on their gargantuan mobile platforms. It measures 160 m tall, 218 m and 158 m and, with a floor of 32,400 m^2, remains the world's largest single-storey scientific building. Nearby is the Launch Control Center, often known as the 'electronic brain' of Complex 39, which undertakes systems checkouts and testing and controls the countdown from a series of four firing rooms. Nevertheless, 'the Cape', as it is often known, represents far more than just a rocket launching site; it is also a major wildlife reserve, providing a range of habitats, from saltwater estuaries and freshwater impoundments to dunes and scrub, for a variety of plants, fish, amphibians, reptiles, birds and mammals: sea turtles, alligators, ospreys, bald eagles, waterfowl, glossy and white ibises and the 'Florida Panther', an endangered subspecies of the cougar.

In the midst of this environment lay the Kennedy Space Center itself and, within its confines, Pad 39B, from which Stafford, Brand and Slayton would add another chapter to its already chequered history. Roughly octagonal in shape, both the 39B and 39A complexes were originally expected to form part of a much wider network of five pads spread along the marshy coastline. As early as 1963, blueprints called for the construction of 39A, 39B and 39C, running sequentially from north to south, with sites for 39D and 39E available for possible future use. However, 39A was not built and the 'original' 39C site was renamed '39A'. Tom Stafford would become the first man to launch from 39B *twice*, having also flown the Apollo 10 mission from there in May 1969. Over the last four decades, 39B has carved its own niche in the annals of space exploration, with three Saturn IB launches to the Skylab station and more than fifty Shuttle flights.

Mounted atop the Saturn IB for ASTP was the final spacecraft of Project Apollo, an effort which had been born very soon after NASA's own creation in late 1958 and whose primary focus for much of its lifetime was reaching and landing on the Moon. Even its name, conferred in July 1960, honoured the ancient Greek god of music, prophecy, medicine, light and – perhaps above all – progress and therefore seemed an ideal representation of the grand scale of the endeavour. In its final form, the 'core' spacecraft comprised two main components. The 'command module' was a conical structure, 3.2 m high and 3.9 m across its ablative base and provided its three-man crews with a habitable environment and control centre during missions to the Moon or in Earth orbit. Both it and the 'service module' were built by North American Aviation of Downey, California (which merged with Rockwell Standard in 1967 to become North American Rockwell, became Rockwell International in 1973 and is currently part of Boeing). The service module was an unpressurised cylinder, 7.5 m long and 3.9 wide, housing six pie-shaped 'sectors' for propellant tanks, fuel cells and scientific instrumentation, arranged around a central cylinder which contained the main engine and pressurant tanks. Major manoeuvres were performed by the giant Service Propulsion System (SPS) engine. Minor manoeuvres and attitude control were achieved using four 'quads' of thrusters, arranged at 90-degree intervals around the main body of the service module. The SPS, some 3.8 m long and fed by a mixture of hydrazine and unsymmetrical dimethyl hydrazine with an oxidiser of nitrogen tetroxide, provided the impulse for inserting Apollo into and removing it from lunar orbit. It was a critical component without a backup and this demanded that its design be as simple as possible: its propellants, pushed into the combustion chamber by helium, were 'hypergolic', meaning that they would burn on coming into contact, eliminating the need for pumps and an ignition system. The propellants would flow as long as the valves were held open by nitrogen gas and those valves were designed for extreme reliability. The two components of Apollo remained connected throughout a mission, with the service module jettisoned just prior to re-entry and the command module parachuting to an ocean splashdown.

On the afternoon of 24 June 1975, Tom Stafford, Vance Brand and Deke Slayton entered a three-week period of medical isolation, known as the 'Flight Crew Health Stabilisation Plan', during which only previously screened personnel come could into direct contact with them. Whilst in Houston, they were quartered in mobile homes, close to the Johnson Space Center's astronaut gym, and in early July they flew to the Cape for final preparations. These included a countdown demonstration test and the only minor issue of note was a problem with Brand's space suit, which was quickly rectified. For Slayton, this period of quarantine was welcome for two reasons: firstly, after waiting 16 years for a flight into space, he did *not* want to be grounded again on the basis of illness, and secondly, he needed the time to brush up on Russian!

In truth, of course, the language skills of all three men were sufficient for them to perform their duties in space. They were not expected to be conversational, or even fluent, but *functional*. At a press conference some months earlier, when asked about the language barrier by a curious journalist, Stafford had responded to the question *in Russian* and Alexei Leonov, seated next to him, had translated *in English*. The only niggling problems were mispronunciations of some words: to the Soviets, the English word 'manoeuvre' sounded like 'manure', whereas to the Americans, the Russian word for 'separate' was similar to 'strangulate'. As already mentioned, accents did not help the situation, particularly Stafford's strong Oklahoma drawl. During their final months of training, the astronauts and cosmonauts spoke in the other language as much as possible.

For Slayton, though, the medical issue had not gone away and even restoration to flight status in 1972 and two and a half years of intensive training and instruction had not silenced all of the naysayers who doubted that he was suitable for a mission. NASA's medical staff insisted on a new flight ruling: if Slayton developed a heart fibrillation during the countdown, the clock would be held at T-4 minutes and he would be extracted from the command module. "Chris Kraft was just livid at this idea," wrote former flight surgeon Chuck Berry in Slayton's autobiography. "He called me and asked me what this was all about – [because] he thought Deke was fully qualified to fly." Berry, who had left NASA the previous year, went to the Johnson Space Center and assured the doctors there that Slayton *wasn't* going to fibrillate on the pad. The rule was unnecessary, he explained, and even if anything untoward *did* occur, he was a slow fibrillator and would not be adversely affected. Even so, the 'Slayton rule' crept again into the countdown procedures. Incensed, Kraft fired two flight surgeons at the Cape in the final days before launch and had Berry supervise instead. Tom Stafford was kept informed of what was happening, but elected not to tell Slayton until after the mission.

When Deke Slayton returned to Earth, he gave Berry a gift of thanks. It was the cardiac monitor from his medical harness in the command module, mounted onto a piece of tracing paper, on which was printed the readout of his heartbeat. The beat was steady, with no fibrillations, throughout the flight . . .

All three astronauts were fast asleep in their crew quarters at Cape Canaveral when Soyuz 19, carrying Alexei Leonov and Valeri Kubasov, underwent its final preparations for launch, several thousand kilometres away in Tyuratam. Chris Kraft, George Low, Glynn Lunney and almost everyone else at the Johnson Space Center who had poured their blood, sweat, effort and tears into ASTP, listened to the voice commentary from the Soviet Union. Compared to the excitement associated with the countdown to an American launch, the final seconds of Soyuz 19's time on Earth seemed almost anti-climatic:

" . . . This is Soviet Mission Control Centre. Moscow Time is 15 hours, 15 minutes. Everything is ready at the cosmodrome for the launch of the Soviet spacecraft, Soyuz. Five minutes remaining for launch. On-board systems are now under on-board control. The right control board . . . opposite the commander's couch is now turned on. The cosmonauts have strapped themselves in and reported that they are ready. They have lowered their face plates. The key for launch has been inserted ... "

Five minutes later, the ethereal stillness of the barren steppe was broken by the retraction of the fuelling tower from the rocket and a steadily increasing din as the engines of the R-7's central core and the four strap-on boosters roared to life. Liftoff occurred at 3:20:10 pm Moscow Time, which was shortly after 7:00 am in Florida and 6:00 am in Houston. Almost five years since its inception, the first part of ASTP was officially underway. The ascent to orbit was almost flawless, with the only problem being the failure of one of the on-board television cameras.

"For a mission whose significance was to demonstrate to a watching world that co-operation in space is possible," Leonov wrote in his autobiography, "this was a problem that had to be solved quickly. We had no choice but to dismantle a major part of our orbital section in order to gain access to the wiring for the system of five cameras connected to the switchboard and fix the problem by disconnecting the switchboard from the circuit. It took us many hours, during which we had been scheduled to sleep."

It was fortunate that Kubasov – renowned throughout the cosmonaut corps as something of a handyman – was aboard, although their efforts to bring the failed camera back to life were ultimately fruitless. This upset some of the Americans, in particular Bob Shafer, NASA's deputy head of public affairs, who was concerned that there might now be no images of the Apollo spacecraft. For the cosmonauts, though, their efforts to fix the camera almost assured them a second career after landing. "On our return to Earth," Leonov wrote, "this prompted a hilarious mail bag of requests from fellow Soviet citizens wanting Kubasov and me to come and fix their television sets!"

Fixing television sets was the last thing on the minds of most of the NASA flight controllers that morning. Despite all the reservations and the mistrust and the lack of transparency and openness in the early days, the Soviets had well and truly stepped up to the plate and launched precisely on time. Barely three months after Lazarev and Makarov's near-death experience, the R-7 performed with perfection, inserting Soyuz 19 into an initial orbit of 186 x 222 km with an inclination of 51.8 degrees. (This inclination was deliberately designed to allow Soyuz to pass well to the north of China, lest an abort should cause the mission to inadvertently land there.) Now it was America's turn. In a coastal region of Florida, prone to electrical storms and severe weather phenomena, in mid-July 1975, there was a 23-percent chance of Mother Nature throwing a spanner in the works; and for this reason, Apollo had no fewer than *five* launch windows – on 15, 16, 17, 18 and 19 July – before it would be necessary to stand down the ground crews. Of course, if more than one or two of these chances were missed, the joint mission might be ruined: for Leonov and Kubasov would have to return to Earth and their backups, Filipchenko and Rukavishnikov, would be launched in their stead. At the very least, such an outcome would prove hugely embarrassing for NASA.

The first order of business on the morning of 15 July was chief astronaut John Young tapping on the bedroom doors of Stafford, Brand and Slayton, giving them

Half a world away from Tyuratam, and several hours after the Soyuz launch, the Apollo crew arrives at the Pad 39B 'white room' for insertion into their command module. Tom Stafford is leaving the elevator, followed by Vance Brand. During this quiet time, as America prepared to launch its final Apollo mission, the men must surely have wondered *when* another crew would follow them into space. They could never have imagined that the *next* manned launch from 39B – that of Challenger in January 1986 – would bring about one of the darkest periods of the American space programme.

the news that Leonov and Kubasov were in space and a long day lay ahead. Seeing Young – a veteran of four space missions himself – reminded Slayton of the role that *he* had fulfilled for more than a decade, waking each crew up on launch morning, joining them for a breakfast of steak and eggs, helping them to don their pressure suits, engaging them in last-minute banter ... then watching impotently as they headed off on their missions and he stayed behind. The tables were now turned. "It was unusual," he wrote, "for me to be on the other end of *that* little bit of business."

Out at Pad 39B, propellant loading of the Saturn IB was nearing completion – liquid oxygen and RP-1 kerosene for its S-IB and liquid oxygen and hydrogen for the S-IVB. Thunderstorms were in the vicinity, but were not expected to disrupt the launch, which was scheduled for mid-afternoon. Glynn Lunney spoke to Konstantin Bushuyev by telephone, congratulating him on the successful launch of Soyuz 19 and advising that events were proceeding normally at NASA's end. Shortly after noon, fully-suited, Stafford, Brand and Slayton left the Operations and Checkout Building and were bussed out to the pad. In his autobiography, Stafford wondered what thoughts were passing through Slayton's head ... and in *his* autobiography, all Slayton could describe was that it felt pretty good to be crossing the swing-arm to the spacecraft. "What the hell," Slayton wrote, "it was only 13 years overdue. I never planned on being the world's oldest rookie astronaut, but I wasn't going to complain."

As he waited to be inserted into the command module, Stafford took a final look at the skeletal framework of Pad 39B. He was keenly aware that this would be the last American manned launch for at least three or four years – ultimately it turned out to be *six* – and he wondered when the next piloted spacecraft would leave from this particular complex. Little could he possibly have known that it would be more than a decade before Pad 39B would again see service ... and on *that* occasion, it would witness one of the greatest calamities of the space programme so far: Challenger.

As rookie astronaut Bob Crippen helped Stafford connect his electrical, oxygen and communications umbilicals on the left side of the command module, Slayton prepared to take his seat on the right-hand side. Although his title implied that his sole responsibility was the docking module, Slayton's primary tasks during ascent and re-entry were to monitor the ship's electrical systems. Last to enter was Brand, the command module specialist, who took the centre seat. Checking in with spacecraft test conductor Clarence 'Skip' Chauvin and the blockhouse capcom, Karol 'Bo' Bobko, Stafford asked if the countdown would be in English or Russian. "Oh, I figured I'd give it in English!" responded Bobko, with a chuckle.

A little under eight minutes before launch, with all 556 switches, 40 event indicators and 71 console lights checked and cross-checked, Stafford asked Bobko to tell the Soyuz crew to get ready for them. The launch itself, at precisely 3:50 pm Eastern Standard Time – shortly before 1:00 am on 16 July in Moscow – was perfect and Stafford found the ascent to be very gentle and smooth, like an elevator. As a veteran of two Gemini launches and a bone-rattling ride on the Saturn V, he was not expecting the rise from Earth to be quite so serene. Even the period of staging, as the S-IB burned out and the single J-2 engine of the S-IVB took over, seemed much calmer; it was "noticeable", he wrote, but nowhere near as violent as the Saturn V.

Slayton, a man of few words, described the experience of his first launch into space as being louder than he had expected, but otherwise not surprising. He had spent most of the previous decade debriefing other astronauts following their own missions aboard Atlas, Titan and Saturn rockets, so his inherent knowledge of what to expect is perhaps not entirely surprising. "Deke felt as if he were riding an old pickup truck slam-banging down a rutted, country dirt road back on his Wisconsin farm," he wrote in third-person narrative in a joint autobiography with Al Shepard. "Eight engines blazed away at full thrust with a cacophony of noises – propellants pounding through lines from turbopumps spinning at tremendous speed, pressures surging with booming thuds throughout the stage, all to the accompaniment of teeth-rattling, eye-blurring shaking." Slayton's desire to sit back and enjoy his ride into orbit yielded to a sensation of balancing at the end of a long rubber balloon which was fighting its way through wild winds.

"I *love it*!" he yelled as the Saturn's first stage burned out and the S-IVB picked

With Alexei Leonov and Valeri Kubasov safely in orbit, the American crew spears for the heavens aboard the final Apollo spacecraft, riding the final Saturn rocket. This perspective provides an impressive view of the marshy landscape of the Kennedy Space Center.

up the thrust for the final boost into orbit. "Man, I'll tell ya … this is worth waiting 16 years for!" It had, after all, been 1959 when he was chosen by NASA for astronaut training.

"You liked that, huh?" grinned Stafford.

"I'd like to make *that* ride about once a day!"

What was *really* surprising, though, was the instant at which he became weightless for the first time; the instant of feeling as light as a snowflake and totally unencumbered by the shackles of gravity. "*Whoppo!* Shutdown [of the S-IVB] was pretty abrupt," he wrote. "You went from being pushed back in your couch to hanging in your straps. We were in zero-G!"

From the centre couch, the only window was directly above Vance Brand's head and his first glimpse of Earth from the edge of space was the glorious, bright blue vista of the Atlantic Ocean. They were, in fact, 1,800 km north-east of Florida and their more northerly azimuth – which, like Soyuz, was inclined 51.8 degrees to the equator – would carry them as far north as England. It would also carry them over areas never before seen from an American spacecraft … including parts of the Soviet Union. "I kept trying to sneak peeks out the windows," wrote Slayton, "and wished there was a bubble turret on the spacecraft, so I could just sit up there and watch the world go by." Yet for both Slayton and Brand, the novelty of where they were was overpowering. As objects began to move around the cabin and his own body floated freely, Brand could only blurt out excitedly in Russian: "Miy nakhoditsya na orbite! (We're in orbit!)" Indeed, they were.

"GLAD TO SEE YOU"

Forty minutes after leaving Florida, and by now on the opposite side of the planet, Tom Stafford radioed Capcom Dick Truly in Houston to advise him that the crew were preparing to perform the 'transposition and docking' manoeuvre to extract the docking module from the S-IVB. In a similar manner to the removal of the lunar module on earlier missions, the astronauts uncoupled Apollo from the S-IVB and the panels of the conical spacecraft adaptor were explosively jettisoned, exposing the docking module to the environment of low-Earth orbit for the first time. With Stafford at the controls, the first problem arose: all he could see was the blinding glare of the sunlit Pacific Ocean ahead of him. "I sat there, a few metres from the DM," he wrote, "and began to sweat. Finally, I decided the only choice was to use the Mark I eyeball – lining up on the cross-shaped target mounted atop the truss behind the DM, and, once the DM had drifted toward a darker background, thrusting closer." Stafford achieved a perfect docking, aligning the two vehicle to an accuracy of better than one hundredth of a degree. It was the best accuracy ever achieved with the Apollo docking system. At the instant of capture, however, the crew were out of direct radio communications with the ground and had to wait some minutes before they could tell Truly that all was well. Shortly afterwards, during a five-minute communications pass over the tracking ship USNS *Vanguard*, the crew successfully extracted the docking module from the S-IVB.

Everything was now in place for Apollo to begin a complex series of manoeuvres over the next two days to precisely match its orbit with that of Soyuz, undertaking the active role in the rendezvous. Shortly before bedding down for his first night's sleep in orbit, Brand set to work disassembling the bulky docking probe from the connecting tunnel between the command and docking modules. It was his intention to open the docking module's hatch and store an experimental freezer – needed for several joint experiments – there overnight. Very soon, however, he realised that he could not properly insert the tool to unlock and collapse the probe. Back home in Houston, Bo Bobko was the duty capcom.

"Okay, Bo," radioed Brand. "Everything in the probe removal checklist on the cue card ... has been going great [but] Step 12 is 'capture latch release, tool 7'. You insert it in the pyro[technic] cover [and] you turn it 180 degrees clockwise to release the capture latches. Well, here's where the problem is and let me explain it to you. Do you have somebody there that knows the probe that can listen?"

Bobko quickly found a probe expert. "Roger," he verified. "Go ahead."

"Okay," continued Brand, "as I look in the back of the ... pyro cover, I'm looking with my flashlight through the hole where I insert this tool and there's something *behind* the pyro cover that's preventing me from putting this tool all the way in. It's actually one of the pyro connectors. This tool has to go down through the pyro cover in between ... some pyro connectors, but one of these pyro connectors has rotated such that it's in the way ... "

Flight Director Neil Hutchinson would later tell reporters that Brand and Bobko spent around 18 minutes troubleshooting the problem with the probe, then decided to delay moving the freezer into the docking module until the following morning. When Brand then tried to close the hatch, he found that the partially-removed probe stopped him from doing so. It was already over an hour past the crew's scheduled bedtime, so Mission Control recommended that they postpone further work. In the meantime, the astronauts slightly raised the command module's cabin pressure to provide additional oxygen and thereby compensate for the nitrogen coolant which was boiling off the freezer.

Sleep proved more awkward than anticipated. The crew folded up Brand's centre couch to gain additional space, with two men sleeping in the side couches and the third in a sleeping bag strung across the command module's lower equipment bay, but even that did not work very well. From the second night, Slayton opted to curl into the narrow, cylindrical docking module (which he had earlier jokingly called "the world's fastest, highest-flying sewer pipe") and sleep there. Brand bedded down in the transfer tunnel and Stafford remained aboard the command module.

Next morning, they returned to the probe issue, completing each of the 11 steps from the previous night in order to re-engage it in its fully locked position. They then removed the pyro cover, straightened out the misaligned pyro cap, proceeded through the 11 *disassembly* steps and finally inserted the key to unlock the capture latches. It was, Neil Hutchinson told journalists, more of an annoyance than anything else and it warranted hardly a mention in Stafford's and Slayton's autobiographies: the former simply making reference to "a balky docking probe", the latter to "some little problem with the hatch". The real deal was the rendezvous.

Two burns of the SPS engine served firstly to circularise their orbit and later adopt an elliptical path to gradually close the gap with Soyuz.

Alexei Leonov and Valeri Kubasov were by no means inactive, despite their unsuccessful attempt to repair the television camera, and by the morning of 17 July the distance between Apollo and Soyuz was a little over 900 km. Shortly after 8:00 am Houston time (5:00 pm in Moscow), Tom Stafford fired the SPS for a third time to lower his apogee – the 'high point' of his orbital path – in order to complete the next stage of the rendezvous. From his station, Vance Brand spotted the Russian craft as a bright speck in the darkness, and five minutes later Deke Slayton managed to contact the cosmonauts in Russian on VHF radio:

"Soyuz, Apollo. How do you read me?"

"Very well," replied Kubasov in English. "Hello everybody."

"Hello, Valeri," Slayton continued. "How are you? Good day, Valeri."

"How are you? Good day," came the response.

Pleasantries completed, it was time for business. Half an hour later, Kubasov switched on the range tone transfer assembly aboard Soyuz to establish accurate ranging data between the two spacecraft. By now, their separation distance had closed to just 220 km. At 9:12 am Houston time, Stafford executed another manoeuvre with the SPS which inserted Apollo into a 210 x 209 km orbit, to effectively catch up with Soyuz. A short terminal phase burn a little over an hour later brought the American craft to within 35 km of its Soviet counterpart. Shortly thereafter, at 10:46 am, Capcom Dick Truly relayed two important messages to the crews: "Moscow is Go for docking; Houston is Go for docking. It's up to you guys. Have fun!"

The two craft drew nearer and nearer. By now, Leonov and Kubasov had retreated into their descent module and closed the hatch to the orbital module; similarly, the Apollo crew had sealed the docking module and were in their couches within the command module. Eight hundred metres of empty space lay between them. At Stafford's call, Leonov rolled the Soyuz some 60 degrees to place it into the proper orientation for the final approach. In Houston, Mission Control was packed: Administrator Jim Fletcher and his wife were there, along with Soviet ambassador Anatoli Dobrynin and chief astronaut John Young, together with Bob Gilruth, former Apollo head Sam Phillips and a gaggle of astronauts, both veterans and rookies: Dave Scott, Joe Allen, Owen Garriott, Bruce McCandless, Story Musgrave and Rusty Schweickart.

Apollo and Soyuz were coming up on the coast of Portugal as Stafford called out the closing distance ... five metres ... three metres ... one metre ... until, finally, at 11:09 am, the two craft met in a metallic embrace and he called: "Contact!" A second or two later, Leonov rendered his own acknowledgement: "Capture! Soyuz and Apollo are shaking hands now!"

The instant of docking, Stafford later wrote, was exceptionally smooth and he promptly retracted the guide ring, actuated the structural latches and compressed the seals to achieve a rigid configuration. "Tell Professor Bushuyev," he radioed, "the docking was very soft." Leonov congratulated him on a good show. It was so good, in fact, that when they checked the alignments later that evening, the centre of the alignment sight sat right on the very *centre* of a bolt that held the *centre* of a target on Soyuz. In other words: Stafford had hit his mark, *dead centre*.

Each of the crew exchanges between the two craft had been worked out months in advance and it was planned for Slayton and Stafford to pass through the docking module for the first meeting with Leonov. "Given the different pressures in the two spacecraft," Slayton wrote, "you couldn't just open the hatches on both ends of the docking module and go through." For that reason, Brand remained sealed inside the command module for the first historic handshake. However, when Slayton opened the hatch to access the docking module, he and Stafford were hit by an unpleasant odour. It reminded them of burned glue. Quickly, they radioed Leonov that they had "somewhat of a bad atmosphere", but the odour dissipated within minutes. It was subsequently blamed on one of the experiments in the docking module. Playfully, Stafford floated to the far end of the tunnel and rapped his knuckles on the hatch leading into the Soyuz orbital module. Leonov rapped back. In Russian, Stafford asked: "Who's there?"

Shortly before opening the hatch into the Soviet craft and the historic, televised handshake, a message was read over the ground-to-space radio link from Leonid Brezhnev. "To the cosmonauts Alexei Leonov, Valeri Kubasov, Thomas Stafford, Vance Brand, Donald Slayton," it read. "Speaking on behalf of the Soviet people, and for myself, I congratulate you on this memorable event ... The whole world is watching with rapt attention and admiration your joint activities in fulfilment of the complicated programme of scientific experiments ... " In conclusion, Brezhnev's message expressed the fervent and sincere hope that Apollo-Soyuz would represent a true forerunner of "future international orbital stations". As will be shown later in this chapter, it is a pity that those heady days of the mid-1970s could not have led to a much broader foundation of co-operation between the two old foes and an even greater tragedy that it would be almost two full decades before anything like ASTP would be attempted again.

For now, though, on the afternoon of 17 July 1975, neither the Apollo astronauts nor the Soviet cosmonauts could have envisaged that the difficult and tumultuous path they had trodden together over the past several years, would be forgotten in just a few months. Nor could they possibly have foreseen that both sides would shortly begin the steady slide back into cold war and, by the early 1980s, the language of conciliation would have reverted back into the language of warning and tough words of evil empires and totalitarianism would be met by equally tough responses of imperialist world domination and the bellicose lunacy of the West. For now, though, on this midsummer's afternoon in 1975, it seemed that the two old foes were setting the seal on a bright future. At 2:17:26 pm Houston time, high above the French city of Metz, Tom Stafford tugged open the hatch and squinted as he peered into the Soyuz orbital module. There, surrounded by a snake-like collection of umbilicals, were their old friends Alexei Leonov and Valeri Kubasov, both beaming.

The meeting was awkward in the peculiar weightless environment, as Stafford and Leonov warmly shook hands and tried clumsily to hug.

"Glad to see you!" Leonov exulted in English.

"Ochen rad! (Very good!)" replied Stafford in Russian. "Tovarich! (Comrade!)"

The five crewmen of Apollo-Soyuz – from left to right, they are Brand, Stafford, Leonov, Kubasov and Slayton – pose beneath a mockup of the docked vehicles in February 1975. The Apollo spacecraft is visible on the left, with the docking module in the centre and the orbital and descent modules of Soyuz to the right.

No sooner had they adjusted to their new environment than the call came from Houston: President Ford was on the line and wished to speak with them. He had been a strong supporter of the mission – or, at least, the measure of détente that it afforded – and asked questions and offered congratulations for nine whole minutes, almost twice as long as scheduled. Responding to the president's queries was not easy. "It was kind of tricky," wrote Slayton, "since we kept having to hand headsets from Tom to Alexei to Valeri to me to hear the questions and give the answers." Ford echoed many of the themes raised by Brezhnev: it had taken the United States and the Soviet Union many years to open this door to useful co-operation in space and he asked how useful the androgynous docking mechanism might be for future missions. Stafford replied that it had performed beautifully.

After the president signed off, the commemorative exchange of gifts began, with Stafford presenting a quintet of tiny American flags and Leonov reciprocating with Soviet flags. Other exchanges included a United Nations banner, launched aboard Soyuz and returned home aboard Apollo. Both crews signed a Certificate of the First International Docking for the official aviation record books, kept by the Fédération Aéronautique Internationale, and a joint plaque (half of which launched aboard

Apollo and half aboard Soyuz) was ceremoniously connected by Stafford and Leonov. "They had a little difficulty making these pieces match," Slayton told journalists in Houston on 9 August. "Fortunately, the docking system worked better than *this* did!" It was then time for a joint dinner, without Vance Brand, around a green metal table in the Soyuz orbital module, and with the whole world watching a lump undoubtedly jumped into Stafford's throat ... for Leonov presented his new American comrades with tubes of *vodka*! Ten years earlier, Leonov had met a group of astronauts – including Deke Slayton – and shared a collective hope to someday drink a toast together in space. Even after ten years, Leonov never forgot the pledge.

Aware that the whole world was watching on live television, as was President Ford, Stafford seemed reluctant to consume anything alcoholic, but his counterpart persisted; it was a Russian tradition, said Leonov, to drink before eating. Unknown to Stafford, the canny Leonov had peeled the labels from a few tubes of borscht and blackcurrant juice before launch and replaced them with labels for various Russian vodkas. In his autobiography, Stafford later recalled his mild disappointment as he slurped ... not vodka, but *borscht* ... through the tube. As Leonov later wrote, "I told him it was the thought that counts!" Leonov's second surprise was a set of pencil sketches of Stafford, Brand and Slayton, which he had drawn during their two years of training together.

Shortly before 6:00 pm Central Standard Time, after the dinner of reconstituted strawberries, Roquefort cheese and sticks of apples and plums, Stafford and Slayton bade the Russians farewell for the night and floated back to the command module, closing and sealing the docking module hatch as they went. Leonov and Kubasov followed suit, securing the hatch of their spacecraft. The two Americans rejoined Brand, who had spent a lonely afternoon keeping an eye on the systems. Right from the start, both sides had insisted on having one crewman aboard their respective craft at all times; this was a point raised by Glynn Lunney at a press conference immediately after the Nixon-Kosygin summit in May 1972. Years later, Brand described his solo work as "kind of minding the store ... holding the attitude for the 'stack' of vehicles, which consisted of Soyuz and Apollo and docking module".

Brand's turn would come during their second day of docked activities, 18 July, when he floated across to join Kubasov in the Russian craft and Leonov came over to the American side. Aboard Apollo, American fayre included potato soup, bread and grilled steak. For four and a half hours that day, Kubasov and Brand worked together in the cramped Soyuz, which they nicknamed the 'Soviet-American TV centre in space', after performing a broadcast from it. Diplomatically, Kubasov wondered aloud to his US audience which of their two nations was more beautiful ... and then concluded that neither possessed the full majesty of "our Blue Planet". Meanwhile, Stafford gave Leonov and the Russian audiences back on Earth a televised tour – in their own language – of the Apollo spacecraft; for the American people, these guided shows from outer space had been commonplace for several years, but in the Soviet Union, such 'live' events from beyond the atmosphere had never been seen before.

The interest of the Russian populace in the joint mission had even inspired them to adopt some of the trappings of the capitalist West. Midway through the mission,

Time reported that several Soviet perfume factories had created a new scent, with the rather unimaginative name of 'EPAS' ('Experimental Project Apollo-Soyuz'), which they intended to sell for $50.75 per bottle in Russia and a mere $10 in the United States! Other efforts to cash-in included a new brand of cigarettes, burdened with the equally unoriginal name of 'Soyuz-Apollo', which Moscow's Yava factory hoped to sell in America ...

Aside from the public relations side of the mission, there was 'real' work to be done, too, and it took the form of five joint experiments, including an experimental multi-purpose electric furnace, located inside the docking module. Measuring 10.1 cm in diameter and 29.2 cm long and weighing 5.2 kg, the furnace was tended by Stafford, Leonov and Slayton. It carried a series of experiments, which focused on the melting and mixing of paired alloys to analyse the effects of convection in the weightless environment, observing the behaviour of specific materials (such as aluminium-antimony, known to have promise for high-efficiency solar cells) and melting and re-solidifing magnetic and semi-conducting crystals. Earth studies were also undertaken, with Slayton photographing ocean currents off the Yucatan Peninsula and in the Straits of Florida; likewise, Brand filmed his own travelogue, covering part of the United States' eastern seaboard, although he was hampered by cloud cover for much of the time. They also kept a keen eye on their hometowns. For Slayton, the farmland of Wisconsin was perhaps more familiar to him than anywhere on Earth. "A hundred and sixty acres of that land," he wrote in third-person narrative in his joint autobiography with Al Shepard, "was where he had spent his boyhood years, farmland with a heart-touching similarity to his family's origins in Norway. For more than a century, his family had lived on that land over which he now sped at five miles *every second*." With such a view 'beneath' him, sleep was difficult and his constant enemy was the clock; indeed, he wrote that he begrudged every moment that he could *not* look at the Home Planet and outwards into the Universe ... for he knew that he would never come this way again.

One of the primary focuses of the Earth studies, geologist Farouk el-Baz told a science briefing on 19 July, was to ascertain the precise *colours* of terrestrial targets. For this reason, a 'colour wheel' was housed aboard the command module and proved particularly effective in the studies of arid desert regions. "We have attempted to get as close as possible to the colour of the deserts as described to us by the crew in the training they did," explained el-Baz. "They flew in T-38s over many deserts in the US and Mexico ... [with] prototypes of the colour wheel." Deke Slayton was astonished that, with new orbital sunrises and sunsets every 90 minutes, the hues and intensities of the clouds altered and even smoke and dust and moisture in the atmosphere constantly changed this kaleidoscope of colour. During orbital darkness, the twinkling lights of cities glimmered below him, as did the fires of desert oil and gas wells and the occasional flash of a meteor enduring a fiery death as it plunged through the atmosphere. Meteors intrigued him; his naked eyes could scarcely gauge their tremendous velocities and yet the results were astonishing, as the invading bodies of rock and dust and ice blazed like welders' torches and trailed streams of particles and superheated gases. Slayton found himself gulping for air and then holding his breath, as if it might somehow freeze the moment to help him savour it. Perhaps, he mused, *this* was the kind of view that angels enjoyed ...

Many space travellers have returned from their journeys, somewhat depressed and disappointed that their lack of ability as wordsmiths or poets render them incapable of truly describing the enormity of what they saw. There are exceptions. Some, such as Mike Collins and Gene Cernan, have described their experiences cogently and with literary vigour. For Alexei Leonov, aboard Soyuz for that week in July 1975, he had his own idea. As a gifted artist, he had taken along with him a collection of crayons and pencils to sketch some of the colours of Earth, as seen by the human eye. "The Black Sea, for instance," he wrote, "really is the darkest sea in the world and I was able to record its true colour."

Other work included studies of aerosols in the stratosphere, observations of the effect of cosmic rays on fungi, the effects of weightlessness on small mice and fish and the retina of the human eye and astronomical and solar physics experiments. They exploited a planned undocking and re-docking exercise, performed on 19 July. After 44 linked hours, the two spacecraft separated at 7:12 am Houston time and Slayton performed an almost flawless re-rendezvous and re-docking with Soyuz. During their period in individual flight, the Apollo crew placed their craft directly between Soyuz and the Sun, such that the diameter of the service module created an artificial eclipse. This enabled Leonov and Kubasov to photograph the solar corona, in conjunction with ground-based observations to compare the effects from instruments located both 'inside' and 'outside' the atmosphere. A second experiment in ultraviolet absorption was also carried out during this independent period, enabling researchers to more precisely determine the quantities of atomic oxygen and nitrogen at that altitude. Instrumentation mounted on the Apollo service module projected monochromatic laser-like beams of light to retro-reflectors attached to Soyuz; when these were reflected back, they were analysed by a spectrometer.

Re-docking did not go entirely as planned and the seriousness of the situation seems to vary, depending upon whose autobiography one chooses to read. With Slayton at the command module's controls, the rendezvous proceeded normally, at first, but then he found it difficult to see the cross-line reference on the alignment sight, due to the glare of the sunlit Earth. It was precisely the same kind of blinding 'wash-out' that Stafford had experienced during the transposition and docking a couple of days earlier. "He proceeded with the docking," Stafford wrote, "which appeared to go smoothly ... but the moment after contact and capture, both vehicles oscillated. It only happened for a few seconds and was probably due to a slight misalignment. The docking was well within limits, so I didn't sweat it." Slayton himself made brief reference to the incident, but did not dwell on it: "I tweaked the hand controller the wrong way, once we had captured Soyuz again, causing the two spacecraft to shake a little."

From Slayton's perspective, the incident was a new lesson in fuel management and provided valuable data for future space rescue scenarios. In his autobiography, though, Alexei Leonov related that, during the contact and capture at 7:33 am, Slayton "inadvertently fired one of Apollo's side roll thrusters, which had the effect of pushing both vehicles off-centre, folding them towards one another". Certainly, television views from inside the command module confirmed that this was a harder docking than the first one. There had been a "real threat" of damaging the joint docking mechanism, Leonov added, and the possibility of a "catastrophic depressurisation of our orbital module". To conclude, the cosmonaut noted that no serious damage was done and even that mission controllers in Moscow received an apology from Houston for the mistake.

Three hours later, at 10:27 am, the two spacecraft undocked for the second and final time. After they had gone their separate ways, they remained from time to time in radio contact and Stafford took the opportunity to gain revenge on Leonov for the vodka/borscht incident. Vance Brand had brought a cassette tape of girls giggling in the shower and Stafford radioed Leonov to ask what he was doing.

"Tom, we are resting, because we've worked so hard."

Stafford replied that *his* crew were still working.

"Why?"

"Listen," said Stafford and motioned to Brand to play the tape, whilst Slayton held down the microphone button. As the sound of running water and giggling girls crossed from ship to ship, Leonov was aghast. "Tom," he asked, "what are you *doing* over there?"

"Working hard!"

With images of scantily clad women frolicking around the command module's cabin now in his head, Leonov was still uncertain. "You *are* kidding, aren't you?"

GRIM REALITY

Two days later, at 1:09 pm Moscow Time on 21 July, Leonov and Kubasov made their final manoeuvre in orbit: a four-and-a-half-minute burn of their retrorocket to begin a ballistic descent through the atmosphere and a landing back on Soviet soil. Within half an hour, recovery pilots could hear the cosmonauts chattering on the radio and a helicopter-borne television crew showed the descent module, hanging beneath its single parachute, as it floated down towards at 1:50 pm touchdown. The landing occurred 87 km north-east of Arkalyk, in central Kazakhstan, and shortly thereafter both cosmonauts extricated themselves from the vehicle and waved at photographers.

Since it had long been evident that ASTP would be the final American manned mission for some years, NASA opted to take advantage of the Apollo spacecraft's consumables and keep it in orbit for a full nine days of research work. It would be the last time that astronauts would perform such tasks for the better part of a full decade. Earth observations were aided by the fact that the orbit for this part of the mission – just 217-230 km – was a few dozen kilometres lower than most previous missions and Stafford made reference to this effect in a Houston press conference on 9 August. "Deke made a comment that it looked like the thunderstorms came a quarter of the way up in altitude to *our* altitude, but, for example, on a clear day over El Paso International Airport, where we usually land going to the west coast, right there were the runways, the taxiways and the hangars. It was a clear day and you

Viewed through the aperture of the open hatch, Tom Stafford, Vance Brand and Deke Slayton undergo a simulation of their mission tasks in March 1975. This image clearly demonstrates the cramped confines of the Apollo command module … and allows the reader to visualise the horror which unfolded in these close quarters during descent and splashdown, when all three men were exposed to potentially lethal nitrogen tetroxide fumes.

could see them, just visually, with your eyes." Another point, Stafford added, was the earlier higher orbital latitude, which permitted them to see much broader swathes of Earth ... and a realisation that "so much of this world ... is just desert and mountains".

Major work on 23 July included Doppler tracking and geodynamics experiments, which employed the Applications Technology Satellite (ATS)-6 in a geostationary orbit above the equator to assess techniques for monitoring terrestrial plate tectonics and mass anomalies. ATS-6, which was launched in May of the previous year, had already seen significant service during ASTP as a communications relay, providing voice links between the spacecraft and Mission Control for up to 55 percent of the time. Its use marked the first occasion on a manned flight in which such a satellite had been employed for voice communications. The docking module, jettisoned in these latter stages of the mission, provided valuable data in support of the Doppler experiment and would eventually re-enter the atmosphere to destruction in August 1975. As they packed up the last of their equipment on splashdown day, 24 July, Capcom Bob Crippen told them that the weather conditions in the primary recovery zone were good: visibility was 16 km, winds at 31 km/h, scattered cloud cover at 600 m and wave heights of 1.1 m. The de-orbit burn was executed at 3:37:47 pm Central Standard Time and, six minutes later, the service module was jettisoned, leaving the command module alone for a fiery plunge back to Earth.

During re-entry, Brand assumed the left-hand seat, with Stafford in the middle and Slayton over on the right side. The computer used a series of tiny reaction control thrusters to control re-entry and aim for the recovery point. One of Stafford's tasks was to deactivate these thrusters, whose propellants included a particularly noxious chemical, called nitrogen tetroxide, at an altitude of around 25 km. After descending another 10 km or so, the drogue chute and finally the main canopy were supposed to automatically deploy, followed by the opening of a vent valve to admit fresh air into the cabin. The command module would then splash gently into the Pacific Ocean, a few hundred kilometres north-west of Hawaii. It did not work out that way ... and almost led to the death of the entire crew!

In their respective accounts of what happened, neither Brand nor Stafford could be entirely sure of who did what – or who did *not* do what – but certainly the re-entry phase was considerably more dynamic than anticipated and an irritating squealing noise in their headsets momentarily distracted them. For whatever reason, Stafford did not throw the switch to deactivate the reaction control jets. "The noise made it impossible for us to hear each other or Houston," he wrote. "In order to be *heard* in the cockpit, we had to *shout*. Either the noise kept Vance and Deke from hearing me or I was too distracted to give the command."

Fifteen kilometres above the Pacific, as intended, the drogue chute deployed and the vent valve duly opened, but the reaction control jets were still spurting and the valve admitted not fresh air ... but a lethal concoction of nitrogen tetroxide. "The vent valve was located right below the ... thrusters," Stafford wrote, "and it sucked some of the nitrogen tetroxide into the cabin." As soon as he saw the yellowish-brown mist and sniffed its pungent, acrid odour, he knew instantly what it was. So did Brand and Slayton. Moreover, all three instinctively knew what it could *do*.

Nitrogen tetroxide is one of the most deadly chemicals used in the manned spaceflight business; highly toxic and extremely corrosive, if inhaled at concentrations of just four hundred parts per million, it can kill. In the same vein, it is one of the most important chemicals ever used in the field of spacegoing rocketry: it is hypergolic with various forms of hydrazine, meaning that it can burn on contact, without a separate ignition source, and recent research has explored its usefulness in future power generation systems as a so-called 'dissociating gas'. As Stafford saw this noxious stuff steadily filling the cabin, he immediately flipped a pair of switches to shut off the propellant gate valves and cut the fuel supply ... but since some residue remained in the lines, it gave little relief. As the fumes set to work irritating eyes, burning faces, noses, mouths and throats, Stafford, Brand and Slayton began hacking and choking uncontrollably.

Four hundred parts per million ... could kill ...

The main parachute canopies deployed as scheduled and the command module struck the Pacific at 4:18:24 pm Central Standard Time – "a real bone-cruncher," Stafford remembered, "nearly 10 positive Gs" – and promptly flipped into the 'Stable 2' position, with its nose submerged and the crew hanging from their harnesses. Brand was seated closest to the vent valve and quickly passed out, his hands clenched. Slayton, too, was feeling nauseous, and it was Stafford who loosened himself and grabbed three sets of oxygen masks. "For some reason, I was more tolerant," he explained later. "I knew that I had a toxic hypoxia ... and I started to grunt-breathe to make sure I got pressure in my lungs to keep my head clear." He secured a mask over the unconscious Brand's face and held it there until he revived, thrashing his arms about for several seconds. At length, with all three astronauts breathing on masks, they were able to inflate airbags on the apex to right the command module into a 'Stable 1' position in the water and Stafford fully opened the vent valve. With a sudden influx of fresh air, the remaining nitrogen tetroxide fumes quickly disappeared.

They were by no means out of the woods yet.

"Then it was *my* turn to screw up," Slayton wrote. "We were in the water a few minutes, still hacking, when they dropped the frogmen. One of them appeared in the window and like a dumb shit I gave him the *thumbs-up sign*! Everything's okay. Well, of course, it *wasn't* ... but everybody outside thought it was, so there was no special effort to get us out of the command module."

Splashdown came a few kilometres from the recovery ship, USS *New Orleans*, and within minutes Stafford, Brand and Slayton were aboard a helicopter, still coughing, but not thinking too much about what they had just endured. Only when they were seated at a press conference on the ship's deck – and, ironically, whilst speaking to President Ford on the telephone – did they inadvertently drop out that the mission had gone well, *except for the final few minutes.* As soon as chief flight surgeon Arnauld Nicogossian learned about the nitrogen tetroxide, he stopped the conference and whisked all three men down into the *New Orleans*' medical bay. Cortisone was pumped into each of them to reduce lung inflammation and, in Slayton's words, it was a good thing: they had felt fine during the helicopter ride, but within three-quarters of an hour, they suddenly began showing the symptoms of full-blown pneumonia! "The next day," Stafford recalled, "we saw X-rays and our *lungs*, where they were completely clear before and right after landing, the next day they were all *white*." It was, said Nicogossian, a classic case of infiltration: an accumulation of abnormal substances in the body. The astronauts were hospitalised at the Tripler Army Medical Center in Honolulu for two weeks, during which time it was found that they had inhaled three hundred parts per million of the toxic gas. Had Stafford not reacted as he did and applied the face masks, all three would have been dead in a matter of minutes ...

If the return to Earth had been grim for the Apollo crew, its grimness would be duplicated in the realisation that ASTP would not live up to one of its promises: it would *not* be the first in a string of ambitious US-Soviet joint missions. In fact, more than a decade and a half would elapse before negotiators again returned to the table to seriously discuss future co-operative efforts in manned space exploration. The astronauts and cosmonauts of ASTP went their separate ways. Stafford retired from NASA later that year to take command of the test pilot school at Edwards Air Force Base in California, whilst Slayton was assigned to manage the Approach and Landing Tests of the Space Shuttle. As a bitter footnote, whilst in Honolulu, a pre-cancerous lesion was discovered on Slayton's lung. Thankfully, it was benign and had actually turned up in a pre-flight X-ray, but had been overlooked. Had it been spotted before launch, he would have been grounded yet again, which would have been cruel luck, indeed. Vance Brand, meanwhile, was also assigned to the Shuttle programme and would go on to command three of its missions in the 1980s and 1990s. Alexei Leonov – promoted to general during ASTP – took charge of the cosmonaut corps and in May 1980 Valeri Kubasov became one of only a handful of civilians to command a Soyuz mission of his own.

Détente between the Soviet Union and the United States would veer sharply off course and, as the 1970s drew to a close, relations were once again at Cold War levels. The Hensinki Accords, signed in the summer of 1975 under Gerald Ford's brief presidency, attempted to improve relations between East and West, including agreements on the inviolability of frontiers, the peaceful settlement of disputes and non-intervention in internal affairs, but superpower involvement in conflicts in the Middle East, Central Asia and the Americas served only to sour relations: the Yom Kippur War, the Chilean coup d'état, the Ogaden War in Ethiopia, the Angolan Civil War, the Nicaraguan Civil War, the Islamic Revolution in Iran ... and, perhaps most fundamentally, the Soviet invasion of Afghanistan.

By the spring of 1977, a new president was in the White House. Jimmy Carter, whose attitude towards the Soviets was much less sympathetic than Nixon or Ford, very quickly signed into law Presidential Directive No. 18 on National Security to reassess the United States' position on détente. At the same time, in an effort to curtail the manufacturing of nuclear weapons and the development of new missile arsenals, Carter and Leonid Brezhnev undertook the second round of Strategic Arms Limitation Talks (SALT II) and would lay down their signatures in Vienna in June 1979. Within six months, however, the Soviets had invaded Afghanistan, prompting an "open-mouthed" Carter and the CIA to begin to arm the mujahideen insurgency. As the decade drew to an end, fears grew that the Soviets were seeking to expand

Recovery swimmers from the USS *New Orleans* prepare to open the command module's hatch after splashdown. Despite a thumbs-up from Deke Slayton, all was *not*, in fact, well. Little could anyone have realised that the three astronauts had inhaled near-fatal quantities of nitrogen tetroxide fumes during their final descent. The triumph of the first US-Soviet joint space mission almost ended in tragedy.

their sphere of influence into Pakistan and Iran and even that they were positioning themselves for a takeover of oil in the Middle East. For his part, Carter refused to permit any outside forces to gain control of the Persian Gulf, terminated a 'wheat deal' with Russia and made the unpopular move of prohibiting American athletes from participating in the 1980 Summer Olympics in Moscow. Together with his national security advisor, Zbigniew Brzezinski, Carter started a $40 billion covert programme to train Pakistani and Afghan insurgents to foil the Soviet invasion. The 'hawkish' foreign policies of Brzezinski have even prompted mutterings over the years that he and Carter had begun arming the mujahideen *before* the invasion, as a way of drawing the Soviets into a protracted and gritty conflict; in essence, creating their own Vietnam.

When one views these important geopolitical events through the looking-glass of history, it is not difficult to understand why virtually no progress was made on the topic of co-operation in space after the return of the Apollo-Soyuz crews to Earth, more than thirty-five summers ago. Today, in the era of the International Space Station and genuine co-operation in space between not merely Russia and America, but other nations, too, it is saddening to consider the possibility of what might have been. Certainly, ASTP Flight Director Neil Hutchinson once commented on how well he worked with his Soviet counterparts, to such an extent that he wished for another such mission. "It's like going to the Moon once and never going back," Hutchinson said. "Ninety percent of the battle is over with ... getting all the firsts done ... I could run another Apollo-Soyuz ... with a heck of a lot less fuss. Though some of the worry in both Houston and Moscow had been in vain, the two teams had confirmed that they could work together in analysing an unforeseen problem."

Others saw it differently. Robert Hotz, then-editor-in-chief of *Aviation Week & Space Technology*, thought the *real* tragedy was the fact that ASTP was a one-off stunt ... that NASA and America had bet everything on a 'political fanfare', when it could have invested the money into a second Skylab orbital station – already built and waiting to launch – for greater long-term scientific return. "Now that it is over," Hotz editorialised, "it is apparent that the decision to fly Apollo-Soyuz, instead of another Skylab, was as foolish and feckless as those other facets of the Nixon-Kissinger détente, the SALT talks, the trade deals and that great treaty that brought peace to Vietnam."

Still, to gain some idea of what may have been, it is necessary to return to the high-watermark of US-Soviet relations: the time in May 1972 when Nixon and Kosygin put their signatures to ASTP. At a press conference in Houston, NASA Administrator Jim Fletcher had responded to a journalist's question by stating that Apollo-Soyuz was merely "a first step in international co-operation" and, moreover, that co-operation in manned programmes "to save duplication of effort between the two countries" was his great hope. Only now, five decades since Gagarin and three and a half decades since ASTP, are we beginning to see the realisation of some of Fletcher's vision. Genuine co-operation was the vision of his deputy, George Low, also. During his visit to the Soviet Union in May 1975, Low spoke of the future with several of his NASA colleagues and his Russian counterparts, including Konstantin Bushuyev. A rendezvous and docking between a Salyut orbital station and the Space

Shuttle was one possibility, as was Soviet participation in a Spacelab flight. Although the latter option was not seen in a particularly favourable light, the idea of a Salyut-Shuttle mission and also the joint development and construction of an 'international' space station were of interest to both sides, a possibility which *Time* told its readers in its summing-up of ASTP on 4 August 1975. Unfortunately, the enthusiasm of Fletcher and Bushuyev and Low and other like-minded colleagues lay at the mercy of the political climate ... and in the late 1970s and early 1980s, that mercy and that climate deteriorated dramatically.

It was kept alive for a time, however. Informal discussions continued between the Americans and the Soviets and culminated in a series of talks in October 1976 at NASA Headquarters in Washington, DC. These established "a meeting of minds" between the two sides on future manned co-operation, with two primary foci: a scientific venture involving the Shuttle and the Salyut space station or the development of "a space platform...bilaterally or multilaterally". By May of the following year, this meeting of minds had crystallised further, when NASA Acting Administrator Alan Lovelace and Anatoli Alexandrov of the Soviet Academy of Sciences explored the topic of a Shuttle docking with a Salyut in greater depth. The result was a document, rather ponderously entitled 'Objectives, Feasibility and Means of Accomplishing Joint Experimental Flights of a Long-Duration Station of the Salyut Type and a Reusable Shuttle Spacecraft', which highlighted the benefits that both sides could bring to such a venture: the Soviet system could achieve long-term missions and the Americans could carry large scientific payloads into orbit.

On 18 May 1977, Soviet Foreign Minister Andrei Gromyko and US Secretary of State Cyrus Vance signed the space co-operation agreement, which took effect six days later...exactly five years to the day since ASTP had been formalised. The new deal would run for a further five years. "This agreement," noted a December 1982 document, produced at the behest of Bob Packwood, then-chair of the US Senate's Committee on Commerce, Science and Transportation, "established the basis for Soviet-American space co-operation through the early 1980s. It was a very important political instrument, because it [ensured] continuity in Soviet-American space relations." The Soviet press, in particular, wrote glowingly of the plans and in November 1977 a meeting in Moscow began to discuss the technical aspects. By April 1978, when follow-on meetings were scheduled to take place in the United States, *Flight International* mentioned the joint Shuttle-Salyut venture, with a rendezvous scheduled for 1981 and a docking a few years later. At around the same time, in the spring of 1978, NASA's Associate Administrator for Space Science, Noel Hinners, testified before the House Science and Technology Committee that future US-Soviet co-operation was crucial, not least because "they have a station in Earth orbit now [Salyut 6] that may be capable of lasting 1.5 years to two years [and] *we* have nothing on the horizon approximating that staytime duration in space". Sadly, this glimmer of future co-operation on the horizon ultimately was nothing more than a glimmer. The plans did not come to pass.

The cause was chiefly political. Issues of human rights violations and the repression of political dissidents, including Anatoli Shcharanski, who was accused of treason and collusion with the CIA, had long bothered the Americans and the implementation of a new Soviet constitution – the 'Brezhnev Constitution' – in the summer of 1977 brought with it worrying signs that new guarantees of individual liberties were a mockery of justice. 'Exercise by citizens of rights and freedoms *must not* injure the interests of society and the state and the rights of other citizens', read one proviso of the Brezhnev Constitution. "Obviously," *Time* told its readers on 13 June, "this statement gives legal sanction for the KGB to proceed, without having to manufacture pretexts, against dissidents exercising the right of free speech, assembly or religion." The situation steadily worsened. When the Carter administration re-established formal diplomatic ties with the Soviet Union's sworn enemy, China, in January 1979, the Brezhnev regime responded with undisguised anger, delaying the planned second round of Strategic Arms Limitation Talks until June. Within the year, the Soviets had invaded Afghanistan and relations had deteriorated still further. Reluctantly, the Americans agreed to abide by the SALT agreements, but were determined to exact punitive action in other areas. Space co-operation turned out to be one of them. Even before SALT and Afghanistan, in February 1979, the new NASA Administrator, Robert Frosch, spoke of a hypothetical joint Shuttle-Salyut venture in the distinctly more frosty language of *if*, rather than *when*. By February of the following year, as Soviet tanks and troops established a forcible toehold in Afghanistan, the situation had scarcely moved. Frosch told Congress that the American and Soviet working groups had "been in abeyance for something over a year". A further meeting was scheduled for October 1980, but nothing was ever formalised and it never transpired. Senator James Exon of Nebraska described the relationship as having devolved into an "arm's length arrangement that we'll more or less continue" and noted, tellingly, that "the direct scientific activities may be affected, but not immediately, since there was no immediate action to be taken anyway".

In some minds, Jimmy Carter and Zbigniew Brzezinski have been seen as the key obstruction to the implementation of any joint plans. In a January 2002 oral history for NASA, Arnold Frutkin, deputy head of the agency's international affairs office until 1978, related that a breakthrough for more advanced co-operation with the Soviets may have been just around the corner. "It [seemed] so logical to continue ... because [ASTP] was so successful," he said. "It seemed to me the thing to do next would be to move [on] into a space station, but that was a huge undertaking at the height of the Cold War with the Soviet Union and it had to be done in such a way that we weren't transferring technology." The Soviets were interested in conceptual studies and developed a draft agreement with NASA, "to the point where they [actually] signed it! There was a *signed agreement* from them for a joint space station programme, but with this careful, limited, step-by-step [procedure, whereby] you would never proceed from one to the ... next ... unless there was complete comfort and satisfaction in the prior [phase]."

Perhaps Jim Fletcher foresaw or knew the sensitivities of the incoming Carter administration about space, for Frutkin noted that he held off from signing the agreement until he had consulted with the new president. "I don't know whether I knew Brzezinski was going to be [his national security advisor] or not," Frutkin continued, but the opinion of the White House was that the plan should not go ahead. "I felt pretty embarrassed, because we [had] led the Soviets into it and then we couldn't follow through, but there [could be] no argument about it. The administration had the right to call it. They didn't want to do it and, certainly, Brzezinski [would oppose it] ... Maybe it was not a good idea. Maybe it was premature, but if the Soviets were willing to get into it, I think it was *not* premature." It would be another decade and a half before Russia would again approach the negotiating table with a view to a joint manned project with the Americans and opening up their space programme to a wider world. Glynn Lunney is not alone in his conviction that ASTP was a vital stepping-stone towards the co-operation which eventually spawned Shuttle-Mir and today's International Space Station. "People would have had a difficult time," he told NASA's oral historian in October 1999, "embracing the level of co-operation that is inherent in the International Space Station without the experience that we had in Apollo-Soyuz. It probably would have been a staggering thing to think about in terms of *never* having had any experience before." Still, there remains one fundamental, enduring tragedy of ASTP: it is devastating that this project, which carried so much promise for the future, should have taken so long to bear lasting fruit.

2

A home after Apollo

OF $10,000 SCOOPS AND $400 FINES

Seven hundred kilometres south-east of Perth, in Western Australia, lies the town of Esperance. Today, it is a large place, with around 15,000 residents, and its spectacular stretch of Southern Ocean coastline makes it a popular destination for swimming, surfing and scuba diving; indeed, in recent times, it has been commended for having Australia's whitest sands, its best beaches and the fearsome 'cyclops' wave. July and August are its wintertime months, when temperatures dip into the low sixties Celsius, and it was at this time of year, more than three decades ago, that the spotlight of international attention and publicity arrived suddenly at this isolated place. The 'spotlight' in question came from the very skies above Esperance ... and *kept coming*, bringing with it a tremendous fireshow and a cacophony of noise and vibration which rattled windows and shook doors. Many residents feared that the end of the world was nigh.

Dorothy Andre recalled that the night of 12 July 1979 was overcast, but the clouds broke when 'it' came through, breaking the sound barrier with a string of sonic booms, "one after the other", and a terrific sheet of flame. Other witnesses to the peculiar happening would add that the object seemed to be shedding fragments as it plunged to Earth. Many hundreds of pieces of debris were found in the next few days and weeks, some in the ocean and others on roofs and in backyards; ranging in size and composition from bits of foam and shreds of insulation to a hatch weighing 45 kg and a huge oxygen tank the size of a Volkswagen van. Today, thankfully, Esperance is quiet once more and Andre and her husband, Mervin, run a museum, which holds a collection of the pieces on prominent display.

The cause of the strange incident: the final death throes of Skylab, America's first space station, launched into orbit six years earlier ...

Remarkably, there were no deaths or injuries on the ground when the fragments fell to Earth. Perhaps even more remarkably, at least one Australian managed to make money from Skylab's demise ... and, three decades after the event, the shire of

Esperance managed to recoup a tongue-in-cheek fine which it had imposed on the United States.

Skylab's demise was by no means unexpected. The station had been in orbit since May 1973, but stronger-than-normal solar activity had inflated Earth's atmosphere and rendered its orbit increasingly unstable. A plan to execute a controlled re-entry was set in motion by NASA, targeting a vast 'footprint' of ocean, some 1,300 km south-east of Cape Town, but Skylab did not burn up as quickly as expected . . . and a calculation error of just a few percent led to its debris raining down over Western Australia.

In the days preceding re-entry, the impending death of the station attracted international interest, with merchandising taking place and wagers laid to predict the time and place at which the first remnants would fall. Chicken Little's apocalyptic cries of "The Skylab is falling! The Skylab is falling!" prompted the appearance of party outfits of beaks and feathers, whilst other booze-fuelled farewells were hosted, not surprisingly, in basements and underground bunkers. Radio stations offered bright yellow T-shirts bearing bull's eyes to their listeners, sales of plastic helmets soared and free legal advice for people hit by debris was offered by New Hampshire attorney John Ahlgren. Others hung mock bull's eyes on the flanks of buildings, confident that "if you give the government a target to shoot at, it's bound to miss!" Buryl Payne, head of MIT's Institute for Psychic Energetics, even used a Fort Lauderdale radio station to tie in with 150 other networks to encourage 40 million listeners to psychically 'push' Skylab into a higher orbit. The *San Francisco Examiner* offered a $10,000 'prize' to the person who delivered the first piece of Skylab to its offices, whilst the competing *San Francisco Chronicle* bettered that with a black-bordered front page notice offering $200,000 if one of its subscribers could prove either personal or property damage caused by the re-entry. NASA's own predictions that the odds of a fragment hitting someone were in the order of 152-to-one and the chances of anything reaching a city of 100,000 or more were in the range of seven-to-one made it increasingly likely that both newspapers might get a call. So too might NASA and the federal government, which had assembled rapid-response disaster teams to head to any nation on Earth that requested assistance following a debris strike.

The reply from Esperance, though, was not for assistance; it was for payback.

One of the first residents to contact the *San Francisco Examiner* was a 17-year-old beer truck driver named Stan Thornton, who found a couple of dozen shards of insulation from Skylab on the roof of his home and promptly caught the first available flight to the United States to collect his winnings. The publicity surrounding the event was amplified a few days later, when the Miss Universe pageant in Perth featured a large chunk of station debris on the stage. However, the most famous episode came full circle almost three decades later. "As a bit of a lark," Mervin Andre recalled, the local ranger issued a $400 fine to NASA for *littering* in the area! The space agency found little humour in the fine and declined to pay; after three months, the shire of Esperance wrote it off . . . but never forgot it. Over the years, Andre received many queries about the unpaid fine, to the extent that it almost turned into an urban myth. Almost. Then, at the end of 2008, Scott Barley, a

DJ at the Californian station 'Highway Radio', asked his listeners to raise the money and pay the long-standing debt. To his surprise, *they did.*

"I thought this fact of the unpaid bill was rather funny," Barley explained. "I thought it would be great if I challenged my listeners to contribute to the $400 fee ... I wanted listeners to start off 2009 with a generous offering of goodwill to our Australian friends south of the equator." Some of Barley's audience gave large donations, with one company offering to match each $50 with $50 *and* a free gym membership. Even three decades later, it seemed, the interest and memory of Skylab was still there ... as was the eagerness to make money from it. "Eventually," Barley concluded, "to my utter surprise, we reached our $400 goal!" In the *Esperance Express*, on 17 April 2009, backdropped by a huge poster, which declared 'In 1979, a spaceship crashed over Esperance ... We fined them $400 for littering', Dorothy and Mervin Andre, together with German Ugarte, the shire's head of tourism and culture, proudly displayed a framed certificate which authenticated the cheque.

Over the years, the fall of spacecraft from orbit has thrown several unwanted and unexpected spotlights of media attention onto remote, isolated communities. Only 18 months before Skylab's demise, a Soviet satellite, Cosmos 954, plummeted to Earth and its fragments landed close to Yellowknife in northern Canada. The nuclear-powered nature of the mission immediately sparked a major search for radioactive debris, Operation Morning Light, between Canadian and American military forces. News of the Cosmos re-entry, like that of Skylab, made headlines the world over, particularly when it became clear that an 80,000 km^2 tract of land, to the east of Hay River, had received a light dusting of radioactive material. The Canadian government later billed the Soviet Union – which had initially claimed that the satellite was completely destroyed during re-entry – for more than $6 million in expenses and additional compensation. They ultimately received half of this sum. Canada's claim came closest to invoking the Space Liability Convention, part of the Outer Space Treaty, signed in 1967 and today acknowledged and ratified by more than a hundred nations. The convention declares that countries or states bear international responsibility for spacecraft launched from their territories and therefore are liable for claims from any aggrieved nation. Since there were no injuries and little serious damage caused by the Skylab incident, a formal claim from Australia against the United States was never made, but the question of the convention has arisen several times over the years, notably during the re-entry of the Mir space station in March 2001. As a consequence, to date, the liability convention has never been invoked and stories of $10,000 prizes and $400 fines and delayed payments are now little more than quirky footnotes to history.

However, the end of Skylab would have untold repercussions for the United States, in ways that Uncle Sam could scarcely have imagined; for in the late 1970s, it was their only space station – and would *remain* their only space station, almost until the turn of the millennium – and many critics have argued over the years that more should have been done to preserve it. Skylab's last crew departed in February 1974, after a record-breaking 84-day mission, and this chapter will show that the scientific, technological and engineering accomplishments of this station were, for their time, unrivalled and unsurpassed. Indeed, just a few months before Esperance's residents were shaken by their unwanted fireworks show, NASA possessed advanced plans for Skylab to be visited and restored to life by the Space Shuttle, offering American astronauts a semi-permanent presence in Earth orbit for at least the first part of the 1980s. Whilst it is fortunate, of course, that falling debris in the coastal regions of Western Australia did not produce a human calamity, the greatest disaster for human space exploration is perhaps that Skylab's full potential was never truly realised.

THE NEXT LOGICAL STEP?

Dreams of a permanent human home in space began more than a hundred years ago, with tales of 'brick moons' to help sailors navigate the oceans and enormous metal cities with life-sustaining gardens and scientific laboratories. In the early years of the last century, such visionaries as Konstantin Tsiolkovski, Hermann Oberth, Baron Guido von Pirquet and Hermann Noordung foresaw giant 'wheels' in the sky, capable of generating artificial gravity for their occupants, and others whose orbits were precisely synchronised with Earth's rotation to permit observations of specific areas over long periods of time. Astronomy, geophysics, communications and medicine were seen as key objectives of such ventures. Until the advent of the Space Age, of course, such dreams remained little more than that, but the 1950s brought an increased wave of optimism that a brighter future was just around the corner and one of the fundamental tenets in building that future after the horror of two world wars was the exploration of the heavens. Journals, books, films and demonstrations of 'future homes' brought these ideas to a wide audience and studies by Krafft Ehricke and Wernher von Braun envisaged 'real' orbital outposts and lunar bases, capable of being launched by rockets which were already in existence. Such outposts, it was expected, would benefit from artificial gravity, as a result of rotating several times per minute, and shapes such as hexagonal 'wheels' with spokes radiating from a central core were considered. The size (some 25 m radius) and mass (in excess of 170,000 kg) of such stations were the limiting factors and problems were anticipated if they were to be launched 'piecemeal' into orbit. Inflatable structures would have solved some issues, but were considered vulnerable to micrometeoroid impacts.

When NASA came into being in October 1958, its initial focus was to launch a man into space. However, planning was soon underway for subsequent programmes and a new, more capable spacecraft – Apollo – entered America's national consciousness. Of course, this would ultimately evolve into the cornerstone of the drive to land men on the Moon, but its original mandate was far broader: as early as July 1960, NASA Deputy Administrator Hugh Dryden advised industry leaders that the agency might someday support long-term space stations, circumlunar expeditions, (eventual) Moon landings and perhaps flights to Mars. A Research Steering Committee on Manned Space Flight had already concluded that a station should be pursued *ahead* of a lunar landing. Clearly, the Moon was just one element in a wide-ranging plan.

From the perspective of what would evolve into Skylab, the station concept is of primary interest here. A May 1961 proposal focused on 'Apollo A', which comprised an adaptor section and an equipment/propulsion module, lofted into orbit by a Saturn booster. It would have provided a long-term base from which a plethora of scientific experiments – from astronomy to medicine – could have been performed and would have enabled techniques, such as spacewalking (known formally as 'extravehicular activity' or 'EVA') to be refined. When, in that same month, President John Kennedy challenged his nation to land a man on the Moon, Apollo was the obvious choice to accomplish the feat, but many of the concepts needed to both reach the lunar surface *and* build a space station, including rendezvous and docking, needed to be mastered. For this reason, Project Gemini was born.

Yet the station remained a clear future goal and from the summer of 1962 designs included a rotating station, launched atop the Saturn V, with six-man crews regularly exchanged by Saturn IBs. Identifying the precise *purpose* of such an outpost aroused more debate; NASA's Langley Research Center in Hampton, Virginia, emphasised a laboratory for research into advanced technologies, whilst the Manned Spacecraft Center in Houston, Texas, said it should provide a waystation to reach Mars. On a more basic level, its role was simply to determine whether men could live and work effectively in space for extended periods of time. The condition of weightlessness, which cannot be simulated on Earth for more than 30 seconds in an aircraft flying a parabolic arc, was poorly understood, as were its biomedical effects on the human body.

In March 1963, before the House Committee on Science and Astronautics, Hugh Dryden indicated that the most obvious future direction *after* the Moon landing should be an Earth-circling laboratory and to press the point NASA awarded two contracts that summer to Lockheed and Douglas Aircraft, charging them to devise a system with an operational life span of five years, during which it would sustain a succession of astronaut crews and supplies. Alongside these studies, efforts to transform the Apollo spacecraft into a miniature station in its own right – perhaps by transporting large telescopes into high orbits for three months at a time – were also explored. One key problem was that Apollo's pressurised volume extended only to the command module and the ascent stage of its lunar module, both of which were limited, and for a time the development of an additional pressurised laboratory to support a 120-day mission was considered. At various points in 1961-1963, North American Aviation (later North American Rockwell) investigated three different concepts, capable of a full year in orbit: an Apollo with 'enlarged subsystems', an Apollo with an attached module and an Apollo with a purpose-built research laboratory.

Each option assumed a crew of two or three astronauts. Fuel cells would be replaced with solar panels for electrical power, a pure oxygen atmosphere would be replaced with a more terrestrial mixture of oxygen and nitrogen and the bulky lithium hydroxide canisters would be replaced by a system of compact, regenerative 'molecular sieves' to cleanse the air of carbon dioxide from the crew's exhaled breath. The least costly option would be to fly an enlarged command module on its own, but this would have restricted the crew to two men, which would in turn have posed 'operational liabilities' in terms of running the spacecraft and completing the mission objectives. A laboratory module would not only increase project costs by as much as 30 percent, but would also impose an undesirable weight penalty on the Saturn booster.

Elsewhere, the development of the Saturn itself had inspired a number of other studies for long-duration civilian stations. In February 1964, for example, Lockheed suggested a 24-man complex, with an operational lifetime of five years and the capability of being resupplied every three months by either a six-man Apollo or a 12-man lifting body. Cost estimates hovered around the $2.6 billion mark. At around the same time, Edward Gray, head of advanced planning at NASA's Office of Manned Space Flight, approached the Langley Research Center to prepare plans for a six-to-nine-person Manned Orbital Research Laboratory (MORL). He also asked the Manned Spacecraft Center to define a long-duration 'Apollo X' craft and the Large Orbital Research Laboratory (LORL). Not only were these ideas demonstrative of the fact that orbital outposts were being taken increasingly seriously as a next step beyond the Moon landing, but they also illustrated a growing awareness of the scale of research which their crews might someday undertake. Of primary interest were astronomy and solar physics – with some MORL plans even calling for a large space telescope, to be launched atop a Saturn IB – although the field of space medicine was also gaining momentum.

The military, and particularly the US Air Force, were similarly intrigued by the notion of establishing a semi-permanent human presence in space, particularly as political relations with the Soviet Union worsened in the mid 1960s and concerns intensified about the increased march of communism across south-east Asia. In May 1966, more than $1.5 billion was pledged by President Lyndon Johnson to build a Manned Orbiting Laboratory (MOL), which would enable two-man crews to spend a month in space, performing surveillance with a battery of large optical instruments and cameras and side-looking radar. In its final form, the cylindrical MOL and an attached Gemini spacecraft would have been launched together, thereby eliminating the need for orbital rendezvous and docking. Ultimately, a combination of inadequate funding and limitations on the Air Force imposed by the 1967 Outer Space Treaty proved the project's undoing. In June 1969, it was ignominiously cancelled, after swallowing close to $3 billion.

From the political stage, though, even whilst MOL was still alive, questions were being asked about its similarities with ongoing civilian efforts. In December 1964, Senator Clinton Anderson, chair of the Committee for Aeronautics and Space Sciences, had complained to President Johnson that having both efforts running in tandem reflected a highly inefficient use of "national treasures". In Anderson's mind, each had its own unique contributions to make, but each should exploit the other's capabilities and together they should lead towards a future 'national' laboratory. Moreover, Anderson felt that the cancellation of MOL and the transfer of its funds to NASA could save $1 billion or more over five years. By the spring of 1965, NASA Associate Administrator for Manned Space Flight George Mueller introduced the concept of the Apollo Extension System (AES) to the House Committee on Science and Astronautics, declaring that the technologies then under development for the Moon landing could form the backbone for any subsequent ventures. This would enable NASA "to produce space hardware and fly it for future missions at a small fraction of the development cost".

These 'future missions' were both varied and exceedingly ambitious, at least in their earliest incarnations. They included lunar-circling flights of four to six weeks, 14-day expeditions on the Moon's surface with rovers and flying vehicles and three-month stays in Earth orbit. Research disciplines expanded further from astronomy and solar physics to biomedicine, the study of artificial gravity using telescoping links, observations of living organisms, including chimpanzees, measurements of solids, liquids and gases in the weightless environment, infrared and microwave remote-sensing of the Home Planet, the launching of unmanned satellites, including the Orbiting Geophysical Observatory (OGO), manoeuvring and docking with other craft and perfecting the techniques needed for spacewalking.

As 1965 wore on, considerable effort was being applied to a relatively old concept of using the 'spent' upper stage from a rocket as a makeshift orbital station. Known as the 'wet workshop' concept, it provided a cost-effective means of reusing existing hardware which would otherwise have served no further purpose. Early proposals from Wernher von Braun had called for the Saturn V's large S-II second stage – measuring some 25 m in length – to provide a core for such a wet workshop. Attached to the top of the S-II, in place of a third stage, would be a large cylindrical module, fitted with life-support apparatus and other supplies for a human crew. After reaching orbit, the S-II would vent any residual propellants through a system of grid-like floors and piping at the base of its tanks, after which a forward hatch would be opened and the cylindrical module would be hydraulically inserted, sealed and repressurised with an oxygen-nitrogen atmosphere to create a large living and working area. Power would come from a set of conformal solar cells. An Apollo crew would then rendezvous and dock with the workshop and spend between six weeks and four months aboard. However, later in the 1960s, when it became increasingly evident that the Saturn V production line would be shut down after Kennedy's challenge had been achieved, other proposals were devised to employ the smaller S-IVB stage from a Saturn IB booster as a wet workshop instead. It is ironic, therefore, that when several Apollo lunar landings were cancelled in 1969-1970, a supply of Saturn Vs *did* become available ... but by that time, so much work had already been done on the design of an S-IVB workshop that it was decided to continue with those plans and they ultimately formed the core of what became Skylab.

Von Braun's relationship with the wet workshop went back even further, to his time at the Army Ballistic Missile Agency in the late 1950s, when he proposed 'Project Horizon' as a temporary shelter for a small crew. In orbit, the astronauts would dock onto the spent rocket stage, purge residual hydrogen from its tanks and prepare it for occupancy using equipment carried aboard their own craft. As the Horizon effort progressed, it was expected that other spent stages would be added, with perhaps two dozen in its final configuration. For a time, this Army-designed outpost went no further, but the transfer of von Braun's group to NASA later led to a revival of the wet workshop. Also interested in the concept was Douglas Aircraft. Keen to establish its own toehold in the manned spaceflight business, Douglas had displayed a mockup of its wet workshop at an exhibition in London, England, in November 1959, which envisaged the use of its Saturn S-IV stage – and later the enlarged S-IVB – as a small orbital station. Neither option would require major modification in order to make it habitable: a wrap-around micrometeoroid shield, fitted on the ground before launch, and an attached storage module to temporarily house equipment which could not survive immersion in liquid hydrogen. In fact, one of the fundamental arguments in favour of the wet workshop was its perceived benefits in getting a long-term station into orbit cheaply and quickly. Douglas' early plans called for an S-IV to be visited by a two-man Gemini crew, who would dock, drain the residual hydrogen, repressurise it with an oxygen-nitrogen mixture and move equipment from the storage module over to the stage itself. This would provide them with all that they would require for a 100-day occupancy, during which time their focus would be upon several dozen physiological, technological and scientific experiments, including a small centrifuge to provide temporary artificial gravity and help to recondition their bodies before returning to Earth.

Subsequent plans, conducted in conjunction with the S-IVB, envisaged an Apollo spacecraft docking with the spent stage and its crew testing spacewalking techniques, fully-suited, in the safety of its unpressurised interior. Alternatively, the stage could be pressurised and its 280 m^3 volume fitted with living quarters. Late in August 1965, three options were examined by NASA: a 'minimum' configuration, in which the empty tank would be equipped with a docking port and no life-support apparatus, an 'intermediate' configuration, with an airlock, power and oxygen, which would enable astronauts to operate inside pressurised suits, and finally a 'baseline' configuration with environmental and attitude controls, experiments and power supplies. Of these, the minimum and intermediate configurations could support missions as long as two weeks, whilst the baseline configuration could sustain a crew for 28 days. As these discussions continued, on 10 September, the project received a new name: 'Apollo Applications Program' (AAP).

From the perspective of the Manned Spacecraft Center, a measure of artificial gravity was highly desirable and they requested a minimum of 0.1 G to be provided by rotating the station on a radius of 20-30 m. This requirement continued until later in 1966, when schedule, funding and design problems gradually eliminated it. Other difficulties centred on the removal of hazardous materials from the spent stage before it could be inhabited. The hydrogen tank firstly had to be 'passivated' – in other words, its propellants totally vented, its high-pressure helium bottles emptied and its pyrotechnics disarmed – and all other potential dangers had to be removed. Provision had to be made for equipment to be fitted onto its internal walls, together with foot restraints and mobility aids, and changes to the size of the 'manhole' hatch in the forward dome of the spent stage needed to be made to allow a fully-suited astronaut through. Micrometeoroid impacts remained a constant worry and preliminary tests on S-IVB skin patches, reported in February 1966, suggested that penetrations could ignite the polyurethane insulation. Efforts to find a solution occupied the remainder of the year. The risk of fire in a pure-oxygen environment was also significant and in December 1966 – just a few weeks before Apollo 1 demonstrated this danger in catastrophic fashion – a recommendation emerged for an atmosphere of 69 percent oxygen and 31 percent nitrogen, pressurised at one-third of terrestrial sea-level. However, it was decided that other compositions were acceptable and, with the addition of external shielding, this would also help to mitigate the micrometeoroid risk.

For the astronauts who would visit this wet workshop, the problems would be immense. Major difficulties in spacewalking had been encountered in June 1966, when Gene Cernan struggled to complete even relatively straightforward tasks, fighting against a fogged-up visor, inadequate environmental controls and other mobility problems caused by the lack of handholds or an effective propulsion device. Simulations on the ground indicated that it would take a pair of suited astronauts at least *six hours* simply to remove the 72 bolts from around the S-IVB's manhole. This was considered intolerable and plans were set in motion to develop a quick-opening hatch in its stead. These changes stalled later that year, when NASA received just $5.01 billion for 1967 – some $250 million *less* than the agency's lowest request to the Bureau of the Budget – and the available share for manned missions was cut by $222 million. Since the Apollo lunar project had first priority, the loss was absorbed by its unwanted stepchild, AAP. The $42 million which ultimately entered AAP's coffers was barely enough to keep it alive.

To be fair, such appropriations were not entirely unexpected. America's involvement in Vietnam had caused troop numbers to swell to almost half a million and military expenditure had peaked at two billion dollars *per month*. The optimistic and heady days of the early 1960s had been replaced by an ominous sense of pessimism for the future. President Johnson had already committed his administration to a $100 billion budget, of which the Medicare and War on Poverty initiatives took centre stage. NASA Administrator Jim Webb's entreaties to Johnson to preserve the agency's budget in December 1966 fell on deaf ears. This was a pity, for AAP had already started to expand into something more ambitious than a makeshift wet workshop and might have included a series of important scientific missions. Among the plethora of proposals by AAP Director William Taylor were the establishment of telescopes, lasers and other optics in high Earth orbits, as well as long-duration missions to the Moon.

Unfortunately, none of these concepts whetted the appetites of congressional critics. When Deputy Administrator Bob Seamans testified before the House Committee for Manned Space Flight in the spring of 1966, he was met by little enthusiasm: compared to landing on the Moon, anything else represented 'busy work' – "boring holes in the sky" was one derogatory judgement – and even Bob Gilruth, head of the Manned Spacecraft Center, was sceptical. Seamans' talk of 45 launches (26 by the Saturn IB and 19 by the Saturn V) between April 1968 and the mid-1970s was audacious, to say the least, averaging six missions per annum and lofting three Saturn IB-launched wet workshops, three Saturn V-launched 'dry' workshops and four flights of an ambitious piece of experimental hardware known as the Apollo Telescope Mount. Gilruth was in firm agreement with the need to continue using Apollo hardware and facilities, beyond the Moon landing, but he felt that AAP lacked a clear direction and its objectives were much less obvious. More importantly, without a significant thrust towards research and development, it

would be intrinsically difficult to maintain anywhere near the kind of momentum that the tightly-focused Apollo lunar effort had enjoyed.

In an eight-page letter to Associate Administrator for Manned Space Flight George Mueller in March 1966, Gilruth expressed worries about other aspects of AAP, including plans to extensively overhaul the lunar module as a platform to support missions for which it was never intended. One such design would have seen the attachment of a 'stack' of *two* lunar module ascent stages – their ascent engines removed to permit the installation of a connecting tunnel – onto the nose of the command module to serve as a laboratory for biomedical, meteorological, engineering, Earth resources or astronomical research. To Gilruth, creating a makeshift space station or conducting a series of ill-defined 'scientific flights' was not enough; a *major* new undertaking, leading, perhaps, to a manned expedition to Mars, was acutely necessary in order to reinvigorate NASA and its political masters and inspire an increasingly fickle public after John Kennedy's goal had been achieved. In the shadow of the Moon, a series of one-off research flights and wet workshops in Earth orbit were seen as just 'marking time' and offered little to advance America's new-found dominance in space. More pragmatically, Jim Webb felt it was highly unwise to start negotiating for costly new programmes until the task of putting a man on the Moon had been accomplished. Nor was it wise to cede to congressional desires to merge AAP with MOL; Webb felt that nations which hosted tracking and other facilities might refuse to co-operate if NASA projects were being used for military purposes, particularly with the war in Vietnam growing uglier.

Still, one element of AAP which *did* survive the budget cull and achieve long-standing respectability was the Apollo Telescope Mount (ATM). This project had been extensively analysed by Ball Brothers of Boulder, Colorado, since September 1965, and offered a three-axis astronomical and solar physics facility for ultraviolet, X-ray and visible-light observations. Although the ATM would later become part of Skylab, it was originally to have flown aboard an Apollo spacecraft, housed inside a bay in the service module, with a launch anticipated during a period of maximum solar activity in 1968-1970. Shortly after achieving orbit, its three-man crew would have deployed the ATM by means of a pneumatically-driven tripod and at least one astronaut – presumably a physicist or astronomer – would have served as its primary operator throughout a scheduled 14-day mission. The role of the crew would have been to ensure the proper orientation of the Apollo spacecraft in order to achieve the

Standing on Pad 39A on the morning of 14 May 1973, the final Saturn V looked quite different to its lunargoing cousins: for it was a two-stage vehicle, comprising the S-IC and S-II segments, but with the inert Skylab workshop and a bullet-like payload shroud taking the place of the S-IVB. Skylab was thus the result of a critical decision, made in the summer of 1969, to adopt the concept of a 'dry workshop', rather than a 'wet workshop'. In doing so, Skylab eliminated the risks and technical worries associated with astronauts having to 'passivate' a spent stage before they could equip it for long-term occupation.

scientific objectives. On at least one occasion, an astronaut would venture outside to retrieve exposed film cassettes and replace them with new ones, in a manner not dissimilar to the 'deep space' excursions performed during the last three Apollo lunar missions.

Although the Ball study was broadly supportive of the ATM concept, its authors cautioned that before the hardware could be committed to other missions, it would be necessary to investigate disturbances of the instrument's alignment induced by the movements of the astronauts and the contamination of the optics by urine dumps, sublimation from the command module or faulty camera film. Early in 1966, Homer Newell, head of NASA's Office of Space Sciences and Applications, and George Mueller concurred that the ATM *was* important, but differed in their opinions of how it should be executed. The former favoured the Apollo service module, since it was cheaper in the longer term and more likely to achieve the mid-1968 launch schedule, whilst the latter encouraged housing the instrument in a modified lunar module, since this would create lower costs from the outset. Each had its drawbacks, of course. A service module-based ATM could only be used once, since its carrier would burn up during re-entry into the atmosphere. Although the lunar module offered a potentially reusable package, it was suffering development problems and was totally untested. Further, using the lander to fulfil a purpose for which it was not designed aroused consternation in Houston and it was felt that the modifications would cost more than $100 million and require two or three additional years of development. Ultimately, Houston's objections were ignored and in mid-May 1966 the decision was taken that the entire ATM, except for the instruments themselves, would be designed, built and integrated into an octagonal 'rack' (essentially a modified descent stage, minus its landing legs). For almost three years, from initial talks with prime contractor Grumman in early June until the cancellation of the Apollo-borne ATM in July 1969, the mission evolved into something based on the lunar module. A modified ascent stage would house the crew and controls, whilst the modified descent stage would house the instrument and a windmill-like arrangement of four solar panels, in a cruciform configuration, to provide electrical power. The spacecraft would be 'held' with the instruments and solar panels facing the Sun. "Eight solar observation telescope systems," noted a Bellcomm report on the ATM, "with cameras, film cassettes, telemetry sensors and six TV monitoring cameras are mounted on the experiment spar, which serves as a common optical bench. The experiments are loaded with cameras and film at launch; each film load is calculated to last about 14 days under nominal frame exposure rates. Replacement sets of cameras and film for each experiment as required will be carried in the [lunar module cabin] for subsequent exchange with the exposed film cassettes by an astronaut performing EVA."

Of course, the ongoing budget cuts, which worsened in 1968 and beyond, would steadily cull more AAP missions, to such an extent that only the workshop and the ATM would survive ... and these would eventually morph into a single programme. The December 1966 plans envisaged Missions AAP-1 and AAP-2 as a manned Apollo, followed by the workshop with an attached 'airlock module' and multiple docking adaptor with five ports. After being launched into a 510 km orbit, the

Apollo would rendezvous and dock and the astronauts would spend 28 days turning the workshop into a rudimentary, two-storey habitat. (At a press conference, one journalist asked George Mueller if the workshop's size was comparable to an average ranch house. "A *small* ranch house," Mueller had retorted with a grin. "The kind *I* can afford to buy!") Medical experiments, including studies of bone and muscle changes and vestibular function, would have occupied the crew, together with engineering tasks, which included a remarkable pair of jet-propelled 'shoes' proposed for possible use during future EVAs. Six months after the return of the first team of astronauts, a second Apollo crew would be launched, followed by the unmanned ATM, and after these had docked they would rendezvous with the workshop. As Bellcomm put it, "The terminal phase of the rendezvous manoeuvre, as well as docking will be accomplished by the use of the [lunar module] and [command and service module] reaction control system. During the final portion of the terminal phase, the [lunar module/ATM] with two astronauts aboard will separate from the [command module], which is under the control of one astronaut." The command ship would then dock with the 'axial' port of the workshop's multiple docking adaptor and stabilise the station. The lunar module/ATM would then dock to a second port, using its own reaction controls for manoeuvrability; consequently, the multiple docking adaptor would be of primary importance in such a mission. "To minimise difficulties associated with the originally planned double rendezvous," continued the Bellcomm report, "recent mission plans required the unmanned [lunar module/ATM] to make a remotely controlled docking … after the manned [command and service module] has docked." In whatever guise the docking took place, it would permit the astronauts to perform 56 days of research aboard the first fully-fledged manned orbiting solar observatory. Other objectives included simply demonstrating the ability to re-activate and re-use the workshop after six months in orbital storage. These missions (AAP-3 and 4) were tentatively scheduled for June 1968, at the start of the two-year period of maximum solar activity and Mueller told journalists that it would represent "the most comprehensive array of instruments that has ever been assembled for observing the Sun". Other plans included the launch of a second workshop in 1970 and adding its ATM in 1971, a year-long mission and a six-man Apollo craft.

At the same time, the President's Science Advisory Committee (PSAC) was reaching its own conclusions about the future of America's space programme in the aftermath of the Moon landing. Published on 11 February 1967, the committee's report endorsed a balanced programme with an "eventual" goal of manned planetary exploration, but required, among other things, a full exploitation of the ability to explore the Moon and qualification of long-duration manned operations in space. In PSAC's collective mind, the ultimate aim of such a programme would be to begin to find answers for the most challenging questions in space science: the existence of extraterrestrial life, the evolution of the Solar System and the origin of the Universe. Although PSAC felt that Apollo-Saturn hardware should be devoted to more intensive manned exploration of the Moon, rather than ATMs and wet workshops, the committee also felt that a semi-permanent, Earth-circling station *was* an essential requirement for qualifying men for long-duration missions in space, in addition to undertaking fundamental research, and that the creation of such a station should be a goal for NASA in the mid-1970s. The wet workshop went some way towards fulfilling this requirement, but PSAC advised closer links between NASA and the Air Force in the MOL project and expressed grave misgivings about the notion of a man-tended ATM. "The heaviest demands on the man [in the ATM project] are to do things which ideally should be done on the ground," the report noted, "or by electromechanical systems ... which do not have to override the angular momentum of the man's movements." In conclusion, the ATM was far from being ideal, but from the point of view of astronomers, keen to fly a high-resolution instrument, it was better than nothing. A delay of a year or so, the committee recommended, was advisable to iron out as many flaws from the man-tended system as possible.

Tragically, on 27 January 1967, astronauts Gus Grissom, Ed White and Roger Chaffee were killed when a flash fire swept through their command module during a 'plugs-out' ground test in Florida. This seriously damaged NASA's reputation, raised questions about its competence ... and left AAP out in the cold. As schedules slipped and more budget cuts were imposed on other programmes to fund the modifications to the Apollo command module, a standing joke quickly developed within the space agency that the workshop was *always* at least two years away from launch. The launch of the workshop was postponed to December 1968 and the ATM to June 1969. When the High Altitude Observatory, the Naval Research Laboratory and the Harvard College Observatory announced that their payloads would not be ready in time, the schedule slipped further. Even the assumption that enough Apollo spacecraft would be available to support these missions was called into question. In the wake of the fire, NASA decided that future flights would only carry those experiments which *directly* aided the lunar landing and a new Apollo Applications flight (AAP-1A) was inserted into the manifest for late 1968. It would run for 14 days and test a piece of hardware called the Lunar Mapping and Survey System (LMSS) in Earth orbit. The original plan was that the LMSS would undertake a polar-orbiting cartographic study of the Moon's surface with infrared and microwave sensors aboard a converted lunar module 'lab', ahead of the first landings. When it became clear that the data from the unmanned Lunar Orbiter spacecraft had provided more than adequate imagery, the LMSS was cancelled by Bob Seamans in August 1967 and AAP-1A was turned into a dedicated Earth resources flight, with half a dozen specialised cameras and four infrared sensors. Circling the globe at an altitude of 260 km, inclined 50 degrees to the equator, it would make six daily passes over the United States and, as 1967 wore on, this seemed an ideal inaugural mission for Apollo Applications, demonstrating that the hardware could be applied to something of direct use to the nation ... but by late December, it was gone, the latest victim of the congressional budget axe. By the summer of 1968, the ATM was itself facing cancellation, but this threat receded when Apollo 8 orbited the Moon that Christmas and Tom Paine arrived as NASA Administrator the following March.

The salvation of the ATM might have been viewed in a positive light, except for one thing: the constantly shrinking budgets had steadily pushed it back from an anticipated 1969 launch to sometime in 1971 ... two years *after* the period of maximum solar activity would have ended. In recognition of this fact, Leo Goldberg, then-head of the Harvard College Observatory, requested the cancellation of a scanning ultraviolet spectrometer from the ATM; he felt that its usefulness for monitoring large flares during the solar maximum was now compromised and instead recommended reinstating an earlier instrument, an ultraviolet spectro-heliometer. NASA's AAP office told Goldberg to continue working on the scanning ultraviolet spectrometer and assured him that the first ATM mission would go ahead in June 1971. In a letter to the Office of Space Sciences and Applications, Goldberg reminded Associate Administrator John Naugle that Harvard had agreed to fly the simplified experiment as a favour to NASA and subject to two conditions: that Harvard would be able to fly its original instruments on the second ATM and that the first ATM would be in orbit in 1969. "I think it is time to face up to the realisation that our participation in the ATM project has been guided more by circumstance and expediency than by the requirements of first-rate science," he told Naugle. "If we do not jointly take the firm action now to reverse this trend, we shall be doing astronomy and NASA both a great disservice." In due course, Goldberg's judgement was accepted, the scanning ultraviolet spectrometer was scrubbed and the old ultraviolet spectroheliometer was duly reinstated. However, these concerns from members of the scientific community underlined a steadily growing realisation that budget limitations and schedule slips were preventing AAP from delivering on its pledges.

It is a sad state of affairs that at least a few of the Apollo Applications proposals drawn up in the months preceding the deaths of Grissom, White and Chaffee – long-duration Moon bases and orbital surveys, surface rovers and fliers, Earth-circling bioscience sorties with modified lunar modules and geophysical mapping expeditions – did not come to pass. One adventurous plan even discussed the possibility of a 400-day Venus flyby mission, launched atop a Saturn V booster in October 1973, with a converted S-IVB for habitable crew quarters. Today, many of these proposals seem as astonishing and unrealistic as were the outlandish planning schedules for lunar bases in the late 1950s, but to be fair the timing of Bob Seamans' words to the House and George Mueller's talk of "embryonic" space stations came at a time when funding for NASA had reached its zenith. "The agency," wrote Deke Slayton, then head of flight crew operations in Houston, "would never have as much support as it had in 1966" ... and, indeed, the funding dropped precipitously from the end of 1967 onwards until nothing of AAP but the ATM and the workshop remained. By June 1968, plans for follow-on lunar missions, which might have provided for extended, two-week stays on the Moon's surface, had also disappeared from the manifest.

For now, though, with the maiden voyage of Apollo upon them and a fervent hope that the promise of AAP would somehow come to pass, more astronauts were needed to crew this plethora of missions. When the NASA selection panel asked Slayton in the autumn of 1965 how many new astronauts he needed out of 35 finalists, he responded, "As many qualified guys as you can find." Nineteen were ultimately picked and reported for training in April 1966: Vance Brand, John Bull, Gerry Carr, Charlie Duke, Joe Engle, Ron Evans, Ed Givens, Fred Haise, Jim Irwin,

Don Lind, Jack Lousma, Bruce McCandless, Ken Mattingly, Ed Mitchell, Bill Pogue, Stu Roosa, Jack Swigert, Paul Weitz and Al Worden. Yet even that was not enough, since the very nature of AAP demanded a large cadre of *scientist-astronauts*, too, including astronomers and physicians. NASA had selected a handful of scientists in June 1965, but the flight projections suggested that a dozen or more were now needed. "We announced that we were accepting applications," Slayton wrote, "then asked the Academy of Sciences to do a rating of the applicants based on their scientific credentials. NASA would handle the rest. We hoped to select 10-20 new scientist-astronauts and expected them to be ... on missions beginning in 1970 or 1971." A little more than nine hundred applications arrived at the National Academy of Sciences by the beginning of January 1967 and, after two months of evaluation, sixty-nine names were submitted to NASA.

However, the budgetary wind of change was blowing fiercely. In July, Administrator Jim Webb had been informed by the Senate Committee on Appropriations that the following year's authorisation bill *would* be severely curtailed and his options were stark: to make cuts to AAP or to unmanned planetary projects, including the Voyager Mars lander. Webb told the committee that, since *both* were essential for America's future space programme, neither option was palatable. It mattered little, for the cuts came rapidly and the agency's funding continued to drop. As the selection process for the new scientist-astronauts proceeded, Slayton suggested reducing the number to five, but the process had already started and eleven were ultimately picked. One day in late September, at a motel called the King's Inn in Houston, Slayton sat down with the new astronaut candidates – Joe Allen, Phil Chapman, Tony England, Karl Henize, Don Holmquest, Bill Lenoir, Tony Llewelyn, Story Musgrave, Bob Parker, Brian O'Leary and Bill Thornton – and told them that there would be few missions for them to fly. "Gents," he told them, in a recollection paraphrased by Joe Allen in his January 2003 NASA oral history, "I've got some bad news for you and that is [that] we've been *told* by the government to take you, but we don't have a job for you, *not one of you*. We've had to make this announcement, but if any of you ... feel that you have more important work to do elsewhere, you'll make no enemies by resigning!" It was typical Deke Slayton; honest, direct and straight to the point. Moreover, the new scientists would need to undergo jet training with the Air Force ... and the ongoing demands of the Vietnam War on combat pilots meant that places would not become available until at least the spring of 1968. Indeed, most of the group – who wryly dubbed themselves 'the Excess 11' or 'XS-11' – would wait for almost two decades before finally reaching space.

Under this menacing cloud of uncertainty, and scarred by the effects of the fire, few decisions about the future, beyond Apollo, could be made. The nature of the orbital workshop was also changing. Towards the end of 1967, reviews began to call for a 'dry' workshop – that is, a modified S-IVB, prepared on the ground with crew quarters and equipment and launched atop the much larger Saturn V. This satisfied a number of lingering concerns. Firstly, launching a dry station was better for human habitation (the astronauts would not need to convert a used, hydrogen-fuelled rocket stage *in orbit*) and secondly, since it reduced the time needed to outfit the workshop with equipment, electrical lines and experiment racks from five or six days to just a couple of days, there would be more time to conduct 'real' science. Still, there remained much support for the wet workshop and vacuum chamber tests at the Marshall Space Flight Center as late as March 1969 confirmed that all measurable traces of hydrogen had disappeared within four to six hours of the onset of passivation. For the astronauts who might someday fly the wet workshop, though, it was not enough. Even the successful venting of propellants from an S-IVB during the unmanned Apollo 5 mission in January 1968 did little to allay their sense of scepticism. Tom Stafford, who knew the dangers and unknowns of working in space better than most, was "appalled" by the very notion of turning a just-used rocket stage into his new home. Al Bean, who would command the second manned Skylab mission, credited fellow astronaut Walt Cunningham – for a time the office's AAP representative – as having been instrumental in recommending a change from a wet workshop to a dry, Saturn V-launched version. Also in favour was scientist-astronaut Joe Kerwin, who noted that the limited amount of detail about the wet workshop's experiments and the precise role of the crew made it even less attractive.

In a NASA oral history interview from May 2000, Kerwin recalled being asked by Al Shepard, then chief of the astronaut office, to go to the Douglas Aircraft plant and take a look at one of the S-IVB stages. Kerwin and fellow scientist-astronaut Curt Michel duly took a T-38 jet and were pleasantly received by the Douglas staff and shown along the production line and the interior of the stage, which was fully 6 m in diameter. "We went on in, in our stocking feet and our white coats and all," Kerwin recounted, "because this was a flight stage . . . and there's this *chemical* smell from the material they used to cure the sides! It was a fibreglass lay-up with some kind of protective coating over it and it smelled funny."

The astronauts posed the obvious question: How were they going to get rid of that smell when they had opened the stage up in orbit?

"Oh, it will leach out . . . *given time!*" came the response.

"What are we gonna *do* in here?"

"Well, we haven't figured that out yet. Maybe you could put on your suits and sort of . . . *fly around* and check out the EVAs in here. It's a nice big area."

"Are you going to pressurise it?"

"Well, we haven't decided, but we think we probably ought to pressurise it with oxygen."

Kerwin and Michel returned to Houston, overwhelmed by the chemical smell from the pristine stage, but decidedly concerned that the wet workshop as a concept was extremely primitive and undeveloped. "Solar arrays," read an AAP Weekly Progress and Problem Summary report, dated 26 April 1968, "would be deployed on the first stateside pass, since the liquid portion of the passivation would have been completed. Gaseous passivation was expected to require approximately 24 hours. The meteoroid bumper would not be deployed until crew arrival, because it would interfere thermally with the passivation." After the liquid hydrogen had fully leached its way out of the stage, the astronauts would have to close the vents, backfill with oxygen and then undo the 72 bolts around the forward dome. All this was before they could even *begin* outfitting it as workplace. Kerwin wrote a memo to Shepard, telling him that "they had a long way to go", whilst Michel added that he "didn't see anything of any scientific value in this whole operation". Their response steadily turned the tide of feeling in the astronaut corps against the idea of a wet workshop.

Unfortunately, Jim Webb vehemently opposed the idea of moving from a wet to a dry workshop. As a dry workshop, by its very nature, would be unable to boost itself into orbit, it would serve as a payload, not an upper stage, and consequently would need the greater lifting capability of the Saturn V, rather than the Saturn IB. Webb had spent much of his tenure as NASA Administrator fighting to persuade Congress and the Bureau of the Budget that the agency *needed* at least 15 Saturn Vs for the lunar landing effort and in August 1968 the direction to terminate further production of engine hardware for the booster began. Now, even with his retirement from NASA imminent, Webb categorically refused to tolerate any suggestion that the Saturn V be used for anything other than the drive to reach the Moon. From the perspective of the Apollo Applications office, using an electrical analogy, Webb was "putting strong impedance in the system", postponing anything to do with AAP unless there was a compelling reason to pursue it. In January 1968, he told George Mueller to adopt a cautious, step-by-step, 'wait-and-see' approach before going ahead with contracts to adapt the Apollo craft for AAP missions and in June he even told centre directors that AAP was nothing more than "a surge tank for Apollo".

When Tom Paine, a scientist and engineer, succeeded Webb at the helm of NASA in the spring of 1969, he promptly indicated from the outset that AAP contracts should go ahead and was warmly enthusiastic of the idea of a long-term space station. His enthusiasm was matched by Congress – perhaps unsurprisingly, following the success of Apollo 8 – but the attitude of the new Nixon administration was extremely cool to the matter and AAP funding was cut by $57 million to just over $250 million ... again, barely enough to keep it alive. Nevertheless, in February 1969, the outstanding performance of the Saturn V prompted AAP Deputy Director John Disher to propose the removal of one booster from the Apollo lunar landing effort to loft a dry workshop into orbit instead of a wet workshop atop the Saturn IB. The proposal was shelved when it became clear that there would be no inherent cost saving and, in late April, George Mueller told the Senate that changing from a wet to a dry workshop was "inefficient and only marginally effective in advancing space technology". Many others agreed, although Wernher von Braun and Max Faget were enthusiastic about the dry workshop. They felt that the Saturn V's launch record was outstanding and to send the workshop aloft in a fully outfitted state would offer huge benefits in terms of its habitability, its ability to carry large pieces of research hardware, better redundancy and the ability of the space agency to undertake a checkout of the complete article on the ground. Bob Gilruth also endorsed the proposal to change to a dry workshop. However, many in the AAP office felt there were "substantial reasons for not changing from the present core programme", not least because it would require many contracts to be re-written and re-negotiated and a large workforce to be re-distributed – particularly that of the Marshall Space Flight Center, which had led the wet workshop studies. "The sheer inertia," wrote David Compton and Charles Benson in their seminal work *Living and*

Working in Space: A History of Skylab, "of a programme as far along as the wet workshop was formidable."

Within a matter of weeks, however, senior managers were gradually won over to the dry workshop camp, including George Mueller himself, although it soon became apparent that an 18-month delay would be inevitable. In their 2009 book *Homesteading Space*, David Hitt, Owen Garriott and Joe Kerwin related the sequence of events which finally changed Mueller's mind: a hands-on session, fully-suited, on a mockup of the workshop in the water tank at the Marshall Space Flight Center in the spring of 1969. This practical experience convinced Mueller that the tasks expected of the astronauts, from loosening bolts to holding themselves steady in bulky pressure suits, were both incredibly difficult and highly dangerous. "From that moment," wrote Hitt, Garriott and Kerwin, "the future was set." Much of May 1969 was spent refining the available options and it was proposed that if the dry workshop option *was* adopted, a quick decision would be necessary from NASA Headquarters and it would be imperative to resist changes further down the line. By the end of the month, four alternative options had been drawn up by Mueller: alternatives 1 and 2 would require a Saturn V to launch the dry workshop and a Saturn IB to launch the crew and the ATM, whilst alternatives 3 and 4 would call for the Saturn V to launch *both* the workshop and the ATM and a Saturn IB to launch the crew. In the subsequent discussions, the first and last options received the most favourable response. Most of the field centre directors concurred, sensing that Option 1 was also more likely to win the approval of both Tom Paine and Congress. After receiving detailed descriptions of all four options on 27 May, Paine was sold on the dry workshop, launched atop a Saturn V, and the decision to formally drop the wet workshop was set in stone from the end of June.

Aware that taking a Saturn V would effectively eliminate a future lunar landing, Paine deferred from signing the change request from a wet to a dry workshop until 18 July, just a couple of days after Neil Armstrong, Mike Collins and Buzz Aldrin had launched aboard Apollo 11. Four days later, with humanity's first Moonwalk triumphantly accomplished, the NASA field centres were formally directed to implement the dry workshop concept. "That was a big day," recalled Joe Kerwin, "when we went from the wet workshop to the dry workshop. Instantly, we had what space programmes up to that time had never had: *weight margin* and *volume*. We had all the weight and volume in the world. They didn't have to cut corners. We could load all the consumables for *three flights* on ... we could put experiments of all kinds up there: medical, solar physics, Earth resources, without worrying too much about weight margins." The immense lifting power of the Saturn V meant that everything could be hauled into orbit at once, all pre-installed and checked out. Launch was tentatively scheduled for July 1972. Apollo 20, the tenth planned lunar landing, disappeared from the 'internal' manifest, as 'its' Saturn V found a new customer, Skylab, and this was announced publicly by George Low in January 1970.

CLEAR ROAD TO LAUNCH

Despite widespread satisfaction that after many years in apparent limbo, the workshop had finally crystallised from "a surge tank for Apollo" – a kind of 'Wednesday's child', one aerospace journalist remarked, full of woe for so long – into a real space station with a real future, there remained much dissatisfaction over its name, for it still retained the rather uninspiring monikers of 'Dry Workshop' or 'AAP'. Shortly after Paine added his signature to the change request, authorising the transition from a wet to a dry workshop, the first steps were set in motion to give it a more appropriate name. More than a hundred suggestions were made, ranging from lofty titles such as 'Socrates' through to the somewhat more questionable 'LSD', and these were whittled down to eight finalists; four drawn from mythology and four from American history. Early in the spring of 1970, NASA's Project Designation Committee mulled over these eight 'possibles', aware that the final choice "could enhance the public identification with the programme and hopefully provide a more manageable term for everyday use". Could 'LSD' do that, perhaps? Ultimately, of the eight submitted options, the committee chose ... *none of them*. The final choice came instead from a name submitted by Lieutenant-Colonel Donald Steelman, an Air Force officer on detached duty to NASA since 1968. His proposal to name this, literally, as a 'laboratory in the sky' ... hence, 'Skylab' ... seemed perfect, rolled smoothly and cleanly from the tongue and summed up the project in a very straightforward fashion.

By this time, the launch of the station was scheduled for late in the summer of 1972, although Tom Paine hoped to achieve an earlier date. The cancellation of two more Apollo lunar landings in the summer of 1970 freed up another pair of Saturn Vs, which in turned opened the possibility of flying a second Skylab and perhaps later establishing a 'space base' with a crew as large as 50 men to carry NASA into the 1980s and beyond. "His main concern," wrote David Compton and Charles Benson of Paine's actions, "was to have 'a major mission of new significance' by 1976 ... something more than just another Skylab ... but he was clearly out of step with the Nixon administration." The agency received a little more than $3.25 billion for its 1971 appropriation. When George Low succeeded Paine as Acting Administrator in September 1970, he found himself with a fight on his hands to defend the interests of the Apollo lunar programme and the future space station, but recognised that, if a choice *had* to be made, then "the weight of evidence seems to favour Skylab". More than a year since the first manned touchdown on the Moon, America had already benefitted from the Apollo investment, and could afford to 'lose' a few additional landing missions from its budget. Cancelling Skylab, on the other hand, would produce absolutely zero return.

Meanwhile, the hardware was steadily taking shape. Under contract from NASA's Marshall Space Flight Center, McDonnell Douglas was responsible for building and testing the main workshop and in April 1969 the S-IVB was removed from storage to begin modifications. When the dry workshop decision was made later that summer, designers from the Manned Spacecraft Center implemented a series of changes to the crew quarters, adding a viewing window to the station's 'wardroom' and extensively upgrading the food storage and preparation facilities. The attitude control system was altered to accommodate an expanded battery of Earth resources instrumentation and the sheer magnitude of the effort – one writer likened it to building Concorde from scratch – meant that in 1971 only 25 percent of the qualification tests required for the workshop had been accomplished, compared to around 80 percent in the case of the airlock and multiple docking adaptor. At length, on 7 September 1972, Administrator Jim Fletcher and Associate Administrator for Manned Space Flight Dale Myers formally accepted the workshop for NASA and it began its journey from California to Florida.

Physically, it measured 14.6 m in length and 6.7 m in diameter, essentially taking its size and shape from that of the S-IVB, with the rocket stage's large hydrogen tank (8.9 m long) now serving as the pressurised habitable section and the smaller oxygen tank providing an unpressurised storage area for rubbish and other unneeded items. In total, it weighed 38,380 kg and offered a habitable volume of 275 m^3. On either side of the workshop was a solar array 'wing', measuring 9 m long. Shortly after orbital insertion, a wrap-around shield would be deployed by torsion bars to 'stand' 15 cm away from the hull to break micrometeoroids into showers of harmless fragments. Within the workshop were two compartments, separated by a grid-like, perforated 'floor'. The rear one included the wardroom for eating, a sleeping area, toilet and a working area for biomedical studies, and the forward compartment offered a larger space in which to undertake scientific experiments. The grid floor, recalled Caldwell Johnson in a May 1998 oral history for NASA, allowed for the design of new footwear to secure the astronauts. "Some of the shoes had ... 'cleats' on them that could go down through the grid and, if you turned your foot slightly, it would lock into the grid. It was quite effective, but they were heavy and clanky and clumsy. Some of the crew would learn to almost use their toes in this grid to kinda halfway-grab-it with the toes. They wore out their little booties ... in no time, with trying to grab hold of the floor." A pair of scientific airlocks, one directed towards the Sun ('solar') and the other in the opposite position ('anti-solar'), allowed for the deployment of experiments into space. An instrument unit attached, as usual, to the forward end of the stage, would control Skylab after orbital insertion, performing in particular the deployment of the aerodynamic shroud, the micrometeoroid shield, the ATM, the solar arrays and the pressurisation of the habitable section.

Also attached to the forward end of Skylab were the airlock and multiple docking adaptor, which formed two components of the same module. Although these were built by different contractors, Martin Marietta and McDonnell Douglas formed a close collaboration when NASA's production plans called for the items to be joined and tested together. Originally, the airlock was intended as little more than a simple tunnel between the Apollo spacecraft and the workshop, but by the summer of 1969 its role had expanded to encompass communications equipment, electrical power facilities and environmental controls. Similarly, the multiple docking adaptor offered services for which it was not initially designed, including displays for the ATM, hardware for the Earth resources experiments and a number of crew stations. The change to a dry workshop concept also led to a reduction in its number of docking ports from five to two; a 'forward' port and an 'axial' port. Both the airlock and the

multiple docking adaptor measured 5.3 m long and 3 m wide. Collectively known as 'the cluster', these three pressurised components of Skylab would represent the largest habitable object yet inserted into space, enclosing a total volume of 347 m³, more than 60 times that of the Apollo command module.

The airlock and multiple docking adaptor also provided a mounting location for the struts of the 11,180 kg ATM, which had itself evolved considerably since its genesis initially as a telescope inside the Apollo service module and then as a lunar module-borne instrument intended for the wet workshop. In its Skylab configuration, the ATM comprised an octagonal structure, 3.6 m long and 3.3 m wide, which carried an experiment canister for eight telescopes and auxiliary systems. However, despite its size, using fine sensors it was able to aim these instruments very precisely at any point on the Sun and hold them steady; in fact, it could be pointed to within 2.5 arc-seconds of the desired target and maintain the position for 15 full minutes, without drifting more than 2.5 arc-seconds. Since conventional thrusters were insufficient to achieve this sort of precision, a trio of control moment gyroscopes – each containing a 9,000 rpm rotor, weighing 65 kg – provided long-duration attitude management and minimised the effects of environmental disturbances and other contaminating effects. During ascent, the ATM would be positioned 'axially' in line with the rest of Skylab and the Saturn V, but after insertion into orbit and the jettisoning of the shroud, a series of electrical motors would rotate it 90 degrees into its operational 'radial' position, thereby freeing up the main port on the multiple docking adaptor. Its 'windmill' of four solar panels would subsequently be deployed, joining those aboard the workshop to supply Skylab with electrical power.

Despite earnest efforts in the summer of 1969 to prevent the station from undergoing the expense of further change, by October of that year more than a dozen 'adjustments' and $100 million of additional costs had been incurred, including the possibility of an extended, 120-day mission for the final crew, the inclusion of a new package of Earth resources experiments and steepening the orbital inclination to 50 degrees in order to improve the surface coverage. It was also recommended that the ATM should be able to acquire solar data in an 'unmanned' capacity, so that it could be operated from Earth, either when there were no astronauts aboard or whilst a human crew was otherwise occupied. Ultimately, however, Skylab would not suffer significant cuts – somewhere in the range of $50 million or five percent of its requested amount – which was absorbed by slowing down the pace of its operations. New launch dates for the last Apollo mission

Photographed by the approaching final crew of Gerry Carr, Ed Gibson and Bill Pogue in November 1973, this image reveals the product of the decision to pursue a 'dry workshop': the Skylab space station. The windmill of arrays belonging to the Apollo Telescope Mount (ATM) are clearly visible at the top left of this image, with the multiple docking adaptor directly ahead.

(December 1972) and for Skylab itself (originally March, then late April, 1973) were announced and would remain, thankfully, essentially unchanged.

In the meantime, other efforts had been ongoing for a 'second' Skylab, which might allow astronauts to fly year-long missions, provide for artificial gravity and permit the installation of a complex array of Earth resources instrumentation in the place of the ATM. The general response from most NASA centres was positive, although the ability to support a mission of a full year would require significant hardware changes and artificial gravity provision would double or triple the overall cost of the project. Internally dubbed 'Skylab II', the project faltered during the latter part of 1970 when its proposed payload weight forced a rethink of ways in which the Saturn V's S-II second stage could be modified. Costs were also high, ranging from $1.3 to $1.5 billion. Nonetheless, by the end of the year, Skylab II had gained strong political support from the House space committees, but its funding would demand either a much larger NASA budget or lengthy delays to a follow-on programme, the Space Shuttle. Since the former was an exceedingly unlikely prospect and delays to the latter were unacceptable under the Nixon administration, Skylab II's future very quickly turned bleak.

For the 'first' Skylab, though, the road to launch seemed clear. However, as the largest *habitable* object yet placed into orbit – and the emphasis certainly was placed on 'habitable' – it was considered vital that some degree of creature comforts should be incorporated. Already, a series of papers presented at a space station symposium in 1960 had suggested that even for hardened test pilots, life aboard an orbital outpost would be "humdrum", analogous to keeping watch at a lighthouse, and even George Mueller had been appalled when first shown a barren, totally mechanical mockup of the wet workshop. "Nobody could have lived in that thing for more than two months," he said later. "They'd have gone stir-crazy!" Caldwell Johnson, assigned as principal investigator for the habitability 'experiment', agreed. "It was very stark," he recalled in his NASA oral history, with "no pleasing carpets, no nice walls, no sofas, no beds. In one of the arrangements, it *did* show some beds, and they were bunks, like one would find on ship-board, except they were folded up against the wall and they hung down on chains! This is in *zero-gravity*, now! It doesn't make a lot of sense. The place was kinda like living in a barn or, worse yet, in a blacksmith's shop." In response, industrial designers Raymond Loewy and William Snaith were hired to conduct a habitability study. Their report, in February 1968, stressed that NASA's plans were poor and suggested a simple working area, with enclosed and open spaces, "flow[ing] smoothly as integrated elements ... against neutral backgrounds". The cylindrical shape of the workshop clashed with rectangular elements, the grid-like floor pattern was "harsh" and the overwhelming atmosphere of the place was "forbidding". Lighting was both random and scattered and colour schemes were dismal and depressing. A neutral background of pale yellow, Loewy-Snaith advised, was advisable and lights with a warmer spectral range, rather than cold fluorescent tubes, should be localised at respective work areas. Over the years, popular opinion of spaceflight has created the maxim that 'up' and 'down' have little meaning or importance in weightlessness. Caldwell Johnson disagreed in the case of Skylab. "It didn't take long to appreciate that things don't work so well if everybody is [in] all kind of funny positions," he recalled. "People don't like to look at each other *upside down*. They don't look the same. If you're going to write directions and placards around the wall, it's nonsense to put some this way and some [that] way, so ... using an 'up' and 'down' in a spacecraft is very important. That way, you can put all the lights in the ceiling, where they're supposed to be, you can put things you've got to do around the side [and] you can put things that you've got to 'walk' on and grab with your feet down here. You end up just like it is on Earth. Besides that, the body – if you ever notice – if you stand up, you can [twist] the trunk and the head almost 180 degrees and reach things around you, but you cannot [bend forward and backward] very far. It's very difficult to do things this way, so all this reinforced the idea of maintaining an 'up' and 'down' in the system." The issue of 'habitability', which has since assumed high importance, particularly aboard the International Space Station, was in its infancy with Skylab and it steadily became apparent that a quite different *modus operandi* would be required for astronauts spending several months in space, as opposed to just a few days in the relatively cramped confines of an Apollo craft.

In the days of the wet workshop, before July 1969, the range of finishes which could withstand even short-term immersion in liquid hydrogen was limited, although the floor plans were addressed (including the installation of removable panels) and efforts were made to incorporate a wardroom for eating, relaxation and for routine 'office work'. By the time the dry workshop was approved, habitability concerns were allayed somewhat ... although little was done at first. The Manned Spacecraft Center criticised its interior as being "austere", like a canvas tent city; moreover, the areas for eating and sleeping were poorly sized and the floor plan made little sense. Many within NASA, including the astronauts, had paid little heed to such aesthetic considerations, in true 'right-stuff' fashion, but George Mueller and others felt that the workshop would be an experiment in the design of future space stations ... and both he and Skylab Director Bill Schneider placed habitability high on their list of priorities. Several improvements had been made by October 1969, including the large viewing window, although concerns were raised about how much it would cost and the extent to which it might weaken the structure. When Mueller asked Loewy-Snaith for their opinion, the response was clear: it was "unthinkable" not to provide the crew with a window. Its recreational value alone would be worth its cost and its implementation was approved by Mueller on the spot.

Other habitability issues included clothing and a decision was taken to replace the one-piece Apollo coveralls with a two-piece uniform and jacket. This was made from a fireproof cloth, with the near-unpronounceable name of 'polybenzemidazole', which came in a golden-brown colour. "Skylab ... was going to have three crews," explained Johnson, "and virtually everything for all three crews had to be put aboard on the first flight. No one knew who the crewmen were going to be. They're not all the same size, so how could you give them all these single-piece garments? Some of them would be too big and some would be tight and I can't imagine staying there three months with a jumpsuit that's not long enough between the crotch and the shoulders! We finally convinced them to use the separate shirt and pants and all, like everyone else does." Other clothing suggestions, such as 'beanie hats' ("to protect their heads from being bumped") and anti-perspiration undershirts ("which smelled awful") did not catch on quite so well . . .

Proposals to include an entertainment centre in the wardroom for watching movies and playing cards, darts or board games yielded little enthusiasm, with many astronauts preferring to spend their free time watching Earth, listening to taped music, reading or exercising on the bicycle ergometer. Personal cassette players were provided, but it is hardly surprising that with the grandeur of the Home Planet and the novelty of their environment, 'recreation' did not pose a significant problem for the astronauts. In Johnson's mind, there was another reason why the astronauts avoided the games, too. "I think now, in retrospect, any psychologist could tell you that they'd want to avoid those," he told the NASA oral historian, "because three men for three months, probably the worst thing they could do was get involved in *competitive games* [and] see who can beat whom! That might just aggravate some situation and I think they intuitively recognised that that's not the thing to do, so they didn't play cards, they played music and they looked out of the window. They liked to exercise; they used the old bicycle. They'd get on that thing and just buzz away."

The need to keep clean led to a decision to create a lightweight, low-cost 'whole body bather' – a bath or shower – but the multi-million-dollar price tag dissuaded many senior managers. Ultimately, a simple shower *was* incorporated into Skylab, but the reviews from the astronauts would be mixed. Improvements in food were definitely needed for two-month missions and, certainly, the standard at the end of the 1960s provided nutrition, survived launch without disintegrating, lasted almost indefinitely . . . but only worked if a way could be found to induce the astronauts to actually *eat* it! In fact, in May 1969, Don Arabian, head of the test division at the Manned Spacecraft Center, tried the Apollo fayre for three whole days . . . and lost the will to live. The sausage patties tasted like granulated rubber which left a sickening taste and by the third day the very process of eating for this self-described "human garbage can" had turned into a chore. Studies of the kind of foods which might be suitable for long-duration missions ultimately concluded that astronauts should be pampered on flights of more than four to eight weeks and both George Mueller and Bill Schneider campaigned vigorously for the inclusion of meals which could be eaten as 'normally' as possible, with a spoon or fork, rather than having to squeeze pastes from tubes or gobble cold, granulated cubes. Indeed, when the Apollo 8 crew ate turkey and gravy with spoons during their return journey from the Moon in December 1968, the impact on their morale had been remarkable. By the time that the workshop turned dry the following summer, the situation had improved significantly: the station would include both a freezer and an oven and foods would be provided in five varieties – dehydrated, intermediate moisture, 'wet-packed', frozen and perishable – to allow them to be better stored and protected from pressure changes. Teas and coffees and fruit juices would be available, but no alcohol.

This may seem unsurprising, but wine and sherry *had* earlier been raised as an option. It was not just for pleasure, but because the wines and sherries under consideration contained little or no protein or controlled electrolytes and were essentially 'empty' calories which would not hamper the results of the medical experiments. Unfortunately, during a lecture, Gerry Carr happened to mention the presence of wine on the menu "in the interest of crew morale" ... and triggered a flurry of complaint letters to NASA. At length, aware of the negative publicity, NASA quietly removed all mention of wines and sherries from the menu. In a memo dated 10 August 1972, Ken Kleinknecht, the Skylab manager in Houston, told Chris Kraft that "there was no basic requirement for including wine", that it was "not necessary for nourishment", that it would "involve an unnecessary expense" and, in the eyes of the press, "would result in adverse criticism for the programme". This had not, however, precluded several wines and sherries from being 'sampled' in the autumn of 1971 at Joe Kerwin's house. Ordinary table wines would most probably have gone bad during the course of a year aboard Skylab and it seems likely that Taylor Cream – a deep-amber-coloured sherry, flavoured with toffee and roasted

Skylab's first crew – from the left, Joe Kerwin, Paul Weitz and Pete Conrad – prepare and eat food in a mockup of the wardroom in the spring of 1973. Although much progress had been made to render the space station more 'habitable' than previous spacecraft (the large window is clearly visible in the background), the result was still very *functional* in design and appearance. For nine test pilots and scientists, however, it mattered little; for Skylab would offer them more than enough work and unforeseen challenges to keep them occupied.

nuts – would have been the astronauts' preferred choice. Still, the issue of 'normal' food and drinks highlighted the need for a sense of terrestrial normality in an 'extraterrestrial' environment. This requirement extended to the design of the wardroom itself, which included a central pedestal to hold the astronauts' food trays. "These trays had heaters in them and recesses that one could put cans of prepared food," recalled Caldwell Johnson, "and it would *heat* the cans with a timer, so it could be set ahead of time and the meal would be ready."

In the meantime, Skylab itself progressed steadily towards launch. In the spring of 1970, the plan was for it to be placed into orbit in November 1972 and then operated by three crews over an eight-month period. A four-week 'activation mission' would launch the day after the workshop had reached space. The later crews would launch in January and May 1973 and each spend 56 days aboard. By the end of 1970, however, further cuts to NASA's budget had pushed those dates to the right and eventually the workshop was rescheduled for launch on the last day of April 1973. This plan would then have called for launching the first crew on 1 May, the second at the end of July and the third late in October. Although these adjusted dates were announced in April 1971, it is remarkable that they proved fairly realistic. By the end of January 1973, minor delays in preparing hardware had enforced a slight postponement of the workshop until 14 May, producing the first 'formal' launch target date. "For nearly two years," read a NASA news release, dated 4 April 1973, "the Skylab team had used April 30 and May 1 as planning dates for the launches [of the station and the first crew]. The planning dates were moved to the month of May when checkout work was running about two weeks behind the pre-launch test schedules. The on-board experiments and the major spacecraft elements have never flown before and require exhaustive first-time testing." Under this revised plan, the second and third crews were rescheduled for launch in August and November, respectively.

With only a few months to go before America's first space station was to take to the skies, three prime crews and their backups had been training feverishly for more than a year. Ironically, the man who would command its 28-day maiden mission – and receive a Congressional Space Medal of Honor in 1978 for helping to 'save' it from an ignominious failure – had actually been frowned upon by NASA just a few years earlier. In fact, when Pete Conrad first underwent the gruelling physical and psychological evaluations to become an astronaut, his application was rejected ... on the grounds that he was 'unsuitable' for long-duration spaceflight!

PILOTS AND SCIENTISTS

"That's a vagina," quipped Charles 'Pete' Conrad Jr. "Definitely a vagina." The psychiatrist noted his response without a word, maybe realising, maybe not, that he was the victim of yet another wisecrack from the gap-toothed, balding Navy lieutenant. Yet, despite his comments about each of the Rorschach cards which he was shown, Conrad was not entirely obsessive about the female genitalia. He had actually been tipped-off the night before by another astronaut candidate, Al

Shepard: what the NASA psychiatrists were really looking for was *male virility*. "I got the dope on the psych test," Shepard had assured him. "No matter what it looks like, make sure you see something sexual." So Conrad did.

His key concern, though, that spring in 1959, had been the impact that this crazy 'Project Mercury' idea might have on his career. Instead of logging hours in the Navy's new F-4 Phantom fighter, he spent a week at the Lovelace Clinic in Albuquerque, New Mexico, serving up stool, semen and blood samples and collecting 24-hour bagfuls of urine. On the evening before a major stomach X-ray, told not to drink alcohol after midnight, Conrad had sat up until 11:57 pm draining a bottle with Shepard and fellow naval aviator Wally Schirra, to loosen himself up for the next day. Conrad doubted that Lovelace's invasive tests had anything remotely to do with flying in space: the physicians, he told Shepard, seemed far more interested in "what's up our ass" than in their flying abilities, which was typical of military flight surgeons. Shepard warned him to be careful, to give the right answers to questions and to remain vigilant that members of staff were watching their every move.

In spite of his frustration, Conrad persevered. He followed Shepard's advice, saw the female anatomy in every Rorschach card, deadpanned to a psychologist that one blank card was upside down, pedalled a stationary bicycle for hours, sat in a hot room for an age, dunked his feet into ice-cold water and argued with one of the physicians that he considered it pointless to have electricity zapped into his hand through a needle. However, all this torture, Conrad felt, would at least give him the opportunity to lay his entire naval career on the line for just one chance to fly something even faster: to ride a rocket, outside Earth's atmosphere, "at a hell of a lot more Machs than anything he was flying right now". Flying higher and faster, and pushing his own boundaries, had been the story of Conrad's life.

Born in Philadelphia, Pennsylvania, on 2 June 1930, the offspring of a wealthy family which made its fortune in real estate and investment banking, Conrad's father insisted that he be named 'Charles Jr' – "no middle name" – although his strong-willed mother, Frances, felt that this tradition of Charleses should be broken. Frances liked the name 'Peter', wrote Nancy Conrad in her 2005 biography of her late husband, and although it never became his official middle name, Charles Conrad Jr would become known as 'Peter' or 'Pete' for the rest of his life. His fascination with anything mechanical reared its head at the age of four, when he found the ignition key to his father's Chrysler and reversed it off the drive. Later, in his teens, he worked summers at Paoli Airfield, mowing lawns, sweeping and doing odd jobs for free flights. Aged 16, he even repaired a small aircraft single-handedly.

"Most people I meet are doing eight things," fellow astronaut Al Bean once said. "Some people can do that. Pete Conrad, my best friend and my mentor, could do that. We could be talking about space and he'd get a phone call about something like the transmission on his race car. 'We've gotta get the transmission changed because the gear ratio's not working on this track.' Suddenly, all his effort would be on that. The next day, someone would call about motorcycle racing. He could do *all* of them well." At heart, Conrad was an engineer and tinkerer.

Education-wise, he would partly follow in his father's footsteps: the private Haverford School, from which he was expelled, then the Darrow School in New York, where Conrad's dyslexia was identified and where he shone. Although his father intended him to attend Yale University, he actually enrolled at Princeton in 1949 with a Reserve Officers Training Corps scholarship from the Navy to pay for his studies in aeronautical engineering. Graduation in 1953 brought him not only his bachelor's degree, but also a pilot's licence with an instrument rating, marriage to Jane DuBose and entrance into naval service.

He breezed through flight training, earning the callsign 'Squarewave' as a carrier pilot. In *Rocketman*, his second wife Nancy wrote that Al Teddeo, executive officer of Fighter Squadron VF-43 at Naval Air Station Jacksonville, Florida, had doubts on first meeting the young, seemingly-wet-behind-the-ears ensign one day in 1955. Those doubts were soon laid to rest when Teddeo discovered that Conrad could handle with ease any manoeuvre asked of him. Tactical runs, strafing runs, spin-recovery tests; Ensign Conrad did it all. "Hell, we refuelled three times till I just had to get back to my desk," Teddeo recalled years later. "It was like telling a kid at the fair that it was time to go home."

Next came gunnery training at El Centro, California, and transition from jet trainers to the F-9 Cougar fighter. In 1958, Conrad reported to Patuxent River in Maryland to qualify as a test pilot. In late January of the following year, he received, along with over a hundred others, instructions to attend a classified briefing. Conrad would not make the final cut for the Mercury selection, but his day would come three years later. His failure proved a little ironic. "Unsuitable for long-duration flight," asserted the explanatory note. He had evidently shown too much cockiness and independence during testing; characteristics which went against the panel's notion of a good, all-rounded astronaut. During his career, Conrad would fly four space missions in total, two of which would set new world records ... for long-duration flight!

When Conrad returned from his third mission, Apollo 12, at the end of 1969, in which he became the third man to set foot on the Moon, he made no secret of the fact that he wanted a second shot at commanding a lunar landing. Both Slayton and chief astronaut Tom Stafford quickly assured him that it would not happen. For a start, dwindling budgets and numbers of flights meant that there were far more astronauts-in-waiting than there were available missions. Conrad was assigned to take charge of the astronaut office's Skylab branch in August 1970, with a high probability that he would command its first crew. It was a post that he held until the end of the Skylab programme in the spring of 1974. Having been formed under Apollo Applications, this office had changed almost beyond recognition since the summer of 1966, when Al Bean took the helm as its first chief.

The very nature of flying for long durations aboard Skylab had already prompted many in the medical community to recommend a commander and *two* scientists on each crew, although Deke Slayton quickly nixed this as impractical. "Skylab was a whole new deal for us," he wrote. "I figured the first crews ... should have at least two people who had some experience in troubleshooting." Slayton found support in Bob Gilruth, who, in June 1970, told Dale Myers that reliability studies indicated "a

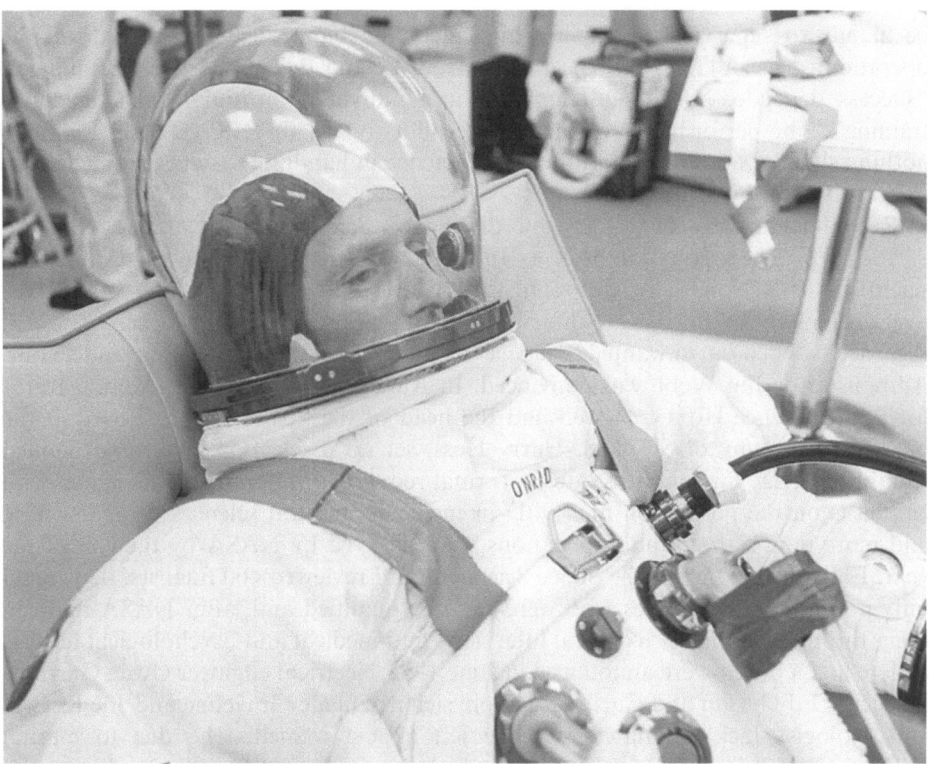

The label "Unsuited for long-duration spaceflight" must have hung around Pete Conrad's neck like a vice ... but he dispensed of it just as quickly. Not only is he clearly *suited*, in this image, for a spaceflight, but by this point in his career he had already flown one record-breaking duration mission, performed an intricate orbital rendezvous and commanded an adrenaline-charged lunar landing. However, in Conrad's mind, it was Skylab that would propel him to his greatest heights and his greatest personal accomplishment. When he received the Congressional Space Medal of Honor in 1978, contrary to popular opinion, it was *not* for his lunar landing ... but for saving Skylab.

high probability of systems problems ... during the mission". Consequently, much of the Skylab crew training would be devoted to systems management and procedures to handle malfunctions. Added to this was the understanding that all crew members would be 'cross-trained' on different tasks, thereby making academic credentials relatively unimportant. The two key positions on each crew were labelled 'commander' and 'pilot' and Dave Shayler has noted that the former would ideally have previous flight (and rendezvous) experience, but the latter could be a rookie. The commander would be responsible for the overall conduct of the mission and specifically for the Apollo systems, the docking and undocking and the flyaround of the station. The focus of the pilot would be upon the electrical and mechanical systems aboard Skylab itself. The third position, known as the 'science pilot', would be mainly to supervise the experiments, including biomedical equipment and the operation of the ATM. Retrospectively, wrote Leland Belew and Ernst Stuhlinger, "success depended more on teamwork and individual attitudes than academic training". The performance of the science pilots on all three Skylab missions was nothing short of admirable and their competence in handling systems problems was in no way inferior to the pilots.

Recommendations to NASA by the Space Science Board of the National Academy of Sciences as far back as July 1962 urged the selection of professional scientists as astronauts. Although the reality that the astronaut office was overwhelmingly manned by military test pilots and a leadership hierarchy which regarded science as unwanted baggage in an exclusively engineering endeavour, Administrator Jim Webb was convinced. In August 1964, Associate Administrator for Space Science Homer Newell and the head of the Space Science Board at the National Academy of Sciences, Harry Hess, set up a selection panel and defined suitable criteria. Two months later, a formal request for applicants was made, with emphasis on the possession of a PhD in medicine, natural sciences or engineering and more than a thousand applications were received by NASA by the end of the year. Early in 1965, when the Space Science Board reviewed 400 finalists, they could only produce 16 candidates who were suitably qualified and, after NASA had run them through the mill of technical interviews and medical and psychological testing, a mere half a dozen were announced in June 1965: electrical engineer Owen Garriott, physicists Ed Gibson and Curt Michel, physicians Duane Graveline and Joe Kerwin and geologist Jack Schmitt. Graveline left almost immediately, due to marital problems, and very quickly the others learned that their position in the pecking order for space missions was by no means assured and all would endure a particularly raw deal.

The intrinsic 'unfairness' of the astronaut office's stance against the scientists quickly became obvious to the 1965 selectees when test pilots recruited a year behind them started picking up flight assignments – including key positions on lunar crews. In a sense, Deke Slayton's explanation that most of the scientists required advanced jet training, which was not complete until shortly after the 1966 class of pilots came on-board, is understandable and as a consequence the two groups were merged as one large unit. However, there was resentment for several years and some resignations. A flashpoint came when Jack Schmitt, the only professional geologist in the astronaut corps, appeared to have been overlooked repeatedly for an Apollo lunar landing and the scientific community began to pressure NASA to fly him. Complaints from the scientist-astronauts themselves to NASA Headquarters in October 1969 that the test pilots were being favoured for crew assignments was greeted with a rebuttal from Slayton that spaceflight was a hazardous business. Eventually, pressure led by the National Academy of Sciences caused the removal of test pilot Joe Engle from consideration for Apollo 17, in favour of Schmitt. This triumph, however, opened up a can of worms to fly more scientists on the future Skylab project.

Having only one scientist on each crew prompted consternation in the scientific community and Dave Shayler has noted that a paper delivered in 1970 by L.B. James of NASA's Marshall Space Flight Center to the 13th International Science School in Sydney, Australia had confidently expected crews of three scientist-astronauts on *each* Skylab mission; indeed, James wrote that crews "may consist of one astronaut who is an astronomer and one who is a medical doctor, as well as a third ... who may have some other scientific speciality". When one considers the apparent disinterest of the pilot-dominated astronaut corps in adding a geologist to one of the Apollo lunar missions, it is easy to understand the scientific community's irritation. As 1970 wore on, NASA found itself under increasing pressure not only to send a geologist to the Moon, but also to add *at least one* scientist, and preferably two, to each Skylab flight.

From Slayton's point of view, it was imperative to have an experienced astronaut in command of the first mission to the workshop and in August 1970 it seemed likely that Conrad, as chief of the Skylab branch, would receive this assignment. Worries about 'safety' and 'flight operations' as a reason for having only one scientist aboard, though, did not wash – for many of them had now achieved their jet-pilot credentials, some even had operational military back-grounds and there would be very few phases of a space station mission which would *require* advanced flying skills. The desire for two scientists progressed, nonetheless, and a recommendation was even made to the Manned Spacecraft Center for two scientists on at least one mission. This eventually came to nothing, particularly when the 'safety-over-science' argument resurfaced in the aftermath of three cosmonauts losing their lives whilst returning to Earth. The final decision came from NASA Headquarters on 6 July 1971: there would be a commander, a pilot and *one* scientist-astronaut, per crew.

By this time, of course, this arrangement was already clear. Earlier that same year, Slayton had identified 15 astronauts for Skylab's three-flight programme: Joe Kerwin would join Pete Conrad and Paul 'PJ' Weitz, Owen Garriott would join Al Bean and Jack Lousma and Ed Gibson would join Gerry Carr and Bill Pogue. The assignments were logical, firstly since – with Schmitt out of the running, seeking a lunar mission – Kerwin, Garriott and Gibson were the most senior scientist-astronauts in the office, having been picked by NASA in June 1965, and secondly their skill sets matched the respective focuses of each mission: the first long-duration flight required the presence of a physician (hence Kerwin), whilst the emphasis of the second and third missions on ATM studies would benefit from the expertise of Garriott and Gibson, both of whom had worked extensively on the instrument during its genesis. Possible crews for hypothetical fourth and fifth missions included physician Story Musgrave, joining Walt Cunningham and Bruce McCandless, and electrical engineer Bill Lenoir joining Rusty Schweickart and Don Lind. Interestingly, had these last two crews flown, they might have come closest to the scientific community's perception of an 'all-scientist' mission. Both Lenoir and Lind possessed doctorates in engineering and physics and, despite a military background, Schweickart's history in academia, particularly as an atmospheric researcher in MIT's experimental astronomy laboratory, had led many of the test pilots to regard him as more of a scientist than a pilot. Similarly, Cunningham had completed doctoral research, save for the act of submitting his thesis, and Dave Shayler recorded a conversation with Curt Michel, which noted that McCandless also held advanced academic credentials (a master's degree in electrical engineering) and so was "informally lumped in with us as scientist-astronauts". When Cunningham realised that he would not receive command of either the first or second missions and, even worse, lost out to a team of rookies on the third mission, he resigned from NASA in August 1971 and was replaced by Schweickart, who, in turn, was replaced by Vance Brand.

In his oral history, Kerwin recalled learning of these assignments. "Slayton came into a pilot's meeting on a Monday morning. [Holding] a sheet of paper in his hand, he said 'The following people are now formally assigned to crew training and mission development for the Skylab programme' ... He didn't say who was prime, who was backup, who was what mission or anything else. All he said was that Conrad was going to be Sky King [head of the group]. We had no idea what that list meant. There was a lot of speculation going on about who was going to be on what mission. There were fifteen of us, which meant there were three prime crews, but only two backup crews, so somebody was going to have double duty as a backup crew, unless the first prime was going to be the last backup. Deke didn't say. Deke was not a man of many words! It turned out, again in retrospect, that the way he had read that list was first prime, first backup, second prime, second backup, third prime." For a time, though, no one really knew. In fact, in a December 2000 oral history for NASA, Ed Gibson recalled that he and Garriott felt sure that *Kerwin* might be assigned to the *final* Skylab mission, owing to a perceived greater usefulness of his medical expertise on a significantly longer flight. In Joe Kerwin's mind, though, it was not until the formal announcement that "[Weitz and I] actually knew for sure that we were going to go with Pete on the first flight".

By the time the Skylab crews were formally announced by NASA on 16 January 1972, therefore, many of them had been training for at least a year and in some cases – particularly the scientists, who had worked ATM and Apollo Applications issues – for over three years. Not surprisingly, Conrad would command the first mission, with Joe Kerwin as science pilot and PJ Weitz as pilot, backed up by Rusty Schweickart, Story Musgrave and Bruce McCandless; the emphasis on biomedical research was illustrated by the presence of a physician on both crews. "Everybody wanted the first science pilot to be a physician," Deke Slayton wrote, "since the mission would be lasting 28 days. Joe Kerwin was the best candidate." The second crew to fly would then consist of Al Bean, Owen Garriott and Jack Lousma and the third would comprise Gerry Carr, Ed Gibson and Bill Pogue. Carr would become one of only five American astronauts in history to command his first spaceflight, having been recommended for the position by Conrad, for his outstanding work on the Apollo 12 support crew. Owing to Carr's military seniority – he was a lieutenant-colonel in the Marines by the time he flew – the decision to assign him to a command position made complete sense to fellow Marine Jack Lousma. "That's the way Deke worked," Lousma recalled. "I think Gerry Carr would have flown to the Moon before me. Gerry was senior to me in the Marines, so I'm not going to fly before him. The way it worked out on Skylab was I ended up flying before Gerry ... and to offset that, Deke assigned Gerry to be the commander, not just the 'ride-along guy' on the

It is not surprising that the first long-duration Skylab mission should have featured a physician, Dr Joe Kerwin, as its science pilot. The adaptation of human beings to the peculiar weightless environment was – and still is – imperfectly understood and part of Kerwin's role was to observe its effects with his own trained mind. Although the Soviets had launched a vestibular specialist, Boris Yegorov, into orbit in October 1964, he did not actually receive his doctorate until after his flight. This made Joe Kerwin the first qualified medical doctor ever to be launched into space.

third Skylab mission. That's the way his mind worked. So to some extent, he was kind of predictable for people who were coming up through the ranks." The second and third Skylab missions were both scheduled to run for 56 days. By the time each crew rode the elevator up the Pad 39B gantry on launch morning, they would have spent almost two and a half thousand hours apiece training for their individual missions.

This is not surprising, in view of the immense number of scientific experiments assigned to Skylab. Already, in the spring of 1970, Richard Tousey of the Naval Research Laboratory had approached NASA with a view to establishing a solar physics course for the astronauts, to enable them to be as prepared as possible for ATM operations. In Tousey's mind, they *could* operate the instruments without an understanding of solar physics, but the data would be inferior, and furthermore he was convinced that two years was realistically needed; a 'cram course', just a few months before launch, would not do, because by that stage the astronauts would be spending a large portion of their time undergoing systems training. At length, the Manned Spacecraft Center agreed in June 1970 to run a ten-week, 60-hour solar physics course, later that autumn, involving all ATM principal investigators and all Skylab prime and backup astronauts. Of course, in the cases of Garriott and Gibson, this was a case of taking a duck to water, but several of the pilots found the training challenging: one remarked that he was "right up to my eyeballs in trouble the whole time, trying to keep up and understand what was going on".

Elsewhere, a much broader training programme was being devised, which would call for 450 hours of briefings, 450 hours of experiment work and almost 700 hours in the spacecraft simulator; almost half as much again as the Apollo lunar crews had required. Training reached fever-pitch in the spring of 1972, with the large water tank at the Marshall Space Flight Center being used by at least one crew for EVA training almost every month. Familiarisation with the ATM equipment took centre-stage and the astronauts spent in excess of 200 hours apiece studying solar activity and responding to mock 'normal' and 'abnormal' conditions. Rendezvous and docking with Skylab were practiced exhaustively in the command module simulator. By September 1972, individual training had morphed into whole-team training, through so-called 'mini-sims', with the crews woken at six each morning, just as they would be in orbit, and advised by teleprinter message what their plans were for the new day. Voice contact from the 'ground' was limited to the schedule expected in flight and the exercises generally allowed both the astronauts and mission controllers to iron out the final planning wrinkles. Even so, the training was insufficient to master Skylab's many functions and by January 1973 Conrad's crew was routinely clocking up 60-hour working weeks in the simulator alone. The workload was so intense and the length of time away from home was so much that Kerwin's daughter, Sharon, later joked: "This would explain why none of your children recognised you after the flight!"

Training, naturally, developed alongside the demands of the experiments. Certainly, biomedical studies had long been one of the primary justifications for a long-duration workshop in orbit and key focuses included the measurement of fundamental changes to the human body in weightlessness and the need to derive a better understanding of the kind of methods needed to aid astronauts' adaptation to the strange environment. Since the earliest days of AAP, metabolic studies, assessment of cardiovascular function and changes in bone and muscle structure were the major areas of study and had led to the creation of a heavily instrumented bicycle ergometer (complete with exhaled gas analyser) and a lower-body negative pressure device to monitor blood pressures, heart rates and leg volumes. Effective medical monitoring would also require the crew to collect, measure and store samples of their own urine and faeces. In the early days, following the approval of the dry workshop, the medical community was undecided about whether the urine samples should be stored in a purpose-built freezer or dried. After a series of studies by contractor Fairchild Hillier, reinforced by parallel work at the Manned Spacecraft Center, it was decided to freeze samples. Stan McIntyre, project engineer for the urine-collection system at the Marshall Space Flight Center, recalled that collecting, sampling and handling fluids in weightlessness had always been a complex 'grey area', but in their account of the programme Compton and Benson wrote that "rather than tackle the job they elected to avoid it and their contractor's scientific advisor assured them that drying would satisfy the medical objectives". Chuck Berry, head of life sciences in Houston, was outraged that the engineers of the urine system were arguing with his *medical* experts on what was essentially a *medical question*. Conversely, even after its approval, many Skylab engineers continued to view the freezer with scepticism and reckoned that a drying process would have been more effective. Elsewhere, the development of a faecal collection device was relatively straightforward: the stools would be collected in plastic bags, weighed and vacuum-dried for return to Earth and analysis. The collector could not be conclusively tested under low-gravity conditions, but a pair of parabolic aircraft flights in November 1969 yielded several good data points. A plan to test it on Apollo 14 was dropped; apparently, the astronaut office was indifferent to the test and the flight's commander, Al Shepard, was vigorously opposed to it.

By the time Skylab reached orbit, it would house no fewer than 19 life science experiments, supporting 28 studies into general habitability, the physiological effects of weightlessness on the crew, fundamental biological processes and biotechnology. These included measuring mineral balance, through daily monitoring of body mass, food and water intake and urine output; exploring the duration of cardiovascular adaptation through the use of the cylindrical lower-body negative pressure device; investigating changes in the function of the heart during long-term spaceflights; comparing blood specimens before, during and after the mission; recording changes in cellular immunity through concentrations of plasma and blood-cell proteins; measuring vestibular function and attempting to chart the susceptibility to 'space sickness' using a rotating chair; analysing sleep patterns with instrumented skull caps; evaluating general adaptation processes; determining the effectiveness of the metabolic system, through the use of the bicycle ergometer; and monitoring the circadian rhythms of pocket mice and vinegar gnats.

For the pilots, much of this work created a considerable amount of head-scratching and Bill Pogue would later remark that he never quite understood what some of the investigations were trying to achieve. "Some of these medical experiments," wrote Deke Slayton, "as they developed over the years, had been pretty wild, if not downright humorous. For example, at one point, the doctors thought it would be great if they could do a biopsy from a Skylab crew member during the mission ... a bit trickier than giving yourself an injection! We were even training some guys to do emergency dental extractions." To ensure that they were as prepared as possible for minor medical emergencies, the astronauts even spent weekends at a Houston A&E department, conducting simple sewing-up procedures and minor operations, and even went out 'on call' with the emergency services. The relationship between the pilot astronauts and the medical community had always been a shaky one – there were only two ways in which they could walk out of a doctor's office: either *fine* or *grounded* – and one particular incident must have appealed to their collective sense of humour. A DIY biopsy kit was developed for Skylab, to which a sceptical Slayton insisted that the *designer* should be its first test subject. "Sure enough, the guy just yowled with pain as this thing tried to rip a hunk of his thigh," recalled Slayton. "Back to the drawing board ... "

Of course, life sciences studies formed only part of the scientific schedule for Skylab and the presence of the ATM pointed to a clear emphasis on solar physics. Its fundamental advantage over earlier (unmanned) Orbiting Solar Observatories was its size and its capacity to store a significant amount of data on photographic film, since the presence of a human crew permitted the retrieval and replacement of cassettes during EVAs. These cassettes were loaded aboard Skylab before launch and extra ones were also aboard the command module for the third mission; in total, the ATM would yield more than 150,000 solar exposures and a vast body of fundamental data. Several discrete instruments formed the ATM, including an X-ray spectrographic telescope to determine temperatures and densities within the 'corona', an ultraviolet scanning polychromator spectroheliometer to record temporal changes in ultraviolet emissions from several solar regions, an X-ray event analyser to obtain data in that part of the spectrum and an extreme-ultraviolet spectrograph and spectroheliograph to photograph the Sun sequentially over long periods in selected wavelengths. Other equipment was employed to 'boresight' these instruments. The control station for the ATM, which included switches, indicators, displays and controls to direct power to the instruments, open and shut their aperture doors, adjust speeds and record data, was housed inside the multiple docking adaptor. A five-channel television screen provided the astronauts with 'safe' views of the solar disk at different wavelengths, not only making them the first people to use live video of the Sun, but also offering them the most comprehensive glimpse of the star which gives us life.

The significance of these instruments – part of the largest manned observatory ever placed into orbit – was highlighted by the fact that more than 300 solar scientists, several satellites and many observatories on the ground supported it in some manner. Daily meetings would be held with a senior representative for each instrument to review the observations and plan the next session. As a consequence, far from the essentially automated role envisaged for the original concept of the ATM in a bay of the service module, the Skylab ATM would require the crew member to take control and perhaps adjust the observation plan to respond to

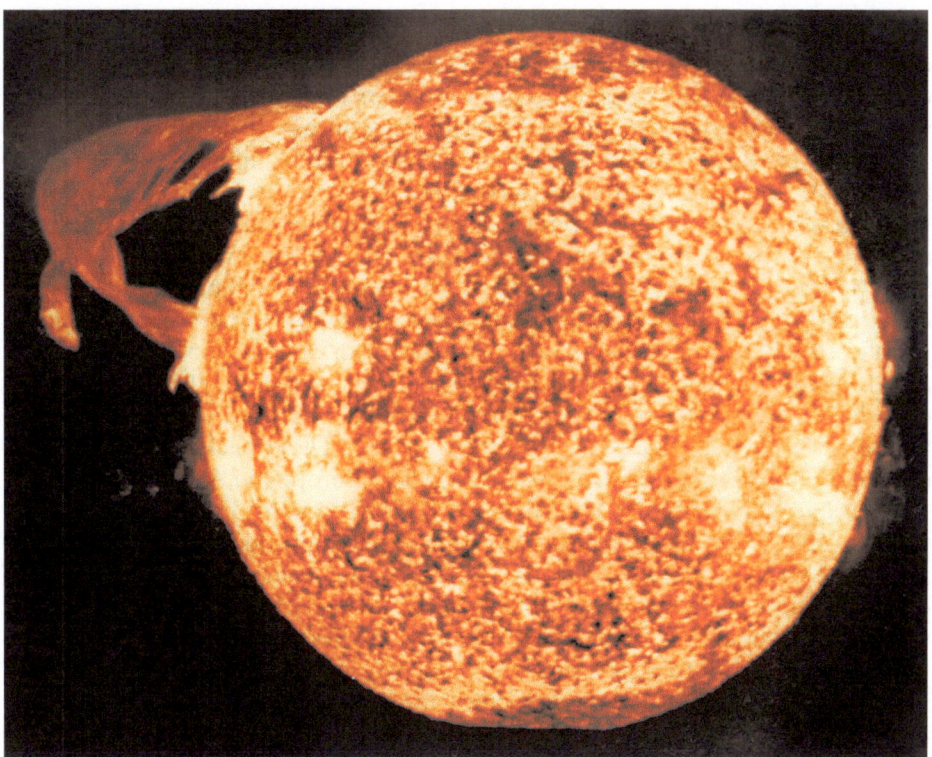

Acquired by the crew of the final Skylab mission, using the ATM's extreme-ultraviolet imaging equipment, this impressive flare extended more than *half a million kilometres* out from the Sun's surface. The solar observations conducted by the three teams of astronauts truly revolutionised our understanding of the sheer majesty and terrifying power of our parent star and Ed Gibson's textbook, *The Quiet Sun*, is still used today.

serendipitous opportunities. All three crew members, and particularly the second and third Skylab teams, were highly trained in its operation. Yet, like the medical experiments, some scientists questioned the competence of the pilots in being able to meet experiment objectives or properly understand what they were trying to do. Gerry Carr once commented that his solar physics training flowed more smoothly when he gave up trying to become a solar physicist and instead looked for ways to become a more competent observer. "The desire to deliver the best performance," wrote Dave Shayler, "could sometimes pressure the crews into trying to learn more than they could handle." In time, the standard of these ATM observations would literally rewrite the textbooks on solar astronomy.

Biomedical research and solar physics were two strands of Skylab's three-pronged remit of scientific enquiry; the third was an in-depth programme of monitoring the Home Planet, using the Earth Resources Experiment Package (EREP). On the first mission, this battery of instrumentation was one of the main responsibilities of Paul Weitz and he found himself frequently travelling to Martin Marietta's facility in

Littleton, Colorado, to follow the development and testing of the cameras. EREP contained six remote-sensing instruments: a multi-spectral photographic facility, comprising a specialised 'terrain camera' with filters to observe at wavelengths from near-infrared to visible; an infrared spectrometer to assess the applications and 'usefulness' of remote-sensing from an orbital platform; a multi-spectral scanner to identify the signatures and map agriculture, forestry, geology, hydrology and oceanography across various spectral bands; a microwave radiometer, spectrometer and altimeter for near-simultaneous measurements of land masses and bodies of water and finally a radiometer to measure the 'microwave brightness' of the surface along the ground track. Supplementing these studies from the ground would be remote-sensing aircraft, co-ordinated, of course, with the 'real-time' flight plan, to correlate with the Skylab data and the 'repetitive coverage' of the station's orbit every five days or so would enable researchers to gather data over a period of months and so track seasonal changes and weather systems. The orbital inclination would cover the United States, Africa, Australia, southern Europe and Asia, much of South America and large swathes of the Atlantic and Pacific Oceans. The astronauts would typically operate the EREP instruments from the multiple docking adaptor.

Astrophysical studies would focus on Earth's outer atmosphere and a range of celestial sources, including the interstellar medium. A nuclear emulsion experiment would record cosmic-ray fluxes and the relative abundances of high-energy primary heavy nuclei; an ultraviolet stellar astronomy instrument – whose principal investigator was Skylab Program Scientist (and astronaut) Karl Henize – would photograph large regions of the Milky Way, devoting particular attention to the breeding grounds of young, hot stars; an ultraviolet air-glow horizon experiment would monitor thermal absorption of Earth's atmosphere; a gegenschein and zodiacal light experiment was to measure the brightness and polarisation of the visible-light 'sky background' from above the sensible atmosphere; a galactic mapping instrument was to survey faint X-ray sources; an ultraviolet panorama instrument would measure the brightness of a large number of stars; an experiment would measure cosmic radiation; and there was a cometary physics investigation, which, as events transpired, would be used during the course of the third mission to make unique observations of Comet Kohoutek.

Elsewhere in the multiple docking adaptor, a materials processing facility was to conduct experiments in the melting of metals, exothermic brazing and flammability trials, whilst a multi-purpose electric furnace focused on re-solidification and crystal growth.

Little training was done on any of these experiments until 1971, with the science pilots initially working with principal investigators and vendors to ensure that the hardware was compatible with the flight equipment and that operating procedures were workable. However, after the formal announcement of the three crews in January of the following year, specialised training got underway, as did simulations of specific parts of a mission, including spacewalks to retrieve and replace ATM film cassettes or recover material samples from the exterior of Skylab. By November, six months before the launch of the workshop and Conrad's crew, the astronauts were routinely running 'integrated sims' – simulations which involved not only themselves, but also the entire flight control team for the mission. Other sims involved the control teams alone. One such exercise, for example, ran for three days in mid-January 1973 and picked up the flight plan on the tenth day of Skylab's mission, as the station made its way across the Atlantic Ocean, heading for the communications relay on Ascension Island. It assumed that Conrad, Kerwin and Weitz were still 'asleep' and enabled flight controllers to simulate a change of shift and prepare a summary plan for the next day's activities.

As Skylab's launch drew nearer, Joe Kerwin, at least, felt that the entire group were better trained than any previous crew, in view of the complexity of the scientific payload and the length of preparation. In fact, selected in 1965, he had been detailed to AAP, and later Skylab, almost since the start of his astronaut career. If anyone deserved a spot on the first mission, it was Commander Joseph Peter Kerwin of the US Navy. Born in Oak Park, Illinois, a suburb of Chicago, on 19 February 1932, he entered the world with the weight of the Great Depression on his tender shoulders. "I'm the seventh child in the family," he told NASA's oral historian, "[and] was accused by my big brothers and sisters ... of having either *caused* or *perpetuated* the Depression by my birth!"

As a child, 'little Joey' read science fiction and was taunted by his siblings over his fascination with the writings of Robert Heinlein and A.E. von Vogt. Graduation from the private Fenwick High School in 1949 was followed by a bachelor's degree in philosophy, with a minor in 'pre-med' to combine his interests in literature and science, from the College of the Holy Cross in Worcester, Massachusetts. On one occasion, he wrote a sophomore English paper about the possibility of life on other planets, which earned him a B-minus grade. The marker's comment in red ink on the front page advised him that, whilst his grammar and phrasing were fine, the subject matter was a bunch of garbage! "I still have that paper," Kerwin recalled years later, "and I chuckle about it every now and again!" He was, however, drawn to the pre-med work and after receiving his degree in 1953, he went directly to Northwestern University Medical School in Chicago, graduating four years later and serving as an intern at the District of Columbia General Hospital in Washington, DC. The Soviets' launching of Sputnik had little impact on him at the time, but he was intrigued by the possibility of the so-called 'doctors draft' to coax health professionals into military service for a period of active and later reserve duty. "They gave you exemptions from the draft," Kerwin recalled, "as long as you were enrolled in a legitimate medical educational programme, including internship and including residency and as soon as you finished it, your name went in the hopper. If it got pulled out, you were up for two years' service." With a residency in paediatrics, he received a military deferment, but when he realised that it was not for him and pulled out, he received "a letter from Uncle Sam" in January 1958, calling him up for the Navy.

His first assignment was to the very last place in flight surgeon school, from which he qualified in December of that same year, receiving about 20 hours' worth of flying instruction. Kerwin then moved to Marine Corps Air Station Cherry Point in North Carolina as a medical officer and enjoyed his tour "immensely", partly because the Marine pilots "allowed me to start and taxi their fighter aircraft around". It was a little unconventional, he admitted, "but the Marines were like that" ... they would teach him how to taxi, as long as he *didn't* take off!

Kerwin's experience at Cherry Point whetted his appetite for naval aviation and in 1961 he was accepted for flight training, transferred from the reserves into the regular service and the following year received his wings at Naval Air Station Chase Field in Beeville, Texas. Consequently, he became one of only a dozen Navy flight surgeons who could combine medical credentials with jet-pilot qualifications. Whilst attached to the air wing at Cecil Field in Florida, he encountered "a couple of friends among my pilot patients" – Jim Lovell and Al Bean – both of whom asked him to assist in filling out their medical forms before submitting their applications to NASA. Both were subsequently selected as astronauts and sometime in 1964 Kerwin recalled sitting at home, in front of the television with his wife, Lee, when he heard an announcement that the space agency was hiring a team of scientists. Ironically, he had been intending to resign from the Navy after his next tour to take a residency in ophthalmology at Northwestern University, but the idea of space travel intrigued him. He promptly applied, through the Navy, and "since there weren't many physicians who could pass the physical *and* had 2,000 hours of time in single-engine jet aircraft, that turned out to be enough, and I was accepted into the astronaut programme in 1965".

Although each of the scientists was overjoyed to have been picked, the reality was that spaceflight remained a risky (and test-pilot-dominated) endeavour and all would have to be qualified in jets before they could even begin training. Fortunately for Kerwin, his naval training and experience as a flight surgeon already allowed him to tick this box, but of the other five scientists, four had either never flown before or had received an insufficient number of hours and were sent off to the Air Force for a year of intensive instruction. In the meantime, Kerwin and Michel (who had accrued 900 hours in Air Force jets during his own military service) were unable to start their own astronaut training and would be merged with a new group of pilots, scheduled for selection in 1966. "They didn't start us into formal training," Kerwin recalled, "because two people don't make a class. They decided to wait for the next class to come along". Duane Graveline left NASA almost immediately, due to marital problems, and the other scientists found themselves embraced into the military ranks of the astronaut corps with varying degrees of acceptance. Years later, Kerwin did not recall whether, as the only active-duty military officer amongst them, he was treated any differently by the hard-nosed test pilots in the office; in fact, "we merged and subordinated our scientific interests to those of the programme". In some cases, this proved difficult. Kerwin recalled approaching Al Shepard at one stage to ask whether he should keep his hand in medicine by going to the clinic for a couple of days each week. Shepard frowned on the idea; although it was important to keep abreast of medical research, he told Kerwin, he was an astronaut, first and foremost.

Initial technical work on spacecraft environmental controls and later space suits was quickly overtaken by assignment in December 1966 to the AAP's experiments branch. Of course, the pivotal change came in the summer of 1969, only days after Armstrong and Aldrin walked on the Moon, when the decision to formally change from a wet to a dry workshop was made by Administrator Tom Paine. To this day,

Kerwin believes that the change was the right decision. With the wet workshop, "I think there would have been a great paucity of in-flight medical experiments and in-flight solar physics experiments that would have impoverished the return from the programme a great deal." Summing up, he felt that a wet workshop probably *would* have accomplished the basic goal of sustaining men in space for three months, but precious little more. Now, almost four years after the dry workshop decision, early in May 1973, Joe Kerwin stood at the Kennedy Space Center, backdropped by the impressive sight of the last Saturn V – its payload covered by a one-off, bullet-like shroud – on Pad 39A, ready to blast America's first space station into the heavens.

Alongside Kerwin and Pete Conrad in Florida that month was Commander Paul Joseph Weitz of the US Navy, nicknamed 'PJ'. Born in Erie, Pennsylvania, on 25 July 1932, Weitz was not unlike many pilot-astronauts from the Apollo era in that he had a natural affinity for aviation; and with a chief petty officer father who fought at the battles of Midway and the Coral Sea, he described himself as "an impressionable young lad during World War II" and the Navy drew him like a magnet. Graduation from high school led to a Reserve Officers Training Corps scholarship from the Navy to study aeronautical engineering at Pennsylvania State University. Whilst there, his instructor advised him to go to sea aboard a destroyer before entering flight training. Weitz would later regret this, "because it put me a year and a half behind my contemporaries". When he finally entered flight school in Jacksonville, Florida, he found himself a classmate of Al Bean, who would subsequently also become an astronaut and command Skylab's second crew.

Weitz' first application to attend the coveted Naval Test Pilot School at Patuxent River, Maryland, was rejected and he moved instead to an air development squadron at China Lake in California's Mojave Desert. His second attempt to enter the coveted school at Patuxent River, Maryland, was accepted – and then *rejected*, because "I had just been moved from the East Coast to the West Coast and they weren't going to move me back to the East Coast again!" Nonetheless, Weitz enjoyed his time at China Lake, serving as a project pilot for five different types of aircraft and helping to develop tactics for delivering air-to-ground weapons. With a degree under his belt, his next (unsolicited) set of orders was assignment to the Naval Postgraduate School in Monterey, California, to study for a master's credential in aeronautical engineering. "I did not apply for it," he told NASA's oral historian in March 2000. "I didn't want to go, because I had what I thought was a good job." At the time, Weitz had no desire to return to full-time education and believed it to be the wrong choice; however, years later, he would admit that "since I did not have a test pilot background directly ... the reason I got selected finally for the astronaut programme was because I did have a master's degree". By the time of his graduation from Monterey in 1964, he had met three other astronauts-to-be: Gene Cernan, Ron Evans and Jack Lousma.

In fact, by completing 'extra work', thanks to the input of a kindly professor, both Weitz and Lousma were able to complete their master's degrees ahead of schedule, in two years, rather than the standard three. Weitz then moved to Whidbey Island in Washington State's Puget Sound to fly the A-3 Skywarrior strategic bomber, undertaking his first combat mission in Vietnam in 1965. It was during this period that "another one of those strange forks in the road" came up and he received a message from the Bureau of Naval Personnel, informing him that NASA was looking for astronaut candidates. Lacking test pilot credentials, Weitz assumed that his chances of selection were slim and he was surprised to make the cut as one of 19 new astronauts in April 1966. His association with AAP began early – in his oral history, he recalled spacewalk training with Jack Lousma to practice taking plugs down and putting drains into the S-IVB hydrogen tank to make it habitable as a wet workshop – and certainly his performance on the support crew for Pete Conrad's Apollo 12 mission in 1969 caught the attention of his superiors and cannot have harmed his chances of receiving the pilot's slot on the first station flight.

On the evening of 13 May 1973, with the launch of Skylab scheduled for 1:30 pm Eastern Standard Time (and 12:30 pm Central Standard Time in Houston) the following afternoon and their own launch set for just after midday on the 15th, Conrad, Kerwin and Weitz were filled with excited anticipation for a mission to which they had been devoted, both professionally and personally, for several years. If everything went according to plan, nine and a half minutes after lifting off, the Saturn V would release Skylab into orbit. The conical payload shroud would be jettisoned a few minutes later, splitting into four sections and drifting away, and then the deployment of the ATM and its windmill of solar panels, the workshop's own pair of solar arrays and finally the micrometeoroid shield would follow in quick succession. Ninety-six minutes after launch, Skylab would be ready for the arrival of Conrad's crew late the following evening. They would dock with the station about seven and a half hours after their own launch, pressurise the connecting tunnel and enter Skylab early on the morning of 16 May to begin their four-week residency. One spacewalk was planned for two and a half hours on 10 June, during which Conrad and Kerwin would retrieve film cassettes from the ATM. A textbook return to Earth and a splashdown in the Pacific Ocean, a few hundred kilometres south-west of San Diego, would follow on 12 June.

Little could they possibly have realised that an unexpected chain of events would unfold only minutes after Skylab's launch, changing their own mission markedly. In time, their performance would win Conrad the Congressional Space Medal of Honor and offer Kerwin, and Weitz, too, the chance to venture outside on their first career spacewalk. More than that, however, America's first mission to a space station would physically and mentally tax all three of them, and NASA, too, to the very limit.

A SPACE STATION ON EARTH

As the launch of Skylab drew closer, it was increasingly obvious to mission planners that a great number of unknowns existed about how the human body would respond to the weightless environment for periods of weeks or months. The longest American mission thus far was the 14-day Gemini VII in December 1965, during which Frank Borman and Jim Lovell underwent a particularly unenviable experience which could only be described as a 'slog'. Living and working in a volume equivalent to the front seat of a Volkswagen Beetle for a fortnight caused even the novelty of flying in space to lose its lustre after a time. The first Skylab crew were to spend four weeks in orbit and then the plan was for the second and third teams to attempt eight weeks apiece. To prepare physically and psychologically for such expeditions and to understand the impact on the crew members who would be embarking on space 'marathons', rather than 'sprints', a Skylab medical briefing in August 1970 identified the need for a ground-based simulation of a full-length mission. Fundamentally, physicians were concerned about changes to the microbial population when three men were confined in close quarters; a 'flare-up' of bacterial infection had to be avoided at all costs.

Dale Myers and Bob Gilruth outlined a plan for such an exercise and sketched out a facility which might also support advanced microbiological research after Skylab. Within weeks, a test plan had been formulated, which addressed flight experiment procedures, a functional evaluation of the Skylab equipment, obtaining 'baseline' medical data and the verification of the analysis, interpretation and reporting plan between the four-week and eight-week missions. Original hopes to use two full-size Skylab mockups proved too expensive for NASA Headquarters to authorise and it was decided to run the test in the Crew Systems Division's altitude chamber, located in Building 7 at the Manned Spacecraft Center. Early plans called for a 28-day test in September 1971 and initiating a 56-day run in November. Each was to feature the operation of Skylab-type experiments, primarily medical in nature, including lower-body negative pressure, blood volume and body mass measurements, vestibular function, metabolic activity and mineral balance. Two teams of volunteers (they did not necessarily have to be astronauts) were to participate in the tests, which were designed to mimic conditions aboard Skylab as closely as possible. Following the extraction of the first 'crew' from the chamber, a 30-day debriefing was planned and its results would help to prepare the second crew. These mock 'missions' would be known as the Skylab Medical Experiments Altitude Test (SMEAT).

As for the chamber itself, it had already been extensively man-rated. Seven metres in diameter and seven high, it was a two-floored stainless steel structure with more than a dozen viewing ports, several penetration bulkheads and a pair of airlocks. It contained closed-circuit television equipment and a closed-loop system to heat or cool its oxygen-nitrogen atmosphere, together with channels for communications, lighting, automatic and backup oxygen supplies and fire suppression systems. For the purposes of these tests, only one of the floors would be used, although the other was available for relaxation. The wardroom and waste management facilities were partitioned, as they would be on Skylab itself, but the remainder of the chamber was modified to accomplish the medical objectives of the workshop. Simulated lighting, caution-and-warning devices and communications (through a capcom and, matching the limitations of tracking, for only 20 percent of the time) would further enhance the sense of realism.

In December 1970, the 28-day test was dropped and it was decided to perform a single, 56-day evaluation. However, the option was retained to run an additional test after the final Skylab mission. It was also decided that astronauts should perform the test and in June 1971 pilots Bob Crippen and Karol 'Bo' Bobko and physician Bill Thornton, were selected for a scheduled start date in July of the following year. In

NASA's formal report on SMEAT, published in October 1973, it was pointed out that the "use of astronauts was considered desirable, since this would [ensure] a general comparability of background, skills and motivation ... [with] subsequent Skylab crews". In a NASA oral history interview in February 2002, Bobko recalled that the 'selection' actually came from a drawing of straws, whilst Crippen remarked that although he had learned never to volunteer, SMEAT sounded "like the best job available" to an astronaut who was years away from any real flight. For his part, Thornton had already worked on Skylab issues – he was one of the principal investigators for its small-mass measurement device, to be used for weighing specimens in flight – and his two comrades had previously worked on the Air Force's Manned Orbiting Laboratory project before joining NASA in September 1969. Consequently, all three had experience in the station design process. During the year before the test, Crippen, Thornton and Bobko participated in the design and layout of the SMEAT chamber and began practicing with the medical equipment in the spring of 1972. In a very real sense, therefore, their training regime matched very closely the actual training of a Skylab crew, with Crippen, Thornton and Bobko receiving equipment and maintenance briefings, bench checks for storage, crew compartment fit and function tests and summaries of emergency procedures. A series of field trips to Air Force medical installations and regional hospitals offered them a thorough understanding of the diseases of the eyes, head, cardiovascular, pulmonary and musculoskeletal systems and even gave them the chance to practice treatments. During the course of that year, each man spent in excess of 500 hours training for SMEAT, which was more than 100 more than scheduled.

Their preparation for 'medical emergencies' was virtually identical to that of the Skylab crewmen and included, notably, dental work. "The plan was ... to make sure that the crews could deal with ... minor medical emergencies," recalled Crippen in a May 2006 NASA oral history, "and part of that was to send us off to dentistry school. We ended up in San Antonio at the Air Force hospital there. In fact, Bo Bobko and I did extract several teeth. I recall this one young man came in ... and his teeth were in terrible shape. He had a couple that needed to be extracted and we asked the dentist, 'Well, are you going to come in?' He said, 'No, you guys can do it.' The kid was probably about late teens or 20 years old and he was a little bit nervous, but I did the novocaine with the needle and all that stuff, and Bo flipped the tooth out, because it wasn't in very firmly. I remember the kid [said] 'You guys are the best dentists I've ever been to.' We didn't let him know the real truth!" Other practice was less satisfying, including CPR on a 'Resuscitation Annie' dummy. When Crippen's turn arrived, he duly thumped the palm of his hand down hard on the dummy's chest ... so hard, in fact, that he broke his fifth metacarpal! "Everybody concluded they didn't want to have a heart attack with *me* around," he mused later.

On the morning that the SMEAT test began, 26 July 1972, Crippen, Thornton and Bobko underwent a medical examination and were required to undergo a period of pre-breathing to purge nitrogen from their bloodstreams. (Crippen later recalled that these preparations were actually *worse* than the test itself.) Inside the chamber, one of their earliest observations was that the reduced-pressure atmosphere caused sound to appear further away and somewhat softer and by shouting to compensate

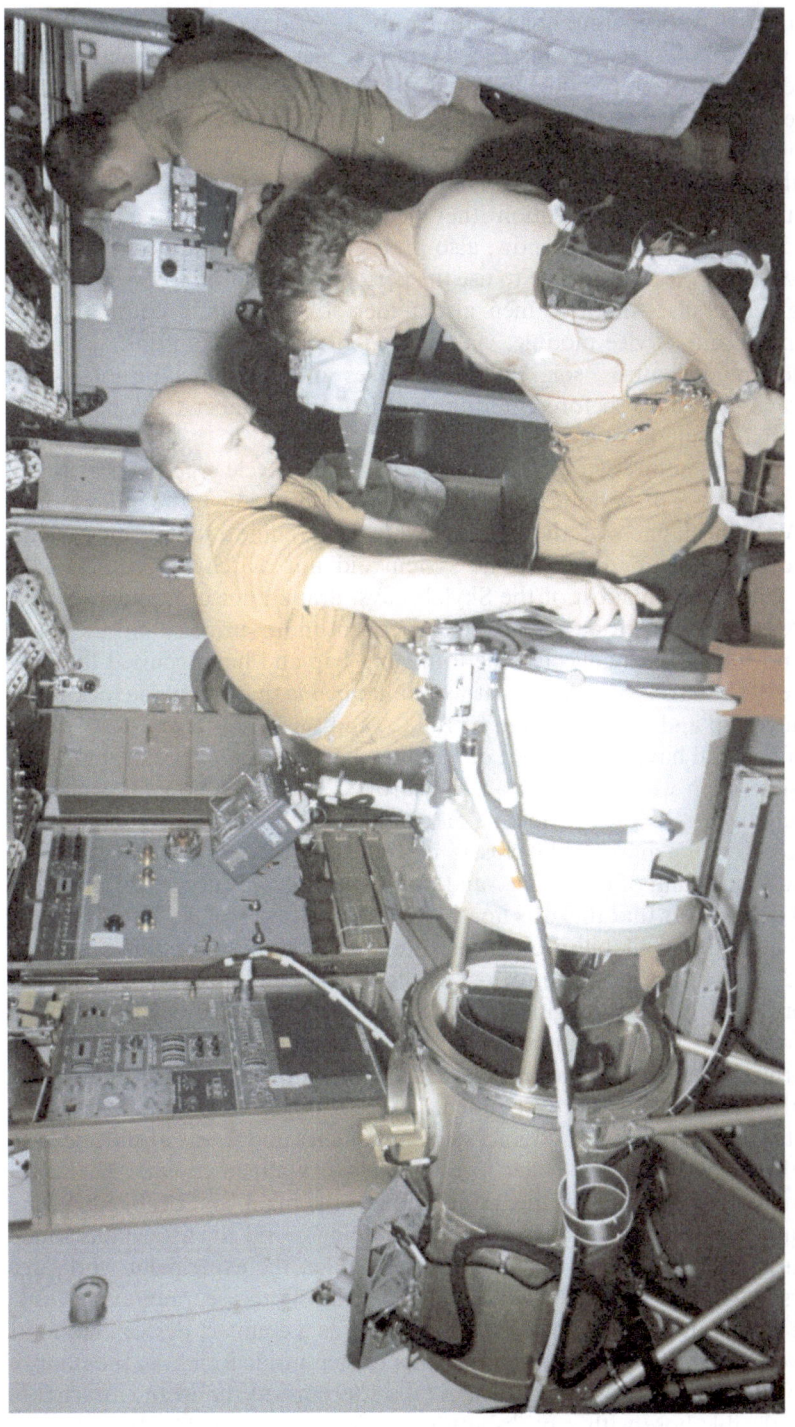

Six weeks before the start of their 56-day mission, the SMEAT team train for their many tasks. Here, Bill Thornton prepares Karol 'Bo' Bobko for a session in the lower-body negative pressure device, which would be used by the Skylab crews to measure blood pressures, heart rates and leg volumes as part of ongoing investigations into the physiological effects of long-term exposure to weightlessness. In the background is Bob Crippen. Note that the SMEAT chamber has been configured to almost exactly replicate Skylab.

they found themselves becoming hoarse after only a few days. Other results of the low pressure were an inability to whistle, a temporary increase in abdominal gas and severe flatulence and a noticeable mildness of sneezing. Typical daily routines involved reveille at seven in the morning, work from nine through to one in the afternoon, followed by a second work period from two until seven and then dinner. A review of their activities and 30 minutes of housekeeping capped off each day and after two hours of 'personal recreation' they bedded down at eleven each evening. Crippen and Bobko, who were, by now, also preparing for their support role in the Apollo-Soyuz Test Project, were able to use the period of recreation to brush up their Russian language skills. All three men also undertook a training course for the Apollo command and service module, via a closed-circuit television link, and even completed electronics courses, solar physics courses and commercial pilots' study courses. It was their choice before entering the SMEAT chamber that they would thus fill their spare time in a 'useful' way ...

Not all went to plan, however, and problems with the bicycle ergometer caused it to quickly break down, prompting the crew to pass it out through the main airlock for repair. With the launch of the 'real' Skylab barely nine months away, such problems had to be rectified. The urine system did not match its requirements. "Most of us, and probably most of the Skylab guys, drank a lot of liquids," Bobko told the NASA oral historian, "and so produced more urine than they had anticipated and ... [so they] collected the urine in a bag and homogenised that and took a sample off and froze it. Well, what would happen is that at two o'clock in the morning, you get up and the bag would be full ... and so, here you are, just kind of groggy out of bed and have to change it out and the bags were *not* very strong. There was an occasion when we dropped one and it broke. You know how messy it is to drop a half-gallon of milk on the floor. You can imagine!" Humour aside, with the Skylab launch drawing inexorably closer, cleaning up the urine spills took the SMEAT crew a minimum of one hour – and *this*, of course, was in *normal terrestrial gravity* – and it was clear that this was unacceptable, not least because hand-washing facilities and even the availability of disinfectants aboard the workshop would be insufficient. A watchful Pete Conrad lost all confidence in the urine system and began to work with Houston engineers to adapt the device already due to fly on Apollo 17. Conrad told senior managers that he was quite prepared to do away with the Skylab urine system entirely and replace it with a simpler, condom-like Apollo device. Shortly after the chamber test ended, late in September 1972, an agreement was reached to expand the urine system's storage capacity and it was also decided to incorporate the Apollo device in order to offer the astronauts a choice.

Life aboard SMEAT was by no means luxurious – the main comfort came from a set of sun-loungers, which were assembled each evening – and the astronauts found themselves surrendering much time to cleaning the toilet, the experiment work area and the floors. A simulated Skylab shower was "delightful" to use, once every week, with all three men showering on the same day. This was a compact device and, in its space-going version, would require the astronauts to step inside a ring on the floor of the workshop and draw up a fireproof curtain on a 1.09 cm-wide hoop and attach it to the ceiling. A flexible hose with a push-button nozzle could spray 2.8 litres of water during each shower and all waste water would be vacuumed from the enclosure and transferred to disposable bags and from thence into Skylab's waste tank. Meals were typical Skylab fayre: sealed in small cans which were inserted into food trays with built-in heating elements. Body intakes and wastes and blood samples were logged methodically for the medical experiments. The collection of similar data, for several weeks before, during and after the test, was remarked upon by Crippen, Thornton and Bobko in their report. "It was our experience," they wrote, "that collecting samples was a rather alien and unpleasant task. It should be made as easy and unconfusing as possible. This was especially felt in the pre-test period, when we were expected to perform our normal activities, including flying and a large number of other tasks in preparation for the test. It seemed that during the pre-test period, one had to think of every action, however basic and simple, because information related to these actions were required for the test." They recommended that a hand-held 'carrying case' to keep liquids and specimens chilled should be made available to Skylab crews for sampling when away from their normal work areas.

Yet, psychologically, they managed, by a strategy of 'gotchas', to maintain good relationships with each other and with the support staff. One dastardly plan, which they were unable to execute, due to time limitations, was recorded by Dave Shayler. "The crew had found that the Skylab clothing shed a considerable amount of lint," he wrote, "and the vacuum [unit] could not cope with the influx of material. The crew therefore redesigned the pickup brush to cope with the problem. The collected balls of lint also gave rise to an idea for a practical joke that the crew wanted to play on the test conductors outside. The serious manner of the controlled experiments and the medical protocol of Skylab ... was a little tiresome on a daily routine and the crew thought that it would be a break in this strict regime if they gave the impression that they were not alone in the chamber! They were in contact with their families once a week, by way of private telephone conversations, and they planned to ask one of the members of the family to smuggle in some fast-food, meat-free chicken bones to a friend in the chamber, who would pass them through the airlock without discovery. The crew were then to place, on a used meal tray, the chicken bones and a couple of hairballs from the suits, with a note saying how much the 'cat' enjoyed the fresh meat!" Had this practical joke transpired, the look of horror on the test conductors' faces at having weeks of medical protocol seemingly ruined would certainly have been a sight to behold ...

Departing the chamber on 20 September, with Crippen and Bobko both sporting full beards, the men expressed glowing praise of the usefulness of the facility ... although the return to the outside world brought its own challenges. Crippen's family boxer dog had given birth to a baker's dozen of puppies whilst he had been away. For his part, Thornton had missed the opportunity to cook his trademark gourmet meals and, although he looked forward to a meal of his *own* choice, he admitted that it would probably be *gluttonous*, rather than *gourmet*. Beneath the veneer of humour, though, Thornton had uncovered a genuine problem with the strict requirement to measure everything that the men ate, drank, urinated and defecated. It was a problem shared by their counterparts training for the real missions.

"We just knew this one was going to make our life miserable," Joe Kerwin recalled. "Don Whedon [the National Institutes of Health physician in charge of the Skylab mineral balance study] ... had done intake and output experiments in medical labs and he knew how to do it, but the thing he wanted to do was to put us all on the same standard diet, pre-flight, in-flight and post-flight: *Everybody* gets 2,400 calories, *everybody* gets 800 mg of calcium, right on down the line, protein, carbohydrate, everything, exactly measured. We said 'This isn't going to work'. We had Jack Lousma, who was a great big 6'1" Marine, who routinely eats 3,200 or 3,500 calories a day and is trim and fit ... and we had Al Bean on the other extreme, who's a 1,900-calorie-a-day person and *he's* trim and fit. You're not going to feed them *both* 2,400 calories a day for what amounts to three months and get away with it. We were rescued by the SMEAT crew. Bill Thornton was built like and exercised a lot like Jack Lousma and he was in there on 2,400 calories a day. He was losing weight like mad and was very seriously irritated about all this. He said 'The only thing that kept me from killing [Whedon] was the two inches of steel between us!'" As a consequence, the astronauts got Deke Slayton and Bill Schneider on their side and convinced Whedon to feed them what they needed to operate and in return they would stick to the monitoring regime and carefully measure their intakes and outputs. Kerwin continued: "We just wanted enough to eat, so we changed the routine and we went through a one-week period of eating all our meals at the laboratory ... being fed carefully-measured portions. They got to know what everybody's dietary intake was and tailored our meals to that dietary intake."

When the formal NASA report on SMEAT was published just over a year after the test was run, it included as one of its key recommendations that the "selection of diet must be more carefully tailored to individual requirements and preferences of crew members than was thought to be the case". In his section on the mineral balance investigation, Whedon took note of Thornton's penchant for "particularly vigorous" exercise and added that "energy requirements for the SMEAT crew were estimated on the basis of age and body weight" and therefore "menus meeting these energy requirements ... were formulated". In addition to the question of dietary intake, the opportunity for 'team training', including the use of mock 'flight controllers', was also judged by the report to be vitally important by the report. Indeed, Crippen, Thornton and Bobko even produced their own humorous patch: with the dog Snoopy as its centrepiece, restrained by a particularly tight leash to signify their medical torment. (Their original idea, rejected by *Peanuts* creator Charles Schultz, actually had the poor pooch ensnared by a fish-hook ...)

Discomfort and unpleasantness aside, the provision of 'baseline' medical data, together with handling techniques and ironing-out hardware problems, had made SMEAT a more than worthwhile exercise. Almost four decades have now passed since Crippen, Thornton and Bobko, clad in masks, entered the chamber to rehearse a space station mission ... but more than one Skylab astronaut acknowledged that by enduring two uncomfortable months inside SMEAT, their colleagues helped to make their own stays in orbit more bearable.

A REWRITTEN SCRIPT

The launch of any new spacecraft cannot be regarded as 'routine'; nor, indeed, can its inaugural checkout in orbit. Skylab was an entirely new concept for the United States and a totally different spacecraft, larger, more spacious and in many ways far more complex, than any that had gone before. Yet on the morning of 14 May 1973, a sense of optimism pervaded the Kennedy Space Center, as the last in a generation of Saturn V boosters was readied for its journey into space. Visually, it looked somewhat different to its lunar predecessors, for, instead of possessing three stages, it had only *two*, and in place of what would have been the final propulsive stage was the inert Skylab space station, capped-off by a conical, bullet-like aerodynamic shroud.

The mere mention of the name 'Saturn V' implies power. From a height, weight and payload-to-orbit standpoint, it remains the largest and most powerful rocket ever brought to operational status. It evolved from a series of vehicles, originally dubbed Saturn 'C-1' through 'C-5', with NASA announcing in January 1962 its intention to build the latter as a booster capable of supporting lunar missions. Soon afterwards, it was renamed the 'Saturn V'. For each of its Apollo missions in support of the lunar programme, it was a three-stage vehicle, with five F-1 engines on its first stage, five J-2 engines on its second stage and a single J-2 on its third stage. Both types of engines, built by Rocketdyne, had shattered the windows of nearby houses when tested. The Saturn V stood approximately 110 m tall – only a few centimetres shorter than St Paul's Cathedral – and 10 m wide across its first stage. For the launch of Skylab, the first and second stages – the S-IC and the S-II – were fed by RP-1 refined kerosene and high-performance liquid hydrogen respectively, with liquid oxygen as the oxidiser. Dozens of railcars'-worth of each propellant were needed to fill a single Saturn V.

The S-IC, built by Boeing, was 42 m tall and its five F-1 engines, arranged in a cross pattern, produced over 3.4 million kg of thrust to lift the vehicle to an altitude of 61 km above Earth. The centre engine was fixed, but the four 'outboard' engines could be gimballed to steer in flight. The S-II was constructed by North American Rockwell, had the same diameter as the S-IC and was 25 m in height. It gained its place in history as the largest cryogenic-fuelled rocket stage ever built and, in some of NASA's earliest space station sketches, might have formed the basis for a wet workshop. On lunar missions, the third stage, the S-IVB, also provided a kind of 'garage' to house the lunar module; *now*, in its place, was America's first space station, comprised of the orbital workshop, the instrument unit, the airlock and multiple docking adaptor and the Apollo Telescope Mount. In total, the Skylab 'stack' measured 36.1 m long and weighed an impressive 90,260 kg.

Today, more than three decades since its final flight, the Saturn V has achieved respectable renown as one of the safest and most successful manned rockets: in 13 flights between Apollo 4 in November 1967 and Skylab in May 1973, it never once failed to fulfil its mission – although on a couple of occasions, its performance was less than optimum. In the case of Apollo 6 in April 1968, the S-IC suffered severe thrust fluctuations, which caused the entire booster to 'bounce' violently, like a pogo

The Saturn IB for the first Skylab crew inches its way from the Vehicle Assembly Building (VAB) to Pad 39B in February 1973.

stick, after which two of its S-II engines shut down prematurely and the S-IVB refused to restart. Then, on Apollo 13 in April 1970, the S-II's centre engine shut down earlier than planned, but the other engines burned for longer to achieve orbit. Since then, four picture-perfect Saturn V liftoffs had delivered a dozen men to the Moon and eight to its surface. On the morning of 14 May 1973, there was every expectation that the final Saturn V would perform with its usual perfection.

Elsewhere, the Saturn IB, carrying the command and service module for Pete Conrad's mission (CSM No. 116), had already been in position on Pad 39B for almost three months, having been installed there on 26 February 1973. The flight readiness test in the first week of April passed without a hitch, as did a countdown demonstration later that same month and at the stroke of midnight on 8/9 May, the launch countdown began, with liftoff scheduled for the 15th. Like the ASTP stack, discussed in the previous chapter, the boosters for the three manned Skylab missions looked peculiarly out of place on the oversized facility, requiring the framework of the 'milk stool' to elevate them to the proper levels of servicing and access arms and gantries. Other than that, the command and service module which Conrad, Kerwin and Weitz would ride into orbit looked little different to the dozen or more that had flown before. However, there were several fundamental differences, for *this* craft – and the pair which would follow it – would spend more than twice as long in space than any of its siblings. Whilst docked at the station, it would be powered down, with the exception of its communications, thermal control systems and instrumentation, but in terms of consumables and propellants it was greatly enhanced. It had double the load of reaction control propellant as previous command and service modules, thanks to a storage module housed inside the bay of the service module previously utilised for the SIM package, which provided for increased manoeuvrability and a much-needed backup facility for the de-orbit manoeuvre at the end of the mission. Additional thermostatically-controlled heaters would maintain temperatures above freezing for propellants, components and propellant lines and an insulating and heat-reflecting thermal paint was added to the sides of the service module. Extra water storage tanks eliminated the need for waste and urine dumps, thereby reducing the risk of such disruptive and contaminating effects on Skylab. However, one of the command module's three fuel cells was removed. Moreover, whilst docked at the space station, Apollo's fuel cells would be shut down and it would draw power from Skylab's own solar arrays and batteries. Larger lockers inside the command module provided greater storage capacity, both to transfer equipment to the station and later to return experiment data and film cassettes back to Earth.

The Saturn V was rolled out to Pad 39A in mid-April 1973 to begin a four-week series of checks and two simulated countdowns. Stacking of Skylab onto the booster had met with delay early in January, when tests of the airlock and multiple docking adaptor fell behind schedule and the EREP control console had to be removed. By 12 April, the Kennedy Space Center's test office verified that "the internal [workshop] is closed out for flight", the aerodynamic shroud was installed and rollout occurred on the 16th. After fluid, electrical and environmental control utilities had been plugged in, the first countdown test got underway. These tests had been a staple of Saturn pre-launch processing, but on *this* occasion, with the requirement to qualify the stack as an integrated system, they were particularly important. NASA's senior management was keenly aware that the very success of Skylab as a project was wholly dependent on the launch of the workshop itself; if something were to go awry, it was unlikely that Congress would release the $250 million needed for a second attempt. A heavy downpour on 4 May and several days of sporadic thunder and lightning caused some concern, especially because the shroud leaked and allowed rainwater to fall onto the ATM. High winds and still more rain delayed attempts to properly seal the shroud until the 10th and when lightning then struck the launch tower this required that the booster's systems be retested.

In fact, the ominous presence of a brewing storm front seriously threatened the launch itself on 14 May. Conrad, Kerwin and Weitz were on hand to observe, along with 25,000 spectators. Precisely at 1:30 pm Eastern Standard Time, the five mighty F-1 engines in the first stage roared to life, raising the question in everyone's mind: had the rocket *risen* or had Florida actually *sunk*? Witnesses to Saturn V launches have employed a range of adjectives to describe its naked power, dazzling light show, intense roar and crackling staccato vibrations, but all agree that it was one of the most awe-inspiring experiences imaginable. On its maiden flight in November 1967, the vibrations of the launch almost shook Walter Cronkite's CBS news trailer *to pieces* ... more than *five kilometres* from the pad! Astronauts Tom Stafford and Mike Collins, who both rode this beast at the start of their own missions to the Moon, and watched others do the same, described the initial sensation as something which one could *feel* through one's feet, rather than *hear* through one's ears. The ground-shaking cacophony on 14 May 1973 as the last of these behemoths headed for the heavens was no less impressive ... or terrifying.

Pete Conrad had flown one of these monsters in November 1969, commanding Apollo 12, and at the moment of ignition to launch Skylab, he instinctively braced himself for the colossal wave of sound, which washed over him and the rest of the crowd as the stack lumbered away from Pad 39A and straight into a deck of thick, iron-grey cloud, trailing an immense tongue of orange flame. "It looked great," Joe Kerwin recalled, despite the poor visibility. Owen Garriott and Jack Lousma were also impressed and left the launch site a few minutes later, stopping briefly at their motel to change into their flight suits, then heading to Patrick Air Force Base to pick up a T-38 jet and fly back to Houston. As they were walking to their rental car for the short trip to Patrick, they happened to encounter Rocco Petrone, newly-appointed head of the Marshall Space Flight Center, who advised them that the ascent had turned up a few telemetry glitches. Still, as Garriott and Lousma took off from Patrick and reached their cruising altitude, they had every confidence that Skylab was operating normally and their future home in space was primed and ready to accept its first visitors in just a few days' time.

That progress, at first, seemed pleasing, with the vehicle going supersonic within a minute of leaving the pad. Shortly thereafter, it passed through a period of maximum aerodynamic turbulence – referred to as 'Max Q' – when atmospheric forces on the vehicle were most severe. Seconds after entering this dynamic period of

The final Saturn V delivers all the light, noise and naked power for which it was already famous as it propels America's first space station – Skylab – into orbit. With an unbroken chain of successes, stretching back to November 1967, there was every reason to expect that the venerable booster would perform to perfection for its swansong . . . but only a few minutes after this photograph was taken, disaster would strike Skylab and a frantic effort to save the $1.5 billion project would begin.

the flight, the telemetry downlink in Mission Control in Houston showed the first indications that *something* was amiss. As Dave Shayler wrote in *Skylab: America's Space Station*, "The data, which went almost unnoticed, indicated a premature deployment of the protective micrometeoroid shield and the No. 2 workshop solar array." If it was *not* simply an instrumentation error, this signified *very* bad news for Skylab. If the micrometeoroid shield and one solar array had indeed deployed during the initial boost to orbit, they were as good as lost and the very future of the space station – and the missions of Conrad and those scheduled to follow him – would hang by a thread. For now, though, the assumption was that the glitch represented nothing more than a spurious signal.

Everything else about the climb to orbit seemed to proceed normally. The Saturn flew its pre-programmed ascent profile, with the second stage taking over when the first stage burned out; in fact, the only problem seemed to be a telemetry indication that the inter-stage failed to separate on time. The five J-2 engines of the second stage were automatically commanded to burn for a little longer than normal in order to compensate for the additional weight. Within ten minutes of leaving the Cape, the S-II shut down crisply and released Skylab at an altitude of 436 km, some 1,800 km downrange, high above the Atlantic Ocean on its steeply inclined orbit and heading away from Newfoundland. The next milestone was for the instrument unit atop the workshop to ready Skylab for orbital operations. The shroud separated and then, at 1:47 pm, electric motors rotated the ATM out 90 degrees. After it had locked itself into place, the windmill of solar arrays was deployed. The orientation seemed to be right on the money and Skylab soon adopted a 'solar inertial' attitude, with the ATM directed towards the Sun. In the euphoria of those first few minutes, the mysterious piece of telemetry about the micrometeoroid shield and the workshop's own solar arrays almost went unnoticed.

Almost . . .

Within an hour of liftoff, Flight Director Don Puddy reported erratic signals. The main solar arrays should have been deployed, under command from the instrument unit, some 41 minutes after launch, when Skylab passed beyond the Madrid tracking station in Spain. Tensions began to rise in Houston, as NASA managers listened for news from the next tracking station, at Carnarvon in Western Australia. The data was confusing. Controllers expected that their monitors would show the two large solar panels fully deployed and producing about 12.4 kW, some 60 percent of the electrical power required to pursue the mission. It was with surprise and dismay, therefore, that the data indicated that power levels were much, *much* lower . . . at a mere 25 *watts*, in fact! The Carnarvon data suggested that the arrays had released for deployment, but had not fully extended, whilst temperature signals from the workshop implied that one array had either been torn away or had suffered severe structural failure, whilst the other had been released, but had not properly deployed. This latter inference seemed consistent with the virtual absence of voltage signals. The data from the next few orbits confirmed a failed micrometeoroid shield and a power outage owing to a solar array malfunction. These concerns were amplified later in the afternoon when telemetry pointed to an electrical short in the pyrotechnic relay needed to release the arrays.

In Mission Control, off-duty flight director Phil Shaffer set to work implementing a malfunction list to handle the myriad problems which were now flooding in from Skylab. "Puddy didn't have time for it," Shaffer related, in an interview quoted by Hitt, Garriott and Kerwin in *Homesteading Space*. Within an hour or two of starting the list, Shaffer found that it already ran to nearly 50 mission-critical items! "At that point," he continued, "we stalled out on the post-insertion activation sequence ... and stuff just kept failing and we could see it was beginning to get hot inside Skylab."

This rising heat offered other, more immediate, worries. The micrometeoroid shield had a secondary function to provide the workshop with thermal control. Its exterior face carried a black and white pattern to absorb the desired amount of heat, whilst its interior face and the hull of the workshop itself were coated in gold foil to regulate the flow of heat between them. As long as the shield remained in place, the system would keep Skylab on the cool side of the comfort zone, but now that it was most likely *gone*, or at least disabled, the gold coating on the workshop would rapidly begin to absorb heat, making its interior uninhabitable. Within hours, sensors on the exterior began to indicate temperatures of 82°C and those *inside* the habitable area moved steadily above 38°C. Thermal engineers at the Marshall Space Flight Center estimated that the exterior would most likely rise to 165°C and the interior to 77°C. This would endanger the astronauts' food stores, their camera film and possibly even the structure of the workshop itself.

Furthermore, engineers feared that materials in the workshop would 'outgas' under the high temperatures, with the resulting contaminants posing an extreme risk to an astronaut crew. The interior walls (originally designed, of course, to house liquid hydrogen, *not* humans) had been insulated with a thick layer of polyurethane foam and fibreglass. One of the constituents of the foam was a particularly aggressive chemical known as toluene diisocyanate – today listed by the New Jersey Toxic Catastrophe Prevention Act as one of a dozen extremely hazardous substances to human health – which tests had shown would begin to break down from heat at temperatures of about 199°C, potentially releasing its toxicity into Skylab's atmosphere. "We identified and procured gas-sampling tubes which would measure the levels of toluene," explained Chuck Ross, the flight surgeon for Conrad's crew, "and adapted them to suck gas through Skylab's hatch equalisation valves. That was so that we could do a 'sniff test' before anyone entered Skylab. We procured two activated-charcoal [filter] masks for the crew to use while first entering the [workshop] ... and we prevailed upon the control team to evacuate all the gas out of Skylab and refill it. I think they did that *twice* before the crew arrived ... "

That evening, at a press conference, Walter Kapryan, NASA's head of launch operations at the Kennedy Space Center, gloomily told the assembled journalists that the agency could only wait for more conclusive telemetry before making a judgement on what to do next. In such a dire predicament, a 28-day manned mission was out of the question; nor was it realistic for Pete Conrad's crew to manually deploy the arrays and rescue Skylab during an EVA. "They [the arrays] either go out or don't go out," Kapryan said glumly. "That is, the astronauts can't do anything about it."

What *was* certain was that Conrad, Kerwin and Weitz would *not* be flying on 15 May and their launch was scrubbed within eight hours of Skylab reaching orbit. The countdown clock, which had been halted at T minus 14 hours and 35 minutes, was recycled to 59 hours and held there. Based on the station's orbital geometry, launch opportunities occurred every five days and the mission was tentatively rescheduled for no earlier than 20 May. "Our first duty was to our families, who were having the pre-flight cocktail party at Patrick Air Force Base", a few miles south of the Kennedy Space Center, Kerwin recalled, with more than a hint of humour. "We called them up and said: 'You can keep having your party, but we're not launching tomorrow! You can go on home'." Next day, the astronauts flew back to Houston in their T-38 jets to develop a new flight plan, which called for a 17-day 'nominal' mission, then 'minimal activity' for a further 11 days in order to gather the medical data for a full four-week residency.

However, as this plan began to crystallise, the situation aboard the station steadily worsened. In order to produce electricity, Skylab needed to remain in a solar inertial attitude, with the Sun's rays perpendicular to the ATM solar panels, but this exposed the full length of the workshop to excessive overheating. For a time, Mission Control ameliorated the problem by pointing the front 'end' of the station directly at the Sun. This lowered temperatures...but also reduced power levels. The best compromise, it was found, was for Skylab to be pitched 'upwards', about 45 degrees, towards the Sun. This permitted just enough sunlight to illuminate the ATM arrays and charge their batteries for the next period of orbital darkness, whilst also stabilising internal temperatures at around 42°C. Conversely, and somewhat ironically, temperatures in the airlock actually *dropped* precipitously and threatened to freeze heat exchangers and coolant loops by 18 May. Manoeuvres to warm the airlock succeeded, but at the expense of overheating the rest of the station. Therefore, the problem of maintaining this fine balance between temperature and power was extremely difficult. Of the nine rate gyroscopes needed for basic attitude management, several had overheated on the day of launch, whilst others produced random errors and sent spurious signals to the three massive control moment gyroscopes; also, the station's attitude propellant had to be used in far larger quantities than originally intended. For his part, Joe Kerwin had nothing but praise for the flight controllers' valiant efforts to keep Skylab on the straight and narrow. "They played this attitude-control game for ten days," he said, "and hosed out a lot of the precious propellant on the workshop, which could not be replenished. I think we had used 60 percent of it by the time we arrived, after *ten days*. That's scary, because this was a *one-year* mission, but they did the job and held it long enough."

'Long enough' was certainly an apt statement, for in the last few days leading up to the launch of Conrad's crew, Skylab's situation was becoming increasingly dire. In its unmanned state, its power requirements were around 4.5 kW, just below the ATM system's maximum output, but the high-angle manoeuvres during its second week in orbit imposed excessive electrical demands and caused several batteries to cease to function, with some failing to revive after the station resumed solar inertial attitude. The arrival of Conrad, Kerwin and Weitz could not come soon enough.

If overheating and power were problems, so too was the potential ruin of the food supply and the astronauts' equipment, most notably the film for the Earth resources experiments. Ground tests verified that canned food could withstand temperatures of 54°C for at least a fortnight and dehydrated food would last even longer, but to be on the safe side the crew was given a quick course on food inspection. A list of medical supplies which might be ruined was drawn up, with the intention of restocking them aboard the command module, and Kodak engineers were approached to assess the potential damage to the film cassettes. The problem was dryness, since heat emulsion on the film would dry out in the low humidity and salt packs in the vaults were not expected to provide sufficient moisture for more than a few days. As a precaution, more film would be carried aloft in the command module's lockers. (In fact, when Conrad's crew finally arrived, they would find that the high temperatures inside the workshop had ruptured their tubes of toothpaste and hand cream and had hardened their shaving cream!)

Although it was clear that some sort of repair was critical, there was one saving grace: not all of Skylab's exterior required protection. In fact, covering the part of the workshop's exterior which directly faced the Sun would serve to bring temperatures within satisfactory limits and, furthermore, such a 'shade' would not need to be tied down or composed of strong or rigid material. In the hours after the accident, options for developing this material were exhaustively brainstormed throughout NASA and the proposals came thick and fast, ranging from spray paints, inflatable balloons and wallpapers to window curtains and extending metal panels. At length, ten options were short-listed for closer inspection, within the guidelines that they must be lightweight, fit inside the cabin of the command module for transportation and were fairly straightforward to deploy. These options were ultimately winnowed down into three finalists: (1) the extension of a sunshade across the exposed hull of the workshop, erected by means of a long pole affixed to the ATM, (2) a sunshade deployed from the command module's hatch whilst station-keeping or (3) a sunshade deployed through Skylab's solar-facing scientific airlock.

Of these, Option 2 was the least technically complex, although its key obstacle was that Pete Conrad, at the command module's controls, would be forced to hold position alongside the workshop, whilst his colleagues opened the hatch to put the sunshade in position. The first option to erect such a shield would require additional EVA training. The third option would require the development of a shade which was capable of passing through a 20 cm^3 opening and then unfurling to cover an area of 7 m^2. It also meant that the scientific airlock would have to be sacrificed. Conrad's crew had already done extensive EVA training on the ATM and felt that with the availability of suitable hand-holds and foot restraints, they could complete Option 1. Similarly, Option 3 was also 'doable', because they could at least work from within the pressurised – but *very* hot – confines of the workshop itself. At length, since no one knew if the scientific airlocks were cluttered with debris, Option 3 was ranked last and the Johnson Space Center set to work on Option 2 and the Marshall Space Flight Center explored Option 1.

The Houston group envisaged a scenario in which an astronaut (probably Paul Weitz) would perform a stand-up EVA (SEVA) in the open hatch of the command module and attach the sunshade in two places to the aft section of Skylab. Conrad would then manoeuvre his spacecraft to the forward end of the station, deploying the shade in the process and finally Weitz would make a third attachment at the ATM. This sunshade very quickly gained the moniker of 'SEVA sail' and its development was conducted under the auspices of Caldwell Johnson, head of the local spacecraft design division. For ten days, his staff worked feverishly on the shade, seamstresses stitched the orange material, parachute packers folded it for deployment, engineers attended to its various fasteners ... and a steady stream of public tours gawped from a mezzanine gallery at what was going on. Meanwhile, in Huntsville, the plan was to perform an EVA from the ATM itself. Their sunshade closely resembled a window blind and its design was completed on the evening of 15 May. Joe Kerwin and the backup commander, Rusty Schweickart, flew to the Marshall Space Flight Center to participate in underwater EVA tests of both sunshades. Like Houston, there was an ever-present sense that they were residents of a goldfish bowl, with journalists and members of the public ever watchful. Late on 15 May, the two men donned training space suits and entered Huntsville's neutral buoyancy simulator – a huge tank, 23 m long and 12 m wide, holding almost 6,500 m^3 of water *and* a full-scale mockup of Skylab – and set to work.

"We had to answer certain very basic questions," Schweickart recalled in a March 2000 oral history for NASA. "Could we get physically around to where we had to be? Could we see certain things? These were questions which you couldn't answer just looking at drawings. We had to get into the water, get on the *real* vehicle and see whether certain things could be done." The pair also had to verify that the hand-holds and foot restraints were adequate on the parts of the structure which would be visited and to ensure that there were no sharp edges on which pressurised suits could snag or tear. Just as the first test was about to start, with Schweickart submerged to his waist and Kerwin about to enter the water, a telephone rang. It was Deke Slayton, the head of flight crew operations.

"Tell them to get out," he told the test operator.

The news was conveyed to Schweickart and Kerwin. "Slayton says to get out of the water."

"I don't care what he says," Schweickart retorted indignantly. "I'm not getting out of the water. We've got to answer these questions. He doesn't understand. Let's go."

After several more attempts to contact the men, Slayton was able to speak directly to Schweickart. "The problem," Schweickart explained, "was somehow related to the fact that the television networks had gotten word that we were doing this over at Huntsville. Of course, everybody was starved for what was going on [and] who was doing what. They wanted *news*, naturally. Somebody had confirmed that, yes, there were a couple of astronauts getting in the water tank, so apparently they were coming ... out there with all their microphones and television cameras and they were going to record this stuff and either Deke didn't like it or Headquarters didn't want that to happen." Schweickart refused to stop the test, telling Slayton that they had barely four days to build and evaluate the repair mechanism from scratch, and so he and Kerwin went ahead. "The media never did show up," he recalled. "I don't know whether they kept them off at the gate, but in any case, it never got there and Deke

Nine days after the accident and two days before the launch of Pete Conrad's crew, on 23 May 1973 the jury-rigged parasol, with its telescoping rods and parachute canopy, is prepared for flight on the floor of Building Ten at the Johnson Space Center. Images such as this exemplified the 'can-do' spirit of NASA and, against many odds, triumph would be snatched from the jaws of defeat.

never said a word and Headquarters never said a word, but that was the kind of craziness that was happening." Slayton kept the men's morale elevated by visiting them one night in their motel room … bearing two large pizzas and a bottle of wine! "Best pizza and wine I ever tasted," Kerwin recalled, despite being officially under pre-flight quarantine …

During their time underwater, the two men evaluated the two sunshade proposals and were able to determine the practicalities of what a spacewalking astronaut could physically see, taking into account the restricted field of view of the suit's helmet. A post-simulation debriefing was then held with the prime crew and some 75 engineers, all clad in blue face masks to uphold pre-flight quarantine rules. "One by one," recalled Schweickart of the exhaustive two-hour-plus session, "we eliminated things and by about midnight ... we basically had the outlines of what we were going to do." The Huntsville sunshade needed further work and the design which steadily evolved was a configuration of two 14 m long poles, which would be 'cantilevered' from the ATM. The poles would be assembled from a dozen smaller sections, allowing them to fit inside the command module, and a rope would run along their length, through a series of eyelets. The 7 × 6 m sunshade would be unfurled by tugging on the rope in a similar fashion to hoisting a ship's sail. This design came to be known as the 'twin-pole' sail. An underwater test by Schweickart and Kerwin on 18 May showed that it would work, but also indicated that its pole sections might separate under stress. A locking nut was modified, the shade's weight was reduced and Teflon inserts were placed into the eyelets to reduce friction. Thereafter, the remainder of the work ran without a hitch.

Meanwhile, the option to deploy a sunshade from the scientific airlock had been revived and was steadily gaining momentum, with a concept that came to be known as 'the parasol'. Tests showed that a combination of coiled springs and telescoping rods could fit inside a standard airlock experiment canister *and* could be deployed smoothly. Jack Kinzler, chief of the Johnson Space Center's technical services division, a close friend and neighbour of Pete Conrad, developed the system by jury-rigging it from a parachute canopy and telescoping glass-fibre fishing rods in hub-mounted springs. Referring to Kinzler, Dave Shayler wrote, "Using a canister similar to those used to place experiments in the airlock and deploying the device by pulling strings tied to the telescoping rods, he clearly demonstrated that as the poles extended and locked, the parachute formed a smooth canopy over the area." For Kinzler, who never gained a degree, but whose experience, according to Hitt, Garriott and Kerwin, "was worth at least a master's in engineering", it was straightforward: with all the talk about doing complex and risky EVAs, no one seemed to have considered doing it the simple way and utilising the scientific airlock.

A technician bought the fishing poles in Houston and Kinzler himself requested a tube from the sheet-metal shop and a large section of parachute material. The result was described by the man himself: "The machine shop fastened the four fishing rods to my base. I fastened the base to the floor of our big high-bay shop area. We fastened the cloth to the rods and long lines to the tips of each rod. I lowered the big overhead crane to floor level and swung my four lines over the crane hook. Then I called [Bob] Gilruth and everybody came over for a demonstration ... I raised the crane back up, letting out excess line, 'til I had enough clearance, then let the crane pull all four lines simultaneously. It looked like a magician's act because our came these fishing rods, getting longer and longer. They're dragging with them fabric. They get all the way to where they're fully out and all I did was let go and it went *sshum*. So the springs were on each corner and they came down and laid out right on the floor just perfectly. *Everybody* was impressed!"

It was decided to use standard space suit material – layers of nylon, Mylar and aluminium – for the shade itself, although very little data existed on the performance of nylon when exposed to long-term vacuum and solar ultraviolet radiation. A decision was taken to cover all three sunshade types with an ultraviolet-resistant material, known as 'Kapton', but this proved problematic, at first, not least because its additional weight might make it more difficult to stow and deploy properly.

Senior managers were in disagreement. Bill Schneider felt that Huntsville's twin-pole sail was most likely to succeed, although Chris Kraft thought it was too heavy. Kraft felt the development of the SEVA sail should continue, in case the twin-pole should fail its tests. At length, during a final review at the Kennedy Space Center on 19 May, Kinzler's parasol was accepted as the primary method – he would later receive a Distinguished Service Medal from NASA for his work – and on the 24th, the flight readiness review endorsed it. Having an astronaut standing in the hatch on an EVA was undesirable, since it would come at the end of a long, 22-hour day for the crew and the contamination effects of the command module's thrusters on the ATM were unknown. Equally, the twin-pole concept did not meet with the approval of the flight surgeons, who were aghast at the prospect of such a complex task so early in the mission, before the crew had properly acclimatised to weightlessness. They also felt that such work might jeopardise Skylab's medical objectives. For his part, Pete Conrad felt that Kinzler's design was the simplest, safest and quickest method ... and hence most likely to succeed. It was expected to be more than sufficient for the first 28-day mission, although the twin-pole and SEVA sails would be carried aloft as a backup and deployed at a later date if the condition of the parasol deteriorated. The review also postponed the launch by five days, until the 25th, thereby allowing Houston and Huntsville engineers to apply finishing touches to their hardware.

Rusty Schweickart considered the delay to be the right decision. "Although there was a lot of inter-centre stuff going on," he told the NASA oral historian, "the logic which supported that was that the crew didn't have to go EVA in order to get the parasol out and try to protect the lab that way, whereas to use the one we developed at Huntsville, you had a whole separate thing of going EVA, which added a level of complexity right at the very start of the mission, which was seen to be undesirable." Years later, Schweickart glowingly praised the efforts of the industrial and NASA workforces to save Skylab during those frantic few days of May 1973. "I probably got a little bit of sleep," he recalled, "but most of the team who worked with me at Huntsville never slept for four days! It was totally round-the-clock and it was not just the resources of the centre; it was *all* of the resources of the *whole* aerospace industry. Anything that we wanted, you simply called somebody and they turned inside out. Three different suppliers would manufacture some thermal material or some device overnight. They would work on it 24 hours themselves. It would be there on the company's private Learjet the next morning. It was unbelievable how hard people worked." 'Borrowing' aircraft – and cars – got a few engineers and branch chiefs into hot water. Bob Schwinghamer, head of the materials lab at the Marshall Space Flight Center, remembered lending the keys of a centre director's car to a colleague ... and then promptly forgetting to return it and receiving a severe verbal roasting the following morning. On another occasion, Schwinghamer was working late, until after the security staff had locked the perimeter gates. "If I call these damn security guys, they'll be here in two hours," he recalled. He decided to climb the fence, almost twice the height of a man and topped by an ominous overhang of barbed wire. "I cut a big gash in my butt," he concluded, "and fell off the fence and fell to the ground. Just when I hit the ground, two headlights came on. These darned security guys drove up and slammed up the brakes and jumped out." One of the guards, who had previously nailed Schwinghamer for speeding, recognised him, grinned and let him go. To sum up: "We didn't let *anything* deter us. A lot of funny stuff happened on our way to the Skylab ... I was getting in hot water all the time and it was day and night. We did all kinds of stuff like that at that time, but we got [the sail] built."

Such comic anecdotes did not detract from the physical consequences of such a punishing workload. Ed Smylie, head of the Crew Systems Division, remembered one of his branch chiefs literally collapsing with exhaustion as he left work late one evening. Yet the sense of teamwork and camaraderie was unmistakable. Joe Kerwin felt the same. "It was a great team," he reflected. "I look on Apollo 13 as the supreme test ... for the Mission Control team. The Skylab problem was the supreme test for the engineering team. Both the contractors and the civil servants joined together, as one, and they figured out what the problem was." To further illustrate his nostalgia for the good old days, Kerwin recalled the motel accommodation during their time at Huntsville, which charged seven dollars per night for a room with black and white television and *eight dollars* for colour ...

As NASA approached resolution on the question of how to repair the workshop, a formal inquiry into the cause of the mishap was set in motion by Administrator Jim Fletcher on 22 May. He asked Bruce Lundin, head of the Lewis Research Center in Cleveland, Ohio, to lead the investigation-board. Reviews of launch data had already shed light on what had transpired. Sixty-three seconds into the Saturn V's ascent, at a point, ironically, when it was obscured from the tracking cameras by a thick deck of overcast cloud, the micrometeoroid shield had prematurely deployed, 'standing out' just a few centimetres from the hull of the workshop, and had very quickly been ripped off, like the skin of a banana, in the supersonic flow. "At this time," noted Lundin's report, published on 30 July 1973, "vehicle dynamic measurements, such as vibration, acceleration, attitude error and acoustics indicated strong disturbances. Measurements which are normally relatively static at this time, such as torsion rod strain gauges, tension strap breakwires, temperatures and [solar array] position indicators, indicated a loss of the [shield]." As a result of *this* failure, others followed: the separation of the micrometeoroid shield caused part of it to wrap around the No. 2 solar array and break the latches on the No. 1 array. A little under ten minutes into the flight, as planned, the burned-out S-II separated, firing its four retrorockets to withdraw from the payload ... and the *plumes* of those retrorockets quickly impinged on the No. 1 array, breaking its hinge and totally *shearing it off*. This incident was referred to in Lundin's report as 'the 593 Second Anomaly'. Telemetry from this point showed a sudden loss of temperature readings and seemingly inexplicable voltage dropouts, which the board took as indicative that the array had *physically separated* from the workshop. "The effect of retrorocket plume impingement was observed almost immediately," the report continued, "on the [No. 2 array] temperature and on vehicle body rates." Under normal circumstances, the two arrays would have been freed from their attachments by a small explosive charge and spring-loaded hinges would have caused them to automatically unfurl. Unfortunately, the No. 1 array was gone and its companion was so clogged with debris as to be effectively 'pinned' to the side of the workshop and hence able to only partially open.

In its concluding remarks, Lundin's report settled on a number of possible causes for the failure of the micrometeoroid shield and felt that the most likely occurrence was an internal pressurisation of its 'auxiliary tunnel' – a tunnel which served as a wiring conduit and was designed to vent pressure as the Saturn V rose through the atmosphere. Due to imperfect seals and fittings, its pressure may have become high enough, about a minute after launch, to slightly raise the shield into the supersonic flow, ripping it off, with the result that it broke the latches on the No. 1 array and part of it became wrapped around the No. 2 array. When the S-II's retrorockets fired, their exhaust finally tore the Number 1 array from its hinge. The fundamental 'human' cause, Joe Kerwin explained to the NASA oral historian, was that the designers of the micrometeoroid shield did not communicate effectively with the aerodynamicists and properly protect it from the supersonic airstream. In fact, failure to recognise such issues during half a decade of development was attributed to a decision to treat the micrometeoroid shield as a *subsystem* of the S-IVB, based on a flawed presumption that it would be structurally integral to the fuel tank. "As a result," noted Hitt, Garriott and Kerwin, "the shield was not assigned its own project engineer, who could have provided greater project leadership. In addition, testing focused on *deployment*, rather than performance during launch." Indeed, one of the recommendations of Lundin's board was for better, more effective and more comprehensive project oversight.

Of course, the state of the arrays and the reason for the No. 2 array being unable to properly unfurl, could only be speculated until the arrival of Pete Conrad's crew and the presence of three sets of eyes to physically *see* what was amiss. If debris *was* the problem, a repair method was acutely needed and engineers from the Marshall Space Flight Center set to work to adapt a cable cutter (not dissimilar to a heavy-duty tree lopper) and a universal tool with prongs for prying and pulling to open the jammed array. Both tools came from A.B. Chance Company of Centralia, Missouri, a manufacturer of equipment for power companies, and were modified to operate from the end of a 3 m pole. In *Homesteading Space*, Hitt, Garriott and Kerwin note an interview with NASA systems engineer Chuck Lewis, who made the initial contact with A.B. Chance and secured the rapid delivery of the tools from Missouri to Alabama aboard a light aircraft. When the A.B. Chance product manager, Cliff Bosch, arrived at the Marshall Space Flight Center, he was joined by Lewis and backup crew members Schweickart and Story Musgrave to begin sorting out the most effective tools. "The one thing they had that was really neat," Lewis recounted, "was this scissor-like cutter that they used to clip electrical cables. The guys at Marshall re-engineered that in about a day and a half to provide some extra mechanical advantage because … we knew what sort of load it was going to take for those jaws to get through that [debris] and whether they were going to be able to pull on it."

On 19 May, the tools were successfully tested in Marshall's neutral buoyancy tank, with the Skylab mockup specially 'modified' with fragments of metal wire bundles, shards of bolts and other objects representative of a failed micrometeoroid shield. Conrad, Kerwin and Weitz took their turns underwater, evaluating the tools, practicing prying the debris away from the array and completing the whole procedure safely. The tools had already left for the Kennedy Space Center when a certification review ruled that the pointed tips of the cutter were hazardous. New heads with blunt tips were quickly prepared and the change was made at the launch site. Now, however, the time for talking was over. Years later, Nancy Conrad related that her late husband's response to the seemingly endless testing was typically to the point: "Just *get me up there*, goddamn it!"

Before the rescue could get underway, of course, the habitable environment of the workshop had to be depressurised and repressurised with nitrogen on no fewer than *four* occasions over a three-day period to purge any toxic outgassed products from its atmosphere. One of the most hazardous materials was a heavy layer of polyurethane foam, bonded to the internal walls, and although the refreshment was expected to be sufficient, it was mandated that Conrad, Kerwin and Weitz would wear masks upon entering Skylab and perform a series of atmospheric analysis checks to establish full confidence in the workshop's environment. As their scheduled 25 May launch date neared, the storage lockers of the command module itself were steadily being filled with additional items: extra cameras for the fly-around inspection of Skylab, new medical supplies and film cassettes to replace those aboard the station, tools with which to free the partially deployed solar array and equipment to detect the levels of poisonous gas. For Gene Kranz, who had supervised the lunar landing of Armstrong and Aldrin and had sweated out the traumatic few days of Apollo 13 from the flight director's console, Skylab was no less seat-of-the-pants exciting than going to the Moon. Years later, he related that people often refused to believe him. "But Skylab *was* as exciting to me as Apollo ever was," he told a NASA oral historian in April 1999. "Skylab to me was a different type of focus; [a] focus as a leader and focus as a team. The Apollo missions were all short, on the order of ten days or so, and it's one thing to hold a team together and … keep the quality for ten days, even though it's very intense. It's another thing to keep this team together for the best part of a *year* and to hand over not tens, but literally *hundreds* of problems, *every shift*, without a glitch!"

With their launch scheduled for the stroke of 9:00 am Eastern Standard Time, the morning of 25 May was particularly peaceful for the three astronauts. "This was the least well-attended Apollo launch in history," Joe Kerwin recalled, "because everybody had to go home and put the kids back in school. We arrived at the command module and looked inside and it was a *sea* of brown rope under the seats and under the brown ropes were all these different umbrellas and parasols and sails and also the equipment that we had selected to try and free up the solar panel, which was a pretty eclectic collection of aluminium poles that could be connected together, and a Southwestern Bell Telephone Company tree-lopper with brown ropes to open

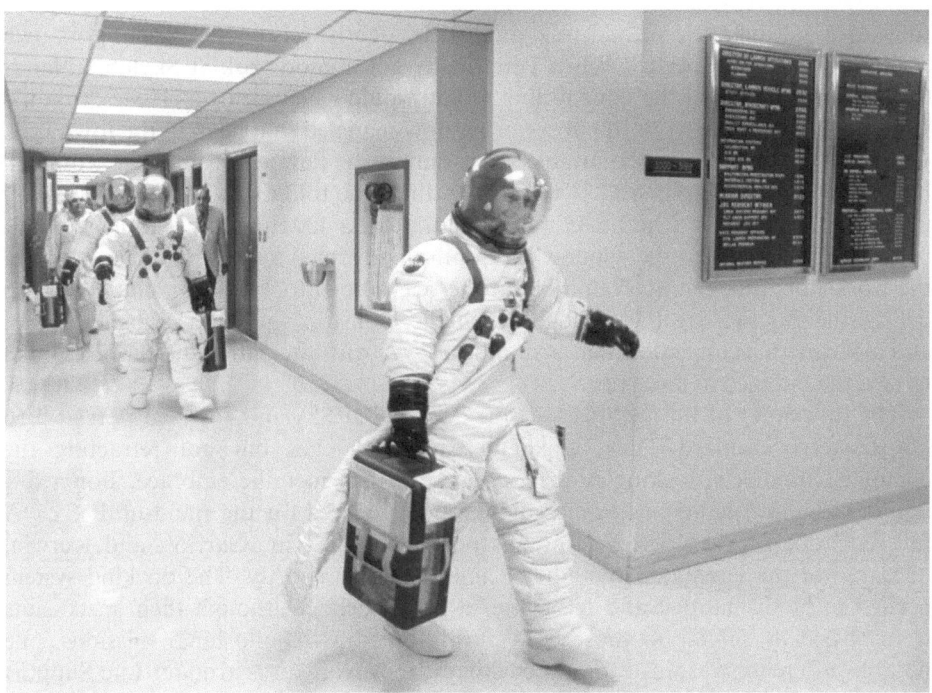

"Just *get me up there*, goddamn it!" Ten days later than planned, Pete Conrad seems buoyant as he leads his men out of the Operations and Checkout Building on the morning of 25 May 1973.

and close the jaws. They handed us the checklist and said 'This is how to operate that stuff.' Some of it we'd seen, some of it we hadn't!" The astronauts were unperturbed. Indeed, as the Saturn IB cleared the Pad 39B tower and roared into the clear morning sky, Conrad declared that *his* crew could fix *anything*. Launch came precisely on time and kicked off an eight-hour orbital ballet to rendezvous with the crippled Skylab later that same afternoon. An initial orbit of 357 × 156 km was gradually refined by a series of four manoeuvres to fly a near-circular, 424 × 415 km path to intercept the space station on their fifth orbit.

"Up until this time," recalled trajectory and rendezvous specialist Cathy Osgood in a November 1999 oral history, "our rendezvous [procedure] had been to rendezvous the first day, after about five orbits. If you didn't rendezvous by about the fifth orbit, well, the astronauts' work day had been so long that they'd have to be facing a docking situation when they were just absolutely exhausted; so, all up until this time, the astronauts themselves wanted to get it all done the first day." Little did anyone know that Pete Conrad's crew's first day would run for no fewer than 22 *hours*.

Conrad's call of "Tally-ho the Skylab!" as a steadily brightening star on the horizon drew closer masked, at first, the seriousness of what the astronauts were about to face. The micrometeoroid shield was indeed gone, as was one of the two solar arrays, whilst the second was jammed by debris, seriously restricting its own ability to deploy properly. "When Pete finally got a good look at Skylab," Nancy Conrad wrote, "he got the same feeling as you would when seeing a classic car you'd invested four years of your life in restoring now mangled in a heap in the town junk yard. The commander's indomitable spirit sank as he flew ... around the wounded bird." As Weitz took pictures – "very frustrating," he told the NASA oral historian in March 2000, since "I had a 35 mm camera, with a 400 mm lens, and it barely fit between the couch and the window" – Conrad performed a flyaround inspection of the workshop, quickly ascertaining that the scientific airlock was *not* cluttered with debris, thereby making the deployment of the Houston parasol a realistic option, and asserting his conviction that a stand-up EVA with the cutter should be enough to free the jammed solar array.

The first order of business was a 'soft docking' at Skylab's forward port at 5:56 pm Eastern Standard Time, engaging capture latches but not retracting the command module's docking probe to ensure a firm metallic embrace. For a few minutes, the astronauts ate a quick dinner and prepared for the stand-up EVA. "A full hard dock wasn't desirable at this point," wrote Hitt, Garriott and Kerwin, "because of the likelihood that they'd undock again shortly. The docking system needed to be dismantled and reset after a hard dock." Although their space suits were physically similar to those used during previous Apollo lunar missions, one notable difference was, instead of a backpack, they wore an Astronaut Life Support Assembly (ALSA) – a kind of 'belly pack', supported by a leg-mounted emergency oxygen bottle. During Skylab-based EVAs, the ALSA would be supplied with water, oxygen and electrical power through an 18 m life-support umbilical. However, for the stand-up task on the evening of 25 May, a shorter umbilical was plugged into the command module's own life-support system. With all three men fully suited, Conrad undocked from Skylab at 6:45 pm, depressurised the cabin and opened the side hatch. With Kerwin hanging onto his ankles to provide stability, Weitz was to reach out and use the modified tree lopper and a kind of 'shepherd's crook' to attempt to free the jammed solar array. It should have been an occasion of euphoria for Weitz, who had not been scheduled to perform a spacewalk *at all* on this mission. Unfortunately, despite his sterling efforts, it did not go well.

At first, Weitz positioned himself with his upper body poking through the hatch into the ethereal blackness of space. Kerwin passed him three sections to assemble a 4.5 m pole with the lopper on the end, whilst Conrad kept the spacecraft steady. "We had seen ... that there was a piece of bolted L-sections from the thermal shield that had been wrapped up around the top of the solar wing," Weitz recalled in his oral history, "and apparently the bolt heads were driven into the aluminium skin. We thought maybe we'd just break it loose, so we got down near the end of the solar array and I got a hold of it with the shepherd's crook. What we really hadn't thought about was – in heaving on it, trying to break the thing free – what I was doing, in effect, was pulling the command module *in towards the workshop*." Weitz could quickly ascertain that he was physically moving Skylab, because its thrusters were spitting and spurting to maintain its attitude and correct the oscillations. Meanwhile, Conrad had the unenviable task of keeping the spacecraft as close as just 60 cm from the station, whilst at the same time trying to prevent the unwanted oscillations from causing a collision. His task was compounded by the fact that a *third* of his field of view was blocked by the command module's open hatch. "It made for some dicey times," Weitz recalled. As the two vehicles entered orbital darkness, he paused in his work, then resumed as they flew within range of the tracking station. The shepherd's crook was getting him nowhere and the torrent of four-letter words from all three members of the crew even prompted the capcom to advise them to modify their language; for they were on an 'open mike'.

The main problem, Conrad told the ground, was that a strip of metal had become wrapped across the solar array system during the separation of the micrometeoroid shield. Its metal bolts had tangled themselves in the array, thereby jamming it, and none of Weitz' actions to cut the strap, even with the lopper, were having any effect. "Rather than cutting it across the short way, we were trying to cut along the long way," Weitz explained, "and just didn't have enough muscle with that thing, because it was six or eight feet out ahead of me and I was pulling on a line to try to do it." The metal strap, ironically, was only a few centimetres wide, but it was riveted fast and Conrad saw they did not have a hope of breaking it using the tools in the command module. The attempt was called off and after 40 minutes or so the astronauts were instructed to close the hatch and re-dock with Skylab. Even this proved easier said than done: getting the pole, lopper and shepherd's crook back inside in a safe and speedy manner led to an inadvertent thump to Conrad's helmet and an accidental kick to Kerwin, whose own language, wrote Dave Shayler, was decidedly "unscientific"!

Docking, wrote Nancy Conrad in her biography of her late husband (probably selecting a few of his own recollected words), "was a bitch". On their first attempt, the probe did not engage with the drogue and no fewer than *three* further attempts were also fruitless. "Pete gave Weitz the controls," Nancy continued, "depressurised the command module and opened the tunnel hatch. He and Joe dove head-first into the bank of circuits and gizmos, Pete cussing a blue streak as he sorted through wires, cutting and splicing like a pissed-off Maytag repairman trying to get a dryer to work again. Kerwin had to hide a smile, despite the seriousness of the situation. The only thing missing from the cussing and grunting wire-snipping Mr Fix-It was the low-creeping work pants revealing more *derriere* than anybody would want to see ..." Weitz then set about rewiring in the right-hand equipment bay, removing the electrical interlock which prevented the main latches from actuating until the capture latches were secure. After an hour or so of re-routing and connecting wires, bypassing electrical relays for the capture latches on the tip of the probe, skinning knuckles and another handful of undesirable vocabulary, Conrad used the service module's thrusters to bring the two collars into direct contact, mechanically triggering the dozen capture latches in a cacophony not dissimilar to machine gunfire.

They were at Skylab to stay.

In their oral histories, both Kerwin and Weitz paid tribute to a conversation with their rendezvous instructor, Jake Smith, in February 1973, in which the minutiae of how to accomplish just this kind of 'backup docking procedure' – which wires to cut and splice – were explored. Had it failed, there was a very real possibility that the mission would have been cancelled and the astronauts brought home. Now, though, with the Apollo spacecraft safely docked, an ecstatic Mission Control were able to advise the crew to grab a bite to eat and some sleep before entering Skylab the following morning. It was 11:50 pm in Florida and 10:50 pm in Houston. Their first day in space had been a long one – more than 22 hours since they were woken up in the crew quarters, prompting Paul Weitz to quip that "union rules wouldn't allow us to work *that* long anymore!" Surprisingly, neither of the rookies had gotten sick. Joe Kerwin had taken a pill shortly after reaching orbit, determined that he would *not* throw up and would *not* screw up the mission ... but, despite the tense nature of those 22 hours, both he and Weitz adapted well to the weightless environment. Making his fourth space mission, Conrad had no problems whatsoever.

Next morning, the crew opened the hatch into the multiple docking adaptor and Weitz, as the systems specialist, was the first to enter Skylab. Pressure checks were quickly followed by air sampling to test for the presence of noxious toluene and carbon monoxide, both of which gave the workshop's atmosphere a clean bill of health. The adaptor and, indeed, the airlock, too, were relatively cool, at around 13°C, but both Weitz and Conrad knew that the unprotected interior of the workshop itself would be hotter and considerably less comfortable. As a consequence, the two men stripped "down to our skivvies" and opened the hatch into the darkened workshop at 3:30 pm, Florida time, to deploy the parasol through the scientific airlock. "It didn't take very long until we figured out why people in the Sahara Desert wear a lot of loose clothing," reflected Weitz, "because ... soon we had long trousers on and we had shirts and jackets and hats and gloves and the whole thing." The temperature was a balmy 55°C – "rather warm", the two men noted – and although there was no evidence of toxic gas and the air was safe to breathe, the workshop's humidity level was very low and they found that they could only remain inside the stifling oven for around 15 minutes at a time, before retreating back to the relative comfort of the multiple docking adaptor.

At length, after several hours, the parasol was assembled and at 7:30 pm Eastern Standard Time, its rods were delicately threaded through the aperture of the scientific airlock into vacuum. Next, the parasol itself emerged, folding out like a big patio umbrella. However, all was not right: one of its four folded arms did not swing out properly and Joe Kerwin, watching from inside the command module, expressed dismay when he saw it had only deployed to cover two-thirds of its required area. "It's not laid out the way it's supposed to be," a dejected Conrad told Mission Control, as it became clear that the parasol was askew and somewhat crinkled. Nevertheless, the ground team in Houston assured the astronauts that the wrinkles had probably set in during the coldness of the lengthy deployment, which took place during orbital 'night-time', and, as the material heated up in sunlight, it would spread out fully. "I think the ground noticed the temperatures coming down," Weitz recalled. "Within an hour, they could tell." Indeed, overnight on 26/27 May, the temperature on the exterior of the workshop dropped by 55°C and its interior by 11°C. Eventually, the interior temperature stabilised at around 30°C (close to a normal day in downtown Houston) and the astronauts could even 'feel' where the

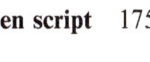

The sacrificed scientific airlock, through which the telescoping rods and the canopy of the parasol would be extended.

parasol lay on the outer hull, by running their hands along the inner wall and *feeling* the differences in temperature ...

For the first couple of nights, Conrad and Kerwin elected to sleep in the command module, whilst Weitz tied his sleeping bag in the multiple docking adaptor, "because it was cooler there and you had a lot more room besides". The process of 'workshop activation' took several days, since the majority of the equipment inside Skylab had been 'hard-mounted' to survive ascent loads and had to be removed and transferred to the required locations. By now, Conrad, Kerwin and Weitz considered themselves bulletproof to weightlessness; they had ridden to orbit, undertaken a complex repair task within their first day aloft and had not suffered the slightest hint of debilitating space sickness. This proved something of a relief, particularly in the light of one of their assigned experiments, the M-131 vestibular function study, provided by Ash Graybiel, head of the Naval Aerospace Medical Research Laboratory in Pensacola, Florida. This study included a rotating chair, part of a 'motion sickness susceptibility test', in which the astronauts would tilt their heads forward and sideways and back, both in and out of the plane of rotation, in a manner designed to bring them to the brink of malaise and vomiting.

Before launch, the astronauts had objected strongly to Graybiel's plan to implement so many head movements – perhaps 70 or more – in the study to purposefully trigger a complex of symptoms known as 'Malaise III'. This would virtually guarantee that they would throw up and Graybiel was obliged to back off to the marginally less severe 'Malaise IIA', which was 'only' expected to cause pallor, sweating, 'stomach awareness' and nausea. "To our amazement," Kerwin told the NASA oral historian, "by the time we had gone through all of our adventures on the first few days of the mission and gotten to the point of doing Ash's experiment, we found that you *couldn't* make yourself motion sick. Once you had adapted, you were bullet-proof! Eventually, all three of us spun the chair up to the *maximum* rotation it was capable of, which was 30 rpm, just about like a long-playing record, and make the *maximum* number of head movements you were allowed to make, which was 150; no malaise, no sweating, *none* of those pre-vomiting symptoms, at all!"

Full-scale scientific operations began soon thereafter, although electrical power was still in short supply. In fact, total consumption during those first few days came very close to the 4.5 kW that was being generated by the observatory's windmill of solar arrays ... which actually represented only around 40 percent of the power which should have been available. The early activities with the ATM drew somewhere in the region of 750 watts. Weitz and his comrades found themselves using the station's lighting sparingly, eating their food cold, not heating water and getting work done as quickly as possible in order to minimise the power loads. On 30 May, another crisis arose: during Earth resources manoeuvres, explained Kerwin, the crew had to orient the station into a so-called 'Z-local vertical' attitude, which took the ATM arrays out of direct sunlight, leaving Skylab reliant on battery power. After one such manoeuvre, four batteries went off-line and ground controllers could revive only three of them, thus reducing the electrical output by *another* 250 watts. "It took us the rest of the day to recover," Kerwin recalled. It was clear that for the

As Conrad, Kerwin and Weitz prepared for their own critical EVA to install the twin-pole sail over Skylab's wounded hull, the *next* crew pressed ahead with their own preparations. Here, Jack Lousma is pictured in June 1973 preparing for the installation of the 'six-pack' of replacement gyroscopes.

mission to survive and succeed, it would be necessary to release the jammed array, and *soon*, and plans were set in motion for an EVA on 7 June.

Under the direction of backup commander Rusty Schweickart, a scenario was put in place and tested in the neutral buoyancy tank at the Marshall Space Flight Center. It required Conrad and Kerwin to move to the antenna boom at the forward end of Skylab and attach an 8 m cable cutter to the debris; this would serve as a makeshift 'hand rail' to enable them to reach the jammed solar array and cut the metal strap. Once there, they would have to break a frozen hydraulic 'damper' on the array. The damper's purpose had been to prevent the array from deploying too fast, but when it partially unlatched after launch, the hydraulic fluid very quickly froze solid. Conrad and Kerwin would connect a Beam Erection Tether (BET, effectively a nylon rope with hooks on both ends) between the solar array and the airlock. When the fouling debris had been removed, it was hoped that by pulling on the tether they would be able to break the damper. During a series of simulations in the Marshall tank on 2 June, Schweickart and fellow astronaut Ed Gibson practiced the procedure and found that, aside from a lack of foot restraints, it was workable. The lack of foot restraints, and lights, hand-holds and visual aids, was not an oversight; it reflected the fact that there had never been any intention to perform maintenance on Skylab. There *were* several *planned* EVAs, for example, to replace film cassettes in the ATM, but actual repair work was considered too dangerous and had not been contemplated. Still, Schweickart and Gibson were perhaps best placed to execute the practice runs on the ground. "I had been working a lot of the spacewalk procedures for the film retrieval," Gibson told the NASA oral historian in December 2000, "so it was natural to then go down to Marshall and start developing procedures for the repair of the station."

In the meantime, aware of the steadily worsening situation, NASA decided that, so long as no more batteries failed, Conrad's team should be able to complete all of their scheduled tasks, including a heavy load of experiments. It also moved the launch of the second Skylab crew forward from 8 August to 27 July, and possibly as early as the 22nd, if circumstances deteriorated further. However, without additional power, the second and third crews would have insufficient resources to complete their own programmes of scientific research. Other managers were jittery about the prospect of attaching a cable cutter to *debris*, and *eight metres away*, at that, but there was little alternative. On the evening of 4 June, Schweickart talked Conrad through the task and, as the astronauts slept, a list of tools and step-by-step assembly instructions were transmitted to Skylab's teleprinter. The crew reviewed these procedures in their spare time and rehearsed the steps inside the workshop on 6 June, with a suited Kerwin, minus his helmet, practicing the movement of the poles and grabbing a mock target with the cutters, before finally venturing outside in the early hours of the following morning.

Since the airlock was right in the middle of the Skylab cluster, with the hatch to the workshop at its aft end and a hatch to the multiple docking adaptor and the command module at its forward end, a fully-suited Paul Weitz had to make sure that Conrad and Kerwin had all of their tools and tethers before he depressurised them. Weitz then retreated into the multiple docking adaptor. The hatch was opened at

10:23 am Eastern Standard Time, just before the workshop entered the dark portion of its orbit. Conrad assembled the tools – six 1.5 m rods were screwed together, the cable cutter was fitted and several metres of rope from the backup SEVA sail were tied to the cutter's pull rope – and then he and Kerwin moved into position alongside the antenna boom. The unlikely contraption thus enabled them to operate the cutter from 8-9 m away ... just far enough from the airlock to the jammed array.

As Kerwin tried to close the cutters against the debris, it became apparent that he was 'slipping', because he was unable to establish a secure position for himself. For half an hour or more, with one hand steadying himself and the other trying to close the cutters, he struggled fruitlessly to complete the work. As his pulse rate began to climb, he decided on an alternative course of action and shortened his own tether, in an effort to steady himself against the edge of the workshop. It worked and after ten minutes or so he was able to tell ground controllers in Houston that the cutters were now securely fastened to the debris. Next, he pulled on the lanyard to operate them ... and nothing happened. Conrad made his way, hand-over-hand, along the length of the beam to see what was amiss, and precisely as he reached the cutter 'end', the jaws *snapped shut*, freeing some of the metal strap at 2:01 pm and hurling the commander "ass over teakettles" into space. Fortunately, his tether restrained him from moving far from Skylab, and the jammed array now stood at 20-degrees-open.

Seventy degrees to go ...

The frozen damper, however, still resisted normal deployment and the holes on the solar array were smaller than on the ground model, so he could only attach one of the two hooks on the BET. The two men heaved on the tether, but without success, until Conrad placed his feet on the frozen hinge, stooped to fit the tether over his shoulder and 'stood up'. Kerwin pulled on the tether and, *this time*, the solar array suddenly released and sprang into its full, 90-degrees-open position. Both astronauts were flung outwards by the catapult-like effect and arrested by their tethers. After three hours and 25 minutes, the two laughing astronauts re-entered Skylab ... and the needles of the electricity meters dramatically *jumped*, signalling success. By the next day, 8 June, solar heating had fully extended the array and it was generating no less than 7 kW of much-needed power. From just 40 percent power, the station's output suddenly increased to around 70 percent.

As a result of this endeavour, NASA's confidence in the abilities of spacewalking astronauts increased substantially. "Before that, there was still the legacy of problems with EVA during Gemini," recalled Rusty Schweickart. "Apollo, of course, made a big difference, but that was sort of running around on the [lunar] surface in gravity again. So, *here* was EVA of a massive scale in weightlessness that we [had] never anticipated; [we] did it with flying colours, everything worked just fine, never had a problem and saved the mission." For Kerwin, his first spacewalk was electrifying – there was *nothing*, he said later, which could possibly compare to looking at the Earth through the visor of a space suit – but also demonstrated that the underwater training on the ground was actually *harder* than doing the real thing in orbit. "No matter how well they weight you out" in the neutral buoyancy tank, he said, "you're not really weightless. If your body is turned upside down, yes, the suit is neutrally buoyant, but you're jammed up against the shoulders and the blood is rushing to your head. If you can do it under *those* circumstances, you're going to find that it's a piece of cake in zero-G."

"DAY 22, FOREVER"

As exciting and dramatic as the first half of Conrad's mission had been, journalists on the ground had other issues on their minds. One of them centred on a clash between NASA's public affairs office and the Office of Manned Space Flight over whether private communications should be considered 'out of bounds'. On the one hand, the space agency did not wish to foster bad relations with the press, whilst at the other extreme, private and unmonitored conversations were essential in discussing issues and opinions and concerns. Some, including John Donnelly, NASA's head of public affairs, felt that 'private' medical conferences would be no more useful than public ones, because pilot-astronauts were so averse to doctors, rarely admitting to medical problems, lest their mission be cut short or they be grounded. When it came to private 'family' conferences, though, it was a different matter and the astronauts felt that it was ridiculous to *not* allow such unmonitored conversations. However, Joe Kerwin reasoned that as active-duty naval officers, their wives knew the score and had long since become accustomed to their husbands being away from home for long periods. The perspective of some journalists was that, as a *public* programme, *everything* should be made publicly available. "Our crew decided before the flight that if we couldn't have a private family conference, we wouldn't have *any* family conferences," Kerwin recalled. "They could tough it out and so could we. So we didn't have any." A compromise would be worked out for subsequent crews, in which the flight surgeon would release pertinent medical information and a public affairs officer would listen to the conversations and tell the press anything which was of relevance to the mission ... but weekly, unmonitored conferences with family members were to be allowed with guaranteed privacy. "On top of that," Kerwin continued, "the crew commander did have the authority to call a private conference with either the flight director or the centre director, if he wanted to. Pete did that once, on the exercise issue."

The 'exercise issue' arose when Conrad asked for a private conversation on the morning of 29 May to clarify a medical matter concerned with the crew's use of the bicycle ergometer. Typically, all three men were required to test their physiological responses twice weekly as part of the metabolic experiment, measuring their blood pressures, heart rates and oxygen consumption in a series of exercise periods. The ergometer also provided an opportunity to evaluate a tool for future long-duration missions. However, the heat in the workshop hampered Paul Weitz' first attempt on the machine on 28 May and this led Kerwin to recommend that these schedules be shortened. Another issue was that fellow astronaut Story Musgrave had devised an elaborate padded harness for them to wear on the ergometer. It fastened around their waist and shoulders and had four bungee-type cords that connected to the floor. The men soon found that this impaired, rather than aided, their exercise. "You don't sit on a bicycle seat in weightlessness," Kerwin explained. "The bicycle seat might as well not be there at all. It's a useless piece of hardware and when you start to pedal, you ride up into this harness ... and it's so tight that it cuts off the circulation in your thighs and your leg muscles get oxygen-starved. They start to *hurt* and you just have to quit before you have reached the aerobic levels that are your goal."

In fact, Weitz curtailed one exercise session, because not only were the waist and shoulder harnesses restricting his movement, but he also found that he was working too much with his hands and not enough with his legs. Kerwin also gave up, but Conrad persevered. "Pete was trying really hard to reach that peak level in spite of the pain and discomfort and he started throwing premature ventricular contractions," Kerwin recounted. "He would throw an occasional odd PVC in his electrocardiogram. That's not that disturbing a thing by itself, but here we are in space and the docs were all very worried about the 28-day duration." At length, on 29 May, Conrad explained during the private conversation that the ergometer could not be ridden as efficiently in space as it could on Earth and requested that the crew should lower the resistance level by up to 20 percent in order to compensate for the difficulty. Mission Control decided that they should only exercise whilst over the United States and fitted with instrumentation to monitor their heart rates and other vital signs. "They were just getting ready to radio that up," Kerwin told the NASA oral historian, "when we were getting ready to radio them down that we had *solved* the exercise problem. We now *knew* how to ride the bike *and* we could get to our max levels without any problem. What we had done was to take Story Musgrave's harness, carefully fold it up and *shove it in the trash airlock* ... "

Humour aside, the demise of Musgrave's harness allowed their exercise sessions on the ergometer to proceed without incident; instead, they would lock the triangular 'cleats' of their shoes into the pedals to stabilise themselves and then grip either the handlebars or the ceiling grid for additional support. Weitz found it was much more effective than strapping himself down and Conrad even pioneered 'arm ergometry', by pressing his feet against the ceiling and then pedalling with his hands. Physician Edward Michel, principal investigator for the metabolic study, was a little irritated that the elimination of Musgrave's harness had invalidated the controls for his experiment, but was nevertheless satisfied that at least the crew were now able to *ride* the ergometer and exercise properly. However, Conrad's premature ventricular contractions, which manifested themselves in a series of heart palpitations and an abnormally high pulse rate, aroused concern in the days leading up to the 7 June EVA. Physician Chuck Ross told Conrad to reschedule his next ergometer whilst over North American tracking stations, enabling the medical team to receive the data quickly, and recommended that he reduce his maximum workload and avoid strenuous activity, including the EVA itself. Conrad was upset by Houston's failure to seek his own evaluation of the situation and assured Skylab officials that *all three* men were fine and in excellent physical condition.

However, these medical conversations raised more than a few eyebrows, not least in NASA's public affairs office, which had regarded them as hardly being worthy of being classed as 'private'. John Donnelly told journalists that Conrad's conversation did *not* reflect any kind of medical or other emergency situation and his demand for

Skylab's first few days in orbit were mired with difficulties. One solar array had been torn away during ascent and the other – seen here in a view from Conrad's craft – was so clogged with debris that it could not properly unfurl.

'privacy' was totally unjustified. He even went so far as to make noises about getting Conrad formally reprimanded for his actions! From the commander's perspective, though, the lack of a private line for any purpose, other than emergencies, actually *inhibited* full and frank communications between the astronauts and the ground. After the mission, Conrad complained that he had been "left in the dark" frequently over plans devised in Houston and, as a case in point, had only learned about EVA plans to free the jammed solar array ... *from his wife*, during a birthday greeting on 2 June!

With the power supply crisis eased, the astronauts were finally able to heat their food, turn on the lights and devote themselves properly to their experiments. In fact, the second half of the mission was able to proceed in a relatively 'humdrum', routine and repetitive manner; so much so, in fact, that during dinner on the evening of the 15th, one of them remarked: "Guys, it feels like it's been Day 22 forever up here!" The primary focuses were upon the ATM, which occupied much of Joe Kerwin's time, and the EREP series of Earth observation experiments, which were supervised primarily by Paul Weitz. Years later, Kerwin would recall that, as a physician, he did not have the same expertise with the solar observatory as electrical engineer Owen Garriott and physicist Ed Gibson, but he had spent many thousands of hours on the ground familiarising himself with its functions and its systems and its malfunction procedures. Weitz, too, would admit in his March 2000 oral history that he had little interest in Earth resources before Skylab. In fact, this led to one rather unfortunate incident, which he and Conrad never managed to live down. "We made one [EREP] pass one time," Weitz recalled. "They had an optical quality window in one side of the MDA, [which] had six multi-spectral cameras; modified Hasselblads and each camera had film in it that had a different wavelength sensitivity. To protect this optical quality window from orbital debris, it had a metal cover over it that could then be, with a hand control, swung out of the way. We made this EREP pass and we were cleaning up afterwards and I lifted up the cameras and the window cover is *closed*." Weitz asked Conrad when he closed the cover. Conrad denied any knowledge. "Well," continued Weitz, "*neither of us* had closed it, because *neither of us* had opened it!" The six-strong complement of high-tech cameras, with their valuable film, had been clicking away, only to produce black images of the inside of the window cover!

Weitz summed the incident up well with a few choice words: Astronauts need equipment "that is simple enough to operate that you can get these dummies to operate it and get good data back!" Privately, the three astronauts devised their own cue card, written in big black letters, and posted it over the switch panel. According to Hitt, Garriott and Kerwin, the cue card read: "Skylab EREP Dumb Shit Checklist". They never forgot again!

Despite both being senior naval officers, Kerwin and Weitz were quite different from one another, the former a scientist and the latter a fighter pilot, and in a sense their respective responsibilities on Skylab were mutually opposite as well. "Doing EREP passes," Weitz recalled, "was incompatible with taking solar data ... [when we] had to be pointed ... toward the Sun. When you took EREP passes, the workshop was what we called 'Local Vertical-Local Horizontal', where it maintained the same attitude relative to the surface of the *Earth*." During this first Skylab mission, the EREP's six remote-sensing systems acquired data of the contiguous United States, the Gulf of Mexico and the Caribbean and were even able to track the progress of Hurricane Ava, which was in full force, just to the south-west of Acapulco.

As the situation inside the workshop stabilised, the astronauts were able to sleep in the wardroom, an area which they found considerably more comfortable than their couches in the command module. For the rookies, particularly Paul Weitz, the notion of attaching a sleeping bag to the wall and resting *vertically* did not seem quite right, and he later admitted to the NASA oral historian that on at least one occasion he tied himself into a horizontal position. The toilet was fine, if noisy, and the shower, which Weitz tried first, left them "smelling good", but proved taxing because so much time was required to dry themselves. The food and drinks were perfectly adequate and a great improvement over earlier types. "It repeated every six days," Weitz explained. "Every other day, for the evening meal, I had prime rib and lobster Newburg one night [and] pork and dressing another. It was relatively juicy. We soon found that they had a Teflon-like material lining these things, so the food didn't absorb any of the metallic taste from the containers. So the liquids would just gradually move, of their own volition, but when they got to the edge towards bare metal now, because we pulled the zip-top off, the fluids would stop. You didn't have to worry about them oozing out by themselves. You had to be careful if you tried to take a spoonful, that you didn't knock something loose. That's how you get splatters around the ward room, careful as you are. I mean, a drop of beef juice would come off a bit of prime rib and you may not know it." Gazing out of the window – either the large one in the wardroom or the smaller ones in the multiple docking adaptor – became a frequent pastime, with the alternatives of darts and playing cards quickly confined to their boxes. Having said this, reading and the playing of music were both popular. Conrad, for one, after eating ice cream and cookies for dessert each evening, would settle down with a book and one of his country music cassettes for an hour or two. After a while, explained his widow, Nancy, it became difficult for Kerwin and Weitz to endure the strains of Patsy Cline and Hank Williams and Willie Nelson: "They were getting a little weary of steel guitars and twang ... "

ENDINGS

The view of Earth from a 50-degree-inclination orbit was remarkable and Paul Weitz recalled seeing enormous swathes of the Home Planet ... and expressed astonishment that humanity *could* have fought over them for so long. "You come down off the east end of Asia, the Kamchatka Peninsula, basically, the Bering Strait," he explained, "and you head south-east then, descending that way down across the Pacific for nearly *half* your revolution, until you come across the southern tip of South America. Then you look at the vast expanse of the Pacific Ocean and you *wonder* how in the world we *ever* fought naval surface engagements there in World War. It's mind-boggling." If Weitz was amazed by the scale of that conflict, he must surely have been equally surprised to gaze down onto a small patchwork of green and brown countryside, irrigation channels and rice paddies, and the meandering finger of the Red River, clinging to the easternmost edge of Indochina, over which he had flown weapon-laden aircraft only a few years earlier. The Vietnam War had begun as an attempt to halt the march of communism across Asia, but, in the spring and summer of 1973, as US military involvement there gradually wound down, it had become a severe embarrassment for the West and a murderous bloodbath which had already claimed tens of thousands of lives.

By the time that Skylab flew, opposition to that war had long since passed its zenith. The sheer numbers of casualties, the notoriously unpopular military drafts, the secret invasions of Cambodia and the ugly revelations about wanton rape and mass murder at My Lai had altered public perceptions of the conflict. Richard Nixon's policy of 'Vietnamisation' – building up the forces of South Vietnam to enable them to better defend themselves against the communist North, whilst at the same time steadily withdrawing US combat troops – had been only partially successful. Early in 1972, the resolve of the communists was amply demonstrated by the so-called 'Easter Offensive', involving a massive incursion into the South and attacks from sympathetic bases in Cambodia. The position of the South had only been saved by American airpower, but when the final US troops withdrew in August, the regime of President Nguyễn Văn Thiệu was left alone.

Efforts were underway to bring both the North and South to the negotiating table and Henry Kissinger – at that time Nixon's national security advisor and later Secretary of State – pressed ahead with secret talks, reaching an agreement of sorts in October 1972. It did not last and negotiations quickly became deadlocked. A US-led bombing of Hanoi in December destroyed much of the North's remaining economic and industrial capability and Nixon also threatened to cut off US aid to the South if the terms of the agreement were not accepted. In January of the following year, the Paris Peace Accords hammered out a string of agreements, officially ending the war, setting a 60-day timescale for US troops to leave the country and calling for national elections in both the North and South. In fact, the only part of the agreement which *was* honoured was the withdrawal of US troops …

The warlike North exploited the absence of American bombing raids to upgrade their logistics infrastructure, putting them in a position to launch a major invasion of the South. In March 1973, Nixon threatened a resumption of military action if the communists disavowed their pledge and violated the ceasefire, but he found little support in either public or congressional circles for troops to be dragged once more into the conflict. In June, Secretary of Defense James Schlesinger reasserted that he would recommend a resumption of bombing if the North tried to launch an attack against the South, but the Senate quickly prohibited any such intervention. By January 1974, the North had made significant territorial inroads into the South and had recovered much of what it had lost the previous year. Two clashes left several dozen South Vietnamese soldiers dead – and, indeed, since the *ceasefire*, the South had suffered more than 25,000 civilian casualties. President Thiệu declared that the North had violated the Paris accords and that his country was once more in a state of war. Richard Nixon's faltering administration, embroiled as it was in the revelations of Watergate, had other things on its mind and even his successor, Gerald Ford, was virtually impotent in trying to drum up more military aid for the South.

By the end of 1974, an increasingly weary Congress cut financial assistance by almost a third, down to $700 million per annum, and pressed its advantage further, calling for more restrictions in 1975 and a final end to funding by 1976. The North Vietnamese, though, remained uncertain as to how far the United States would go to protecting the South: in December 1974, they attacked Phuoc Long Province and captured its provincial capital within a matter of weeks. The US did nothing. Gerald Ford sought funds to resupply the South, but Congress refused. At this point, despite having several times as much artillery, tanks and armoured vehicles as the North and a numerical superiority in the order of two-to-one in combat troops, the South began to crumble under the pressures of increasing oil prices and lack of US support. The North, in contrast, was well-organised, highly-determined and well-funded. In April 1975, Thiệu, by now fearful of a coup against him and blaming the deteriorating situation on the "irresponsible" withdrawal of the US forces, resigned in tears and fled to Taiwan. A few days later, the North Vietnamese overran Saigon. The war was over by the 30th, marking an overwhelming communist triumph. Today, almost four decades later, and despite now possessing many of the trappings of capitalism, Vietnam remains one of the world's last bastions of communist rule.

As the war in Vietnam headed towards its end, in the middle of June 1973, the expedition of Conrad, Kerwin and Weitz similarly entered its homestretch. Although their main success had been the salvation of Skylab from almost certain failure, the three men worked feverishly during their last two weeks to accomplish as many of their scientific tasks – ATM observations, EREP passes, medical experiments – as possible and to leave the station in as good a condition as possible for the arrival of Al Bean's crew at the end of July. In fact, by the end of their third week, against all the odds, they had achieved 81 percent of their ATM observations, 88 percent of the EREP surveys and 90 percent of the medical tests. However, Kerwin found that one of the biggest problems with the control console of the ATM was its flare detection alarm system; it was meant to alert the crew if a solar flare developed, but frequently went off whenever Skylab passed over the eastern portion of South America, where Earth's radiation belts dip much lower than normal in the region known as the South Atlantic Anomaly. This increase in the magnetic field duly triggered the alarm and prompted Kerwin to lambast it as "useless" after the flight. Still, on 15 June, a good-sized solar flare materialised and Weitz managed to track it through two full minutes of its rising portion and the subsequent fall, monitoring several displays at the ATM console, initiating the relevant programs on the computer and pinpointing the target with the solar telescopes. Operating the ATM was a complex task in itself, involving a so-called Joint Operating Program (JOP), whereby sometimes as many as four or even five instruments would need to be employed almost simultaneously. Bob Parker, a member of the 1967 scientist intake, once described running a JOP from the ATM console as like playing an organ, with all of its complexity, whilst Bill Pogue summed up all the manual sequencing and the need for split-second timing and the precise positioning of instrument gratings with just a single word: *frustrating*.

"I'd like you to be the first to know," Kerwin advised Mission Control of Weitz' achievement, "that the pilot is the proud father of a genuine flare!" In this and many other instances over the following months, many of the criticisms of the ATM and concerns about the role of man-tended observatories in orbit evaporated. The crew's work was acknowledged on 17 June, when President Nixon himself called Conrad to compliment them on their superior performance. This performance, it seemed, also extended to physical training, as the enormous habitable volume of Skylab permitted almost unimaginable feats of gymnastics.

Original plans called for Conrad and Kerwin to perform a two-hour EVA, late in the mission, to replace film cassettes in the ATM. However, it was decided to offer Weitz another opportunity to venture outside. Working from the airlock module, he and Conrad began their ascent to the ATM, using five workstations along a route which they had nicknamed 'the EVA Trail'. They moved with the aid of a series of single and double hand rails and their movements were "as easy as driving down a freeway" and then set to work at the observatory. Conrad replaced canisters of film in the six instruments, cleaned the optics of the white light coronagraph using a fine brush to reduce glare, retrieved a set of exposed material samples and returned to the airlock. They also managed to free a stuck electrical relay. The excursion lasted 96 minutes, over an hour less than the time allotted. In total, between them, Conrad, Kerwin and Weitz had chalked up over five hours of EVA time in perhaps the most dramatic fashion ever attempted at that time.

With things going so well, Mission Control considered keeping the astronauts in orbit for another week, extending the mission to 35 days. Pete Conrad and his crew were game, but eventually it was judged unnecessary. "We finished our routine," recalled Kerwin, "got to the day before re-entry, which was basically de-activation day, powering it down, taking a lot of close-out photos [and] getting rid of all the trash." This final disposal of trash almost led to a problem. Since the workshop was essentially a modified S-IVB stage, its hydrogen tank provided the habitable area and its oxygen tank served as a storage area for rubbish. Typically, the astronauts would open a hatch at the base of the hydrogen tank, insert their bags, close the hatch and then activate a plunger mechanism to shove the detritus down into the unpressurised oxygen tank. "It worked fine until the very last day," Kerwin explained. "Weitz put [some trash] in there and he closed the top hatch. He opened the bottom hatch and went to plunge it ... and it wouldn't plunge. It was stuck halfway through the bottom hatch and it looked very much as if we had just destroyed the trash provisions on Skylab." Thankfully, after a couple of hours of jigging the plunger, Weitz managed to push the trash into the tank. The three men cheered. They were going home after all.

For Conrad, he knew it was the end of his astronaut career; although he would continue in his role as chief of the Skylab branch until the spring of the following year, he had already made the decision to retire from NASA after this flight. He was 43 years old and – whilst in orbit – had celebrated his 20th wedding anniversary with his then-wife, Jane. He took one last, long look at the first glimmers of orbital sunrise over the vast bulk of Australia, looming in Skylab's wardroom window, then floated up the length of the station to the command module. The satisfaction of the

After completing the longest manned spaceflight to date, the command module is winched aboard the USS *Ticonderoga* on 22 June 1973.

last four weeks had incomparably eclipsed walking on the Moon; he, Kerwin and Weitz had soundly broken the 23-day space endurance record, set two years earlier by the ill-fated Soyuz 11 crew, had saved this multi-billion-dollar national research laboratory from an ignominious death and – perhaps most importantly – had demonstrated that the presence of *man in space* was acutely necessary. Automated machines *could not* have saved Skylab in the way that his crew had so expertly done.

At 4:58 am, Florida time, on 23 June, Conrad, Kerwin and Weitz undocked from Skylab, thus ending its first period of human occupation. The command and service module performed a fly-around inspection, during which visual observations were made and photographs taken, before the large SPS engine was fired in a separation manoeuvre. This 'shaping burn' lasted ten seconds and shaved about 80 m/sec from their orbital velocity. Three hours later, at 9:10 am, a second burn, lasting just seven seconds, reduced their velocity still further, placing the spacecraft on a trajectory to intercept the atmosphere. Kerwin vividly recalled the experience of sitting through an SPS burn. "We're taking it in the couches," he told the NASA oral historian, "lying 'sideways' ... and two of us *greyed out*. Damnedest thing I ever saw. Your visual field *contracts* like it does when you're pulling Gs in an airplane." The sensation surprised Kerwin, at first – could 28 days in space *really* have affected him *that* much? – and, years later, he remembered wondering if they would be able to tolerate re-entry after such a lengthy period of weightlessness. However, the remainder of the descent did not pose a problem, with gravitational forces moving no higher than about 4.5 G. "I guess you might say it was just sort of settling the blood back where it belonged," Kerwin concluded.

Paul Weitz also remembered experiencing "some tunnelling" in his vision after four weeks in space, "which surprised us mightily", but it only persisted during the first SPS burn; by the time of the second firing, both he and Kerwin were back to normal. Pete Conrad, of course, making his fourth re-entry, was totally unaffected. Weitz, who would also later fly the Space Shuttle and experience its quite different re-entry scenario, remembered looking over his shoulder and seeing a vast 'wake' of ionised material stretching behind the command module as it plunged through the atmosphere ... and *burning pieces* of something hurtling past the window. "I knew that as the heat shield ablated, you've got vapour and hot gas," he recalled, "but *nobody* told me that *burning pieces* came off! I was a little startled, to say the least."

The splashdown, around 12 km from the prime recovery vessel, the aircraft carrier USS *Ticonderoga*, was exceptionally smooth. The 'Tico', in fact, was midway through her final voyage, bringing to a close a career of almost three decades, which had seen service in the Philippines and Japan in the Second World War and later in Vietnam, as well as recovery of the Apollo 17 astronauts in December 1972. She would be decommissioned in September 1973 and sold for scrap the following year. The command module bearing Conrad, Kerwin and Weitz hit the waters of the Pacific at 9:49 am Eastern Standard Time (8:49 am in Houston), wrapping up a record-setting mission of 28 days, 49 minutes and 48 seconds and more than four hundred orbits of Earth. "We landed so close to the carrier," said Kerwin, "they almost had to *move* to avoid our landing on the flight deck!" For the first time, a command module would be picked up by a ship's crane and settled onto the carrier, in order not to subject the astronauts to the stress of egressing into a raft and being hoisted into a helicopter; "they figured they'd go easy on us," he continued, "since we'd been up there for a month."

Conrad had told his crew before re-entry that each and every one of them were going to *walk* out of their ship – there would be no "carrying us out on stretchers" – but the conditions of the three men varied sharply. Immediately after splashdown, Kerwin advised his comrades that they needed to take some fluid. Without further ado, he knocked back a helping of strawberry juice ... and very quickly got himself seasick. "I should have chug-a-lugged the damn drink with *breakfast* that morning instead," he explained, "but I had it on the gently bobbing command module and ... I didn't feel very good!" Weitz thought it would have been better for Kerwin to have remained in his couch. With the benefit of hindsight, so too did Kerwin! Still, recalled Mel Richmond, head of the NASA recovery team, Conrad was spring-loaded and ready to go, paused momentarily on his haunches, when the command module's hatch finally opened.

On the *Ticonderoga*, steaming back towards San Diego, flight surgeon Chuck Ross was midway through his medical evaluation of the three astronauts, when he received a personal call from Richard Nixon himself. The president wanted Conrad and his crew to come to the Western White House for dinner in a couple of days' time, where "a guest" would be "awfully keen" to meet them. The guest concerned was Leonid Brezhnev, who had come to the United States in an official capacity as part of negotiations of the Strategic Arms Limitation Talks and their follow-on treaties. Nixon asked Ross how soon he could release the astronauts from post-flight quarantine. In her biography of her late husband, Nancy Conrad paints a humorous picture: *officially*, the three men should have been undergoing a rigorous, measured and steady re-adaptation to terrestrial gravity, including many days of heavy medical observation, decontamination and a careful re-introduction to normal air and food. In reality, Conrad had already smoked a fat Cuban cigar and knocked back a celebratory tumbler of whisky ...

Regardless of Ross' opinion, NASA had already agreed to Nixon's request and as soon as the *Ticonderoga* docked in San Diego, a Sea King helicopter flew the three astronauts and the flight surgeon to Marine Corps Air Station El Toro, from whence a bullet-proof Secret Service car raced them off the freeway ... and was promptly pulled over by a police patrol! The officer considered it more than a little odd that four men were sitting in the back of a car, hurtling through a sleepy coastal town on a Sunday morning, wearing hospital-style face masks. The incredulous image is conveyed in the words of Nancy Conrad:

"Gentlemen, take your masks off."

"Officer, these men have just returned from space."

"Yeah, right. Drop 'em. *Now*."

"Look, pal," interjected the Secret Service agent, more than a little irritated, "they're the NASA astronauts. You know, Skylab?"

"What the *hell* is Skylab?"

Clearly, the miraculous repair of a multi-billion-dollar national asset, and weeks of wall-to-wall television news coverage, had been lost on at least one representative of the San Clemente Police Department. Fortunately, a quick glance at the balding, gap-toothed Conrad reminded the cop of someone he knew ... and, at length, the identities were revealed. They even signed a couple of autographs for the embarrassed officer's children. When they finally arrived at the Western White House, La Casa Pacifica, on the beaches of San Clemente, they spotted Nixon and Brezhnev approaching on a golf cart. Instantly, Conrad took a command decision – "We're going to be with the President and the head of state of the USSR," he told Kerwin and Weitz, "and I ain't wearing no damn *mask*" – and promptly pulled it off. His two colleagues, of course, did the same.

Interspersed between the giving and sharing of gifts, Brezhnev – a jocular bear of a man, all three astronauts agreed – invited them to dinner in Moscow at some future date. Conrad agreed, but informed the General Secretary that although he would fly *anywhere* for a free meal, he was not prepared to do *anything* until he had spoken to his wife. "Sure enough," said Kerwin, "the following year, we got this invitation to join a group, with our wives, and had a very interesting tour of Russia." At this time, of course, détente was at its peak and the Apollo-Soyuz Test Project was scheduled for the summer of 1975. No one could have foreseen how dramatically US-Soviet relations would freeze in the coming decade.

For now, America had reached the very apex of its adventure in space and the Soviet Union was lagging seriously behind. Almost two years had passed since the deaths of cosmonauts Georgi Dobrovolski, Vladislav Volkov and Viktor Patsayev and, although Russia's manned spacecraft had been extensively improved and was ready to fly again, not a single cosmonaut had ventured into the heavens – or *would* venture into the heavens – until the early autumn of 1973. In the meantime, a dozen citizens of the United States had left their bootprints on the lunar surface, the Saturn V rocket had delivered both them and Skylab with near-perfection and America had proved the importance of having *men*, and not just *machines*, in space. Since the pioneering EVA by astronaut Ed White in June 1965, more than 30 spacewalks or Moonwalks had been conducted ... and all but *two* of those had been undertaken by US astronauts. The Soviets' dream of landing men on the Moon had vanished, its Soyuz spacecraft was technologically far behind Apollo and *now*, it seemed, the Americans were streets ahead in long-duration operations aboard Earth-circling space stations. Behind the smiles at La Casa Pacifica, Leonid Brezhnev's 1969 promise that the Soviet Union would steam ahead with its space stations, as "the main road for man into space", and thereby eclipse the achievements of Apollo on the Moon, now seemed more than a little hollow.

In retrospect, it is a pity that Skylab was so woefully under-utilised, and that none of the Nixon, Ford or Carter administrations adequately funded *both* the Shuttle and a follow-on series of missions to save and continue operating the workshop. The last few decades in space might have been quite different. Instead, the indifference of America's senior politicians to human space exploration – and their short-sighted *need* to make access to low-Earth orbit (supposedly) cheaper and more efficient – would leave NASA's hopes and dreams essentially hamstrung and in the process would leave the door wide open to a series of Soviet space spectaculars. In fact, by the time that astronauts John Young and Bob Crippen rode the first Shuttle into

"I ain't wearing no damn *mask*!" In 1973, détente between the United States and the Soviet Union was at its peak, Apollo-Soyuz was just two years away and a dinner date in Moscow awaited Pete Conrad, Joe Kerwin and Paul Weitz. Best to check with the wives first …

orbit in April 1981, the records set by Skylab would have long since been superseded, two and three times over, by cosmonauts aboard a series of no fewer than *three* Soviet space stations. As NASA and its Washington-based bean-counting masters sought a 'routine' access to space, the Soviets would already be *living* there.

But not yet. Betwixt Conrad's impressive example of triumph over adversity and the remarkable voyage of Apollo-Soyuz in July 1975, two more Skylab crews would further advance America's endurance lead in space. They would spend, respectively, two and three months in orbit and their work would establish Skylab as an unrivalled research laboratory in the sky ... truly a 'Sky Lab', with the emphasis placed most strongly on the latter. The activities of Al Bean and Owen Garriott and Jack Lousma and or Gerry Carr and Ed Gibson and Bill Pogue would prompt a re-evaluation of the abilities, performance and *requirement* for humans in space by literally rewriting the textbooks on solar astronomy, surveying our Home Planet in new ways, fulfilling many of the dreams of the early thinkers of Apollo Applications ... and even bagging a glimpse of a comet to boot. The final two Skylab missions would also bring a wry smile of satisfaction to the faces of the 1965 and 1967 scientist-astronauts. Not for nothing would Bean, in command of the second crew, describe *his* mission as "one hundred percent more productive" with scientist-astronaut Garriott aboard. In fact, on these expeditions to Skylab, the very distinction between 'scientist' and 'pilot' would begin to blur and 'Real Science' would take centre stage.

3

A tale of science, sickness and the Sun

LONG-HAIRED MESSENGER

Since the first recorded discovery of their remains, a few kilometres east of Düsseldorf in western Germany, the Neanderthals have received a particularly harsh press in popular culture. Brutish, grotesque and enormously strong, yet not quite *human*, the old one-liner *Where the men are men … and so are the women* seems to have been invented perfectly for them. Even the workers who excavated their bones from the Neander Valley in the summer of 1856 were convinced at first that they had found a bear, not one of our distant ancestors. Dragging their womenfolk and exhibiting nothing but the most rudimentary elements of humanity, the Neanderthal was the archetypal caveman, wiped out a few tens of thousands of years ago by the greater intelligence, better adaptability, higher technological expertise and strong social cohesion of anatomically-modern people. As with so many things in life, the reality was quite different. More than a decade ago, my fascination with the Neanderthals as a history undergraduate, led me to write my final-year dissertation about them. Fellow students joked that my dissertation would be the *shortest* ever filed in the university archives; what *else* could I possibly write about the Neanderthals, they jested, apart from the word 'Ug'? In the end, I spent a fascinating year researching and writing about them and learned a great deal about this much maligned and often misunderstood predecessor of ours. In fact, I found them to be not too dissimilar to us.

Scientific opinion today, thankfully, has long since begun to shift from that of the snarling caveman to a maker of highly sophisticated stone tools and implements, the guardian of a bigger brain than modern humans and the prime custodian of a vast swathe of Earth for more than a hundred millennia. In fact, the first physical traits of the Neanderthal can be traced back almost a quarter of a million years and the 'classic' specimen in its final form ranged widely across Europe, the Levant and Central Asia from around 150,000 years ago, during the Middle Palaeolithic, when the ice sheets began to recede, the last Ice Age steadily gave way to the present 'interglacial' period and anatomically-modern humans proved more adept at coping with new environmental conditions and problems. Pockets of Neanderthals continued to survive for a while, as evidenced by their bones and tools excavated in the coastal regions of Spain in the west and Uzbekistan in the east, but they were gone from the archaeological record not long after 29,000 years ago. Their intellect, the quality of the implements which they left behind and the complexity of their social structure has long persuaded many anthropologists that these creatures were far from dumb brutes. Remains from Shanidar Cave in Iraqi Kurdistan have suggested that they cared for their elderly and, perhaps, that they placed flowers and left worked-chert points at the graves of their dead, which some scholars have seen as possible indicators of shamanism and ritual. Elsewhere, at the Kebara rockshelter in Israel, fragments of a hyoid bone – a component of the larynx and a key player in the production of speech – have been found, adding to a hefty weight of evidence for sophisticated language skills.

A hundred and fifty thousand years ago, the 'classic' Neanderthal had well and truly arrived on the scene and, maybe, at that time, one of their number happened to set down a half-made chert tool, look up into the night sky from the crackling embers of his fire and behold something *moving* against the ethereal blackness: a strange star *with a tail* – a kind of 'long-haired messenger' which, far into the future, would convince King Harold's men that ill fortune awaited them on the eve of the Battle of Hastings and which, even today, the superstitious among us continue to regard as an omen of cruel luck. They are 'comets' and throughout history, their manifestation has aroused fear and uncertainty: to the ancients, they foretold an imminent catastrophe – a plague, perhaps, or a flood – or the impending death of a king or a nobleman. Plutarch famously related that a brilliant comet shone for seven nights over Rome after the assassination of Julius Caesar in 44 BC. To the biblical Book of Revelation and the Jewish Book of Enoch, such 'falling stars' were believed to represent heavenly visitations and in 7 BC King Herod is said to have been warned that the apparition of a 'hairy star' over Judaea would herald the birth of a boy whose achievements would outshine his own. Centuries later, a comet blazed in the skies of Gaul as Attila the Hun overran one of the last provinces of the Roman Empire and in 1456 Pope Calixtus even *excommunicated* a comet which had appeared after the Turkish conquest of Constantinople. His declaration had little impact. The comet upon which the pontiff vented such fury would return again and again – every three-quarters of a century, in fact – and today it is revered as one of the best-known and most celebrated: Halley's Comet. Even in relatively recent times, though, superstition has taken over from common sense. When Halley reappeared in 1835, it was accompanied by widespread panic that it would crash into Earth and at its next visitation in 1910 people even bought masks and 'comet pills' to protect themselves from the noxious gases of its tail. Novelist Mark Twain, who famously penned the tales of Tom Sawyer and Huckleberry Finn, happened to have been born in the year of Halley's 1835 apparition. Later in life, he became convinced that he would die during its *next* appearance in 1910. His reason: "The Almighty has said, no doubt, here are two *unaccountable freaks*. They came in together; they *must* go out together!" Twain died in April 1910, only days after the comet made its reappearance.

Indeed, until the 16th century, the mainstay of scientific opinion about the nature of comets rested with the Greek philosopher Aristotle, who believed that they could not possibly belong to the 'perfect' celestial realm and so posited that they were a phenomenon of the upper atmosphere, from which hot, dry exhalations gathered and burst into flame. Then, in 1577, the Danish astronomer Tycho Brahe used measurements of a comet taken from several geographical locations, including some by himself, to determine that it *must* have been at least four times more distant from Earth than the Moon. In Brahe's mind, comets *could not* be elements of the upper atmosphere, but must represent something from beyond.

Since the beginning of the Space Age, our understanding of these long-haired messengers has grown ever sharper: we now know them to possess irregularly-shaped nuclei of rock and water-ice and a multitude of other volatiles – prompting the nickname of 'dirty snowballs' – and as they approach the inner Solar System, these volatiles quickly vaporise and stream away, carrying dust particles with them. These streams create a vast 'atmosphere' around the comet (its 'coma') and the force exerted by the Sun's radiation pressure and solar wind cause immense 'tails' to form, sometimes many millions of kilometres long. Our knowledge of their appearance, their composition and their evolution has multiplied, thanks to our extraterrestrial emissaries: from Giotto's impressive photographs of the nucleus of Halley's Comet, with its 'jets' of dust, in the winter of 1985-86 to Deep Space 1's astonishing shots of the hot, dry Comet Borrelly 1 in September 2001 and from Stardust's collection of crystalline, "fire-born" material from Comet Wild 2 in January 2004 to Deep Impact's spectacular effort to blast a man-made crater into Comet Tempel 1 in the summer of 2005. Further flights of exploration are also planned. A second voyage of the Stardust mother craft revisited Tempel 1 in February 2011 to analyse the after-effects of that collision and to test the hypothesis that much of the ice in a comet is stored in subsurface 'reservoirs'. The European Rosetta spacecraft will land its instrumented Philae probe onto Comet Churyumov-Gerasimenko in November 2014. Once there, Philae will use harpoons and drills to physically anchor itself, for the first time, onto the surface of a comet. All of these ventures have carried a substantial monetary cost, but the media exposure which they have afforded and the popular interest which they have garnered tell us that comets are just as fascinating today as they were to our most ancient ancestors.

If a Neanderthal did happen to look up from his fire and his tool-making into the darkened sky, seven and a half thousand human generations ago, he might have seen one such long-haired messenger, faint, but possibly visible, passing overhead. He may have wondered what it was, where it came from and what it meant. What he could not have possibly known, however, was that it would not be seen again by the eyes of a hominid until a year which in the Western consciousness came to be known as Nineteen Seventy-Three, the year after his distant descendants had left their last footprints on the Moon and the year in which humans would begin to routinely live away from the Home Planet for the first time. The comet was spotted on 7 March by a Czech astronomer named Luboš Kohoutek, a professor at the Hamburg-Bergedorf Observatory in West Germany, whilst making observations of minor planets. At the time, it was little more than a diffuse point of light, moving slowly north-westward in the constellation of Hydra (the Water Serpent), but its distance from the Sun – some five astronomical units, or around 750 million km, about as distant as the orbit of Jupiter – quickly whetted astronomers' appetites that it would be a particularly bright comet. Calculations by Brian Marsden of the Smithsonian Astrophysical Observatory in Cambridge, Massachusetts, added that it would reach a perihelion of just 0.14 astronomical units *and* would subsequently pass Earth in January 1974 at a distance of just 95 million km.

All in all, the early discovery of Kohoutek's object pointed to something intrinsically brighter than Halley's Comet and probably something quite large, too. Elizabeth Roemer of the University of Arizona estimated that the nucleus could measure 35 km or more in diameter, whilst other astronomers speculated that it might weigh many billions of kilograms. Spectroscopic evidence for water-ice had been obtained and astronomers had reason to hope that hydrogen cyanide and methyl cyanide – both 'polyatomic molecules', previously only detected in intergalactic space, never in comets – were present in the nucleus. Consequently, in the late spring of 1973, Comet Kohoutek carried much promise. Its hyperbolic trajectory led to theories that it originated in the Oort Cloud, a spherical 'shell', far beyond the orbit of Neptune, on the very edge of the Solar System, and as such it was suspected that this might be the latest in a series of very infrequent visits to the inner Solar System. Unlike the 'worn-out' Halley's Comet, the possibility that Kohoutek might be a 'virgin' comet from the Oort Cloud sparked optimism that it would offer scientists a chance to study material virtually unchanged from the primordial state *and* that it might be visually stunning.

A spectacular display of 'outgassing' of this material and a brilliant, glittering tail as it neared the Sun was also increasingly likely. *Sky & Telescope* magazine predicted in May that it should be a conspicuous, naked-eye object of first magnitude or brighter, whilst other writers expected it to rival or even surpass the best views of Halley's Comet. It was predicted that by early 1974, after perihelion, Kohoutek's tail would be a fully grown and shimmering 'streamer', extending across maybe one-sixth of the night sky. In such an eventuality, exulted Harvard astronomer and comet specialist Fred Whipple, it might "well be the comet of the century".

Not everyone was convinced, however. Many observers acknowledged that comets were notoriously unpredictable and even *The New York Times* warned that Kohoutek might not live up to its billing, but none of this deterred the sky-watchers of 1973. Kohoutek shows were held in planetaria across the world, binocular and telescope purchases picked up and escalated at an exceptionally brisk pace – one company even announcing a 200 percent profit in its sales – and the luxury liner QE2 sailed from New York with 1,700 passengers on a special 'comet cruise'. New York's Hayden Planetarium planned a spectacular, six-day 'Flight of the Comet' aboard a chartered Boeing 747, in time for the January 1974 perihelion, when Kohoutek was expected to reach maximum visibility. With candlelit (and *comet-lit*, it would seem) dinners offered for a total package of more than a thousand dollars per person, it was bound to be spectacular. Yet as one journalist later remarked, the desire for success has a tendency to make us smug, and although Kohoutek *would* indeed be visible to the naked eye and *would* reach a magnitude of about -3, it did not live up to the hype. Today, it is generally thought not to be a pristine Oort Cloud comet, but a rocky object from the Kuiper Belt, a disk of material in the outer reaches of the Solar System, a 'long-period' comet which, after its appearance in Neanderthal skies and in 1973, will not again grace humanity with its presence for another 75,000 years. In fact, perhaps a little unfairly, Kohoutek's name has become synonymous with spectacular duds. As for the unfortunate ocean voyagers, their thousand-dollar 'comet-lit' QE2 dinner tables actually brought them little more than disappointment, cloudy skies and seasickness . . .

Today, Kohoutek's greatest claim to fame is that it appeared when it did: for America's Skylab space station, with its powerful Apollo Telescope Mount (ATM) and battery of sophisticated astronomical equipment, was in orbit and its second and third human crews – both of which included professional scientists among their number – were primed and ready for their missions of discovery. The timing could hardly have been better. Astronaut Karl Henize, himself an astronomer, summed up an excited spirit of optimism in the scientific community as spring burned into summer and summer cooled into autumn in 1973. He saw Kohoutek, potentially, as an astronomical 'Rosetta Stone': an unexpected find, an incredibly fortuitous quirk of serendipity, which might reveal vital clues to unlock the mystery of how the Sun and its attendant planets – and, by extension, *ourselves* – came into being, more than four and half billion years ago.

A SICK CREW AND A SICK SHIP

As June wore into July 1973, Skylab's fortunes had been snatched from the jaws of defeat and a pyrrhic victory of sorts had been transformed into a triumph. During launch, one of the space station's power-producing solar arrays and its micrometeoroid shield had been torn away, creating a very real possibility that the multi-billion-dollar effort was over before it had even begun. Then, in a remarkable example of NASA's 'can-do' engineering spirit and a rapid-fire ten days in May, a practical repair had been conceived, perfected and tested on the ground, then executed spectacularly by Skylab's first crew, Pete Conrad, Joe Kerwin and Paul Weitz. The stage was set for the arrival of the second team of astronauts in late July, who would attempt a record-breaking mission of 56 days in orbit. Then, on 20 July, with barely a week to go before their launch, the flight of Al Bean, Owen Garriott and Jack Lousma was slightly extended to 59 days, in order to provide a better recovery posture for their scheduled splashdown in late September. Whilst in orbit, they would build upon the achievements of their predecessors, more than doubling the time spent aloft and performing over five dozen scientific and medical experiments. Their mission would also include high drama, including debilitating space sickness only hours after launch and the very real possibility of a fully-fledged rescue . . .

Commanding the mission was Alan LaVern Bean, a man who undoubtedly owed his second chance at flying into space to his good friend and mentor, Pete Conrad.

They had flown together on Apollo 12 in November 1969, setting their bootprints into the ancient dust of the Moon's Ocean of Storms, but the opportunities afterwards had been somewhat limited. Conrad's flight experience quickly assured him the headship of the Skylab branch of the astronaut office and he advised Bean – who had been assigned as chief of Apollo Applications back in 1966 – to join him. Bean took the advice. Conrad extended the same gesture to his other Apollo 12 crewmate, Dick Gordon, but the latter wanted to stick with Apollo, hoping to someday command a lunar landing voyage of his own. Unluckily for Gordon, a near-disaster on Apollo 13 in April 1970 and a steady thinning of NASA's budget scrubbed the final few landings from the manifest. Gordon never made it to the Moon's surface. For Bean, however, a seat on a long-duration Skylab mission was assured in January 1972, when he was named to lead its second expedition, planned for the summer of the following year.

Born in Wheeler, Texas, on 15 March 1932, Bean's father worked for the Soil Conservation Service and his mother ran her own ice cream shop. A keen athlete, he received his degree in aeronautical engineering from the University of Texas at Austin in 1955 and joined the US Navy through the Reserve Officers Training Corps. Initial flight instruction was followed by assignment to a jet attack squadron, based at the Naval Air Station in Jacksonville, Florida. From an early age, Bean was fascinated by painting and took night classes in oils whilst in Florida; indeed, after leaving NASA, he would devote his life to conveying the story of his lunar adventure on canvas. Today, his paintings routinely fetch thousands of dollars apiece and he always adds a few unique finishing touches to each one: a smearing of real Moon dust, salvaged from his own space suit patches, or a print made by a real lunar boot or a scratch made by a real Apollo geological hammer or a groove etched by a core tube sampler. Bean has been interviewed many times over the years and has always described himself as "an artist, creating paintings that record for future generations mankind's first exploration of another world". Some astronauts have said that going to the Moon did not change them. For Bean, walking amongst the craters of the Ocean of Storms was a profound part of his life and has guided him ever since.

In his history of the Apollo missions, Andrew Chaikin described how 'different' Bean was compared to most other astronauts, for he exhibited few of their macho qualities and their overarching desire to be first. Instead, he had followed a simple motto: Work hard and *someone* will notice. Since his days in Jacksonville, he had been alternately nicknamed 'Sarsaparilla', because he never touched a drink, or simply 'Beano'. As a squadron pilot and, later, as a test pilot at Patuxent River in Maryland, his sheer determination had enabled him to master weapons delivery systems and turned him into an outstanding aviator. Whilst at test pilot school, he met the one instructor who would most significantly change his life ... Pete Conrad.

All three Apollo 12 crewmen responded to NASA's call for astronaut candidates in the spring and summer of 1962; Conrad made it, but Bean and Gordon would have to wait another year before their time came. Selection to the world's most elite flying fraternity in October 1963 would bring both frustration and triumph for them. Within months, they had received their individual technical assignments. Gordon would head the Apollo branch of the astronaut office, whilst Bean was given duties pertaining to spacecraft recovery systems and range operations. Unsurprisingly, Gordon was first to draw a flight opportunity: he served as Conrad's pilot on Gemini XI in September 1966 and performed a spacewalk. However, Bean wound up as backup commander of Gemini X, which made him the first of his class to receive a command position (albeit in a backup role) on his *very first* assignment. The down side was that Deke Slayton's three-flight system of crew rotations should have pointed Bean to command Gemini XIII ... except that the project *ended* with Gemini XII! Nevertheless, the fact that he had been assigned a backup command implied to Bean that he was highly regarded by Slayton and Chief Astronaut Al Shepard and that a firm flight assignment *must* be just around the corner.

His hopes subsided when, in late 1966, he was detailed instead to work on the Apollo Applications project and any chance of a lunar mission seemed gone forever. In his autobiography, however, Slayton endorsed Bean's initial supposition: that he *was* highly regarded. "Al was just a victim of the numbers game," Slayton wrote. "I would only point to the fact that he was the first guy from his group assigned as a crew commander. I was confident he could do the job." Also brimming with confidence was Owen Garriott, the scientist-astronaut who would accompany Bean on his Skylab mission. "Anything that Alan undertakes," Garriott told a NASA oral historian in November 2000, "you can be sure he is going to give it his absolute 110 percent. He stays focused. He is a 'clean-desk' type of person: you go in at the end of the day [and] his desk is *clean*. Mine gets cleaned maybe once a year, if I'm lucky!" Clearly, Bean's years of experience in the military and as a naval aviator had contributed to his sense of order.

So it was that in the autumn of 1966, when Pete Conrad was asked to pick an astronaut to someday serve as his lunar module pilot, the organised, reliable, dependable Al Bean was his first choice. Slayton turned him down, saying that Bean was currently working on Apollo Applications, and Conrad opted instead for a burly and jocular Marine Corps aviator named Clifton 'C.C.' Williams. Less than a year later, in October 1967, Williams was tragically killed in the crash of his T-38 jet. Within weeks of his death, Conrad caught up with Bean with some good news, at last. He had asked Slayton to reconsider Bean for his crew ... and the request had been accepted. After a year in the relative purgatory of Apollo Applications, Bean's future held greater promise. That promise was delivered in spectacular style in November 1969, when Bean became the fourth man to walk on the Moon. A little more than three years after *that*, in the early summer of 1973, his time spent working on Apollo Applications also paid off and bore its own fruit, for Bean was now to command his *own* mission. This effectively brought him full-circle from being the *first* member of his class to receive a *backup* command position to the *last* member of his class to receive an *actual* command position. His mission would bring more than its fair share of surprises and challenges and would truly put Skylab to work as a *scientific* platform from which astronauts could operate for lengthy periods of time. A handful of words from Bean's oral history, recorded by NASA in June 1998, sagely summed up his rags-to-riches career path as an astronaut: "Life is a dance," he said. "You learn as you go."

That dance seemed to draw ever closer as Pete Conrad's team completed their repairs on the station. Fears about the health of Skylab's stabilising rate gyroscopes and the possible deterioration of the makeshift parasol prompted NASA to get the second crew aboard as soon as possible. On 6 June, the launch of Bean, Garriott and Lousma, which was originally scheduled for early to mid August, was moved forward to 27 July, with an option to fly as early as the 22nd if the situation worsened. Ultimately, mission managers settled on the 28th as the most desirable date, since it offered more suitable conditions for a rendezvous with Skylab during the astronauts' fifth orbit. Moreover, noted Bill Schneider in a NASA news release, the revised launch date would "schedule the mission at a time when the relationship of the Sun to the [workshop's] orbit plane is most favourable ... and will therefore provide the most power for conducting the experiments." As these plans crystallised, the Saturn IB rocket and the attached command and service modules were rolled to Pad 39B on 11 June. A series of inauspicious lightning strikes on the launch tower certainly raised a few eyebrows, but the damaged parts – including instrumentation for the spacecraft – were replaced and retested without incident. The countdown commenced on the morning of 25 July, with launch set to occur at the start of a ten-minute 'window', opening at 7:08 am Eastern Standard Time on the 28th.

In orbit, the station itself had been carefully depressurised and its internal temperature was kept at a controlled level to prevent condensation whilst unoccupied. Solar observations were also pursued, autonomously, using the ATM instruments, although this work was suspended following the failure, on 19 July, of a primary rate gyroscope – one of nine electronic instruments used to provide primary and backup control of the station's orientation in the roll, pitch and yaw axes. Four days later, NASA announced that Bean's crew would carry a 'six-pack' of replacement gyros (properly termed a 'rate gyro augmentation package') to provide redundancy for both its stay and that of the final crew. "At present," the space agency reported, "one [gyro] has been turned off because it malfunctioned, while five others have overheated to some degree at one time or another during more than two months the spacecraft has been in orbit." The news release cautioned that an actual replacement would probably not be effected unless *all three gyros* failed in one axis, requiring approximately 20 minutes of EVA time. Also aboard the command module was an improved version of the parasol. The astronauts had trained extensively to install both this and the Marshall Space Flight Center's twin-pole thermal shield, which had been developed in May to provide backup micrometeoroid protection and supplement the parasol and had been taken to the station by Conrad's crew, but not deployed. Twenty hours before the crew was set to lift off, Skylab's atmosphere was repressurised, in anticipation of the arrival of its second team of human visitors. Despite its very long duration, Bean's mission would be packed to the rafters with scientific and engineering objectives: a preliminary timeline, released by NASA on 12 July, highlighted plans for an EVA on the 31st, in which film canisters would be loaded for the ATM and the twin-pole sunshade would be installed. Two more excursions were also scheduled for 24 August and 19 September for the purposes of retrieving and installing more ATM film.

If Skylab was ready for an influx of new arrivals, then the Kennedy Space Center in Florida was equally ready and received its own fair share of visitors on the day of launch, with no fewer than 35,000 people cramming the press site, public viewing areas and causeways, excited by the drama of the last several weeks and eager to observe the next part of the adventure. The morning of 28 July dawned dreary and overcast, evidenced by NASA television footage from the day, with only a few lights in the vicinity of the pad twinkling through the gloom and revealing the Saturn IB atop its milk stool. Aboard the command module, Bean was positioned on the left, with science pilot Garriott in the centre seat and square-jawed pilot Jack Lousma over on the right. During training, they had convinced themselves that all three of them came from the US Navy – Bean was an active captain in the service, Lousma was a major in the Marines (which "really is just a subset of the Navy," according to Garriott) and the science pilot himself was a veteran of three years of naval duty in his younger days. "My total time in the military," Garriott acquiesced, "hardly compared to the career that both Alan and Jack have, but that was a minor unifying factor and we had a chance to joke about it from time to time."

Owen Kay Garriott had been born in Enid, Oklahoma, on 22 November 1930, his Christian names honouring his father and paying homage to a diminutive of his mother's middle name, Catherine. In *NASA's Scientist-Astronauts*, Garriott told Dave Shayler and Colin Burgess that his ancestors on his father's side were French farmers and his own grandfather had routinely travelled all day to transport goods and supplies by horse and wagon from the closest railway stop to Enid's tiny convenience store. Years later, Garriott – the first of *two* generations of his family who would ultimately break the surly bonds of Earth and venture into the heavens – would express astonishment that, aboard Skylab for a few weeks in the summer of 1973, he would routinely cover the same distance as his grandfather … albeit in barely *four seconds*.

Garriott's fascination with physics and engineering emerged at a young age, thanks to his father, a geologist by training who spent his career working as a chemist and as an oil and gasoline distributor. One day in 1944, he came home and invited his teenage son to attend an adult class on radio theory with him. The class included electronics, an understanding of Morse code and the construction of radio and transmitting equipment and they both quickly passed their Federal Communications Commission exams. Engineering was now in the young man's blood and from high school he entered the University of Oklahoma on a Navy scholarship and received a degree in electrical engineering in 1953. As part of his commitment to the Navy, Garriott – by now married to his school sweetheart, Helen – undertook three years of active military service, acting as an electronics officer and participating in several tours aboard destroyers at sea. After the completion of his military duty, Garriott moved to Stanford University in Palo Alto, California, working in the Radio Propagation Laboratory, and completed his master's degree in 1957. Next came his doctorate, which he gained in 1960 and which offered him an inroad into the Space Age for the first time. "My dissertation," he told Shayler and Burgess, "used the radio signals from Sputnik 3 to study the electron content of the ionosphere." Garriott remained at Stanford for five years as a faculty member, specialising in electromagnetic theory and ionospheric physics and rising to the position of Associate Professor of Electrical Engineering.

A casual conversation with a friend over dinner, early in 1965, prompted Garriott to secure himself a private pilot's licence, with an instrument rating, and to submit an application to NASA for its much-publicised first selection of scientist-astronauts. In his oral history, more than three decades later, he doubted that his doctoral research or even his work as a member of the Stanford faculty was *necessary* for admission into the hallowed ranks of the astronaut corps, but the *credentials* – the degrees and the published papers – served to qualify him. "What NASA wanted then," he explained, "is not a world-class athlete, but somebody who has *nothing wrong with them*. They want average, everything down the middle, in terms of your biological characteristics. You wanted to be as normal as you can possibly be!" One area in which Garriott was 'borderline' was his eyesight: astronaut candidates needed 20/20 uncorrected vision. His left eye was fine, but his right eye was on the edge.

"Maybe 20/25," the optometrist told him.

"Well, let me blink a little bit," Garriott replied. "Let me see if I can't focus a little bit more carefully."

The optometrist retested him and agreed that he had just hit 20/20, although Garriott knew that his right eye at near distances was hazy. Years later, he recognised that if the optometrist had classified him as 20/25, he would never have been accepted by NASA.

Soon after their selection, the new astronauts – among them future Skylab science pilots Joe Kerwin and Ed Gibson – were despatched for flight training. Kerwin had already done a lengthy stint in the Navy and was a qualified jet pilot, but most of the others had never flown an aircraft before and even Garriott's private licence did not suffice. "Even though we were light airplane pilots," he explained, "three of us were not jet-qualified. The other three of us went to flight school at Williams Air Force Base in Arizona." After a year of instruction, the old heads in the astronaut offices must have been a little surprised to see the wet-behind-the-ears scientists coming back to Houston with their jet credentials. All of them – Garriott, Kerwin, Gibson and geologist Jack Schmitt – even underwent advanced helicopter training later in their NASA careers. Upon their return to the heat and humidity of Houston, however, it was back to the grinding business of preparing for future space missions. Jungle survival training in Panama and geological expeditions to Alaska and Iceland and Hawaii occupied their time, as did the steadily growing space station project, which was at that time known as Apollo Applications.

By the time he ascended the tower to Pad 39B on the morning of 28 July 1973 for his first launch into space, Garriott had been training almost exclusively on the station for more than five years and knew his two crewmates not only as colleagues, but as friends. All too often, the perception of Garriott as the 'science pilot' on the mission has led many observers to assume that the duties of his colleagues were exclusively piloting or engineering roles. The reality, of course, was that all three men were cross-trained and each was as competent and internally motivated as the others to accomplish Skylab's scientific objectives. Their deep-seated friendship persisted and, even three decades after their mission, they took family holidays together. Garriott was impressed by Bean's tenacity, attention to detail and unwavering focus on The Mission. Jack Lousma was "almost a conventional Marine" – it would be highly unwise to get in the way of this all-American boy after he had been given a task to complete, "because he is *going* to get it accomplished!" Years later, Garriott would pay tribute to Lousma's loyalty, friendship and an infectious personality which was impossible to dislike.

Jack Robert Lousma was born in Grand Rapids, Michigan, on 29 February 1936, making him unique amongst flown spacefarers, having entered the world on a 'leap day' and therefore becoming the only astronaut 'leapling'. For a member of the elite test-piloting community which populated the bulk of the astronaut corps at the time, it was there that the uniqueness ended, for Lousma grew up with an insatiable desire to fly. As a child, he built model aircraft and was amazed during a visit to his grandparents' farm when his cousin, a fighter pilot in the Army Air Corps, flew overhead. Recalling the electrifying event to a NASA oral historian in March 2001, the memory was still strong for Lousma. "He just came so low between the barn and the windmill, I could almost see his eyeballs!" Indeed, with another relative who became an airline captain, it would appear that aviation was in the Lousma family bloodstream.

Still, the young Lousma's first ambition was to become a businessman, "so I studied those kinds of courses and when I went to the University of Michigan [at Ann Arbor], that's where I started". Eventually, though, his love of aviation caused him to gravitate towards aeronautical engineering and he duly received his degree in 1959. Lousma placed his ambition of a master's credential on a back burner and sought out the military services, with the ambition of becoming a pilot. Unfortunately, both the Navy and the Air Force turned him down for pilot training, because he was married, but one day he spotted a group of Marines on the university campus, "all dressed up with their red stripes and turtlenecks". Engaging them in conversation, Lousma quickly discovered that the US Marine Corps did indeed have a pilot training programme which accepted married candidates.

Graduation from initial training took the now-Senior Lieutenant Lousma to Pensacola, Florida, for flight instruction, followed by Beeville in Texas for six months of advanced work. He received his wings in 1960 and moved to Marine Corps Air Station Cherry Point in North Carolina, serving first as an attack pilot and later a reconnaissance pilot with all-weather fighter and tactical electronic warfare squadrons. This exposure to the realities of military aviation convinced him that this was a career for him and in 1965 Lousma completed a master's qualification in aeronautical engineering at the Naval Postgraduate School. Later that same year, hungry for a new challenge, he noticed a NASA advertisement, inviting applications for a new class of astronauts. Lousma considered the whole thing to be a "don't call us, we'll call you" situation and had little illusion about being selected, but tendered his application regardless. The Marine Corps came back to him with a problem: his height. At 1.85 m tall, Lousma was an *inch* above the space agency's six-foot maximum requirement.

"I'm *not* an inch over six feet," he protested. "I've *never* been taller than six feet. I *know* that must be an error."

"It was on your last flight physical," came the response.

"Well, they must have done it incorrectly, because I'm *not* over six feet."

Whilst his application was being processed, Lousma was posted to Puerto Rico and Cuba for a couple of months, during which time he realised that when he stood next to the door each day, by nightfall he would be half an inch *shorter* than normal. "So I practiced standing all day," he told the NASA oral historian in March 2001, "and found that if I hunkered down just right, I could come *under* six feet. They must've been right, but I wasn't willing to admit it!" On returning to the United States, Lousma received a special measurement from his flight surgeon ... which clocked his height at 1.82 m, just a fraction of an inch *under* NASA's upper limit!

Medical screening at Brooks Air Force Base in Texas followed, after which Lousma returned to Cherry Point. Upon disembarking from his jet after flying a reconnaissance mission one day in the spring of 1966, he was greeted by a captain with a telephone message for him in the ready room. In a few adrenaline-fuelled moments that only the successful astronaut candidate can ever fully appreciate, Lousma was given the immortal words by Chief Astronaut Al Shepard: *You're in.*

A year of basic astronaut training was followed by technical assignments within the office. Amongst Lousma's earliest duties were EVA demonstrations of how to adapt the Apollo Applications wet workshop for human habitation. Like Al Bean, however, he continued to maintain an active role in the unfolding effort to land a man on the Moon, helping to check out Apollo 9 and 10's lunar modules before launch, then serving as one of the capcoms during the ill-fated Apollo 13 mission. In fact, Lousma was not just *any* capcom: he was the astronaut 'on console' on the evening of 13 April 1970 and it was he who received Jack Swigert's urgent call of "Houston, we've had a problem". Even three decades after the event, he remembered that night and the tense days which followed with crystal clarity: "It was just absolutely overnight, a *cataclysmic* change in the demeanour of the *whole* mission!" The safe return of the Apollo 13 astronauts also contributed to sounding the death knell for several later lunar landings, one of which Lousma had reason to believe he might have been aboard. During his support work on Apollo 9 and 10, he had accrued more than 700 hours in the lunar module simulator and in an oral history for NASA he recounted that he knew from around the end of 1970 that he would not fly to the Moon. When ASTP appeared on the horizon, Lousma signed himself up for a one-semester course in Russian, took the exams, informed Slayton and received the post of backup docking module pilot. At around the same time, however, he was told by Slayton of his selection to fly with Bean and Garriott aboard Skylab and noted that in total he and his crewmates spent around two and a half years in dedicated training. The official NASA press release announcing this selection finally appeared in January 1972.

When the astronauts for the second Skylab expedition arrived at the pad for their launch into space, Lousma still had to pinch himself. Clambering into the transfer van, encumbered by his bulky white suit and oxygen hoses, he had to constantly remind himself that this was *not* a simulation or a drill; it was the *real thing*. The flashlights of the assembled press were *for real*, as were the checkpoints, the last-minute examinations by the flight surgeon, the traditional astronaut's breakfast of steak and eggs, the sticking of biomedical electrodes to their chests and the pressure checks of their space suits. "The technicians ... weren't saying much of *anything*,"

Suit technician Al Rochford adjusts Jack Lousma's communications headgear on launch morning.

Lousma recollected, "because I think they were afraid they would disturb you. You could tell that they knew that this was the moment of truth and it was quite clear that this was the day we were all waiting for."

Out on the gantry of Pad 39B, awaiting his turn to board the command module, Lousma noticed that the normal beehive of activity and the dozens of hard-hatted technicians who normally swarmed around were conspicuously absent; the launch complex was eerily quiet, with hardly any sound, save for the hissing airflow inside his helmet. "Boy, this was the *absolute* 2001 experience," Lousma said. "This was *really* spacy." As the science pilot, occupant of the command module's centre seat, Garriott was the last to enter and had a few moments of private reflection, gazing across the slumbering Florida landscape. "There was a long training period leading up to this moment," he recalled. "A fiery rocket would soon take our speed from zero to over five miles *a second* ... in *less* than ten minutes!" It was a sobering realisation, which, by Garriott's own admission, he would remember lucidly, decades later. "We got in the spacecraft," continued Lousma. "They closed the hatch and went away. We're out there on the launch pad, all by ourselves now, and we can feel the [Saturn sway] in the breeze a little bit and we're hoping for good weather. We're there for about two and a half hours before liftoff, helping the ground crew check out the spacecraft. I fell asleep for a little while ... "

From the press site, several kilometres away, CBS anchorman Morton Dean was relaying live commentary to his audience. Next to him sat former Apollo astronaut Wally Schirra, who had been snapped up in 1969, after his retirement from NASA, as the network's 'special expert consultant' and had helped to cover each space mission since. "T minus 40 seconds," Dean intoned, as television images showed the Saturn IB, atop its milk stool, scarcely visible against the murk of the early morning. A few twinkling searchlights glimmered through the pre-dawn gloom. "The spacecraft commander has now made the final guidance alignment," he continued. "That's the final action to be taken by the crew on-board the spacecraft until after the launch. T minus 30 seconds ... the eight first-stage engines will ignite at 3.1 seconds in our countdown. They'll be held down while thrust is built up at the zero mark, at which time we'll get liftoff."

The *absolute* 2001 experience about which Lousma had dreamed was about to become a reality. The satellites about which Garriott had written his doctoral thesis would soon be his travelling companions for the next eight weeks. The oils with which Bean had long experimented at his painting classes could not come close to truly representing the sheer iridescence of colour that he would see whilst looking onto the Home Planet from on high. All three men were about to embark on the ride of their lives.

Morton Dean's voice noticeably notched up an octave as he prepared his audience for the big event and those final few seconds before the Saturn was set to shear its manacles and tear itself loose from Earth: "T minus ten, nine, eight, seven, six, five, four" – then a tongue of bright orange flame, which steadily thickened and expanded into a vast, blinding sheet, erupted from the base of the booster and licked at the metallic trusses of the milk stool – "three, two, we have ignition sequence start, all engines are running" – as the eight H-1 engines roared to near-full power, almost obscuring the rocket from view with their glare – "we have a liftoff ... and the second manned crew has cleared the tower!" As the Saturn rose into the overcast sky and quickly vanished into a deck of gloomy, lowering cloud, Dean turned to Schirra and chattered over the airwaves about his impressions of the launch: the rolling thunder and pulsations of sheer naked energy which now pummelled their bodies.

From his couch on the right side of the command module, Lousma noticed a very definite "heavy vibration" at the instant of liftoff. This sensation very quickly damped out, but the initial 'chugging' of the Saturn IB did little to disguise the tremendous acceleration, which peaked at around four and a half times the force of gravity. As the S-IB first stage expended its load of propellant and separated, the three men were greeted by an eerie, though temporary, silence, as they coasted for a while, interrupted by the detonations of explosive charges, then the seat-of-the-pants push as the single J-2 engine of the S-IVB took over and continued the boost into orbit. Through a command module window, Lousma vividly remembered seeing a vast, circular 'fan' of debris, flying outwards in all directions and glinting in the sunlight. For Owen Garriott, the sensation of being pressed into his seat at several times his Earthly weight to suddenly floating in his harness was electrifying.

Barely ten minutes after leaving Florida, the second Skylab crew were in space, primed for an eight-hour chase of the station and a late-afternoon rendezvous and docking. Also hitching a ride to their new orbital home was one of the largest 'crews' ever launched; for in addition to Bean, Garriott and Lousma, a complement of almost *eight hundred* living creatures were packed into various nooks and crannies and lockers and containers within the command module. Their purpose was to permit researchers to analyse circadian rhythms and closely monitor how they would cope and adapt to the strange weightless environment. The astronauts would share their home with two common cross spiders (*Araneus diadematus*), named Anita and Arabella, together with a pair of 'brackish water' minnows (colloquially known as 'Mummichogs') and 50 of their eggs, half a dozen pocket mice from the deserts around Palm Springs in California and 720 vinegar gnat pupae. The fish, NASA revealed, had been added to the list at Garriott's request and were caught in the coastal waters near Beaufort in North Carolina. Their presence in a specially-built aquarium was part of vestibular studies into the possible disorientating effect of weightlessness, whilst Anita and Arabella – part of a student experiment proposed by 17-year-old Judith Miles – would be monitored to assess their ability to build webs in conditions utterly peculiar to those in which their species had evolved. Lastly, the mice and gnats were kept in a compartment aboard the service module and were part of investigations into changes of circadian rhythms.

In spite of a seemingly perfect launch and insertion into orbit, problems did not take long to catch up with the crew. About three hours after leaving Cape Canaveral, midway through their orbital ballet to rendezvous with Skylab, Bean reported what he described as "some kind of sparklers" streaming past one of the command module's windows after the first firing of the big Service Propulsion System (SPS) engine. Lousma had spotted it first and called Garriott to his window on the right side of the spacecraft to take a look. "I looked out his window," Garriott recalled, "and here came what looked to be the *nozzle* of one of the reaction control thrusters just *floating by the window*! It *couldn't* have come off the spacecraft ... the propellant line to that nozzle had sprung a leak, so when the propellant comes out of the fuel line, it then freezes on the nozzle and then after a certain amount of it escapes, it acquired the shape of the nozzle. What floated away was an *ice sculpture* of that reaction control thruster!" A few seconds later, Bean was startled by the blaring of a master alarm, which indicated low temperatures in one of the ship's Reaction Control System (RCS) thruster quads.

All three men realised that a propellant leak of either hydrazine fuel or nitrogen tetroxide oxidiser had most likely occurred in 'Quad B', one of four sets of RCS manoeuvring thrusters, spaced at 90-degree intervals around the circumference of the service module. It was not good that it had come during the rendezvous with Skylab *and* so early into their mission. They quickly set about securing and shutting down the entire quad on Lousma's side of the spacecraft. Data telemetred to the ground revealed a clear drop in pressure, as both the hydrazine and pressurising helium rapidly fell to off-scale lows.

In *Homesteading Space*, Hitt, Garriott and Kerwin noted that, due to the "reduced authority" of only having three sets of RCS quads at his disposal, Bean was obliged to pulse the remaining thrusters for longer periods to achieve a perfect rendezvous. "It *really* incapacitated us a lot," Bean recalled. "The main effect we had was any time I did anything, we went off-attitude in the other axes." For the entire crew, it was acutely disappointing. During training, their instructors had thrown them hundreds of faults and failure scenarios and they had managed to overcome each one without so much as a blink of the eye or a bead of sweat on the brow. Now, said Bean, "we realised we were lucky we didn't have some sort of explosion and blow that leaky quad thruster right off and *really* have a problem." The situation was further complicated by the reality that, in 1973, devices to precisely measure a spacecraft's closing range rate had not yet been invented. Owen Garriott, tasked with helping Bean to stick to the correct trajectory, had trained to use two range measurements from the on-board radar transponder at two different times, then dividing the *range difference* by the *time difference*. By Garriott's own admission, it was far from perfect, and it quickly became apparent that it was not slowing them down sufficiently to complete the rendezvous.

Conversely, added Bean, "one of the worst things you could *ever* do was slow down *too much*, because then you had to use fuel to get closing again, all the timing's off, you came into daylight too soon – all these things were going on in my mind at that time." He lauded Garriott as "a great 'back-of-the-envelope' guy", capable of making accurate and rapid calculations and recommendations, but doubted his science pilot's constant reminders that they were closing too fast and needed to apply more braking. The difference of opinion led Garriott, at one stage, to head down to the lower equipment bay, and *that* certainly got Bean's attention. They were indeed closing too fast. Had it not been for Garriott's admonitions, they would have closed too rapidly and sailed straight past Skylab, which would have necessitated a re-rendezvous, an unnecessary wasting of precious manoeuvring propellant and, in

Bean's eyes as a naval aviator and the mission's commander, would have been "*real embarrassing*".

The astronauts' first view of Skylab had come at a distance of more than 700 km and docking occurred a little under nine hours into the mission. From Lousma's perspective, as he shot photograph after photograph during the rendezvous, it was both spectacular and unnerving; for the parasol erected by Pete Conrad's crew looked decidedly flimsy. "It was flapping in the breeze ... [of] the exhaust from our thrusters," he told the NASA oral historian. "We thought we were going to blow it off, so we suspended the fly-around [inspection of Skylab] and went in and did the docking." Despite the problems encountered during the final closure, the docking ran flawlessly, backdropped by the spectacular vista of the mighty Amazon River, longest in the world, snaking its way across almost 7,000 km through the South American heartlands of Brazil and Colombia and Peru.

Despite the good health of the station and – it seemed – the Apollo spacecraft, the first major obstacle quickly reared its head, when the astronauts fell victim to space sickness. The first affected was Lousma, who had begun experiencing the symptoms shortly after reaching orbit. He took a scopalomine-dextroamphetamine pill for anti-motion sickness and felt well enough to fulfil his photographic duties during the rendezvous. However, very soon, Bean and Garriott reported stomach 'awareness' and were unable to move quickly around the interior of Skylab. At length, even with the benefit of the anti-motion-sickness medication, it was becoming increasingly difficult for the astronauts to prevent themselves from becoming ill. Years later, Lousma vividly recalled that the arrival in the workshop seemed to exacerbate the problem. "We hadn't spent a lot of time in ... a volume [this large]," he recollected. "It was mostly in confined quarters, where you don't have to move very much, but *now* we were climbing out of the command module and going through the tunnel into this big volume and we had all kinds of room to operate in and move around in."

Space sickness – properly known as 'Space Adaptation Syndrome' (SAS) – is today known to affect around half of all space travellers. Research over the past five decades has generally concluded that it is a kind of nauseous malaise, somewhat akin to motion sickness, which typically lasts no more than the first two or three days of a mission. In the early years of the space age, it manifested itself in sensations of disorientation and discomfort, coupled with feelings of dizziness and recurrent headaches. The first victim was Soviet cosmonaut Gherman Titov in August 1961, who described the peculiar feeling of being unable to determine 'up' from 'down'. Even today, explanations and countermeasures for the condition remain imprecise. It appears – as Lousma, Bean and Garriott found to their cost – to be aggravated by the subject's ability to *move around* freely in the weightless environment, and seems more prevalent in 'larger' spacecraft.

Modern thinking postulates that the influence of weightlessness on the vestibular apparatus – the workings of the inner ear, which control our sense of balance – could represent a possible root cause. This disorientation arises, it is theorised, when real sensations from the eyes and other areas of the body conflict with those false sensations from the vestibular apparatus and the experience of the brain, gained during a lifetime spent in 'normal' terrestrial gravity. Over a few days, a 'repatterning' of the central memory network occurs, such that unfamiliar sensations from the eyes and ears begin to be correctly interpreted and adjustment to the new environment can commence. Today, motion sickness medicines have been shown to help counter it, but are rarely used, with most spacefarers expressing a preference to adapt naturally over a few days in orbit, rather than risk starting their missions in a drowsy state.

By the morning of 29 July, after a particularly unpleasant sleep, the breakfasts of Bean, Garriott and Lousma were but partially eaten and the astronauts quickly found themselves behind with their timeline. A concerned Bean asked Mission Control for the opportunity to allow them to rest for a while and to move their first off-duty day forward from 3 August to 30 July; additionally, it was decided to postpone their first EVA by 24 hours. Originally, according to a detailed timeline published by NASA a few weeks earlier, no fewer than *three* EVAs were scheduled for the mission – on 31 July, 24 August and 19 September – for the primary purpose of installing and retrieving ATM film, but the first excursion was especially critical, since it would also involve the deployment of the Marshall Space Flight Center's backup sunshade. Bean's crew had barely started their two-month mission and were already behind on the timeline. Lousma would later admit to being perplexed at his unpleasant reaction to the space environment. He was a Marine Corps aviator, used to sickening aileron rolls and other stomach-churning aerial manoeuvres and, along with his crewmates, had done a significant amount of training in rotating chairs on the ground, making various head movements to induce nausea and better 'condition' themselves. It all appeared to have been for nothing. "Actually," Lousma added, "on the ground, we were one of the *most* resistant crews to that kind of experience, but when we got in there, we were one of the *least* resistant!"

"Now, Alan and I never got *frankly sick*," Garriott admitted, "but we were feeling what I call *lethargic*, you know … really not up to speed, not ready to charge full-speed like you wanted to do." They coped, nonetheless. The men paced themselves for the first couple of days, limiting their movements and changes of orientation, taking more time to rest, stopping and shutting their eyes whenever they felt ill, then going back to work again when, as Lousma put it, they had allowed their "gyros to unspin". Medical specialists on the ground advised the crew to continue taking their scopalomine-dextroamphetamine anti-motion-sickness pills (which functioned by blocking off nerve endings to the stomach) and recommended several ten-minute sessions of head movements to alleviate the nausea and hopefully overcome the difficulties of adapting to weightlessness. Many of these recommendations came from Ashton Graybiel, principal investigator of the M-131 vestibular study, who warned that remaining still and resting would not improve the situation. His research had convinced him that subjects adapted more quickly in a slowly-rotating room when they made rapid head movements and he advised them to undertake such movements at least three times per day. Graybiel's advice was supported by the astronauts' own NASA physician, Chuck Berry, but the mission management team in Houston remained sceptical and advised the crew

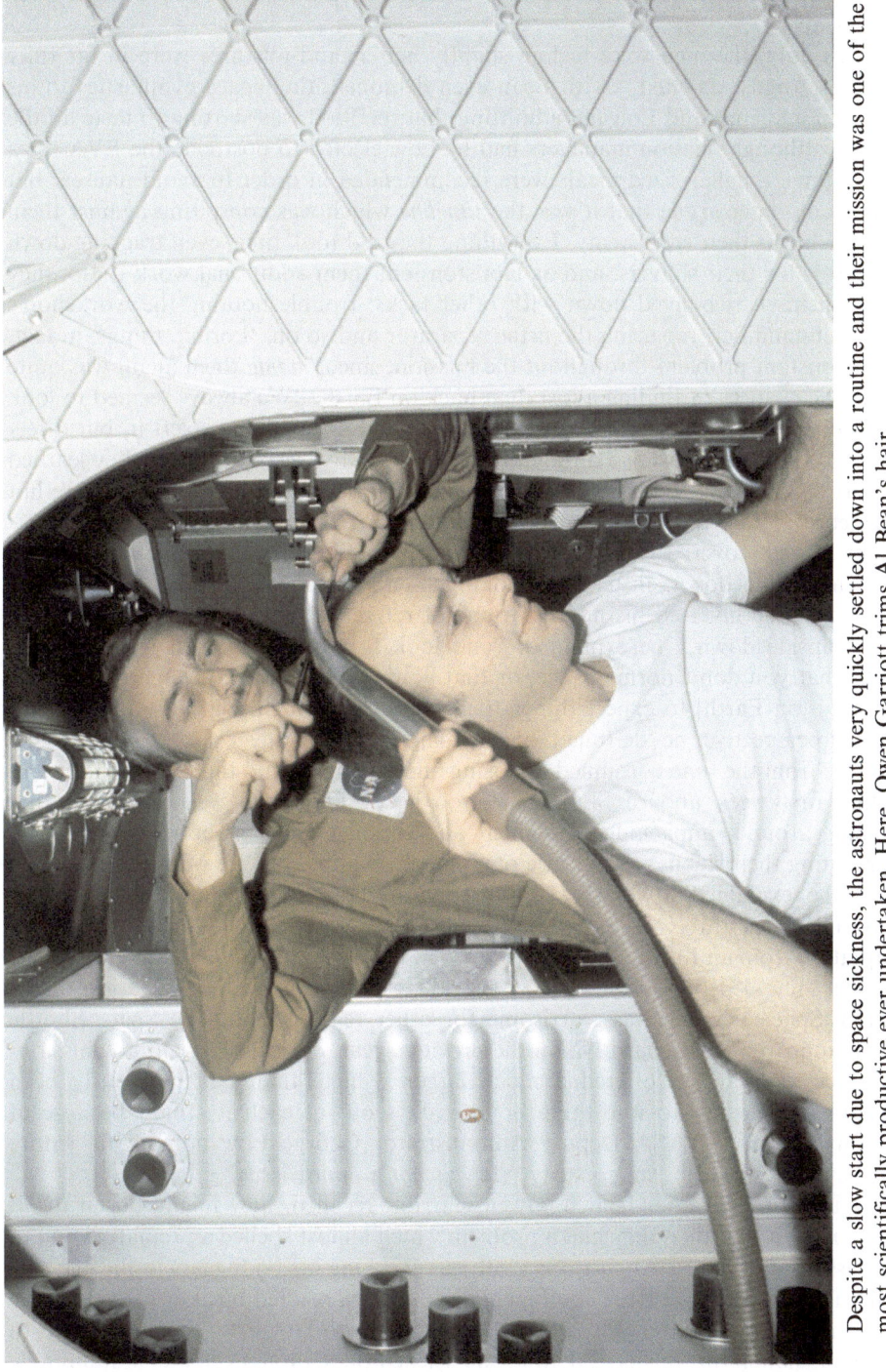

Despite a slow start due to space sickness, the astronauts very quickly settled down into a routine and their mission was one of the most scientifically productive ever undertaken. Here, Owen Garriott trims Al Bean's hair.

to complete the reactivation of Skylab at their own pace, whilst 'trying out' the head movements.

By 31 July, the men were feeling slightly better, and all three were in far finer spirits as August dawned – with Bean even demonstrating space gymnastics to his television audience and Lousma admitting that the food was starting to taste a little better – although mission managers had by now elected to postpone the EVA for a second time. Smaller-sized meals were recommended in order to avoid nausea, but Bean would later argue that it was the *timeline* which was conspiring against them and which was their *real* enemy. Everything they did took time, even tracking down new heads for their shavers, and as Houston sent them additional work to do, they found themselves bogged down with other tasks: troubleshooting the workshop's balky dehumidifier, repairing the urine separator and so on. 'Losing' things, in fact, was a constant problem throughout the mission, since *finding* them again was quite unlike the chances of finding a lost item back on Earth. "We always seemed to look on hard surfaces," said Garriott, "where we would normally have left it, but three-dimensional space is just too difficult to search visually." After a while, they learned to start by checking the intake duct of the station's air-circulation fans, where lint and debris – and, quite often, a few little items, pens, pencils, notes and so on – would be rediscovered. Lousma, on the other hand, turned the familiar Earthly problem upside down: instead of looking on the *tops* of places to find his lost possessions, he inverted his normal frame of reference to look *underneath*. "When you're upside down," he explained, "and look for something, you look at those places that you don't normally see, or that your eye doesn't get drawn to, because you tend [on Earth] to expect things to be sitting on something." By adopting an unusual perspective, he 'de-tuned' his normal search strategy.

Aside from the time consumed by losing and having to find things, the sheer *pace* of that first week aboard Skylab was too hectic and Bean was convinced that zooming around, unpacking equipment and supplies and activating systems, was aggravating their sickness and sense of stomach awareness. "We were not eating on time," he explained, "we were not getting to bed on time and we were not exercising." All three, Bean felt, were absolutely crucial in ensuring a smooth adaptation from an Earthly lifestyle to the new environment and he strongly advised the next crew, led by Gerry Carr, to give them priority over the activation of the station. Space sickness was worrisome for other reasons, too. The Space Shuttle effort, approved by Richard Nixon the previous year, envisaged seven-day missions in a spacecraft which offered a much larger volume than had previously been available. "Were we to lose three or four days out of each ... flight because of motion sickness," NASA Deputy Administrator George Low noted, "the entire Shuttle effort would be in jeopardy." It was *not* a minor issue.

Around six days after launch, Owen Garriott recalled, they finally began to hit their stride ... and then they had a problem which almost spelled a premature end to the mission. Early on 2 August, Garriott was checking his body mass in the rotating chair, as Bean prepared EVA equipment and Lousma readied breakfast for the three of them in the wardroom. All at once, Lousma called Garriott to the window ... to behold a *snowstorm* outside! At first, the science pilot thought it was a stunningly beautiful auroral display, since at that time they were flying high above New Zealand and the southern auroral area. However, he was left open-mouthed as a flurry of snowflake-like particles – "a real blizzard" – flooded past the window. "Well, since it *doesn't* snow very much in space," Garriott deadpanned to the NASA oral historian, "we realised there had to be another explanation and in this case it was obvious: we had *another* fuel leak!" Bean and Lousma headed rapidly for the command module to check its systems and quickly identified a problem with the Quad D set of thrusters, directly opposite the failed Quad B. At first, a link was not made with the earlier Quad B problem. Temperatures in Quad D were falling rapidly and the crew were first advised to activate its backup heaters. An hour later, a further drop in temperature *and* pressure was noticed, suggesting that a *second* leak, similar to the Quad B incident, had indeed occurred; a suspicion confirmed by visual observations from the crew. "We could see [the quantity] going down on the meter," Lousma recalled, "so we shut off all the valves and made sure everything was secure, just closed up all the plumbing and reported it to Mission Control."

With two of the four RCS quads shut down, the situation began to turn ugly, for the quads were a critical component of the spacecraft's manoeuvring capability and the astronauts were *less than a week* into their eight-week mission! The rate of the oxidiser leakage was low, and only around ten percent had been lost, but Mission Control and the crew were concerned, since there was no indication of how rapidly it would increase. They *could* use two, or even one, quads to get home, but it was not acceptable for re-entry, without further investigation.

The astronauts and ground controllers mulled the situation over. Of paramount importance was figuring out *why* the failure had occurred at all. Was it a systemic flaw? Would there be a *continuous* stream of systems failures? Or was it purely a spurious, totally random pair of unrelated incidents, which gave the *appearance* of having a common cause? No one knew. Early supposition was that the batch of nitrogen tetroxide might have been contaminated, which placed a cloud of uncertainty over the *other two* thruster sets, Quads A and C. Moreover, if the leaks persisted, internal circuits aboard the service module might render the craft unusable. Later that morning, Bean spoke directly to Chris Kraft in Houston, imploring him to allow them to remain aboard Skylab to complete their mission, but he was advised that if the situation worsened a rescue mission with a modified command and service module might be attempted. "If we had *another* leak," Lousma reflected, "we sure wouldn't have enough to get home. It was this that caused them to position the rescue vehicle on the launch pad."

Acutely disappointed, since they had just begun to work at full capacity on their scientific tasks after several days of down time due to space sickness, the crew was now faced with the very real possibility of not only a curtailed mission . . . but maybe even the need for *another crew* to be hurriedly primed for launch to dock at Skylab's second port and bring them home, abandoning their own craft. Rescue missions for Skylab had been extensively explored, baselined, analysed and planned by NASA for more than two years; indeed, a pair of astronauts had undergone specific training to accomplish it. Both were experienced pilots and both were convinced that they *could* rescue Bean's crew, if necessary. However, in August 1973, the dangers were evident:

such a rescue mission had *never* been attempted and if the thruster quad fault was a *systemic* flaw, then it was possible that the *rescue craft* might be similarly affected. It would be risky gamble ... but one which two rookie astronauts were more than ready, more than willing and more than able to take.

A SKYLAB RESCUE?

One evening, late in 1969, astronaut Jim Lovell took his wife Marilyn to see a movie which, for them, would prove horribly prophetic. *Marooned*, based on the novel by Martin Caidin and starring Gregory Peck, told the unsettling tale of a three-man Apollo crew returning from a five-month space station mission. Shortly before re-entry, an SPS engine failure and lack of available RCS propellant left them both powerless and helpless: unable to return to Earth, unable to redock with the station and unable to remain aloft for more than a few days. In true Hollywood fashion, of course, an impromptu rescue was orchestrated and, despite the death of one astronaut, most of those involved lived happily ever after. A few months later, Lovell and his Apollo 13 crewmates found themselves in similarly harrowing, though chillingly real, circumstances, the story of which would find its own way to the silver screen. What is perhaps less well-known to today's cinema audiences is that a much closer parallel to *Marooned* very nearly transpired in the summer of 1973.

On 13 August of that year, *Time* magazine told its readers the unsettling news from Skylab: that one of the service module's four RCS thruster quads had sprung a leak shortly after launch and *another* quad had proven inoperable a few days later. Not only was Mission Control concerned that all four quads were of identical design, but their oxidiser – *and* the oxidiser for the big SPS engine – originated from the same batch. If that batch was contaminated in some way, NASA could have a systemic problem on its hands which might affect the whole spacecraft and prevent Bean, Garriott and Lousma from returning home safely. In theory, Apollo could be controlled with just one quad or even using the thrusters on the command module itself, but the risk of further deterioration prompted the space agency to take steps to implement a rescue plan first laid out two years earlier. Of course, Skylab's multiple docking adaptor carried *two* ports (hence its name), which might permit a rescue craft to visit. Yet the effort itself to orchestrate a real rescue was an audacious endeavour, costing more than two million dollars, which called for a second Apollo to be outfitted with five seats; two for its own pilots and three others for Bean, Garriott and Lousma. The stranded crew could then transfer from the station to the rescue vehicle and return to Earth. Assessments of the practicability of such a plan were first made in April 1971 and it was considered possible for NASA and its workforce to prepare and launch the mission at any time between ten and 45 days after being given the go-ahead. It was by no means a simple 'paper' exercise: by March 1972, NASA had committed itself to having a Skylab rescue capability and Navy and Air Force helicopters supported trials at sea later that same year.

With Garriott focused on the scientific side of the mission, Bean and Lousma, as the primary pilots of the Apollo, had involved themselves extensively in designing

An Apollo command and service module docked with Skylab. Two of the four 'quads' of RCS manoeuvring thrusters are clearly visible.

the rescue craft. In fact, said Lousma, they would have provided a rescue capability for the *previous* Skylab crew, ferrying Conrad, Kerwin and Weitz home, in the event of insurmountable problems with their own craft. "Al and I worked with Rockwell and the NASA engineers," Lousma told the NASA oral historian, "in configuring a command module that had two flat couches underneath the three couches on top and that would handle five people. In the centre, between the two people on the very bottom floor – on the 'bottom bunk', so to speak – there was enough room to put some of the experimental data and other kinds of things you'd want to bring back for data reduction." The biggest concern, Lousma added, was the potential 'stroking' of the upper deck of couches. Normally, the standard three couches were designed to 'stroke', or have their supports compress, like a shock absorber, in the event of a rough landing. If that happened to the returning rescue craft, the supports might compress ... *onto* the astronaut in the couch below! However, since no couch had ever stroked during any of the previous dozen Apollo splashdowns, NASA considered the risk a minimal one.

Now, in the first half of August 1973, it seemed that a rescue mission might happen for real. Preparations at Cape Canaveral now shifted into high gear, with efforts to ready Pad 39B for its second Saturn IB launch in a few days. In fact, the booster to be used was the one already earmarked to transport Gerry Carr's crew to Skylab in November. On 3 August, the processing schedule was accelerated, with technicians and engineers working around the clock, seven days per week, to prepare the vehicle for its possible new role. The Apollo (known by its mission designation of 'Skylab-Rescue' or 'SL-R') would be stripped bare of stowage lockers in its lower equipment bay to accommodate the three couches and 'ballasted' with around 450 kg of lead to compensate for centre-of-gravity offsets. Upon receipt of 'the call', the so-called 'field modification kit' to turn a three-man command module into a five-man rescue ship would commence. Launch of the SL-R mission was scheduled for around 5 September, approximately three-quarters of the way through Bean, Garriott and Lousma's baselined mission. On 10 August, less than a month before this scheduled launch, the SL-R command and service module was transferred to the VAB for final checkout. By necessity, this work was abbreviated and it was expected that the spacecraft *could* be readied for installation onto the Saturn IB and rolled out to the pad in as little as *three days*. Flight readiness checks would then be accomplished by 24 August, propellant loading would commence on the 27th and launch would occur a little more than a week later. The SL-R mission was expected to last no more than five days, ending with a splashdown in the Pacific on 10 September. Looking at the numbers, Bean, Garriott and Lousma would thus complete the vast majority of their scheduled mission in terms of duration, landing perhaps 44 days into their planned 59-day residency. They could *still* claim a new world record for a single space mission, but there would be a caveat: they would be forced to leave the results of most of their scientific research behind on the station ... but, hopefully, this would be retrieved by their successors.

Of course, no daring rescue could possibly go ahead without a heroic crew. Enter, therefore, the unsung would-be heroes of Skylab: Vance Brand – named only six months earlier as the command module pilot for the American half of ASTP – and

Don Lind. Both had extensive expertise with both the Apollo spacecraft and the station. Teamed with fellow astronaut Bill Lenoir since January 1972, they had trained as backups for Bean, Garriott and Lousma and also for the final Skylab crew of Gerry Carr, Ed Gibson and Bill Pogue. Additionally, Brand and Lind would pull double duty as the SL-R stand-by crew for both missions, though Lenoir had also trained for the role and was considered equally qualified. ("I was not in the discussion that selected the crew," Brand related. "We just found out. Both were capable of doing that job. Bill was a scientist, but also an excellent engineer and pilot. Everybody cross-trained for everything.") Brand would have commanded SL-R, with Lind – a 'scientist-pilot', holding both military flying credentials *and* a PhD in physics – would serve as the command module pilot. As already noted in Chapter One, Brand had served on the Apollo 15 backup crew and knew the spacecraft intimately. "I did a lot of contingency work figuring out, for example, how you could complete a rendezvous if you lost your inertial navigation unit," he told a NASA oral historian in April 2002, "by gauging things, using charts, using the spacecraft guidance, navigation and control equipment in ways that it was never, ever designed to be used. I thought I could almost fly the spacecraft without thinking at the time, because I had so much exposure to Apollo training."

However, as the commander of the rescue flight, Brand had his own personal doubts. He had become very familiar with the schedule of launch preparations for both the spacecraft and its Saturn IB booster and felt that the 'long pole' to achieving their 10 September target was getting the vehicle ready on time. "I think we would have been lucky to be off 30 days after that," he recalled, "but we were talking about that, aiming for that."

When the trouble with Bean, Lousma and Garriott's spacecraft became clearer, Brand and Lind spent most of August 1973 "figuring out how to rendezvous with them, where we would dock, outfitting our command module so that it had … padding on the aft bulkhead where people could lie … [to] get five people in the spacecraft". In addition to the risk of stroking the struts of the couches, the number of bodies inside the command module might also have posed difficulties after splashdown; if the spacecraft entered a so-called 'stable 2' orientation – with its apex in the water – it could prove extremely disorientating for the crew. "Vance and I had gone through some training on this out in the Gulf [of Mexico] with a real command module," Lind recounted in a May 2005 oral history. For the purposes of the five-man exercise, three other astronauts, including Bill Lenoir, had joined them. "I said to [Lenoir] … 'Now, if we get in 'stable 2', remember you're on top of the vehicle and so when you unstrap, make sure you have a hold of a stanchion someplace'. He looked at me like, 'Lind, how *dumb* do you think I am?' Well, we got in stable 2. The radio called for this guy to unstrap. It was hard to realise that you were now strapped [to] the ceiling, because it was bouncing around in the water. He released his seatbelt and, wham, [Lenoir fell several feet]. Then he looked at me like 'If you say *one word*, I'll kill you!'"

Other considerations included which experiments and samples the rescue mission could bring home, in view of the limited storage space aboard the SL-R command module. Brand assigned Lind the job of 'cargo master' to make these important decisions, based on weight and centre-of-gravity concerns. "I had to inventory all the possible decisions to maximise the scientific return with the limited capability we had to return all that data," Lind recounted. According to contemporary NASA documentation, in addition to the safe return of both crews to Earth, the main objectives were to bring back 'selected' experiment data, to perform a diagnosis of the failure of the original Apollo craft and to configure Skylab for a revisit. In terms of experiment samples to be returned to Earth, the frozen urine specimens and dried faeces – "ironically", Al Bean pointed out – were of primary interest from a medical perspective. However, film tapes from the EREP and ATM and, wherever possible, student experiment results would also be brought home. Aside from the rescue itself, the men also worked a number of what Brand referred to as 'unorthodox' procedures, whereby they could bypass the service module and perform a re-entry with only the command module's thrusters. Years later, Brand would compare the feverish seriousness of their efforts to the ten days in May 1973, during which NASA had miraculously saved Skylab from disaster; many within NASA, including the astronauts themselves, believed that the rescue would most likely go ahead. The mood was lightened a little when further investigation confirmed that none of the RCS oxidiser batches were contaminated and had not contributed to the leak. It also appeared that Quads A and C were unaffected and a subsequent investigation would attribute the failure to undetected loose fittings in oxidiser lines.

Brand and Lind's preparations also, through their effort to identify workaround procedures, included an ability for Bean to manage a safe undocking and re-entry *without* the need for Quads B and D. The men also successfully worked out a method to achieve translation capability and complete a de-orbit burn using only the command module's attitude thrusters. "We were so clever as the backup crew," said Lind, "that we worked ourselves out of a flight! You really didn't want to *have* to go rescue them. You really wanted to bring them back safely with all their equipment, so that was the right choice." Their workaround, together with the engineering expertise of others on the ground and of the crew in orbit, ensured that Bean's crew could complete their flight. Rather than making a two-stage SPS burn for re-entry, it was decided that the men would perform a single burn and that if there were no more leaks, the rescue mission could be stood down. In a strange way, it was disappointing, particularly for Lind, who had long desired a Skylab mission, but he knew that both Lousma and Bill Pogue were "depressingly healthy" and there would be essentially no chance that he would be called upon to take their place on a prime crew. Even the old backup crewman's technique of straw dolls and long pins, it seemed, would have little effect. Flying the short SL-R mission was his last chance to see Skylab up close and in orbit.

As the situation began to improve, with the workarounds and the relief that the leaks did not represent a systemic flaw and the oxidiser batches were not contaminated, plans changed quickly. For the men in space, it was not a serious problem, since they had no increased chance of fire or pressure loss aboard Skylab, they had plenty of food stores – even with a voracious appetite like that of Lousma amongst their number – and their only worry was the risk of having to come home early and curtail a mission which had got off to a poor start due to the sickness. The

The Saturn IB rolls out to Pad 39B on 14 August 1973. By now, it had been effectively 'stood-down' from an active role in the rescue of Al Bean's crew and would instead be retasked to its original role: launching Gerry Carr, Ed Gibson and Bill Pogue towards Skylab in November.

decision to remove SL-R from consideration changed everything, in Brand's words, virtually overnight. On 14 August, only days after the transfer of the spacecraft to the VAB, NASA announced that the launch vehicle was now being retasked for launch no earlier than 25 September; effectively, the agency was removing it from immediate duty as a rescue craft, since Bean, Garriott and Lousma were already scheduled to land at around that time. This decision ties in with Lousma's assertion to the NASA oral historian that "after about ten days or so" into the crisis, the crew was advised that they could stay aboard Skylab and continue their baselined mission. The decision underlined a growing confidence that the crew would most likely be able to complete the planned 59-day flight. On that same day, 14 August, the Saturn – now reassigned to its original purpose of transporting the final Skylab crew of Carr, Gibson and Pogue – was transferred to Pad 39B to begin its own launch preparations. Nevertheless, upon the completion of hypergolic propellant loading aboard the booster on 10 September, the vehicle remained in a 'Launch Minus Nine Days' stand-by status until Bean's crew were safely back on Earth.

A few months later, in January 1974, the SL-R command and service module was transferred to the VAB, where it later saw service as a backup vehicle for the Apollo-Soyuz mission. After that, for many years, it resided in the Kennedy Space Center's visitor area and, in 2007, was 'commandeered' by NASA to aid studies of a similar rescue capability for the Orion spacecraft. It is interesting and appropriate that lessons from the past continue to be learned for the future and, indeed, the post-Columbia practice of having partial crews on stand-by for a rescue is by no means a new one: it had been pioneered, worked out and perfected all those years ago. For Brand and Lind, losing the chance to fly the rescue mission as a direct result of the thoroughness of their own work was a bittersweet experience. "You really feel not just a professional obligation, but also a personal obligation," Lind remembered, "to the fellows on the crew that you know so well to do that job very well. We did the best job we could and were able to convince management that we had enough redundancy to bring the guys home with the quad problems." For Brand, who would fly aboard Apollo-Soyuz less than two years later, given the choice, he would have preferred the joint flight with the Soviets to having to undertake the SL-R voyage. It is also ironic, as the reader will recall, that part of the original rationale for Apollo-Soyuz was to show how a spacecraft from one nation might someday rescue the crew of a spacecraft from the *other* nation. Still, commanding the world's first-ever fully-fledged space rescue mission, he knew, would have stood out as an intensely rewarding professional achievement for the rest of his life. Today, almost four decades later, whilst it would have been 'interesting' to see Brand and Lind fly their mission, we should perhaps be thankful that they did not have to.

MEN AT WORK, REST AND PLAY

The reaction aboard Skylab to the notion of a rescue, not surprisingly, was one of dismay. "Our bias," Garriott recalled years later, "was that we must stay. Let us continue our work up there. We still had some control over the command module reaction control system … so if it really came down to having to get back home, we thought we could do it with the command module reaction control system. We didn't want to leave." Nor did Bean or Lousma. They knew that, even if the rescue craft *was* despatched to pick them up, its arrival would come more than six weeks into their eight-week flight. If the situation worsened, Brand and Lind's rescue mission was in a good posture to recover them, but in the meantime Skylab was a stable platform for them to remain aboard. If they would be staying for six weeks anyway, regardless of whether a rescue attempt was pressed into action, it made sense to keep them aboard for the full eight weeks. "The command module sitting there with a little less propellant was not a threatening thing," Lousma told the oral historian. "We could continue working in the … station, doing our normal daily activities and continue on until the decision was made as to what to do. Even if they were going to launch a rescue ship, we were safe. There was no chance of fire, no increased chance … of losing pressure within the spacecraft. We had plenty of food and water. The biggest concern we had was that we were going to have to shorten our mission and … come home early."

In addition to their own vestibular experiments, the first EREP passes were conducted without incident and Arabella – one of the two spiders, who, like Anita, would become virtually a household name that summer of 1973 – was shaken out of her vial and into the thumb-sized storage cage by Garriott on 5 August. "She immediately bounced back and forth," wrote Bean in his mission diary, recorded in *Homesteading Space*, "front to back, four or five times, then locked onto screen panels at the box edge provided for visualisation. There she sits, clutching the screen." He talked to Garriott about feeding her; no, replied the science pilot, they should wait a while, since she would only begin to spin her web when she became hungry. Garriott knew that both spiders could comfortably survive without food for two to three weeks and the primary focus of Judith Miles' experiment, after all, was to assess their ability to spin webs in weightlessness. At length, Arabella began spinning her webs; scruffy ones at first, but each day their quality improved to a level comparable with Earth-spun ones. After three weeks or so, she was returned to her container and Anita got her chance in the cage. For his part, Lousma admitted to some good-natured jealousy that these two arachnids were receiving far more public attention and interest than the three human astronauts themselves.

Judith Miles had come up with the idea of a spider experiment after reading an article in *National Geographic*. She noted that, on Earth, spiders sense their own weight to determine the required thickness of web material, using both wind and gravity to initiate the construction process. She was fascinated by the possibility of investigating how spiders would adapt to building their webs in the *complete absence* of gravitational force aboard Skylab. Unfortunately, Anita was found dead in her cage on 16 September. Shortly after returning to Earth, Arabella also died and dehydration seemed to be the only visible cause. Examination of the two spiders' webs revealed that their threads were significantly finer than those spun on the ground, before the mission, suggesting that Anita and Arabella both used a weight-sensing organism to size their threads.

Before leaving Skylab, as a 'gotcha' for Gerry Carr's crew, Bean, Garriott and

Lousma planned to set up a huge fake spider and web to place over the hatch to the multiple docking adaptor. It would certainly give Carr and his all-rookie team a momentary scare, or so they hoped. Unfortunately, they mistakenly thought that they had left it behind on Earth and regrettably did not set it up ...

As the days wore on, the crew improved in spirits and set to work on one of the largest complements of scientific and medical experiments ever attempted in space. One additional task was their first EVA, delayed repeatedly due to their sickness and ongoing problems with the service module's thruster quads. When Bean's mission was originally baselined, its three planned spacewalks were scheduled to install and retrieve ATM film; now, the workload of the first excursion had been expanded to encompass the fitting of the Marshall Space Flight Center's sunshade on an A-frame-shaped structure known as the 'twin pole'. The previous parasol, extended through the scientific airlock by Pete Conrad's team, had now been in place for almost three months and was showing visual signs of deterioration. The Marshall device would be installed *over* the top of the Houston parasol. "After the [ATM] film installation was completed," Garriott recalled, "I had to connect eleven sections of aluminium poles, *twice*, forming two long poles. These were then extended to Jack, some 40 or 50 feet away, where the poles were mounted in a 'V' and a large 'sail' pulled across them with nylon lines." Although this kind of excursion had never been planned, the route which Lousma would take from the airlock to the central station of the ATM was a fairly straightforward one and its traverse had been rehearsed repeatedly by the men for the purposes of film installation and retrieval. Additionally, Lousma would have handholds available to assist his movement and, when he had established himself at the ATM, he would install a set of foot restraints to anchor himself in place.

Early on 6 August – ten days into the mission – Garriott and Lousma set out, leaving Bean inside the workshop "to tend the store". With the science pilot located inside the airlock module, Lousma commenced the tricky process of getting into position, on the central part of the ATM, then began fitting a two-slotted adaptor to receive the ends of the long poles. Meanwhile, Garriott used a bayonet-like connector to fit the segments of the poles together. "He fitted each segment into the next," noted the authors in *Homesteading Space*, "depressed a spring, rotated the segment about 20 degrees and latched it into place." Finally, a rubber ring was rolled over the fitting to secure the connection. He then passed the poles, each of which was 17.5 m in length when fully assembled, to Lousma, who fixed them into their slots. The pilot's job was then to deploy the Marshall shade onto the poles, stretching it across and rendering it taut with long 'ropes' to cover the whole exposed section of the workshop and Conrad's old parasol.

Perhaps unsurprisingly, the work took them twice as long in space as it had done in the neutral buoyancy tank on the ground, and both men feared that the narrow diameter of the twin poles in comparison to their extreme length might make them somewhat 'whippy', a little like fishing rods. Lousma did not concern himself too much about that, but their flimsiness was definitely noticed by both spacewalkers as the assembled poles grew longer. Ultimately, it would work very well in continuing to keep down temperatures inside the workshop, but one problem did later arise, evidenced from the photographs. "When you look at the Skylab photos," Lousma said, "the sunshade is kinda brown, but has a white streak in there". During the feverish activity back in May, when the Marshall sail was conceived and built, the adhesive used to hold its pieces together had been given insufficient time to properly cure. When the time came for the astronauts to unfold it, accordion-like, in space, this very adhesive prevented it from deploying in the manner that it was meant to do.

"I had to bring that whole thing back toward myself," continued Lousma. "It was all out of the bag and billowing up all over and, by hand, I had to unfasten *all* of those folds! Then I had to attach the two corners that were nearest me with a long lanyard and drift out to two places on either side of the [multiple docking adaptor] to attach the lanyards. When the large sail was deployed, the twin poles were flopped down on top of the parasol and against the Skylab workshop and the lanyards tightened. It nearly covered the workshop and worked quite well ... but it turned out that I had missed one of those folds and so it was out there like that for a long time and getting browner and browner. Then the Sun did the rest of the job and unstuck that one little piece and so you see that white streak in there – *that* was the one that had remained folded for the longest time."

Overall, the deployment of the Marshall sail demanded almost half of the six-and-a-half-hour spacewalk, with the remainder being devoted to the installation of ATM film, an unsuccessful exploration of the outside of the workshop to identify the source of a mysterious coolant leak, visual checks of the command module's RCS quads and the fitment of small panels to measure micrometeoroid impacts. Garriott and Lousma returned inside the airlock module after six hours and 31 minutes, which – excluding the Moonwalks of the last three Apollo lunar missions – established a new world record for the longest spacewalk to date. Temperatures within the workshop after the installation of the twin-pole sunshade began to drop almost immediately to a comfortable 22°C or thereabouts, a full six to eight degrees cooler than Conrad's crew had experienced during the last few days of their mission with the parasol.

Several memories exist of this spacewalk, not least from the two men directly involved in it, even though Garriott's diary summed up the day in just a handful of words: 'EVA Day Went Very Well'! For Lousma, recounting the experience for the NASA oral historian almost three decades later, each day of his two months in space seemed to blur into one ... *except*, that is, for the day of his spacewalks. "The spacewalks are just absolutely the high point," he explained. "It really is the most memorable part of being in orbit; just an unusual experience, in that when you go outside, it's *different* than being inside. When you're inside, you look through the window and you see part of world." Outside, in the ethereal void, the Earth was an enormous 'sphere', just beyond the helmet visor. Lousma was dazzled by the glare of the unfiltered sunlight and was astonished that, for the very first time, he *really* felt the sensation of speed and motion over Earth. At the same time, however, unlike the related sensations associated with speed in an aircraft, there was *no* vibration and absolutely *no* sound. "It's like gliding along on this magic carpet," waxed the tough, square-jawed Marine, "going into the sunset and sunrise every hour and a half ...

doing that for six hours." From Skylab's altitude, he could see freeways and cities and airports with his naked eyes and, looking further, could spot landmarks: right up into the middle of Hudson Bay and, further on, the twinkling lights of Washington and Baltimore and New Orleans and Denver.

On one occasion, perched on the end of the ATM, installing a film cassette, Lousma remembered moving into orbital night, somewhere over Siberia, he guessed, and being thrown headlong into the pitchest blackness. Through his visor, he could hardly see his own gloved hands and had the profound realisation that "it's just me, God, the spacecraft and my buddies and *that's it*". Fortuitously, in the days before Tracking and Data Relay Satellites and more-or-less continuous communications coverage throughout an entire orbit, the early space explorers, like Lousma and Garriott, were not always over a ground station. "Sometimes," Lousma noted, "when we were out there, we would take our time more than we should. Mission Control didn't know [everything] we were doing. Sometimes, if we were lucky, we could miss *all* of the ground stations for a *whole* revolution!"

Ten days into their mission, though, Bean, Garriott and Lousma knew that they were seriously behind schedule, due in part to their sickness and in part due to the demands of the thruster quad problems ... but they also *knew* that none of them would be going home without completing one hundred percent – or *more*, preferably – of their assigned work. In the days after Garriott and Lousma's EVA to install the twin-pole sunshade, therefore, they hunkered down to work. In fact, in an impromptu heart-to-heart with mission controllers on 12 August, Bean requested *more* work to do – "We've got the ability and time and energy and I know y'all do down there" – and Houston duly obliged, increasing each man's daily schedule for scientific experiments or repairs by four hours to 12 hours. From his own perspective, Owen Garriott felt that it was from *this* point, around 16 days into the mission, that the entire crew really felt that they were turning a corner and making aggressive inroads into completing their scientific schedule. "We're considerably more efficient," he wrote, "and the flight plan may be a little less tight."

Efficiency was the name of the game in Bean's mind. He had been shocked early on, when he learned from Mission Control that after a particularly exhausting work session they had actually only accomplished 50 or 60 percent of their assigned tasks. As a result, they stopped taking meals together, allowing two of them to be working at all times, and even manned the ATM from the moment they awoke, whilst the others occupied themselves with other chores and routines. Each morning, Lousma would head for the teleprinter, on which Mission Control had transmitted the day's activities for each crew member during the night; sometimes, the strip of paper was 10 m long and he would be obliged to cut it into small pieces and give each man their respective piece. To keep themselves on track, they had small notebooks in their back pockets – red for Bean, white for Garriott, blue for Lousma – exhaustively detailing their daily routines: from which experiments they were assigned to complete through to which meals they were scheduled to eat. There were occasional exceptions to the 'separate meals' routine, of course, and Bean would remember some of the best memories from the flight were their ice cream parties, with the three of them together at the wardroom window, gazing at the Home Planet passing serenely 'beneath' them, "just like an atlas", whilst slurping ice cream and strawberries. "I remember those as really nice times."

Since Garriott was the best of the three at operating the ATM, the pilots generally ended up swapping their duties with him to offer the scientist the majority of time at the observatory's controls. The movement of supplies and equipment and trash which needed to be undertaken on a given day was done during the working time, thereby reducing the 'housekeeping' load, until the astronauts were actually *ahead* of the schedule. "We were ahead," Bean noted, "and they would call us and they wouldn't have anything new the next day and we'd be twiddling thumbs! We were ready to go, but they hadn't geared up for us yet!" Mission Control, with Capcom Story Musgrave on console for much of the time, eventually responded by giving them *lots* more work to do. Nor did it seem like a chore. With no outside disturbances – aside from the view, perhaps – there was little else to distract them in Bean's mind and the effort of the past two and a half years was most definitely *not* going to be wasted. At length, ground controllers were struggling to keep up with them. By the end of the flight, this would have paid its own dividends, for Bean's crew would execute one of the most scientifically productive space missions ever undertaken and, despite the slow start, exceeded their pre-flight research expectations by a whopping 50 percent.

With the ATM fully loaded with film and ready for operations, Garriott set to work on 7 August observing the Sun's corona for three hours and successfully photographed a medium-sized flare a couple of days later. Then, on the 10th, astronomers at the Canary Islands Observatory detected an even larger solar event and, even though it was their half-day off, Bean and Garriott sprang to the ATM console to film an enormous eruption of solar radiation which turned out to be a specimen typically only seen two or three times each year. A simple calculation revealed that the *minimum* speed of the eruption – properly termed a 'coronal mass ejection' – at more than 500 km/sec! Two weeks later, on 21 August, a huge prominence was detected on the eastern edge of the solar disk, "like a big bubble", according to Al Bean, which gradually curled and contorted its way, loop-like, across a distance nearly three-quarters of the Sun's diameter. The astronauts earned specific praise through their judicious use of limited film stores in the ATM's white light coronagraph. All told, the men acquired more than 77,600 images of the corona in the X-ray, ultraviolet and visible portions of the electromagnetic spectrum by the mission's close and would spend over three hundred hours conducting astronomical observations.

In addition to the ATM observations and EREP passes and medical experiments, one notable 'corollary' investigation was the M-509 multi-mode manoeuvring device; essentially a large backpack which NASA hoped to develop for future EVA use. Astronaut Bruce McCandless, who had earlier served as backup pilot for Pete Conrad's mission, played an integral role in its development and in 1984 would test-fly its successor, the Manned Manoeuvring Unit (MMU), on its first sojourn in space. Years later, in a March 2006 email correspondence with this author, McCandless admitted that his long wait to fly in space was because he "probably lavished too much attention on scientific and engineering interests, as opposed to the

Solar observations were one of the key foci of each Skylab mission, with operations centred on the control console of the Apollo Telescope Mount. Here, in the spring of 1973, the crew for the final flight – from the left, Bill Pogue, Ed Gibson and Gerry Carr – practice their responsibilities on the console.

flying, flying and more flying." In the late 1960s, he began working closely with David Shultz and NASA's Charles 'Ed' Whitsett to evaluate the validity of a self-contained manoeuvring unit for future spacewalkers. The result was the M-509. "I hoped to be the first to fly it," McCandless told me, "but that was not to be. I was named as backup pilot for Skylab 2 and waved goodbye to being on the prime crew."

When Conrad arrived at the station, together with Joe Kerwin and Paul Weitz, the manoeuvring unit might have been a useful tool in their repairs, but for one thing ... the increased solar heating, it was feared, might have damaged or impaired the M-509's nickel-cadmium batteries and rendered them unusable. Nevertheless, on 13 August 1973, *inside* the workshop, Bean and Lousma had the opportunity to test-fly the backpack and generally gave it glowing reports, although they felt that future models would benefit from more speed and less precise attitude controls. Bean wanted something that flew more like a spacecraft, in order to ensure that the astronaut's intuitive piloting response was the correct one. Later, although Garriott had received limited training on the M-509, Bean offered him the chance to test it. The science pilot was suitably impressed. "With just a modest amount of training," he told the NASA oral historian, "in how to fly a spacecraft with a translation and attitude control thruster, you could fly around that spacecraft, making turns all the way around the ring [of] lockers, holding yourself maybe an inch away from the surface and maintaining attitude control and translation control where you wanted. It was very pleasant to see how easy it was to fly and perhaps useful to the designers, as well, to see that a person with essentially no training could learn to fly it so quickly."

Living in space, Skylab-style, brought its own unusual comforts and, although Bean felt that the station's shower was cooler than he would have liked, it was still strange to feel the water droplets clinging to his body like blobs of jelly. "It built up around the eyes, in the nose and mouth and it gave a slight feeling of trying to breathe underwater," he wrote. He found himself shaking his head violently, only to see the water droplets spraying in all directions, some clinging to the walls of the shower, others to his body and some actually distended and sprang themselves back. Both Bean and Lousma agreed that improvements needed to be made: the shower's vacuum head did not conform to the body too well and additional towels were required to dry themselves, since the water did not drain. For Garriott, the whole process of showering in space was a waste of time; the process involved "floating in an erectable, water-tight cylinder, preparing warm water, spraying yourself to get wet, soaping up, rinsing off, collecting waste water, then reversing the whole process". Typical shower runs could take an hour or more and the results were often little better than a moist wash cloth might have provided in five to ten minutes. In fact, by the end of the mission, Garriott would not have used the shower at all, Lousma only once and Bean twice. Still, showering 'properly' was something that all three men missed acutely. When they returned to Earth, they were picked up by the recovery vessel *New Orleans* in the Pacific and one of their first ports of call was the ship's shower room!

In addition to their scientific workload, the need to conduct regular repairs and maintenance was a constant headache, particularly a leaky dehumidifier. On 20

August, Bean spent virtually his entire day inspecting it: adding nitrogen, checking connections, listening with a stethoscope and applying a soap solution. Coolant loop leaks had been detected two weeks earlier and Garriott and Lousma had actually attempted to identify their source from outside the station during their first EVA, without success. By the end of Bean's mission, Skylab was running on its backup coolant loop, although plans were afoot to devise a means of replenishing coolant in time for the arrival of Gerry Carr's crew in November 1973. The second EVA of the mission, on 24 August, in addition to retrieving and replacing ATM film, was tasked with installing the six-pack of rate gyros. Mission Control *had* managed to maintain one good gyroscope in each axis, and usually a serviceable backup, too, and there was a real risk that an unsuccessful attempt to install the six-pack could possibly signal the end of the mission. However, three days before the EVA, mission managers agreed to go ahead with the six-pack installation. As the commander, it was for Bean to decide which of the three of them would be best suited for the excursion and in his diary, preserved in *Homesteading Space*, he noted a few difficult decisions: "Made up my mind that it would be Owen and I, but after reading the procedures realised that I should stay in because of my [command module] experience. Owen and Jack are just not up on it and it is the best decision. Jack will do the six-pack [installation], as he is the most mechanical. Owen does not do those things as well as Jack; it will be taxing to tell him tomorrow ... "

In his entry for the following day, 22 August, Bean added that both Garriott and Lousma were happy with their assigned roles for the EVA; all three of them knew that the pilot was physically the strongest member of the crew and "if anyone could twist those connectors that had never been designed to be loosened in flight, Jack was the man. He *needed* to be out there".

The EVA on 24 August ran like clockwork, lasting four hours and 30 minutes in total. Although the six-pack of gyroscopes were actually installed *inside* Skylab, the need to turn over attitude control to the new system meant that Lousma and Garriott would have to venture outside to connect a set of cables which would circumvent the original set in favour of the new ones. In effect, Lousma would use a tool to twist the old connectors from their sockets and twist the new ones into place. Unfortunately, working in space proved notoriously more difficult than working on the ground, and Lousma ran the risk of twisting *himself*, along with the connectors, unless he securely anchored himself in place. "So I ended up wedging myself somehow, so that when I turned on these [connections] that were hard to get off, I didn't rotate myself *out of the picture*," Lousma explained. "It took a fair amount of sweat and so forth to figure out how that was going to be done. It was one of those things that the water tank misleads you on. It's *not* perfect in neutral buoyancy."

The two spacewalkers did have the opportunity to glance again at the glorious vista 'beneath' them: Garriott was elated by his panoramic view of the snow-capped Andes, framed by three of the ATM's solar arrays, stretching towards Peru. As he cast his eyes steadily southwards, he beheld high mountain lakes and salt deposits, extending all the way to Tierra del Fuego. Lousma was suitably impressed by the glimmering orange lights of cities somewhere over Russia. At the time, of course, he and Bean had recently been announced as members of the backup crew for Apollo-

Soyuz and were destined to visit the Soviet Union in November 1973 as part of their joint training regime.

The replenishment of the solar observatory allowed Bean to note with glee in his diary on the 25th that it "felt good to have ATM film again". It was apparent that Garriott was the best qualified and the most productive at the observatory's controls, to such an extent that Bean began assigning several of his own and Lousma's stints on the console to the science pilot. "We were journeymen there," Bean noted in his diary. "Jack, maybe, was better than me, but Owen was much superior and we could do the other stuff like heating up the furnace so it'll melt metal. You just turn on the switch and do the checklist and we didn't have to make much in the way of decisions about what we were seeing there. It's just a fact. Owen was just superior at it and it fit him. He enjoyed it and knew more about it and loved it." That is not to imply, of course, that Bean and Lousma were novices; for *all three* had cross-trained extensively for virtually every Skylab experiment and work task and responsibility and more than one of the science pilots – Kerwin, Garriott and Gibson – would note that the *distinction* between the scientists and the pilots became decidedly blurred as the missions wore on. Yet the scientists' value aboard Skylab was unmistakable; after the mission, Bean would laud Garriott as having ensured that the flight was one hundred percent more productive with him aboard and would note, touchingly: "I'm *really* sold on these scientist-astronauts!" Clearly, they had come a long way from the unwanted outcasts forced down NASA's throat a little less than a decade earlier.

As August wore into September, and the astronauts entered their sixth week aboard Skylab, the difficulties of mobility and the problems of space sickness which had plagued them in their first few days seemed to be gone. "It was instinctive for us to *float*," Lousma described. "We never thought about walking. We could spin our bodies and arrive pretty much where we wanted to with our hands and feet ready to stop our motion. If we had to spin a little faster, we just tightened up into more of a ball. It was a very relaxing, pleasant, comfortable feeling. We really enjoyed being weightless." The rotating chair ceased being seen as an instrument of torture and they could perform as well in orbit as they could on the ground. "We got to feeling better and better and better," he continued, "and there came a time after a couple or three weeks when I felt as good in space as I'd ever felt on Earth." Nonetheless, medical specialists on the ground were aware that this crew would be more than doubling the endurance of Pete Conrad's team and elected to extend the mission incrementally, in seven-day steps, keeping a close eye on the astronauts' health and performance. "We did medical experiments every three days," said Lousma, "and they would look at the information and, working with us from the ground, decide whether or not we could stay for longer." Aware that Gerry Carr's mission (at the time) was also scheduled to run to around two months, Bean's crew *really* wanted to make sure that they snared the record and asked for an additional ten days on top of their baselined 59 days. Mission Control deliberated for a few days and finally turned them down, on the logic that they had used up almost all of their assigned reserves of food (and, indeed, Jack Lousma had dipped rather too heartily into the *next crew's* stock of strawberry drinks and butter cookies) and film for the ATM and EREP observations.

Another item of which the crew was in dwindling supply was clean underwear. Ordinarily, Bean, Garriott and Lousma would change their underwear every other day and, since Skylab did not possess any sort of laundry facility, 'soiled' items were deposited into the station's trash airlock. The latter was essentially the depressurised oxygen tank at the base of the workshop. "The ground had the delicate dilemma of deciding how to provide enough sets of skivvies for both crews from a carefully calculated, limited supply," wrote Hitt, Garriott and Kerwin in *Homesteading Space*, "without compromising the duration of the present and next missions, the doctors' hygiene restrictions and, especially, the crews' most personal expectations with respect to living and working in space." Late in the mission, on the last appointed day of the final set of skivvies, Mission Control uploaded their 'solution' to the problem with a tongue-in-cheek teleprinter message. It took the form of a 'Good News and Bad News' joke: the Good News was that all three men *would* get their expected change of underwear that day ... but the Bad News was that Bean would exchange his underwear with Garriott, Garriott with Lousma and Lousma with Bean ...

Golden Rule Number One for long-duration spaceflight: *Never* lose your sense of humour.

The third spacewalk of the mission pleased Garriott particularly, as he became the only member of the crew to have performed all three. This time, on 22 September, he would be teamed with Al Bean on a relatively short excursion – it turned out to last two hours and 45 minutes – to retrieve film from the ATM in readiness for their impending return to Earth. By his own admission, Bean found it difficult to decide who should perform the spacewalk. Original plans had dictated that Garriott would perform only the first and second outings and Bean's decision that he should also participate in EVA-3 was down to the feeling that Lousma would probably get the opportunity to fly again and perform a spacewalk in the future. In his diary entry on 8 September, Bean felt that Garriott probably would not get the chance to do another EVA in his career – and he was *right*; but neither would Lousma – and also that the science pilot had "made this spaceflight much more interesting than it could have been with three operational types". Was it simply a 'reward' for Garriott, therefore? Perhaps. Ironically, both Garriott and Lousma *would* fly again and the science pilot would end up with the longest total flight time out of the pair of them.

For Al Bean, this would actually be his *very first EVA* ... his two previous excursions, of course, were both on the lunar surface, back in November 1969. Consequently, he would become one of only a handful of humans – only four, in fact – who have had the opportunity, to date, of performing both a spacewalk outside their spacecraft *and* a trudge across the dusty terrain of the Moon. Of those four, most considered floating high above the Home Planet to incomparably surpass the experience of walking on the lunar surface; the almost-Godlike feeling of looking down on everything that makes us human was both profound and quite ethereal. Bean agreed. "It was much more 'science-fiction' to go EVA in Skylab than it was to go EVA on the Moon. The EVA on the Moon was much like training: you were in light, the sky was black, but everything else was the same. When you go EVA in space, it's like crawling out the window on an airliner and just going along the wing

Al Bean participates in his first – and only – EVA, just a few days before his crew was scheduled to return to Earth. Note the Astronaut Life Support Assembly (ALSA), a kind of 'belly pack' oxygen supply, at Bean's waist. One of the solar arrays of the Apollo Telescope Mount is visible just over his left shoulder.

and looking in the *engine* ... something that would be *impossible* to do! I think it's the nearest analogue to what we actually do on EVA. We crawl out on the vehicle and go along the side and there's nothing you can do on Earth like that." For Bean, it was truly an unforgettable experience, with its highlight being an entire half-orbit in darkness with no responsibilities. At one stage, whilst working on the ATM, Mission Control told him that, since they needed to test the ATM doors in orbital daylight, he would need to 'wait out' the roughly 35-minute nighttime pass. Several hundred kilometres 'below' his boots, as Skylab drifted serenely into orbital darkness, Bean noticed that they were heading out over the western Mediterranean, with the heel of Italy and the bulk of Sicily clearly visible. Casting his gaze eastwards, the yellowish-brown of Egypt, with the Nile snaking through its middle, came quickly into view, then the slumbering, darkened Middle East. Fellow astronaut Ed Gibson, who would fly on Skylab's final mission, would later remark that taking a *picture* of a rose garden is never quite the same as actually *being there* and the same was true of observing Earth from orbit. The sheer breadth and depth of colour and iridescence of life could only be truly appreciated by those who had actually been there. Some of the most stunning and awe-inspiring images that I have ever seen from space were published a few years ago in *Orbit*, a book co-authored by veteran Shuttle astronaut Jay Apt; yet I know that even those images probably captured a mere fraction of the true majesty of our planet. From his lofty position, high above the roof of the world, Al Bean was exultant. For fun, at one point, he even kicked himself out of his foot restraints on the ATM and spun himself into a handstand. Years later, he convinced himself that he had set a new world record handstand for both height and speed.

One record that he had *already* set was for having achieved more time in space than any other man. The previous record, set only a few months earlier, belonged to his good friend, Pete Conrad, who had amassed something in the region of seven weeks across four spaceflights; by the time that Bean splashed down in the Pacific on 25 September he would have spent a total of 69 days, or almost ten weeks, off the planet. Even *that* record, though, would not last for long; the final Skylab team of Gerry Carr, Ed Gibson and Bill Pogue would smash it in the opening days of the following year and in the spring of 1978 the Soviets would firmly take the lead ... a lead which they still hold, unrivalled, even today.

As their mission entered its homestretch, Bean, Garriott and Lousma continued working feverishly to maximise their scientific output. By the end, they had made more than three dozen EREP passes and obtained a total of 930 minutes' worth of data. Although their daily hours on the ATM were reduced from 12 to just eight in the final days, they would win deserved praise from solar scientists for some of the most astonishing data ever gathered. On 6 September, for example, they observed a major flare which Bean called "Big Daddy" – it spanned 18 times the diameter of Earth – and were lauded for their "intelligent" use of the observatory's instruments. The medical data from the flight – which included a haematology and immunology programme, assessments of mineral balance, evaluations of hormonal changes and associated fluid and electrolyte parameters, extent of bone mineral loss, cardiovascular effects and measurements of metabolic activity – had yielded an unprecedented and valuable harvest of data covering almost every aspect of how the human body was affected by long periods in the strange weightless environment. Less than six months earlier, before the launch of Conrad's crew, NASA could only draw on a 14-day Gemini flight in terms of long-duration expertise. The Soviets had flown for longer, of course, but when their cosmonauts returned from an 18-day mission in June 1970, they had been in a pitifully poor physical state ... and the crew of Soyuz

11, who landed in the late summer of the following year, had died during re-entry. Years later, Jack Lousma was not alone in insisting that *their* mission broke entirely new ground for understanding how humans could live and thrive in space. Garriott agreed and felt that a strict and consistent fitness regime was the key. "It stresses the importance of exercise. If you do not exercise, you're going to come back in the same physical condition that you [would] be in if you'd lain in the bed. You shouldn't expect to come back *without* muscle atrophy [or] orthostatic intolerance; you'd probably faint, until your body learned how to readjust the fluid balance. We learned the importance of exercise and if you have the appropriate *amount* of exercise – namely one to two hours a day – then you're going to come back in essentially as good a condition as when you left."

Integral to living and thriving, of course, was a good sense of humour, keenly demonstrated by the Good News-Bad News debate over the crew's underwear. They also decided, late in the flight, to play a gotcha on Mission Control, in particular the unfortunate capcom, Bob Crippen. Ordinarily, open space-to-ground communications were effected through the 'A' channel, whereas the astronauts could speak directly to mission controllers, capcoms, doctors and managers through the 'B' channel, which was more restricted. Notably, in the days prior to the selection of women astronauts, the voices crackling backwards and forwards across these channels were exclusively male ... and then something strange happened. On 11 September, in the cunning climax of a dastardly plan that they had been cooking for some time, a *female voice* – the voice of Owen Garriott's wife, Helen-Mary – arrived in the headsets of astonished mission controllers in Houston.

"Hello, Houston, are you reading Skylab?"

A female voice? Sat at his console, aware that other controllers were also listening in, Crippen was stunned. Surely this was a wind-up.

"Skylab, this is Houston," he began, then paused. "I hear you alright, but had difficulty recognising your voice. Who do we have on the line up there?"

"Houston, roger," came the sweet, feminine reply. "I haven't talked with you for awhile. Isn't that you down there, Bob? This is Helen, here in Skylab. The boys hadn't had a home-cooked meal in so long, I thought I'd just bring one up. Over."

Crippen must have struggled to restrain a grin. "Roger, Skylab. Someone's gotta be pulling my leg, Helen. Where *are* you?"

No mercy would be granted on the poor capcom, however. The torment continued:

"Right here in Skylab, Bob. Just a few orbits ago, we were looking down on those forest fires in California. The smoke sure covers a lot of territory and, oh boy, the sunrises are just beautiful. Oh oh ... see you later, Bob. I hear the boys coming up here and I'm not supposed to be on the radio ... "

The confused expression on Crippen's face must have been a picture. Helen-Mary Garriott couldn't *really* be up there ... could she? And if she wasn't, then *how* could she possibly know what the forest fires – which had plagued much of California throughout August and into September 1973 – would have looked like from orbit? The transmission from Skylab had no gaps or obvious breaks and *everyone* in Mission Control had heard Crippen having a *two-way* conversation with Garriott's wife. When the science pilot keyed his mike and spoke to Crippen shortly thereafter, the embarrassed capcom claimed complete ignorance ... but privately Flight Director Neil Hutchinson and others burned to know *how* Helen-Mary knew that *he* was on the capcom's console at that precise time, *how* she knew about the fires and *how* had she been able to respond to his questions in real time?

Many years later, in *Homesteading Space*, Owen Garriott finally exposed the truth ... and the role that the 'innocent' Crippen actually played. "It had the flight controllers puzzled for over 25 years," he wrote, with undisguised glee. "My objective was to pretend that my wife ... had come up to Skylab to bring us a hot meal, even though this was an obvious impossibility. Here is how the scheme worked. I recorded her voice on my small hand-held tape recorder before the flight, pretending to have a brief conversation with a capcom, with timed gaps for his replies. The capcom would be my only 'accomplice', but his role would be carefully disguised. It was also necessary to have some recent event mentioned to validate the currency of the dialogue, so it would seem it could not have been recorded before flight. I knew that both Bob Crippen and Karl Henize were going to be capcoms for Skylab, so they were *brought into the planning*, given the script and *rehearsed* on their timing. They kept the short script on a piece of paper ... awaiting the right moment." That 'right moment' came at the end of the previous orbit, when Garriott warned Crippen that he would "have something for you" on the next communications pass. *That* was Crippen's cue to have his script – which told him exactly how to respond to each question and for how many seconds – primed and at the ready.

"No one *ever* worked out how this was accomplished," concluded Garriott. Not until the crew's 25th anniversary reunion, in Houston in 1998, with many of the former flight controllers and managers present, did he reveal what had happened in its entirety. Crippen had loved being part of it. "There was head rubbing," he said later of the reactions of his perplexed colleagues. "We did a good job. It was fun. Working those missions got to be tough. We did all kinds of things to try to come up with levity. That was a nice one that the crew got, that the ground control didn't know about!"

It was not the only time that humour, albeit of a slightly less 'socially acceptable' variety, was exercised between space and the ground. One of the crew's cameras had particularly low intensity and a Polaroid had been packed to permit the photography of faint objects. Early in the mission, one of the solar physicists, Paul Patterson, had asked Garriott if he had managed to set up the Polaroid. Garriott responded that it was next on the list. Each day, Patterson repeated the question, but due to other tasks, Garriott did not get chance to break out the camera. At length, when he did, he decided to take a few pictures to ensure that it was working properly ... and was shocked when the Polaroid produced a centrefold image, straight out of *Playboy*! He and Jack Lousma stared at each other in astonishment. They tried again. Out came *another* image of a scantily-clad young woman, and *another* and *another*. In fact, the *whole film* had been pre-exposed with *Playboy* centrefolds! When Patterson next spoke to Garriott and asked him if he had set up the camera, the science pilot replied simply: "Yes, Paul, and we got ... some *very* interesting pictures!" In Garriott's mind, such moments of levity served to cement the team and boost good feeling and morale.

Morale was high during the final days of the mission and on 21 September, the astronauts began a somewhat protracted checkout of their command and service module; protracted, by necessity, due to lingering concerns about the state of the RCS quads. Already, back in August, the command module had also developed a small coolant loop leak, seeping out at a rate of about one percent per day, but mission managers had long since concluded that the flight would be over before this turned into a serious problem. Additionally, the men deactivated the systems aboard Skylab itself, setting up a portable fan in the multiple docking adaptor to circulate air over the newly-installed six-pack of gyroscopes to keep them cool in preparation for the arrival of Gerry Carr's crew in November.

Bean was pleased that a large chunk of time had been allocated to their duties aboard the spacecraft – six hours from the closing of the Skylab hatch to undocking, then almost two more hours until the de-orbit burn and another 24 minutes before entry interface. "A nice, slow timeline," Bean had noted in his diary, "that will allow us to get set up, double and triple checked for our entry." During the 21 September checkout of the Apollo, Bean had not been surprised that his ship had passed all of its tests without incident – "somehow I knew they would" – and did not anticipate any problems with the remaining quads. After eight weeks of weightlessness and a job exceptionally well done, both he and Lousma were ready and keen to go home, although Garriott, it seemed, wanted to stay longer. "I'll miss old Skylab," the science pilot poignantly wrote that same evening. "Really hate to leave for a variety of reasons. Mostly all the unique things to do and see. A geographer's paradise. Next time!" Privately, Garriott knew that there was a very real chance that there might not be a next time, although with the Shuttle on the horizon, there was reason to hope.

Despite the reasons for wanting to remain aboard, leave they must and the time for departure was approaching rapidly. Now was the time at which the tests undertaken by Vance Brand and Don Lind to develop procedures for a controlled re-entry with only two quads or even the command module's own thrusters might be put to the test. Brand was in Mission Control throughout the entry phase of Bean's crew, confident, but concerned, nonetheless. After procedural adjustments had been made to compensate for the locked-out RCS thrusters, the return to Earth passed without incident, although Garriott felt embarrassed at his difficulty in reading Bean's scribbled checklist notes and Bean lamented his failure to properly position some RCS switches on the instrument panel. From Lousma's perspective, the hypersonic flight back through the atmosphere was exceptionally dynamic; "like flying in the cone of a flame". The occasional roar of a roll-control thruster would startle them and, at 10 km above the Pacific, the loud clang of the command module's apex cover being jettisoned, to make way for parachute deployment, was noticeable, to say the least.

Splashdown, 300 km south-west of San Diego, left the craft in a 'Stable 2' apex-down position. "Hanging from the ceiling in 1 G was uncomfortable," Lousma said later, "after two months of weightlessness. The [command module] is not a good boat, either, especially *upside down*!" Still, it very quickly righted itself and the crew soon saw the welcome face of one of the recovery frogmen in the hatch window.

Aboard the USS *New Orleans*, the process of conditioning themselves back to an Earthly existence was harder than they had ever expected: walking across a room to their bunks inevitably caused them to fall over and only the eyes – and the presence of light – could assure them of which way was 'up' and which way was 'down'. It took several days for the otoliths of their vestibular systems to re-adapt sufficiently to be able to provide them with a good sense of up and down in complete darkness.

Many of the photographs of a televised welcome-home ceremony on the deck of the recovery ship show Bean, Garriott and Lousma sitting on chairs with legs spread wide. The commander, for one, was not feeling well. He had felt fine, albeit a little 'heavy', in the water, and thought that their inflated G-suits were a complete waste of time ... until he stood up. By his own admission, Bean did not feel faint, but he certainly did not feel quite right. All three of them sat through the ceremony, hoping it would soon be over, keeping their legs in a "wide stance because of our lack of stability". On his way down to the ship's sick bay, for standard end-of-mission tests, Bean felt himself swaying involuntarily from side to side; the doctors had to assist him at one stage. It did not make sense to him that *he* was rolling left and right and, at first, he even convinced himself that the *ship* must be in rough seas. At length, in the sick bay, it felt good to lie down, but the doctors tried to get him to stand as much as possible. Garriott and Lousma seemed to re-adapt far more quickly and it was not for several days that Bean felt able or willing to sit or stand. Even in debriefings, he would get out of his chair and lie on the floor, propping his head up in order to talk. Fifty-nine days, 11 hours and nine minutes since leaving Earth, those first few hours and days back on the planet were *no* walk in the park.

Interestingly, Lousma, who had adapted with such difficulty to the weightless environment, was back in pre-flight shape within six days of splashing down. Like Bean, he didn't feel particularly *bad*, but just *lazy*, with the desire to lie down, and a sensation of lightheadedness was always present when standing or sitting. Adjusting to their new home, back on Earth, also made for some humorous episodes. On one occasion, in the *New Orleans*' sick bay, Lousma woke in the night to see a light on outside his room, and was surprised when he was unable to float over to the switch to turn it off. Attempts to toss a shaving cream bottle from one hand to the other were similarly unsuccessful; instead of floating gracefully from palm to palm, it dropped instantly. "Pow, right in the sink," he recalled with a grin. "Smashed the whole bottle!"

Their achievements, however, were extensive. In October 1973, they were hailed by NASA as the 'Supercrew' which had exceeded its mission goals and achieved a 150-percent scientific return. Years later, Bean still kept a copy of the NASA article in his briefcase, exhibiting justifiable pride in the fact that his crew had started their mission behind schedule and had fought fiercely against the timeline and come out on top. Jack Lousma felt more professionally satisfied after his mission than at any other time in his life and Owen Garriott, for his part, paid homage to the indomitable team spirit which had allowed them to achieve all they did. The next crew, scheduled to launch just a few weeks later, would certainly have its work cut out. Bean and his men had shown that, obviously, NASA was scheduling insufficient work for its astronauts. *That* would most definitely change in time for the final

Skylab mission, which, in a press release, dated 26 October 1973, the agency confidently expected to run for at least 60 days and perhaps as long as 85 days. It was taken for granted that with this last roll of the dice in long-duration spaceflight, Gerry Carr, Ed Gibson and Bill Pogue would perform at least as well as Bean's crew. Unfortunately, their mission would begin badly, with another aggressive episode of space sickness and an unfortunate "error of judgement" on the part of the commander, then deteriorate into what Carr himself has confessed was "a mutiny in space", yet it would emerge as a hugely successful, record-breaking space voyage whose duration would stand for almost half a decade.

THUNDERBIRDS AREN'T GO!

Before July 1973, if *anyone* was going to get sick in space, it was *not* expected to be Jack Lousma, the tough, square-jawed Marine Corps aviator, who had proven himself indestructible, both in high-performance jet aircraft and during a plethora of ground-based physiological and vestibular trials. Yet it was he who seemed to have suffered the most when he reached orbit and entered Skylab; even his normally insatiable appetite had suffered for a time. Next on the list of potential 'indestructibles' should have been Lousma's successor aboard the space station: Bill Pogue, a 43-year-old veteran of the US Air Force's famed Thunderbirds aerial demonstration team. If anyone's stomach could be turned upside down and wrung inside out with multiple aileron rolls and other churning manoeuvres, with no ill effects, then surely Pogue was the man. Right?

Wrong.

William Reid Pogue was born in Okemah, Oklahoma, on 23 January 1930, making him, at the time of his launch to Skylab in November 1973, the oldest rookie yet sent into orbit in NASA's history. Interestingly, his parents were both of Native American descent, their family line tracing its origin back to the Choctaw people of the south-eastern part of the United States, whose territorial homelands centred on Mississippi, Florida, Alabama and Louisiana. He could reasonably claim to be the 'first' Native American astronaut, but it was John Herrington, who flew the Shuttle in November 2002, who was an actual *enrolled* member of a tribe.

As a child, Pogue was fascinated by aircraft and aviation, making and flying model aircraft and, aged eight or nine, watching a Ford Tri-motor transport aircraft performing a loop over a field close to his home. In his formative years, Pogue attended primary and secondary schools in Oklahoma and won a bachelor's degree in education from Oklahoma Baptist University in 1951, with the intention of becoming a mathematics or physics teacher, but the outbreak of the Korean War persuaded him to take the examinations for admission into the Air Force for cadet training. Pogue was originally meant to be a gunner, but was instead sent to flight school and ended up flying 43 missions in the straight-winged F-84 Thunderjet, on interdiction missions by bombing trains and providing close aerial support to front-line American troops. He returned to the United States in December 1953 and was assigned to Luke Air Force Base in Phoenix, Arizona, as a gunnery instructor. It was at this time that he was approached to join the Thunderbirds, a newly-created aerial demonstration team, based at Luke.

Pogue loved the two-year assignment as a Thunderbird solo and 'slot' pilot. "It was a lot of fun," he told a NASA oral historian in July 2000. "No paperwork. All we had to do was *fly* and we travelled all over the country." At this stage, the team was very much in its infancy and the first Thunderbird pilots flew the single-seater F-84G Thunderjet, the swept-wing F-84F Thunderstreak and, for VIP and press/narrator rides, the two-seater T-33 Shooting Star. Shortly before Pogue finished his tour with the team, in the summer of 1956, the F-100 Super Sabre was introduced to provide them with a supersonic capability and they were relocated from Luke to Nellis Air Force Base, near Las Vegas, where the squadron remains headquartered to this day. In addition to the flying, another positive result of Thunderbirds membership was that Pogue had the pick of his next assignment. He chose postgraduate school and was enrolled for a master's credential in mathematics at Oklahoma State University, hoping to someday become an instructor at the Air Force Academy in Colorado Springs. He received his degree in 1960 and taught there as an assistant professor for a couple of years, before his attention was drawn to NASA and its astronaut corps. "I had already decided I would like to get into the space programme," Pogue recalled, "and talked to the professor of mathematics. This was kind of a tricky operation, because you had a five-year tour there. A lot of them … really held you feet to the fire, but I went in and explained what it was I wanted to do." The professor was particularly supportive and sent Pogue to the University of California at Los Angeles in the summer months "to special courses" and recommended the young pilot most favourably.

Those 'special courses' were test pilot school, which was – at the time – a prerequisite for any serious astronaut candidate. However, unlike many other astronauts, he graduated from the Empire Test Pilots' School in Farnborough, England, as part of an exchange programme between the United States and Britain. "That was very interesting," he told the NASA oral historian, "working with the Brits. *Very* competent air crewmen." He flew a variety of aircraft during his time in England, from fighters to transports and sailplanes to the Shakleton four-engined patrol bomber. "The thing about it was, if you *asked* to fly it, they would *let you*! They would give you the book; well, actually, it was the *plane*. I recall many times coming in to land an airplane, before I'd make my let-down, the first time I'd flown it, I'd have to look on the back of it and fortunately on the back of these pilots' notes … it listed all the *air speeds* for approaches! So I'd memorise the two or three air speeds and then come in and land fine!" If astronauts-in-waiting needed the 'Right Stuff', it seemed that Bill Pogue had it in spadefuls. He completed his tour in England in September 1965.

His next assignment was to the famed Edwards Air Force Base in California, where he was placed briefly in Fighter Ops and later taught operational mathematics at the test piloting school *and* translated all of the mathematics text into vector notation. "I finished *that*," he concluded, "about three or four weeks before I came to JSC." Pogue was surprised, in the light of his lengthy stint in England, that his application to NASA had even *arrived* on time, but after undergoing the standard battery of physical and physiological tests, he was called to Houston's Rice Hotel, late in March 1966, for his formal interview. "As I understood it," he recalled, "they weren't going to select [more than] about 12 or 15." Eventually, aware that its lunar programme and Apollo Applications would need the manpower, NASA picked 19 fliers ... and Bill Pogue was amongst them.

His relationship with Apollo Applications, the effort which later morphed into Skylab, began early, when he was assigned to work with Al Bean. Subsequent assignments included supporting Apollo 7, in which one of Pogue's duties was to reduce the crew's notes and checklists and simplify the approaches to working with the spacecraft systems. In true support-crew fashion, his general role was very much "sort of a go-fer" and he and his two comrades, fellow 1966 selectees Ron Evans and Jack Swigert, spent much of their time undertaking tests which the prime and backup crews were too busy to conduct. An actual flight assignment for Pogue was a long time coming, but in the spring of 1970 he was placed on what he described as "a phantom backup crew" for Apollo 16, with an anticipated – although not formally announced – future assignment as command module pilot on the Apollo 19 mission. This mission was scheduled to occur in late 1973 or sometime in 1974, after Skylab, and would have been the final lunar landing. Unluckily for Pogue and his phantom-crew mates, Fred Haise and Gerry Carr, the prospects for Apollo 19 did not look good that spring and on 2 September 1970, due to steadily tightening budgets, the flight was cancelled. Haise, Pogue and Carr were on a field geology trip in Arizona when they learned the news and all three were devastated as they saw their chance to go to the Moon evaporate.

Yet their chances of flying in space were still there. Early in 1971, Deke Slayton named Pogue and Carr as prime crew members, along with scientist-astronaut Ed Gibson, to the third and final Skylab mission. Since Pogue had previously trained as a command module pilot, he was assigned as the pilot on the crew, because many of his former duties would translate smoothly across from one role to the other. Gerry Carr, on the other hand, had originally been training as a lunar module pilot and, with no lunar module involved in Skylab, it made sense, in Slayton's mind, to give him command of the mission. He would thus be the first rookie astronaut to command an American space mission since Neil Armstrong on Gemini VIII in March 1966. Slayton had another reason to give Carr this command. Personal recommendation went a long way and Carr's sterling work for Pete Conrad on the Apollo 12 support crew had earned him deserved brownie points. Conrad, of course, was now 'Sky King' and was not only in charge of the Skylab branch of the astronaut corps, but had a big say in who should fly each mission. With all these factors in mind, it is not surprising that Carr received the Skylab command.

Pogue has never openly commented about receiving the junior position on the crew, even though (as an Air Force colonel at the time of the mission, as compared to Carr's rank of lieutenant-colonel in the Marines) *he* was the senior military officer and one can only assume that the two men were far too professional for such an issue to arise between them. In *Homesteading Space*, Ed Gibson referred to Pogue as "an average, mild-mannered mathematician", but added that "he was once grounded for flying too low behind enemy lines ... [and is] a sharp, aggressive guy". Like Pogue,

Carr had received an exceptional education in aeronautics and he was also an extremely competent Marine aviator. In Gibson's mind, that said it all. "When assigned to the mission," the scientist noted, "I *knew* I was in fast company!"

ERRORS IN JUDGEMENT

Gerry Carr, commander of the final Skylab crew, floated silently in his seat on the left side of the command module. It was the evening of 16 November 1973. Every hour and a half, his ship passed from orbital daytime into nighttime and the interior was cyclically flung from blinding sunlight into the pitchest blackness. The cabin had assumed a more spacious demeanour since they had blasted off from Cape Canaveral a few hours ago, since *now*, of course, the astronauts were operating in a three-dimensional volume; ceiling, walls and 'floor' now effectively morphed into one. However, as he floated there in the darkness, Carr had a problem. Nothing was wrong with the spacecraft; nor, indeed, was the docking with Skylab in jeopardy, for it had recently been successfully accomplished. The all-rookie crew had performed exceptionally well so far ... bar one issue. Carr's problem was with his pilot, Bill Pogue, who had just thrown up due to space sickness ... and the very real need to do something about it.

At the time, space sickness was a major issue, both medically and politically, having almost incapacitated Al Bean's crew for a couple of days at the start of their mission. Congress had only recently approved development of the Space Shuttle, a unique winged spacecraft which would routinely carry astronauts into orbit for several days at a time, as much as *fifty* times each year. The Shuttle's pilots would be required to actively guide their craft back through the atmosphere at the end of each mission and perform a pinpoint runway touchdown, like an airliner. Their descent would be entirely unpowered and with no option to go around, they would have just one chance to land. Lawmakers were concerned that if those pilots were incapacitated by space sickness, even for just a couple of days, then the entire project was unworthy of the risk. Others thought the Shuttle was nothing more than a multi-billion-dollar white elephant and were lobbying for it to be cancelled. The dangers of space sickness would offer them precisely the ammunition they needed. Gerry Carr was well aware of the implications; for the Shuttle would effectively be his nation's *only* manned spacecraft for the foreseeable future. As commander of Skylab, he now needed to make an executive decision. He could either *admit* Pogue's sickness and potentially kick off a political firestorm within NASA ... or say *nothing at all* and get rid of the evidence. Carr chose the second option. He would live to regret it.

By his own admission, Carr had been overwhelmed when Deke Slayton chose him, almost three years earlier, to lead this mission and was exceptionally pleased to be flying with Pogue and Ed Gibson, two men he greatly respected. Their training for the mission had increased dramatically following the return of the station's second crew in September 1973; clearly, NASA managers realised, they had set far too little work for the astronauts to complete. They would not make the same mistake with Carr's mission. During a press conference on 2 October, Ken Kleinknecht, the Skylab manager in Houston, enthused over the accomplishments of Bean's crew, gushing that they had shown that mankind "was able to do more than we thought he could do". If Skylab itself was once considered the unwanted stepchild of the Apollo lunar landing effort, then, for a time, Carr, Gibson and Pogue felt that they were the unwanted stepchildren of the training staff, for the bulk of the simulator time in 1973 had been taken by the two prior crews, together with Rusty Schweickart's backup team as they struggled to devise a solution to the solar heating issue in May, *and* by Brand and Lind as they worked against the clock in August to identify work-arounds for the RCS thruster problems. "Thus, we were left doing only peripheral stuff," Carr related. "We'd go wherever they *weren't* getting trained. Whichever simulator or trainer they *weren't* using, we would use, *if* it wasn't down for maintenance." They also spent a great deal of their own time on learning about the ATM – Gibson was the astronaut corps' resident solar physics expert, having recently published a textbook, entitled *The Quiet Sun* – and on becoming "intelligent" observers of Earth. By the time they finally launched, they would have received direct, one-to-one tuition from over a dozen of the world's foremost experts in various terrestrial phenomena: from earthquake faulting to ice formation at northern and southern latitudes and from the dynamics of the oceans to the formation of deserts. At length, with Bean's crew safely in orbit and the need for Brand's rescue mission gone by the middle of August or thereabouts, the final Skylab crew had their much-desired simulator team for three months of relatively intense training.

Integral to their own training was an understanding of how the previous two crews had performed. "When the first crew came back after 28 days," Carr reflected, "they were pretty wobbly, pretty weak. So the second crew and ours decided to bump up the exercise periods. Al Bean's crew doubled their exercise period from a half hour to an hour a day. Turns out that didn't appear to be enough, either, so we increased it again to an hour and a half." One area that *did* cause serious concern – and *would* certainly cause concern in flight – was that Bean's crew had hustled about their work too quickly and it was felt that this course of action might not be wise in a mission lasting three months. "We began telling some of the managers that we didn't think *that* rate of work was wise for an 84-day mission," continued Carr, "because we weren't sure that we were going to be able to sustain it. We thought that the workload should be slacked off some and there should be more rest. Everybody agreed to that and the experiments were slowed and spread out quite a bit."

Unfortunately, *more* experiments were then added, "and we allowed ourselves to get trapped into this new situation". Carr would find that he and his crew had over-committed themselves and, in orbit, after a few days, would be in real trouble. As the days ticked away to their scheduled launch, their chances of getting into orbit diminished rapidly. In late October, Bill Schneider announced 10 November as the target date, with the first of up to five EVAs scheduled for a week later to install the first set of ATM film. Repairs of an antenna on the S-193 experiment would occupy a second excursion, whilst a third – planned for four and a half hours on Christmas Day – would include among its list of tasks photography of the long-awaited Comet

A technician leans precariously from his work platform on 7 November 1973 to examine one of the Saturn IB's stabilising fins. Worrisome cracks had been detected the previous day and more than a dozen would be uncovered in total, resulting in a delay of almost a week in getting the final Skylab crew off the ground.

Kohoutek. A fourth EVA on 29 December would perform additional experiments and retrieve material samples from Skylab's hull and a final outing in mid to late January would collect the final sets of ATM film for return to Earth. Although NASA hoped to extend the duration of Carr, Gibson and Pogue's mission to 85 days, the agency cautioned that any extensions would be decided and implemented on a weekly basis.

In the meantime, the Apollo spacecraft and its Saturn IB continued to be readied for flight. Standard pressurisation and other checks of the booster in October had gone well, until, on 6 November, a management meeting at Cape Canaveral was interrupted by the worrisome news that inspectors carrying out routine structural integrity checks had found a series of hairline cracks in the aft attachments of the eight stabilising fins at the base of the Saturn's first stage. More than a dozen cracks were found in total, the longest of which ran to several centimetres in length. Most likely, the cause was related to age – the rocket stage itself had been delivered to the Cape more than *seven years* earlier and kept in storage for much of that time – but three months spent sitting on the launch pad, in the salty air of the Cape, was also a contributory factor. With the possibility that the compromised fins might be ripped away during ascent, NASA opted to delay the launch so that replacements could be flown in from Michoud, Louisiana, and installed. This was no mean feat. Merely rigging the support platforms on the milk stool in order to remove the *first* fin took 35 hours. Nevertheless, by the early hours of 13 November the work was completed and launch rescheduled for the 16th. In the meantime, the astronauts returned to Houston to participate in revised launch-abort procedures, since the postponement would inevitably affect their launch azimuth.

Carr was particularly upset by the delay, since their original date on the 10th happened to be the official birthday of his parent service, the Marine Corps, founded in 1775. With the exception of John Glenn, who flew into orbit alone, Carr would become the first Marine to command a spacegoing crew. In order to commemorate both this anniversary and this milestone, the Commandant of the Marine Corps himself, General Robert Cushman, and several of his key staff were due to visit the Cape and watch the launch. The bitter disappointment over the delay was quickly replaced by frustration and, as the issue of the cracks was steadily resolved, by dark humour. "We started making some comments about calling the vehicle 'Old Humpty Dumpty'," Ed Gibson told the NASA oral historian. "We were just kinda kidding and then, somehow, that got out in the press and, of course, those guys who were working around the clock, all day and all night, it *didn't* sit too well with some of them." Nothing was ever 'said' ... at least not until shortly before launch. "When we got about 20 minutes before launch," Gibson continued, "we got this message from them: 'Good Luck and God Speed from all the King's Horses and all the King's Men'. It was a neat little comment!"

Not surprisingly, the night before launch had been a sleepless one for Gerry Carr. With liftoff scheduled for 90 seconds past ten on the morning of 16 November, he and Gibson and Pogue had struggled for the past week or so to adjust their circadian rhythms to the routine of bedding down in the Cape's crew quarters at around six each evening, then awakening at the unholy hour of two or three in the morning.

Following routine medical and microbiological checks – "to find out what kind of flora and fauna were living on us" – Carr and his men breakfasted with Deke Slayton, Chief Astronaut Al Shepard and Ken Kleinknecht. Despite having been a member of the astronaut corps since 1966, Carr still found it a little unusual that a launch-day 'breakfast' always consisted of steak and eggs. Of course, the *purpose* of such food was not dietary, but on what might be euphemistically described as 'low roughage', but its unhealthy nature tickled Carr. "In subsequent years," he noted later, "my wife and I have totally modified our diet so that now we don't touch any of those foods, mainly because of their high cholesterol and fat content. It's amazing that dieticians in those days thought that lots of steak and eggs was the best thing in the world for us!"

After emerging from the breakfast room, suitably fed, the astronauts were helped into their pressure suits. Carr recalled attaching a self-winding, counterweighted Movado watch to his ankle. Although he knew that he was not meant to carry any additional items into orbit, he was curious to know how the weightless environment would affect the performance of the counterweight in the tiny timepiece.

Out at Pad 39B, they were greeted by a scene of utter silence and stillness and an overwhelming sense that *today* was the real thing; the event for which each of them had devoted the best part of a decade. At the top of the elevator, the white room personnel, led by Guenter Wendt, who received a complimentary bag of sourballs from Carr, set to work getting them settled into the spacecraft. Fellow astronaut Hank Hartsfield finished the final checks, crawled underneath the couches and was gone. "After Hank exited the spacecraft ... they closed the hatch and it was just the three of us in there," Carr recalled, "and we began the preparations that we had done so many, many times on the simulator." Except, that is, for one small diversion. "I remember we looked up ... leaned forward a little bit and looked over at each other and just *giggled* like a bunch of schoolgirls!" The feeling that this was *really* going to happen, for a few seconds, totally consumed them. It was going to happen *today*.

The launch itself was picture-perfect and all three men would relate lucidly their memories of riding the Saturn IB for the heavens. As propellants flooded into the combustion chambers of the eight H-1 engines, Bill Pogue likened the sound to someone having simultaneously flushed every toilet in the Astrodome. Seconds later, the behemoth left Earth, convincing Ed Gibson for an instant that the *basement* had just exploded. Naturally, for a veteran Thunderbird, Pogue thought he stayed pretty cool and calm throughout the experience, although a subsequent conversation with flight surgeon Fred Kelly assured him that his pulse rate had actually accelerated from 50 to 120 at the instant of liftoff. The vibrations in the command module's cabin were intense throughout the climb; one of Pogue's responsibilities was to follow the launch profile in his procedures book and he quickly found that his hands – and the *book* – were shaking so much that he could hardly read it. The noise was so intense that they could hardly hear each other's voices over the intercom between their suits. "Once you go supersonic," Pogue continued, "the noise *stops*, because the shockwave detaches and you no longer get the air noise. Then, after you get above about 40,000 feet, the turbulence cuts down considerably. By the time you hit 60 [thousand feet], there's none. The sky turns black and so forth and you're on your way, but it's a *real* soft ride after that."

Carr and Gibson would agree with their crewmate that the 'first' part of the ascent, in the lower atmosphere, was definitely the most dynamic and 'exciting'. "You get an awful lot of turbulence and a lot of shaking," Gibson said. "I would equate it to being a fly glued to a paint shaker! There's something massive there that you're sitting in that's really giving an *extreme* turbulence." For his part, Carr likened the first part of the journey uphill to riding a train with *square* wheels! At the point of staging, when the first stage burned out and the single J-2 engine of the S-IVB second stage took over, all three agreed with veteran astronaut John Young that it resembled a train wreck in terms of its rough abruptness: the transition from four times their normal, terrestrial 'weight' to around one and a half times was acutely noticeable. Thereafter, flying the S-IVB for the remainder of the trip into orbit was a dream – a long elevator ride, according to Gibson – and when its engine finally shut down, some eight and a half minutes after lifting off, all three men were astonished to observe that their supposedly 'clean' craft was now filled with debris: dust, particles, paperclips, screws and washers. "People had done their best down there at the Cape to keep it clean," said Gibson, "but you can't keep everything out. All that is soon dissipated or taken out by the airflow in the cabin. It all happened *real* quick."

Arrival in space was greeted, at first, with complete bewilderment. Looking down on Earth, even though the notion of 'up' and 'down' in this peculiar environment was gradually losing its meaning, Carr's months spent studying terrestrial features and landmarks counted for nothing; he could see *nothing* that he recognised. Then, after about half an hour or so, he spotted the heel of Italy – which *really did* look like a boot – and it was this image which would remain with him, indelibly printed on his memory. "I've never forgotten that particular experience," he said. For Gibson, the gradually receding coast of Florida was his first view of the glorious Home Planet, although he, too, saw the heel of Italy and understood for the first time what it felt like to travel at more than 8 km *per second*.

For Pogue, on the other hand, the euphemism known as 'stomach awareness' would very soon take centre stage. In the light of the difficulties encountered by the previous crew, NASA had decided that, after docking with Skylab, Carr, Gibson and Pogue would remain aboard the command module for their first night's sleep in space and enter the station the following morning; the rationale was an effort to assist their adaptation to weightlessness before entering the large, open volume of the workshop. After two unsuccessful attempts, docking was duly accomplished some eight hours into the mission and the crew remained awake late into the night, stowing equipment, when, all at once, Pogue felt sick. *Really* sick. Years later, the beleaguered pilot would explain that the sensation took the form of a severe headache and nausea and that Carr's suggestion to eat something did not help matters. In fact, a mouthful of stewed tomatoes – the only item left in Pogue's evening meal – very soon sent him scurrying for his sick bag.

One of the greatest ironies was that Pogue had actually taken the scopalomine-dextroamphetamine anti-nausea medication before launch and Carr had not ... and

Spectacular liftoff of the final Skylab crew on 16 November 1973.

yet the commander experienced no feelings of sickness whatsoever. Nor did Gibson, the scientist. Under normal circumstances, Carr knew that he was not even allowed to *drive* after having taken 'Scop-Dex', and he most definitely *did not* want his ability to be compromised whilst in charge of a multi-million-dollar spacecraft. Of the three of them, Pogue, the ex-Thunderbird, was considered the *least* likely to get sick. Years later, he would see it as evidence that none of the medical community had a real handle on what caused space sickness and, for the most part, the theories and the prescribed medications were inconclusive and of only limited usefulness. Now, of course, with the pressure on them as a crew *not* to get sick, and with all the medications aboard, they faced a real dilemma: what to tell the ground?

Skylab had drifted out of direct radio contact at this stage and it was Gibson who suggested simply disposing of the 'evidence' – Pogue's sick bag – in the station's trash airlock and keeping quiet about the matter. In so doing, Carr agreed, they could avoid getting the medical community "all fuzzed" and hopefully get their mission off to a smooth start. To Gibson, the desire to avoid space sickness was very much a political one: as already noted earlier, the Shuttle had only recently been approved by Congress and there were lingering worries that if astronauts could get sick and potentially incapacitated for several days, the whole *raison d'etre* behind *having* a reusable winged spacecraft might be compromised.

Things were not looking good. Carr and Gibson tried putting Pogue into the docking tunnel, hoping that air from a cabin fan might make him feel better, but to little effect. Hopefully, 'Old Iron Ears' Pogue would feel better the next morning and no harm would have been done. With regard to the food that the pilot had not eaten, they would say that he was just *not hungry*. Before retiring for the night, Carr read his status report to the ground, admitting to Pogue's nausea and highlighting that he had not eaten all of one of his meals. Unluckily, one of Pogue's responsibilities was the spacecraft's communications system … and, as specified in the checklist, he had left the switch 'on' to the equipment which was recording their in-cabin conversations. Whilst the crew slept that night, Mission Control downloaded the tape and heard *all* of their discussion about concealing the evidence!

Early the next day, 17 November, Pogue felt better, but took things slowly as he and his comrades ate breakfast and watched the Alps and south-eastern Europe drift beneath them. By mid-morning, they were inside Skylab, Carr switched on the lights and the men set to work on their respective duties: setting up communications links, starting the environmental control system and generally reactivating and testing the workshop after its eight-week slumber following the departure of Al Bean's crew. In the meantime, back in Houston, the tapes from the previous evening were being transcribed and their startling contents led to a medical conference being convened that afternoon. Later, Al Shepard himself came onto the capcom's console to address Carr directly. Shepard, the first American in space and the astronaut's direct superior, did not mince words.

"I just wanted to tell you," he said, "that on the matter of your status reports, we think you made a fairly serious error in judgement here in the report of your condition."

Carr accepted the rebuke. "Okay, Al, I admit it was a dumb decision."

Shepard was not to be put off. If Pogue's sick bag *had* been disposed of in the station's trash airlock, it could screw up many of the medical experiments. Shepard pressed Carr to assure him that they had *not* gotten rid of the bag and that its contents would be weighed, as per the crew's training, as part of the mineral balance studies.

The incident, whilst relatively minor, underlined in some managers' minds a fear that Carr's crew were unwilling to engage in frank and open communications with Mission Control. Flight Director Neil Hutchinson did not doubt the crew's integrity, but made certain that any further problems would require flight controllers to take immediate steps to set matters right. The situation would grow markedly worse over the coming days. Even three decades later, Bill Pogue would recall that he got on perfectly well with both Carr and Gibson, but felt it would have been nice to have had an experienced crewman aboard. They were the first all-rookie American spacegoing crew since Gemini VIII in March 1966, during which Neil Armstrong and Dave Scott had experienced problems after docking with an Agena target vehicle. Barely a day into the mission which would make or break Carr's astronaut career, the enormous responsibility and the very real need to make this the best Skylab flight of the series weighed heavily on his mind.

Gerald Paul Carr came from Denver, Colorado, where he was born on 22 August 1932. Most of his formative years were spent in Santa Ana in California, from where he received his secondary education. As a teenager, he and a friend would cycle from their homes to Orange County Airport – now John Wayne Airport – to wash aircraft in exchange for a 20-minute 'hop' in an old Martin Aviation Taylorcraft. The flying bug, in Carr's case, at least, bit early. So too did other bugs: an active Boy Scout, he became an Eagle Scout and was later a junior assistant scoutmaster, participated in student government and played high school football. He became a naval reservist in 1949, during his final year in high school, and remembered that his duties included cleaning the F-6F Hellcat aircraft, checking its oil and fuel, starting its engine and warming it up each day. When offered the chance to obtain a degree, Carr proceeded to the University of Southern California to study mechanical engineering, part of the Reserve Officers Training Corps (ROTC) detachment. Decades later, he recalled a commanding officer who laid out the choice for him: if he wanted to join the Navy, he could go to the strict Naval Academy in Annapolis, but for the Marines, he could pick a university of his own choice.

The bonuses of taking the Marine route, the commanding officer told him, were twofold:

1) "In the Marine Corps, they don't care where you came from, whether you're an Academy graduate or not ... "
2) "Why should you lock yourself up in the Bastille for four years, when you can go to a college with *girls* for four years ... ?"

Carr picked the Marine option.

On graduating in 1954, he was commissioned into the Marine Corps and attended basic school in Virginia, flight instruction in Florida and Texas and was then assigned to the All-Weather Fighter Squadron, VMF-114 – the notorious 'Death

Dealers', famed for having provided close aerial support during the bloody Battle of Peleliu in the Pacific during the autumn months of 1944. Over the next three years, Carr gained experience in the Grumman F-9F Cougar and Douglas F-4A Skyray and undertook a Mediterranean cruise aboard the USS *Franklin D. Roosevelt*, as a preliminary step to qualifying to operate from aircraft carriers.

Partway through his cruise, he was informed of his selection for the Naval Postgraduate School. Carr graduated in aeronautical engineering in 1961 and was sent to Princeton for a master's qualification. The focus of his work was upon aircraft stability and control and in 1962 he was detailed to Marine Corps Air Station Beaufort in South Carolina, where he flew the F-8 Crusader and served as the maintenance officer of VMF-122. He subsequently moved to Japan and was present when the Tonkin Bay crisis escalated the Vietnam conflict. Back in the United States, Carr worked on the design of fighter trajectories to intercept different aerial targets. During this time, he learned that NASA was screening new astronaut candidates and knew that Marine aviators John Glenn and Clifton 'C.C.' Williams had already entered their hallowed ranks. Carr decided to give it a try. In the winter of 1965, as a 'conditional candidate', he participated in the space agency's physical and psychological evaluations, but was convinced that his flat-footedness and susceptibility to hay fever might preclude him from selection. Not so. On April Fool's Day 1966, Carr was in California, working on his trajectory-intercept tasks, when he received the call from Al Shepard which would change the course of his life. He was told to keep quiet, but when he informed his wife that he had been picked to join the most elite flying fraternity in the world, her tongue-in-cheek response was to look him straight in the eye and quip: "*April Fool!*"

Carr's selection was certainly no joke and over the following years he played a pivotal role in America's drive to land men on the Moon: serving as a support crew member for the historic Apollo 8 voyage, during which he helped to develop and maintain the flight data files for Frank Borman's crew and supporting Pete Conrad's team on their Apollo 12 landing mission. Many commentators and historians have remarked that Conrad was so impressed by Carr's performance that, after becoming head of the Skylab branch of the astronaut corps, he personally recommended the Marine aviator to command the station's third crew. Before that, however, Carr's experience with the lunar module gave rise to the strong expectation that he would serve in at least a backup capacity on a later Apollo mission to the Moon. Had the final three landings not been cancelled due to budget cuts in 1970, it is quite reasonable to suppose that Carr might have served as the lunar module pilot on Apollo 19.

Alas, it was not to be. Sometime late in 1970, Tom Stafford, who was then serving as Chief Astronaut whilst Al Shepard trained for Apollo 14, asked Carr if he would be willing to fly with Gibson and Pogue to Skylab. Unsurprisingly, Carr's response was yes. Their names were publicly announced in January 1972.

If Gerry Carr had made an error in judgement by concealing the evidence of Pogue's space sickness, then certainly his science pilot, Ed Gibson, had done the same a few years earlier, in an unfortunate episode involving a helicopter. On 2 August 1969, just a couple of weeks after the historic Apollo 11 lunar landing, he took a small Bell 47 chopper out over the mudflats, near La Porte in Texas' Harris County, to practice vertical landings. Gibson had been chosen as a scientist-astronaut by NASA in the same group as Joe Kerwin and Owen Garriott and was well aware that helicopter training offered one of the closest terrestrial analogues to the behaviour of the spidery lunar module. *Everyone* in the corps wanted a seat on a landing crew, of course, but Gibson was under no illusion that one of those seats might come to him. He knew that NASA's test-pilot-dominated astronaut office regarded the scientists as markedly inferior. Still, Gibson *enjoyed* flying helicopters and the desire to gain a seat on a space mission exerted an incredibly strong pull.

"We all went through the Navy programme at Pensacola and got Navy helicopter certification," Gibson recalled to the NASA oral historian. "I enjoyed it. It was one of those things, again, where it takes a while to catch on, but once you do, you really enjoy it." On that particular Saturday morning, early in August, he wanted to practice so-called 'run-on landings', during which pilots would approach the ground at a rate of maybe 30-40 km/h and slowly set the chopper down onto its skids. The area near Ellington Field was too hazardous at the time and Gibson flew north, finally spotting what looked like a broad patch of dry ground, seemingly perfect for the exercise. He performed several landings and everything seemed fine; the centre of the field was indeed firm. Next, he decided to fly over to the side for another landing. Unluckily for Gibson, what appeared to be solid ground was actually a thin crust of dryness over a metre or more of *mud*. "Of course, that's not a lake bed up there," Gibson continued, "but, nonetheless, it appeared that way, so I went over to the side and made a run-on landing. Unfortunately, the thing was crowned so that on the side it was lower, and relative to the water table, the water was *this* far below the surface, whereas over in the centre, it ... had been a couple of feet below. The centre was firm, [but] the other had just a small layer that looked firm, but when you put the skids on, the weight on the skids broke *right through*."

Gibson's beautifully orchestrated approach ended up with a wrecked chopper, as the tiny Bell got itself stuck in the mud. "The next thing I knew, I was hanging in the straps, looking at the *mud*, with gasoline dripping over my shoulder and one *destroyed* helicopter all around me." The still-spinning rotor blades ripped off the tail. Years later, Gibson would praise his own airmanship ... and woefully condemn his knowledge of local *geology*! Deke Slayton took up the story in his autobiography: "He was a hundred yards from dry land in the middle of what was, essentially, *quicksand*. He had to slog his way through that stuff. The manager at some oil company site on the edge of this threw down a ladder to him. He made it just in time; Ed had been burning up his adrenaline the whole time and was just about to pass out." Slayton would subsequently blame managerial error and Gibson's inexperience – he had, after all, only completed his initial helicopter instruction at Pensacola a few months earlier. The incident did not harm Gibson's astronaut career, but it must have reinforced in some of the test pilots' minds the notion that the scientists simply did not possess the good judgement to perform in such unexpected situations. "It was pretty clear to me," Slayton wrote, "that there weren't going to be many scientist-astronauts landing on the Moon, anyway."

From his perspective, Gibson considered himself lucky not to have been thrown out of NASA. "If they wanted to get rid of me," he recounted, "they had a very good reason to do it right there." But Edward George Gibson was a valuable commodity at the space agency. Not only was he an accomplished pilot, in spite of the helicopter mishap, but he had also established a reputation as the astronaut corps' resident solar physics expert. In fact, in 1973, he would publish a handbook, entitled *The Quiet Sun*, which would be used not only for Skylab ATM training, but is actually still used today as a solar physics text. A physicist by training, Gibson has admitted that his original specialism was in plasma physics and that he knew little about the Sun – "except that it was big, round, yellow and hot" – but decided on his own initiative to brush up his skills, since one of Skylab's three main scientific themes was solar science.

Gibson came from Buffalo, New York, where he was born on 8 November 1936. His father ran a family company which produced marking devices – including rubber stamps, steel dies and stencils – and the young Gibson considered that this was the inspiration for his fascination with science and engineering. "He wanted me to go into business with him," Gibson explained. "He thought that if I'd learned the engineering side, I could always get the business side later, so for lack of any real direction in my life, I went into engineering. Once I got into it, I found I really liked the basic science more. I like physics and then I combined that interest...with my interest in astronomy, got interested in rocketry and space travel." He graduated from Kenmore High School and aspired to go to Cornell University, but his application was rejected. Ultimately, he entered the University of Rochester in New York and graduated with a degree in engineering in 1959.

Academic success was tempered by disappointment: his wish to enter the US Air Force had been scuppered by a bout of osteomyelitis – a bacterial inflammation of the bone – in his youth. Gibson decided, therefore, that if he could not *fly* aircraft, he wanted to involve himself in *building* them. He pursued master's studies in mechanical engineering, with an option in jet propulsion, at California Institute of Technology. During this time, he also served as a design engineer for Buffalo-based Sylvania Electric, performing thermal, vibration and shock tests of electromagnetic countermeasure systems for the Convair B-58 Hustler. A doctorate in engineering physics followed – "I never would have anticipated *that* earlier in life" – from Caltech in 1964. By this time, he was also married with four young children.

Completion of his PhD led to a position as a senior research assistant for Applied Research Laboratories, an adjunct of the Ford-Philco Corporation in California, to focus on theoretical and experimental studies in laser pumping and the operational breakdown of gases within their aeronutronic division. Although he enjoyed the work, Gibson found it somewhat unsatisfying...until, one morning at breakfast, his wife Julie drew his attention to an article in the *Los Angeles Times*. NASA was calling for scientist-astronauts. Gibson initially thought that she was playing a prank on him, but when he realised that the notice was real, he took only a few minutes to decide to apply.

He was shocked as his application progressed further and further...and really began to take his chances of success more seriously when NASA physicians told him that his osteomyelitis had been dormant for two decades and would not be a factor in his selection. Interviews and physiological testing and meetings with the psychologist followed and to his great surprise, in June 1965, Gibson was chosen as the youngest of six scientist-astronauts. Only two were qualified pilots with jet experience and so Gibson was sent, along with Owen Garriott, Duane Graveline and Jack Schmitt to Williams Air Force Base in Arizona in July to complete a year-long undergraduate training programme. With the exception of Graveline, who was obliged by NASA to tender his resignation in the midst of a particularly ugly divorce, the others received their Air Force wings in the summer of 1966.

Very soon, Gibson recognised that not only were the scientists still treated as second-class citizens by many of the test pilots, but they were also unlikely ever to gain a seat on a lunar landing crew either. "During this early phase," he explained, "Al Holt, who was in flight crew support, and I decided that the guys didn't know anything about the Sun or solar physics and we were looking around at how to get people educated, at least *enough* so that they would know what they were doing. So he and I decided to write a solar physics guide and it was strictly just top-level stuff. Some of the books that were out there really didn't address the way – the *simple* way – which guys needed to understand." But which one of them would be dealing with the Sun's 'active' state, with the flares and the prominences and so forth, and which would handle the 'quiet' or 'steady' state? They tossed a coin. In a sense, Gibson lost and his focus was upon the Quiet Sun.

"Al never pushed it along as far," he remembered. "He finally got something in a guide put out, but then I wrote *The Quiet Sun* and it was in a very preliminary state. It was really just a guide." At the time, however, Gibson knew that the forthcoming space station project, which was then part of Apollo Applications, would feature a large solar observatory and he decided to pursue his work further, in the hope of someday gaining a seat on a mission. Having said this, as the 'wet' workshop became 'dry' in the summer of 1969 and the project was renamed 'Skylab' in the spring of the following year, the chances of Gibson getting one of the three station missions seemed to be a no-brainer. With the resignations of Graveline and Curt Michel, there were only four scientists in line. Jack Schmitt ended up as lunar module pilot of Apollo 17, putting his geological expertise to work on the Moon's surface, whilst Kerwin, Garriott and Gibson ended up on Skylab.

The only question seemed to be *which* of them would fly in which order. One assumption was that Kerwin, as the only physician amongst them, might go on either the *first* mission (to establish and understand problems involved with long-duration spaceflight) or the *last* (to spend the longest period of time aloft). Garriott was senior to Gibson in terms of age and experience, which also threw another consideration into the mix, but until 1971 the precise order remained unknown to the three scientists. When Skylab finally reached space and problems with its micrometeoroid shield and solar arrays were discovered, Gibson thought that his place on the third crew had turned into something of a 'short straw' ... for the station might not even *last* long enough for him to get there. In the years since, his opinion mellowed. "When it *did* hold together that long," he explained, "and we ended up going from 56 to 84 days, I came out and *really* got the good deal." In Gibson's mind, he ended up in the right place at the right time.

MUTINY IN SPACE?

None of the Skylab crew – Gibson, Carr or Pogue – felt themselves to be in the right place on the morning of 17 November 1973, as their ears were seared by the stern admonitions of Chief Astronaut Al Shepard. All three men felt bad about it, particularly in view of the fact that there was now an unspoken element of distrust in the relationship between themselves and Mission Control. In hindsight, Gibson acquiesced, they could and should have handled it differently. "We should've just said: 'Hey, guys, your pills *didn't work*. They're wafting across the command module now, along with Bill's tomatoes!' That's probably the most regrettable thing I have about that whole flight ... that we were not smart enough to handle it properly, because it caused everybody a lot of problems."

On beginning this new day, Pogue felt a little better, if slightly under the weather, and Carr admitted to a peculiar, knot-like sensation in the stomach, but the surprise to all was that Gibson – the *scientist*, the one with the least flying time – experienced no problems whatsoever adapting to weightlessness. Their first encounter upon entering the voluminous expanse of the workshop was *three figures*, clad in the golden-brown Skylab flight suits and positioned at various positions – a joke set up by Al Bean's crew. "When we arrived," Carr deadpanned later, "we found three dummies that had been packed and put there by *three previous dummies!*" Humour aside, it was quite a shock to come across three 'people' in a supposedly unoccupied spacecraft ...

Enthusiasm and gusto were definitely the watchwords in those first few days, because, like their predecessors, the new crew were behind schedule and needed to work hard in order to catch up. Finding things in the vastness of Skylab was difficult, too. "One evening," Pogue related, "my flight activity message for the next day directed me to recharge the fluid level in a water loop used to cool an electronics package." The task seemed straightforward enough ... and it *was*, but Pogue did not count on the fact that it would require an age to find all of the required parts. He found a small flashlight to observe the accumulator, then went in search of a long hose which he would hook up between the water tanks to the work site. He couldn't find it. Unfortunately, at the time, Skylab was out of direct communications with the ground and Pogue had to wait 20 minutes until the next stateside pass. Mission Control got in touch with his predecessor, Jack Lousma, who happened to be at home in Friendswood, mowing the lawn at the time. "He wiped some of his sweat off," continued Pogue, "and said he did remember using it, but if it wasn't in the designated stowage location, he didn't have the foggiest notion of where it might be." Eventually, the beleaguered Pogue found two shorter sections of hose, which he connected together to span the distance. In his 84 days aboard Skylab, he never found the hose for which he had so diligently searched. On other occasions, calibration weights would mysteriously disappear, a systems checklist would float away and not be seen again for several weeks. Even Pogue's reading glasses vanished. *They* turned up, thankfully, within three days, floating near the 'ceiling' of Gibson's sleep compartment. As the mission wore on, the station's air duct turned into a sort of 'lost-and-found' department and was the first place the crew would

Suspiciously floating in Bill Pogue's flight suit, one of the three dummies awaits the crew of the final Skylab mission in November 1973.

check for anything they had lost. One morning, Carr and Pogue even lost *Gibson*, who was searching for some old procedures books and had buried himself deep behind the station's freezers.

Aside from adaptation to their new environment, the crew's first week in space included an EVA by Pogue and Gibson to reload film into the ATM and check out an inoperable antenna on an external radiometer/scatterometer and altimeter, known by the experiment code number of 'S-193', which was part of the Earth resources payload. Late on the morning of 22 November, Thanksgiving Day, the two astronauts ventured outside to begin work. The space suits, Pogue remembered, were hard to work with, particularly in view of the reality that the astronauts' spines lengthened slightly after a few days in weightlessness. "Our space suits had been very carefully custom-fitted ... so when we got up in space and our bodies increased in length, it was not only difficult to get into them, but you got back after working six or seven hours outside [and] you'd have *cable burns* on your shoulders because, from crotch to shoulder, you'd *grown* two inches and that put a *lot* of pressure on the shoulders." The excursion itself was highly successful, running to six and a half hours. As the S-193 experiment had not been designed for an EVA repair, there were no handholds or foot restraints. Pogue and Gibson improvised by moving, hand-over-hand, along a rigid dump pipe for the station's molecular sieve, giving it an additional role to its normal job of getting rid of carbon dioxide and water vapour. The S-193 repair was awkward – "a Dixie screwdriver operation," according to Pogue – but the two men took turns and eventually succeeded.

"Removing the cover from the antenna electronics box," explained Gibson, "turned out to be exceptionally difficult. On one side of the box, four screws had to be removed. On the ground, it was easy, but in flight, because the *real* box had a metal lip that closely overhung the screws, it was anything but easy. The small screwdriver that fit the small screws had to be inserted into the slots from the side of the screw heads, rather than from the top, which was extremely difficult in our large, bulky EVA gloves." Even *accessing* the electronics box was tough, as they had to cut away several layers of aluminised Mylar insulation, much of which was highly reflective and floated away, producing a 'cloud' of flashing lights, like fireflies. In addition to the satisfaction of a job well done, Pogue and Gibson greatly admired the astonishing vista of the Home Planet and the Universe. After his eyes had become suitably adapted, Pogue managed to recognise a few constellations. For Gibson, the sensation of *height* was evident: "It's like going up into a tall building where you look out the window and you think it's interesting that you see all the little people down there ... but now, if we *open* the window and take you out to the end of a long springboard, where we get this steel-fisted Arnold Schwarzenegger, who's going to grab you by your ankles and hold your head down ... somehow it *feels* a bit different!" As a physicist, intellectually, Gibson knew that the laws of orbital mechanics meant he was up there to stay, and would not fall, but for a while his stomach told him something distinctly different ...

The only failure on the spacewalk was a camera problem, which prevented Pogue from photographing the amount of contaminating debris around Skylab; the shutter-speed knob spun ineffectually in his gloved fingers and he managed only a few exposures before having to give up on it. The repair of the antenna, though, was a major triumph and its restoration delighted the S-193 investigators.

Working in a space suit in *real* space was quite different to rehearsing their movements in the water tank, at the Marshall Space Flight Center in Huntsville, Alabama. For one thing, the resistance of the water made motions considerably more difficult. In orbit, with the benefit of weightlessness, the tasks were simplified considerably. "If you could do it in the tank," Gibson related, "you could do it in space." The day concluded in fine style, with an ample Thanksgiving dinner. Carr selected prime ribs, Gibson went with turkey and Pogue chose chicken. It was very good food, but they did note a tendency of blandness. Condiments helped a little, although they could be used only sparingly, particularly salt, lest they interfere with the medical experiments.

Whether afflicted by sickness or not, the three men also needed time to get themselves adapted to their strange new environment. Even Gibson, who was untouched by the malaise, remembered looking at his face in a mirror and seeing an enormous pumpkin, pierced by a pair of blood-red eyeballs, staring back at him; it was clear that his heart and arteries, no longer constrained by terrestrial gravity, were pushing blood inexorably towards his head. It felt a little like lying down on Earth, with his feet elevated, a truly peculiar sensation. Sleeping was the ultimate in relaxation, with no pressures acting upon any part of their bodies, and one evening Gibson tried to bed down in the open expanse of the central workshop. "Once I relaxed," he noted, "my knees would bend slightly and my arms would float out straight, just like the position I had assumed floating in water many times on Earth. After a few minutes, I would drift off into a nice … relaxing … quiet … *WHACK!!!*" Gibson had inadvertently floated, whilst asleep, directly into a wall! He made sure to be careful to secure himself after that unfortunate episode …

During the first few days, Carr elected to swap roles with Pogue, "because my job was more sedentary than his". The ailing pilot was thus able to stay quiet and get the commander's work done and vice-versa. After a while, Pogue perked up and began to carry his load with the same enthusiasm as his crewmates. Gradually, this enthusiasm became wearing, for the performance of Bean, Garriott and Lousma had encouraged mission planners to pile more and more work onto Carr's team. If they missed a step or made a mistake with a task, the timeline was so densely packed for all three men that they would end up racing against the clock, fighting a losing battle to keep up. "It was *hard* on morale," Carr recalled. "We were rushed and not able to get things done and experiments completed." The low morale was accompanied by an acute feeling that they were letting down the principal investigators; they pictured the scientists on the ground grinding their teeth as the three astronauts kept reporting that they were unable to complete all tasks on time because they were rushing around too much, making mistakes. Circumstances were not aided by the rather unpleasant attitude taken toward the astronauts by several of the scientists, some of whom regarded the crew as little more than a trio of semi-trained chimps. Early in the mission, Pogue was assigned to photograph a barium cloud, exploded by a Black Brant IV rocket launch from Fairbanks in Alaska. The objective of the NASA-funded study was that the barium vapour would be injected into Earth's magnetosphere and, after ionisation by solar ultraviolet radiation, would illuminate geomagnetic field lines and make them visible to sensitive equipment on the ground. The two principal investigators for the experiment, Milton Peek and Eugene Wescott, were respectively based in Hawaii and Alaska, with other researchers aboard a modified NC-135 aircraft and the mission's backup pilot Don Lind offering additional support from Houston. "The experiment," reported a NASA news release, dated 20 November, "is analogous to using iron filings, sprinkled on a sheet of paper over a magnet to show the magnet's field lines."

However, after struggling to find each piece of equipment needed for the camera, Pogue positioned everything in the correct angle, checked his watch to ensure that all was ready ... and saw nothing. "*Where's* the barium cloud?" he recalled later. "I missed it. I don't know why none of the numbers worked, but in any event I missed the cloud." Despite not understanding the problems that Carr's crew faced in space, particularly during initial adaptation, those first few days of the mission seemed to reinforce this notion. "Then we discovered that we had been scheduled at nearly the same rate that the second crew had achieved at the *end* of their flight," Carr continued. "*That* explained why we were having so much trouble keeping up. But by the time that was finally recognised, we had achieved a skill level that was adequate to get the work done." They ate dinner together each evening, partly for the social contact and partly to cement their cohesion as a crew, but found themselves hurrying off to work on experiments until well into the evening. By bedtime, none of them felt ready to sleep, because there were tools and pieces of equipment to put away or set up for the following morning. Their minds were moving too fast to rest, which impaired their ability to sleep and caused their productivity to suffer. Planners on the ground began to reschedule their exercise sessions, to enable them to catch up on the experiments. There was nothing wrong with that, it seemed ... until it became clear that at least one exercise session was scheduled *right after dinner*! "That's *no* time to be exercising," observed Carr afterwards, "particularly up there, where you couldn't belch, because with your food floating around inside you, you were liable to get it back with your belch."

Not surprisingly, as time wore on, the crew became quite testy with Mission Control. Carr had requested before launch that their first day off, scheduled for 19 November, should be cancelled, but Flight Director Neil Hutchinson felt that they would benefit from the free time and two opportunities were made available in the first few days. Already, Hutchinson had the feeling that the crew had made an error in estimating how long they could get things done and their workload was already too intense. On the evening of their second free day, 24 November, Carr told mission controllers that the week had been "frantic" and that, try as they might, they were struggling to keep to the timeline. Words of wisdom from Ed Gibson's high school track coach summed up the crew's feeling: If you want to win the quarter-mile race, *sprint* the first hundred yards, then just gradually increase your pace. Aboard Skylab, they were moving so fast, and so constantly, that it was impossible to avoid making mistakes. On more than one occasion, Gibson recalled floating through the multiple docking adaptor, muttering creative profanities, little realising that the voice recording link to the ground was open. "The situation," he explained, "was compounded a bit because people had not yet fully come to grips with the fact that Skylab was a different animal than all the relatively short missions to date." For ascent and re-entry, fine, it was *necessary* to have their timelines spelled out in second-by-second detail, or at least in blocks of a few minutes, but on an extended mission of *three months*, such micromanagement was neither suitable or appropriate. One such instance of this came one morning, when they found that Mission Control had sent their daily, minute-by-minute schedule to the teleprinter ... and it ran to a length of over 20 m, virtually the length of the workshop. It would take all day simply to *read* it ...

Nor was there an option to simply have a heart-to-heart with the ground and thrash things out, because all communications had to be open for the whole world to hear, including the press. At one stage, the astronauts decided to make better use of their time by having one of them listening for radio traffic at a time, thereby allowing the others to switch off the distraction of their radios and get on with their work. This worked well for a time ... until, during one particular orbit, they forgot and *all three* of them left their radios switched off for an entire pass. "That caused a lot of concern on the ground," Carr said, "and, of course, the press just thought that was wonderful." In their sensationalist minds, the astronauts were on strike, on the brink of mutiny, turning their radios off deliberately. It was an unfair accusation which led to a stigma that would hang over Carr's crew for decades.

Meanwhile, the men took a stoical outlook, trying to push on with their work and hope that circumstances would improve. They did not. It was time for Carr, as the commander of the crew, to make a stand. Having made it known prior to launch that his crew intended to take the activation of the workshop at a slow pace and ease their adaptation to weightlessness, and already mentioned the frantic nature of the first few days, on 6 December he spent several minutes complaining to Mission Control that the schedule was too full. His crew would not be expected to work for 16 hours a day, every day, for 85 days, *on Earth*, so it was unfair to expect it of them in space. There the issue rested for a time and little more was said during the last couple of weeks of the month.

Christmas – the second one to be celebrated by Americans in space – may have helped to distract the men from their workload, together with the anticipated arrival of Comet Kohoutek. Carr's crew built a crude tree out of packing material from food containers and decorated it with makeshift ornaments. They even crafted a small, long-tailed star from silver foil and put it in pride of place at the top of the tree, in honour of their cometary visitor. Systematic studies of Kohoutek from Skylab had begun on 23 November, when Pogue used a photometric camera to record its intrinsic brightness, followed by analyses of the composition of its coma and tail a few days later. By the weekend before Christmas, more than a dozen such observations had been completed. Another EVA, conducted this time by Carr and Pogue, took place on Christmas Day, one of whose objectives was photography of Kohoutek. "Bill and I were out for seven hours," Carr recalled. "I was amazed when I got back in, because I expected that I'd have to go to the bathroom something fierce, but I didn't. Apparently, I'd gotten rid of a lot of fluids in the form of sweat through my pores. When I got back in, I was really sweaty, but I really didn't have to urinate. I was just amazed that, after seven hours, I wasn't pretty interested in streaking to the urinal!"

Photography of Kohoutek was one of Bill Pogue's tasks and in his interview for the NASA oral historian he remembered floating in the station airlock, surrounded by his cameras, two large film magazines for the ATM and tools which Carr would use to routinely service the solar observatory. "Gerry went hand-over-hand to the end of the solar observatory," Pogue related, "while I got the replacement film magazines ready. I operated an extendable boom to transfer the first film canister to Gerry; he removed it and loaded the exposed canister to the boom; I retracted the boom while Gerry loaded the fresh canister to replace the one he had just removed and when he gave me the okay, I sent the second canister out. We repeated the procedure and were finished in record time." Next came the photography of the comet. Pogue carefully set up his camera, mounting it onto a strut and positioning it such that one of Skylab's ATM arrays barely blocked the Sun. He could not physically see the comet, but Mission Control had earlier sent him a diagram on the teleprinter. "The instructions were clear and it was a fairly easy job," he recalled. "I turned on the camera and I was finished."

Also finished was Pogue's experience of spacewalking, for this second excursion would be his last; the next two EVAs would both be undertaken by Carr and Gibson. The child in Pogue suddenly took over. "I decided to make the most of it. I crawled all over the accessible parts of Skylab. It reminded me of when I was a kid, doing a mud-crawl in a four-feet-deep stock tank used for watering cows and horses." His adventure ended at the solar 'end' of the ATM, offering him a stunning and unobstructed view of Earth; it felt like Pogue was doing a swan-like dive through space. His fun was arrested by a sudden call from Gibson. One of the station's three gyroscopes had failed during the 22 November EVA and now, as Pogue lingered close to the ATM, *he* was actually causing a second one to throw a fit. "Our suits were fed by oxygen from inside Skylab," he explained, "and there was no recycling of the air. It automatically fed in near the back of my head, flowed down across my face and then escaped out the front of the suit near my waist. The outward airflow had acted like a small thruster, like letting the air out of a balloon. Although the force from the escaping air was small, my position at the Sun 'end' of the ATM magnified the thrusting effect because I was about 30 feet from the centreline of Skylab. In other words, this lever arm was giving the force of the escaping air a *lot* of leverage. The airflow from *my* suit was rotating a *one-hundred-ton space station!*"

The experience, though, was more than worth it, particularly when the men returned inside the airlock and were advised that, with a duration of six hours and 54 minutes, they had set a new world record for the longest EVA to date. Yet the relentless march of the timeline continued to conspire against them. Pogue complained on 12 December that the tight schedule had lost him the chance to complete a series of assigned photographs, because he had to set up the required equipment in a hurry, and on the 20th Gibson recorded that managing the crew's time with teleprinter messages which dictated their day on a minute-by-minute basis was "no way to do business". He even described the first five weeks in space as "nothing but a 33-day fire drill". On 28 December, after submitting his status report for the day, Carr told Capcom Dick Truly that he was preparing a special message for Mission Control, to be sent later that evening, before retiring to sleep. He wanted a discussion about the concerns the following day. If it took a couple of hours, well, everything else would have to wait. Any hope that the conversation could be aired on a private loop was gone and the press jumped hungrily on the matter. "We started talking ... as we came up over Goldstone [tracking station in California]," Carr explained. "We had the whole US pass, essentially, for me to tell them all the things that were bothering us. We need more time to rest. We need a schedule that's not quite so packed. We don't want to exercise after a meal. We need to get the pace of things under control." Carr concluded his remarks by asking for Mission Control's response during the next communications pass. Throughout this 20-minute period of recess, he conferred with Gibson and Pogue to put together a summation of their needs. "During the next pass, *they* bent *our* ear with all of the things that we were doing, including our rigidity that made it difficult for them to have the flexibility to schedule us how they needed to."

All told, the conversation ran for almost an hour and covered a multitude of issues. Dick Truly assured them that everyone was happy with their performance. With Deke Slayton and JSC Director Chris Kraft in attendance, Truly felt the need to provide an assurance. "Gerry, let me say one thing," he said. "Dr Kraft and Deke have been here and listened ... and they're very happy with the way you're doing business ... and they think we've made about a million dollars tonight." It was not vain praise. Mission managers had already reviewed the crew's performance after a month in orbit and Bill Schneider had told the press on 13 December that Carr's team had already completed 84 hours of solar observations, 12 EREP passes, 80 photographic and visual surveys, all of their assigned medical tasks and three major repair jobs. Unless something unforeseen transpired, Schneider told the gathered journalists, he was optimistic of a mission lasting at least 60 days, "open-ended to 84". Looking back, Ed Gibson would laud his commander for making a stand on the matter of the schedule, although, tellingly, he noted that Mission Control was by no means blameless, even describing their continuous demands and requests as "obnoxious".

Next morning, the 30th, Houston came back with a list of recommendations. One was that the crew's menial, routine chores – the housekeeping – would henceforth be put on a so-called 'shopping list', to be completed when time permitted during the course of the day. Moreover, the men would no longer be hassled during meal times and they would be given no major assignments after dinner in the evening. The experiments, however, which needed to be 'hard-scheduled' to a particular time slot, would need to remain as they were. Fine, Carr replied, aware that this opening-up of the schedule had already taken much of the pressure off the crew's shoulders. In fact, this opened-up schedule even allowed them to conjure some experiments, suitable for television audiences on Earth. Several of these included demonstrations of the behaviour of water in weightlessness are still used today as part of classroom science experiments. From Mission Control's perspective, Neil Hutchinson admitted that there were indeed several serious scheduling and performance problems in the flight. Indeed, Harvard Business School would later publish a case study about the

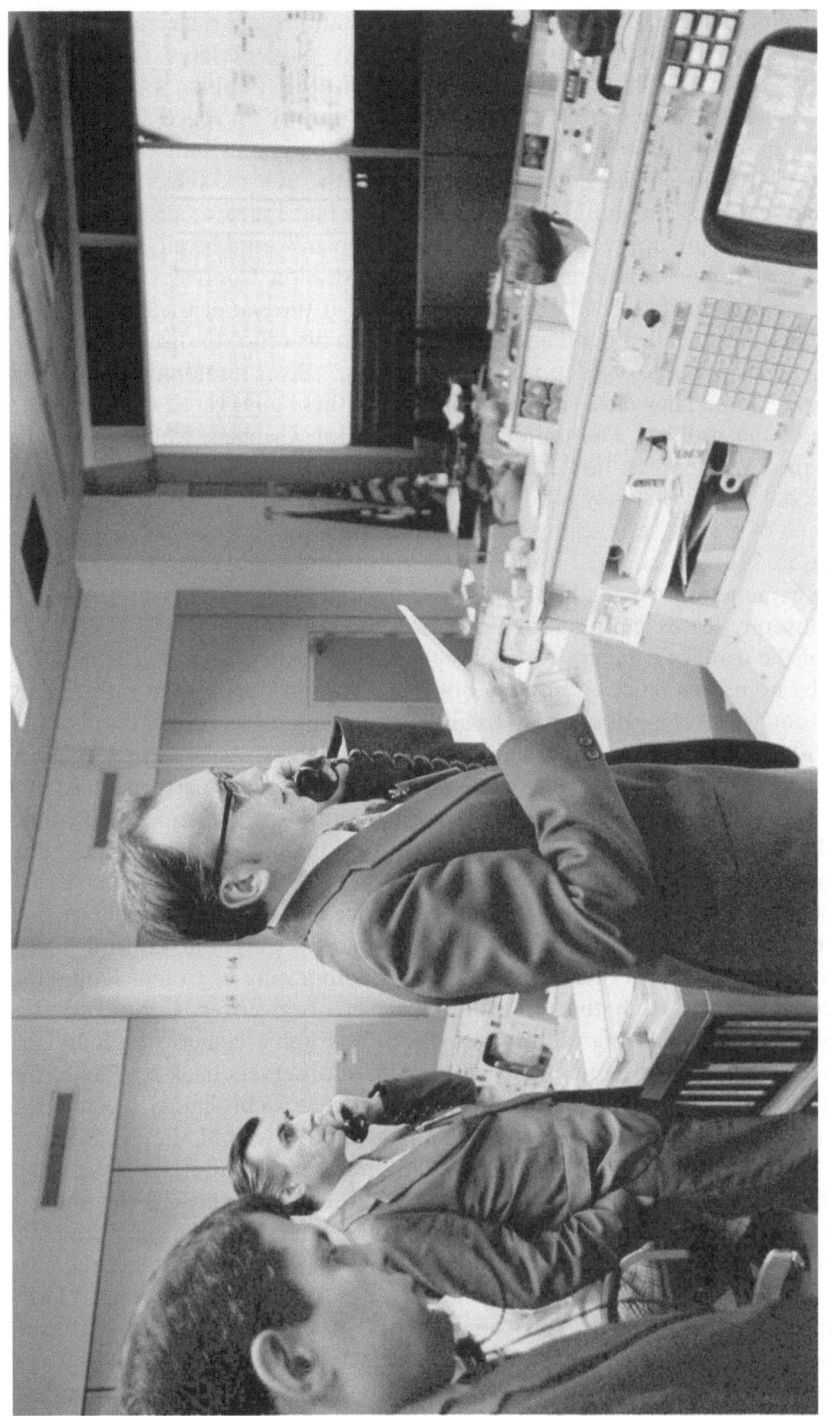

Czech astronomer Luboš Kohoutek, discoverer of the comet which bears his name, visited the Mission Control Center in Houston, Texas, in early January 1974. In this image, he speaks by telephone to the final Skylab crew, only days after Carr, Gibson and Pogue had executed a pair of spacewalks to photograph the comet.

supposed 'strike' in space, detailing unrealistic expectations and miscommunication as part of a flawed management process. To this day, perhaps unfairly, the achievements of Carr's crew are overwhelmed in the popular press by the unfortunate label that they were the first to stage a 'mutiny' in space.

In a sense, the over-performance of Al Bean's team had contributed to the stress under which Carr, Gibson and Pogue had worked for the first six weeks of their own mission. Years later, Bean would accept that NASA had failed to properly switch gears after his flight and appreciate that Carr and his men would be aloft for a much longer period of time. More detrimental was that Mission Control *started* Carr's crew at the same point that Bean's crew had reached at the *end* of their mission. At one stage, Chris Kraft called Bean and Pete Conrad to his office to discuss the issue. "Mission Control plans to lighten up on these guys," Bean implored Kraft, "but they don't ever do it. They *have* to lighten up and let these guys catch their breath!" Also aware of the problem was unflown astronaut Bob Crippen, who had led the SMEAT team a year earlier; he felt that Conrad's crew had spent much of its time repairing Skylab, with relatively little time on the experiments, and Bean's men had the advantage of a slow start, ramping up into a more aggressive final few weeks, when they were properly adapted to the new environment. Carr, Gibson and Pogue, at the other extreme, were being literally burned out right from the outset. It was *not* a good philosophy for executing a long-duration space mission.

As one of the team of capcoms for this flight, Crippen *knew* that the events of 28 December by no means reflected any kind of 'rebellion' on Carr's part; rather, they offered Mission Control a wake-up call to understand the problems they faced. After the heart-to-heart, everything smoothed out and the final six weeks ran much more smoothly and pleasantly. The drive to achieve the highest possible performance from the crew was a hard-won lesson, Gibson recounted, but it was a lesson which would have untold ramifications in the planning of future long-term missions, including those aboard today's International Space Station.

With these terse exchanges going on between space and ground, it might at first glance seem surprising that a *third* EVA – this time by Carr and Gibson – was undertaken on 29 December, acquiring yet more photography of Comet Kohoutek and collecting samples from the airlock's micrometeoroid cover. Central to this photography effort was the far ultraviolet electronographic camera, built by the Naval Research Laboratory. Like many disappointed observers back on Earth, the astronauts could hardly describe the long-haired messenger as brilliantly spectacular, but Carr was impressed, nonetheless. "It was so faint," he told the NASA oral historian, "that we really had to work to find it. Once we did find it, we observed a gorgeous thing: small, faint, but *gorgeous*." Still, although he and Gibson shot as much film as they could during their three and a half hours outside, they privately doubted that it was sensitive enough to record sufficiently good data. Over the next few days, from inside the workshop, the best results came from the ATM's coronagraph ... and, interestingly, from a series of pencil sketches by Gibson himself, today enshrined within the hallowed precincts of the Smithsonian.

With the dawn of the New Year, much of the comet-hunting on the ground had come to an end and there was much disappointment that Kohoutek had not lived up to its media-hyped billing. Journalists started calling it the 'Flop of the Century', ignoring the scientific yield and focusing only on its brightness as a marker of significance, but in reality the fault lay fairly and squarely upon their shoulders; for it was the media which treated Kohoutek as a sure thing, a dead cert, right from the outset. In March 1974, *Sky & Telescope* glumly told its readers that, whilst professional astronomers were jubilant with their observations, "the general public wondered what had happened to the spectacle promised by the news media".

If the media were disheartened at the start of 1974, the crew aboard Skylab considered the last six weeks of their mission to be much more satisfying than the first six. Work aboard the orbiting laboratory was pleasurable, at last, and fleeting moments of joy – and humour, too – came from regular contact with loved ones back on Earth. On one occasion, Gibson described for his wife and their children the beauty of watching fires along the coastline of Africa. He thought they would be hanging onto his every word ... until Julie, his youngest daughter, asked her mother if she could go outside and play! On another occasion, Pogue was disturbed by a message from his wife, alerting him that she had received a letter, advising that her husband's life insurance policy was about to expire. Pogue was particularly irritated, since he had *requested* the possibility of making a three-month pre-payment into his policy, before launch, to cover the time that he would be away ... and was told that pre-payments were not possible, but assured that the policy *would* remain effective throughout his mission.

In the days preceding their launch, Carr's crew was tentatively scheduled to return home in the early part of January 1974, after around 56 days, but the assumption was that extensions would be considered on a weekly basis, depending upon consumables and the good health of the astronauts. These extensions came thick and fast through the month of January. Three days before they were due to break the record set by Bean's team, NASA officially extended them to at least 63 days, with Bill Schneider praising their "good spirits", their "excellent physical condition" and the "good shape" of Skylab itself. It was a dramatic change and remarkable turnaround from the exhausted trio of spacefarers who had struggled to keep up with the timeline a few weeks earlier. Another extension pushed their mission to 70 days, with splashdown rescheduled for no earlier than 24 January, then further into the first week of February and finally settling on the 8th. "The last six weeks of the flight were very pleasant for me for two reasons," Bill Pogue remembered. "One, we'd achieved the skill level sufficient to do the job quickly and accurately, and second, I no longer suffered from the head congestion that had plagued me for about the first six weeks." As they neared the 12-week limit, Pogue even hinted that NASA Administrator Jim Fletcher wanted them to spend an *additional* ten days aloft, but mentally the crew were ready to come home and, besides, the consumables aboard the workshop were almost exhausted. A potential fourth Skylab visit, lasting three weeks and possibly crewed by Vance Brand, Bill Lenoir and Don Lind, was unlikely to ever take place. It had been considered as an option early in the spring of 1973, but, according to Dave Shayler, "proceeded no further than coffee-table discussions". Had it flown, one of its primary functions would have been to mothball the workshop and perhaps conduct a few more experiments. However, the extension of

Carr's mission by a month past its original 56-day mandate sufficiently maximised the scientific return from Skylab and eliminated the need to launch a costly additional flight. On 3 February 1974, Carr and Gibson performed the final EVA from the station, retrieving the last of the ATM film and gathering other material samples for return to Earth and subsequent analysis.

Five days later, the time was upon them to leave. No more crews were expected to occupy the station, although there was a faint hope that a Space Shuttle might someday visit, perhaps to re-boost it into a higher orbit for possible future use. For that eventuality, Carr, Gibson and Pogue left a sort of time capsule – a variety of material samples, held inside the station, but kept in vacuum, since Skylab was depressurised shortly after their departure – which a subsequent crew might collect. However, there was little time to ponder. Frozen urine samples had to be stored aboard the command module, as did the materials samples and the film cassettes from the ATM and other equipment. The men went about their tasks with enthusiasm, finally undocking and performing a flyaround inspection of their old home, then firing the service module's SPS engine to depart for the last time.

With Skylab now gone from their immediate lives, the crew duly jettisoned the service module ... and ran straight into a problem. "I looked over at Gerry as he was moving the hand controller," explained Pogue, "to get the right entry attitude, which we absolutely had to be at for re-entry to avoid landing in the wrong location or being *cremated* before our time ... and *nothing* was happening." He advised his commander to 'go direct' – to go to the hard stops on the hand controller, bypassing all the black boxes and putting the 'juice' directly to the solenoids controlling the propellants in the reaction control thrusters. Fortunately, directness worked and re-entry proceeded without further difficulty. It later became clear that in the gloom of orbital darkness, Carr had mistakenly pulled the circuit breakers for the *command module*, instead of the *service module*. "It was dark," Pogue concluded, "and he just pulled the wrong ones. But it turned out fine."

Descending through the steadily thickening atmosphere was truly spectacular – Ed Gibson likened it to travelling through a purple neon tube, of gradually increasing brightness and intensity – and as the forces of extreme deceleration increased, a white-hot flame enveloped them. They were no longer in a neon tube. They were in a blast furnace. From the centre seat, Gibson could *see* the command module's thrusters pulsing as the computer actively guided them towards their splashdown point, a little under 300 km south-east of San Diego. Less than 5 km away, steaming towards them, was the helicopter carrier USS *New Orleans*, their prime recovery ship.

Splashdown, when it finally came, was into a thankfully calm sea, with virtually no wind, although the command module quickly assumed the inverted 'Stable 2' position. Within moments, though, the flotation bags around the spacecraft's apex inflated and bobbed it over into the Stable 1 configuration required for the recovery operation. Helicopters hovered nearby and frogmen were soon in attendance. Dawn had barely broken in the Pacific and the final Skylab crew were safely home. They had left Earth as NASA's first team of rookie spacefarers in almost a decade ... and had returned as world record-holders, having jointly secured an impressive mission

Bill Pogue grins through a heavy beard and tired eyes after jointly establishing himself as one of the world record-holders for space endurance. The 84-day achievement would not be eclipsed for almost half a decade. Note the charred exterior of the command module.

of 84 days since their November liftoff. None of them felt seasick, although the sensation of *weight* was strong; Gibson could *feel* his head on his shoulders and it took great effort simply to move his arms, but all three men could be thankful and deeply satisfied that their long mission had ended in fine fashion.

"A FEW EXTRA BUCKS"

And what of Skylab itself, the reader might ask? Here, the story turns to alternate possibilities, depressing realities, and eventually winds up in a shower of burning debris, screaming through the sky above Western Australia. Just before undocking, Gerry Carr had boosted the station about 11 km higher in its orbit, such that it circled 433 × 455 km above Earth. In a press release, NASA announced that it confidently "anticipated the orbital life of Skylab [to] be about ten years." Next day, on 9 February 1974, flight controllers remotely vented the atmosphere from the workshop and established it in a gravity-gradient stabilised attitude and shut down most of its systems. It could still respond to telemetry signals, whenever its solar arrays were exposed to sunlight, but with a troublesome control moment gyroscope and erratic behaviour noted on two coolant loops, there seemed little possibility that it would be reoccupied.

A second dry workshop, more or less identical to the first and labelled 'Skylab B', had already been built and paid for, but ultimately was never used, much to the regret of many in NASA, including the astronauts. Even today, it is considered a colossal waste of an enormous financial investment. In the weeks before the launch of Pete Conrad's crew, in late March 1973, the space agency hoped to fly Skylab B sometime after ASTP and before the first launch of the Shuttle. Unfortunately, the risk that assigning additional funding to the second workshop would divert monies from the politically protected Shuttle contributed to its demise. By August, as Al Bean's crew beavered their way through a 59-day flight, NASA quietly announced that Skylab B was being deleted from the manifest. Today, it sits in the Smithsonian and, having cost the taxpayer an estimated $1.5 billion, is perhaps the museum's most expensive exhibit. And today, many astronauts, including Owen Garriott, are still bitter that it was forsaken so easily.

"What should have been done was to put [Skylab B] into orbit and then continue flying crews to gain six- and 12-month experience," he told the NASA oral historian. "The value of the Skylab programme could have been doubled or tripled by simply taking advantage of the flight hardware that we had already built and paid for. The decision was made not to do it, because it would have cost a few extra bucks and possibly competed with the development of the [Shuttle], which didn't fly for another eight years. *That's* my biggest annoyance about some of the foolish decisions, largely political, largely financial, that were made at the end of the Skylab period."

The potential for Skylab B is discussed at length by Dave Shayler. He postulates that it might have been launched as early as December 1975, for possible occupation by several three-man crews for between six and ten months, punctuated by shorter visiting missions to exchange Apollo command and service modules, in a manner not dissimilar to the pattern followed by the Soviets, starting with their Salyut 6 station. Such speculation today, of course, is purely academic, but one thing is clear: Skylab B, had it flown, *could* have sustained American astronauts for missions in excess of the three-month record established by Carr, Gibson and Pogue. Not for more than two decades would American astronauts surpass their time in orbit … and they would do so aboard a *Russian* space station.

As the 1970s wore on, the only chance of saving the orbiting hulk of Skylab was with the Shuttle, originally scheduled to undertake its maiden voyage sometime in 1979. Solar activity and its impact on Earth's upper atmosphere had led NASA managers to think that the station would probably endure until the early 1980s and plans were set in motion for a possible Shuttle re-boost. Engineers analysed the practicalities and concluded that the winged spacecraft would probably be unable to physically dock with Skylab until at least its fifth manned mission, by which time its systems would be sufficiently tested and flight-proven. This would be too late. Rendezvous and docking would have to be abandoned in favour of attaching some sort of remote-controlled booster to the station. In September 1977, NASA gave the go-ahead for such a mission and a two-year programme to build appropriate hardware got underway. By the end of November, Martin Marietta had received an initial $1.75 million letter contract to study the options and plans steadily began to crystallise for a boxy Teleoperator Retrieval System (TRS) – a kind of 'tug', the size of a truck, laden with propellant tanks and engines, to be deployed by the Shuttle on only its third orbital mission. This would be guided remotely by the crew to dock with the workshop. NASA would then have two options: either to fire the TRS engine and cause Skylab to perform a controlled, destructive dive into the atmosphere, thereby ensuring a safe disposal and avoiding populated areas, or to insert it into a higher 'storage' orbit for subsequent reuse, since, in space, a pressurised volume was a precious commodity. By March 1978, the names of the men who might be called upon to complete the mission had been formally announced: Apollo 13 veteran Fred Haise would serve as the commander and old Skylab hand Jack Lousma would be the pilot. Delivery of the $26 million TRS to Cape Canaveral was scheduled for August 1979 for launch late the following month.

Years later, Lousma described his responsibilities on the scheduled five-day flight. He and Haise would have maintained a station-keeping position, around 300 m from Skylab, after which Lousma would have deployed the TRS using the Shuttle's robot arm and manoeuvred it – "like a radio-controlled airplane" – to a docking with the workshop. "The mission had never been planned to begin with," he told the NASA oral historian, "so the Shuttles didn't have rendezvous radar. Fred got busy and started getting that implemented and also developing rendezvous procedures. I worked with Martin Marietta [as] the lead on the development of the [TRS]." In many minds, the effort was pointless, futile and a waste of both time and money. Chris Kraft wanted little to do with it and even NASA Administrator Robert Frosch, who had taken over from Jim Fletcher in 1977 and steered the space agency throughout the Carter years, felt that the chances of successfully pulling off the re-boost were fifty-fifty at best.

What the agency had not anticipated was a marked increase in solar activity during the latter part of the decade and its consequent effect on the upper atmosphere. Early calculations estimated that solar action and increasing atmospheric densities would cause Skylab's orbit to become susceptible to increased drag, dropping by perhaps 30 km in altitude by 1980 and faster within another couple of years. Sometime around March 1983, it was thought, the workshop would finally burn up in the atmosphere. On the face of it, with the Shuttle scheduled to fly in 1979, there seemed plenty of time. However, as the next solar maximum period approached, it was clear that the Sun was considerably more active than anyone had expected. In fact, as early as 1977, the National Oceanic and Atmospheric Administration (NOAA) cautioned NASA that the present solar cycle was the

Had unexpectedly fierce solar activity not taken its toll, there was a very real possibility that the Shuttle might have performed a rendezvous with Skylab to either prepare it for a controlled re-entry or to re-boost it into a higher orbit for subsequent re-use. In this 1978 artist's impression, Fred Haise and Jack Lousma are station-keeping with the orbital workshop. The Teleoperator Retrieval System (TRS) has been remotely docked onto Skylab and is in the process of burning its twin engines. The grim reality was that the Shuttle was delayed and the workshop descended to Earth far sooner than anticipated ... and America's chance to have its own space station throughout the 1980s and beyond was lost.

second most intense in a century-long history of record-keeping. This fact strongly suggested that Skylab would be drawn back to Earth far sooner. In December of that year, just after the contract award to Martin Marietta was signed, a meeting of the American Geophysical Union heard from NOAA's chief forecaster, Howard Sargent, that the current cycle of sunspots was particularly intense. NASA, and particularly the Marshall Space Flight Center, largely ignored the warnings for a while, hoping that the Shuttle would be ready in time to save the workshop. To be fair to the space agency, there really was no single reliable means of predicting sunspot activity which was universally accepted by the whole solar science community; all were based on an analysis of a century of historical data. "Fred and I would come into work every morning," Lousma related, "and they'd have a picture of the Sun and...where the Skylab was on the wall." In this constant battle between Sun and Skylab, there could be only one winner ...

Haise and Lousma were determined to fight, nonetheless, and pursued their training with renewed vigour. "The simulations for the [TRS] were done at Marshall," continued Lousma. "We knew that the Skylab wasn't just sitting there motionless, waiting to be docked with, but it was actually augering through the sky with a motion that made the nose of it wobble in a circle. When we designed the booster package, it had to be capable of being flown so it could match that wobble of the docking port around a centreline. I had to fly that booster over there and match that circular motion and have enough power in the control system to make sure that could be done. We simulated [the] known wobble of the Skylab and made sure we could get to that, then we added to it and made a bigger wobble so we could size the control system. That was part of the development."

As Haise and Lousma readied themselves for their hypothetical mission, NASA redoubled its efforts to regain control of Skylab and bring it out of a years-long hibernation. Beginning with an initial contact from a Bermuda tracking station in February 1978, system by critical system, it gradually came back to life ... but all was not well. The control moment gyroscopes had significantly deteriorated, coolant needed for the solar array batteries was depleted and propellant levels were low. More trouble was afoot with the Shuttle. By the end of 1978, it was obvious that the oft-delayed spacecraft would not be ready for its *maiden* flight until at least the autumn of 1979, thereby pushing Haise and Lousma's mission back until November of that year. If Skylab was still in orbit by then (and *that* was optimistic), a re-boost would be incredibly complex.

The final nail came in December 1978, when two separate test-stand failures of the Shuttle's main engines eliminated any chance that it could fly until at least 1980. The North American Air Defense Command (NORAD), whose data was extensively employed by NASA for the purposes of tracking and re-entry prediction, estimated that Skylab would come down in July or August of the following year ... more than eight months *before* the earliest opportunity for Haise and Lousma to get to it. Ten days before Christmas, Robert Frosch glumly notified President Carter that a practical Skylab re-boost was now out of the question and on the 19th the press was informed. NASA would henceforth focus its energies upon ensuring that the workshop could execute a controlled re-entry, without causing injuries or damage on the ground. At first glance, this seemed unlikely. Population density maps, provided to the agency by the Department of Defense, made it possible to analyse variations in Skylab's orbit and consider techniques for better controlling its descent. The judgement was that there was a one-in-152 chance that a human being might be hit by falling debris. This risk could be minimised to a tenth of that figure if it was brought down at a specific location. "If re-entry could be contrived to happen *east* of North America," wrote Hitt, Garriott and Kerwin, "and *west* of Australia on one of these orbits, Skylab could be safely disposed of." For Jimmy Carter's administration, the political risks were equally as high as the human ones. The remains of the Soviet Cosmos 954 satellite, which fell near Yellowknife, in northern Canada, in January 1978, had polluted a vast area with residue from its nuclear reactor. Although Skylab carried no radioactive material, the episode was still at the forefront of many minds and a major international incident seemed inevitably on the cards if the descent path could not be properly controlled.

The station was therefore inserted into a solar-inertial, high-drag attitude in January 1979. During re-entry, and particularly during the so-called 'terminal phase of decay', at an altitude of around 125 km, its stabilising control moment gyros would be switched off. This would cause Skylab to tumble and the known lower drag effects associated with a tumbling configuration would result in a more predictable re-entry location and impact footprint. Other minor adjustments could also be made to the workshop's attitude to slightly increase or decrease atmospheric drag effects and thus lengthen or shorten its re-entry, but independent studies by the Batytelle Institute predicted that several pieces of Skylab would probably survive the passage through the atmosphere ... and some of them might weigh several thousand kilograms apiece! The impact date, which was steadily refined in the spring of 1979 to sometime around the beginning to the middle of July, was expected to create a debris footprint 6,400 km long and 160 km wide. Finally, as July dawned, it was decided that the station would indeed be placed into an end-over-end tumbling motion during re-entry, thereby offering a more accurate prediction of its impact point. NORAD anticipated impact on the 12th and as controllers watched and listened, Skylab's tortuous six years in orbit finally came to an end. In the official Skylab Reactivation Mission Report, published by NASA in March 1980, the majority of the break-up occurred between the altitudes of 60-95 km: in sequence, the workshop's final solar panel was torn away by steadily increasing aerodynamic forces, after which the ATM and the main part of Skylab were separated and the solar observatory's own windmill of arrays came apart. Final disintegration of the workshop and ATM most likely occurred some 60 km above Earth. One of the earliest visual sightings was from an airline pilot, flying along Australia's west coast, who reported an intense fireworks show, as multiple streaks of debris snaked across the sky.

"The actual vehicle," wrote Hitt, Garriott and Kerwin, "was stronger than the specs required. It held together longer than was calculated, breaking up over the Indian Ocean. Most of the debris fell harmlessly into the water, but some chunks fell in Western Australia, along a line from south to north-east of Perth." A week later, as Perth hosted the Miss Universe pageant, a large fragment of Skylab was set up on the stage. As an interesting footnote, although the attention centred on places like Esperance, with their dusting of debris, the larger pieces probably carried on into the Outback. Perhaps some are still there, awaiting an explorer to trek out into the wilderness and find them. One individual who certainly did find something which made him a small fortune was 17-year-old Stan Thornton, the beer truck driver, whose discovery of a few sizeable fragments in his mother's backyard resulted in a last-minute flight to the United States to pick up his reward from the *San Francisco Examiner*. "The radio people were in Esperance with the help of a Swan Brewery Learjet within two hours," Thornton later recalled. "They had already contacted Qantas to arrange my ticket and passport and the next day I flew out of Perth for the United States. In San Francisco, I was greeted at the airport by Qantas manager Gil Whelan, who had arranged a limousine, and I was taken straight to the downtown *Examiner* office."

More than three decades since Skylab's fiery fall, a space station once again circles Earth, with American astronauts aboard. The International Space Station is the first to exceed Skylab in terms of size, mass and technological complexity. The records set by the Conrad, Bean and Carr crews have long since been eclipsed. Today, it is not uncommon for Americans to spend in excess of one hundred days in orbit and a handful – Peggy Whitson, Mike Foale, Mike Fincke and Jeff Williams, at the time of writing – have totalled around a year of their lives off the planet on multiple flights. In the same way as the Apollo lunar explorers could not have envisaged a half-century or more before humans would return to the Moon, Ed Gibson considered it unimaginable that it would take over 20 years for a fellow American to exceed the duration that he and his crewmates set in February 1974. By his own admission, Gibson hoped that their record would be broken within three or four years. And, in truth, *it was* ... for in March 1978, Soviet cosmonauts Yuri Romanenko and Georgi Grechko, whose exploits will form part of the next chapter, would smash it by spending almost 14 weeks in orbit. Nor was their venture an isolated incident, for the Russians *continued* to set record after record, all the way up to the remarkable 14-month expedition of Valeri Polyakov aboard the Mir space station between January 1994 and March 1995. Interestingly, the end of Polyakov's mission coincided with the launch of astronaut Norm Thagard, who would finally break Carr, Gibson and Pogue's long-held American record. In terms of endurance, for more than three decades, the Russians have held all of the records when it comes to spaceflight. Today, cosmonaut Sergei Krikalev holds the unchallenged world record ... and *his* total space time of over 800 days, spread across no fewer than six space missions, is more than *double* that of his closest American rival.

Of course, one should not judge success by endurance records alone. Skylab was a remarkable example of triumph over adversity and its scientific achievements, which included hundreds of hours of solar and astronomical observations with the ATM, many dozens of EREP passes, several spacewalks, an enormous base of biomedical data and three rapid-fire endurance missions are but part of a much more significant picture. For Skylab provided astronauts with a real *home* in space for the first time and the reality that it was so woefully under-exploited in its final years and that it could not be saved is as tragic as its scientific and technological accomplishments are triumphant. Many of the lessons learned aboard Skylab – how to live and work in weightlessness for extended periods, how best to manage training, schedule and interpersonal pressures, how to minimise physiological effects such as space sickness and how to keep oneself healthy and functional in the most hostile environment ever visited by human beings – continue to be learned and built upon today. Such knowledge *will* be needed for us to take our next steps outward in the next half-century and beyond. The Moon may no longer be America's immediate destination, but if President Barack Obama's vision reaches maturity, then we may see astronauts venturing out to explore an asteroid and Mars before the 2030s. If that does happen, it would not be waxing too lyrical to assert that the successes, tragedies, trials and tribulations of Skylab, its three remarkable crews and the hundreds of thousands of nameless heroes on the ground who made it all possible, have played an immense role in carving our future in space.

4

Red stars in the East

THE ORDEAL OF VITALI ZHOLOBOV

In the silence and eerie stillness of the Salyut 5 space station, cosmonaut Boris Volynov was more than a little uneasy. It was 17 August 1976 and he was struggling to sleep. Together with his crewmate, Vitali Zholobov, he had been in orbit for exactly six weeks. Many journalists on the ground confidently expected the two men to return to Earth in early September, after a mission of around two months. This would not establish any new world records, since America's final Skylab crew had spent 84 days aloft in a mission which ended more than two years earlier. Still, it would mark a major milestone for the Soviets as they strove to make space station living a routine occurrence in the late 1970s and beyond. Since the advent of the Space Age, they had enjoyed a virtual monopoly in the heavens: it was they who placed the first artificial satellite, Sputnik, into orbit, and it was their technological prowess which enabled Yuri Gagarin to become the first man to venture through the thin veil of Earth's atmosphere and into the infinity beyond.

More triumphs followed. In rapid succession, cosmonauts were spending several days aloft, whilst their American counterparts could barely manage a couple of suborbital 'hops' and orbital flights of just a few hours apiece. When the United States pledged that it would put a man on the Moon by the end of the 1960s, the Soviet Union seemed to be far ahead, dropping bombshell after bombshell: with what appeared to be the first rendezvous between two manned craft, the first woman in space, the first three-man crew and the first spacewalk. After that, the monopoly faded and the sands began to shift. America benefitted from an enormous space budget in the middle 1960s and the Soviets not only lost their Chief Designer, Sergei Korolev, but also received a new head of state, Leonid Brezhnev, whose interest in space affairs was inspired only by the ability to score political points. Whilst the Americans walked on the Moon and later occupied Skylab, the Soviets seemed confined in low-Earth orbit with a spacecraft whose capabilities were a mere fraction of those of Apollo. At length, they placed a series of space stations into orbit in the early 1970s – some with a military focus, others with exclusively civilian goals – and cosmonauts steadily began to snatch back some of the ground they had lost to the Americans. In the summer of 1975, Pyotr Klimuk and Vitali Sevastyanov spent an impressive 63 days aloft aboard Salyut 4, missing out on taking the world endurance record from Skylab by just three weeks.

Now, with Boris Volynov and Vitali Zholobov aboard Salyut 5, observers were expecting the duration of the Klimuk-Sevastyanov mission to be surpassed, albeit only by a narrow margin of a few days. An understanding of the landing requirements for Soyuz missions had enabled early space historians to make generally accurate predictions for the likely return of cosmonauts to Earth ... and also generally accurate assessments of which flights had been curtailed, due to mechanical or other difficulties. "A successful mission," wrote Phillip Clark, "is ideally recovered before sunset and not significantly earlier than three hours before sunset." Since the prime recovery zone was in Kazakhstan, Soyuz crews descended back to Earth on a north-easterly track across Africa, preparing for their de-orbit burn in daylight, performing the manoeuvre over the South Atlantic and landing in Soviet Central Asia in the late afternoon. Additionally, the Soviets attempted, wherever possible, to impose a pair of primary criteria for each landing: that it should not occur in the hours of darkness and that the retrofire procedure should happen in sunlight, in order to permit manual orientation of the spacecraft. "The latter condition," continued Clark, "means that the lower limit on the landing time is not strictly speaking three hours before sunset, but the approximation ... to define landing opportunities will not be greatly in error." With this in mind, Clark estimated that Volynov and Zholobov were most likely to remain aloft until early September 1976, completing a mission of around 54-66 days. Using Clark's landing window data, the Swedish space analyst, Sven Grahn, on his website, www.svengrahn.pp.se, has suggested a scheduled return date of 1 September, which would have placed the mission duration at the lower end of this range.

Clearly, this latest mission was but part of a far larger effort to cement Soviet space ambitions in the coming decade: an effort which would see the establishment of a series of space stations, aboard which cosmonauts would spend many months, surpass the endurance record established by Skylab and implement other steps acutely needed to live off the planet for long periods of time. By the close of the 1970s, these outposts in Earth orbit would not only have eclipsed Skylab in terms of duration, but also in terms of a support infrastructure: regular unmanned cargo freighters, rather mundanely called 'Progress', which would haul fresh food, water, propellant, clothes and equipment, and a unique series of visiting missions, some of which would include international cosmonauts. As the Americans dug in for several years on the ground before the first flight of the Shuttle, the Soviets were gearing up for half a decade of space spectaculars.

Kicking off this new endeavour were the crew of Soyuz 21. In command was Colonel Boris Valentinovich Volynov of the Soviet Air Force; a man whose adventures in space had already been thwarted at each turn by his Jewish heritage. Selected as one of Russia's original 20 cosmonaut candidates in March 1960, he served as Valeri Bykovsky's backup on the Vostok 5 mission and was widely expected to receive assignment to a later flight or perhaps command of the first multi-manned Voskhod. By January 1964, however, Marshal Sergei Rudenko, the Soviet Air Force's deputy commander-in-chief, requested Volynov's transfer from the Voskhod to the Soyuz training group. This apparent 'downgrading' was vetoed by Nikolai Kamanin, commander of the military cosmonaut team, who, in July of that year, named Volynov as a candidate to train for the Voskhod 1 command. Despite receiving formal certification to fly in August and seeming to be a strong contender to lead a crew, Volynov was ultimately dropped from the mission only days before launch. So too was fellow cosmonaut Georgi Katys, whose father had been executed in one of Stalin's purges and who had a pair of half-siblings living in Paris. In spite of the solid support of Mstislav Keldysh, head of the Soviet Academy of Sciences, Katys was cast aside, with Kamanin noting in his diary that an "unfavourable background ... spoils the candidate for flight". When Sergei Korolev heard of the decision to drop Volynov on the basis of being a Jew, he was reportedly furious, but was advised by Nikita Khrushchev not to "rock the boat ... it's not worth it!" Unperturbed, Volynov next trained to lead the long-duration Voskhod 3 mission and, after its cancellation, moved over to Soyuz. In May 1968, he was assigned to command one of its early missions and, despite passing all of his exams in September and being commended for his "mastery" of the systems, almost lost this assignment too. When the Central Committee of the Communist Party met on 20 December to discuss the crew selections, further "unhappiness" was expressed over Volynov's background. On this occasion, however, good fortune smiled upon him and his position on Soyuz 5 was accepted.

The man who aroused all this debate was born in Irkutsk in southern Siberia on 18 December 1934. He received schooling in Prokopyevsk and developed a love for the wildness of the taiga which would remain with him throughout his life. He moved to Prokopyevsk before the Great Patriotic War, where his mother, a paediatrician, surgeon and traumatologist, raised him on her own. After graduation, Volynov applied to his local Komsomol Committee for a letter of recommendation to a pilots' school and, with this in hand, experienced a taste of military life: endless drilling, guard duty and kitchen detail. His first solo flight, it is said, gave him little satisfaction, but, thanks to his instructor, Veniamin Reshetov, and his air force squadron commander, Major Ivanov, he eventually developed a love of aviation.

Completion of the military pilots' school in Novosibirsk in 1955 was followed by marriage to his childhood sweetheart, Tamara, and the birth of their first son, Alexei. Then, in the closing months of 1959, with a glowing reference from Major Ivanov, he was one of hundreds of young Soviet Air Force pilots to be interviewed about flying a quite different machine: the Vostok spacecraft. Acceptance in March of the following year marked the beginning of his long wait for a mission into space. It also marked the end of a long and difficult selection process. The search for the world's first spacefarers had begun in earnest in May 1959, when representatives of the armed forces, the scientific community and the design bureaux met at the Soviet Academy of Sciences in Moscow, under the auspices of Mstislav Keldysh, to discuss methods of choosing the most suitable candidates for Earth-orbital missions. Aviators, rocketeers and even car racers were considered, but at length, bowing to the Soviet Air Force, Keldysh agreed to narrow the criteria to qualified pilots from this branch of the military.

Despite an obvious vested interest in wanting to have 'its' fliers take the first manned spacecraft beyond the atmosphere, the logic was inescapable: as already noted, military pilots had proven themselves under exposure to hypoxia, high pressures and varying G loads and had undergone rigorous ejection-seat and parachute training. In addition to their flying experience, candidates would only be admissible if they could meet the height and weight requirements of the Vostok spacecraft: they needed to be no taller than 1.75 m and weigh no more than 72 kg. Moreover, in the expectation that they would be embarking on lengthy careers as 'cosmonauts', the age limit at selection was firmly set at between 25-30 years old.

Throughout 1959, groups of physicians were sent to air bases in the western Soviet Union and by August the selection teams had the records of over 3,000 pilots available for inspection. Most of these were eliminated at a fairly early stage, on the basis of not meeting the height, weight, age or medical criteria – some, indeed, were dropped for bronchitis, angina, gastritis and colitis, renal and heptic colic and pathological cardiac shifts. The remainder were then systematically interviewed from early September, still unaware of exactly what the so-called 'special flights' project entailed. The list was soon reduced to a little over 200 candidates. They were despatched in groups of about 20 for further tests at the Central Scientific Research Aviation Hospital in Moscow. In addition to more interviews, the candidates were spun in stationary seats to assess their vestibular apparatus, placed in low-pressure barometric chambers and spun around on a centrifuge to evaluate their performance under high-G loads. Original plans, it seemed, called for seven or eight pilots, but Sergei Korolev insisted on tripling this number, for no other reason than because he wanted a larger team than the United States' seven-strong group for Project Mercury.

In January 1960, Marshal Konstantin Vershinin, commander-in-chief of the Soviet Air Force, formally signed plans to establish a centre for cosmonaut training in Moscow. Although it was nominally under the control of physicians, the Air Force General Staff eventually assigned Nikolai Kamanin command of cosmonaut affairs. It was he who approved a final shortlist of 20 candidates in late February: Ivan Anikeyev, Pavel Belyayev, Valentin Bondarenko, Valeri Bykovsky, Valentin Filatyev, Yuri Gagarin, Viktor Gorbatko, Anatoli Kartashov, Yevgeni Khrunov, Vladimir Komarov, Alexei Leonov, Grigori Nelyubov, Andrian Nikolayev, Pavel Popovich, Mars Rafikov, Georgi Shonin, Gherman Titov, Valentin Varlamov, Boris Volynov and Dmitri Zaikin. The age criterion was waived in a couple of instances out of respect for their exemplary performance during testing, and ran from just 23 for Bondarenko to 34 for Belyayev. Some of these men would become the most famous names in the history of spaceflight, whilst others would remain anonymous ... and, in a few cases, fall into disgrace.

On 7 March, the cosmonauts were given their welcoming speech by Vershinin at the Central Scientific Research Aviation Hospital. A week later, after settling their affairs at their individual air bases, they began training with Vladimir Yazdovsky's first class in aerospace medicine. The following four months were consumed by a mixture of in-depth lectures and an intense physical fitness regime, the latter of which included two hours per day of intensive callisthenics at the Central Army Stadium in Moscow. Their parachute training was conducted in the Saratov region, near Engels. They jumped from a converted Antonov An-2 aircraft, and within six weeks each man made 40-50 jumps over water and land, from high and low altitudes and in daylight and darkness.

Almost a full decade later, in January 1969, Volynov commanded the crew of Soyuz 5 and participated in a stunning rendezvous with the Soyuz 4 spacecraft and its lone pilot, Vladimir Shatalov. During the mission, fellow cosmonauts Alexei Yeliseyev and Yevgeni Khrunov spacewalked over to Shatalov's ship and returned to Earth with him. A day later, Volynov came home alone ... and came within a whisker of death in the process. The instrument module at the rear of his Soyuz failed to separate properly, causing the spacecraft to adopt an improper orientation as it re-entered the atmosphere and placing the cosmonaut in dire peril. By divine providence, Volynov somehow survived the inferno of re-entry and made a very hard landing, hundreds of kilometres off-course, in the Ural Mountains. Spitting bits of broken teeth as he went, the bruised and bloodied cosmonaut trudged through the snow in sub-zero temperatures, until at length he reached the safety of a peasant's cottage. For some time, the psychological trauma that Volynov had endured was expected by many in the cosmonaut team to keep him permanently grounded. Yet in April 1970, he was identified to command a possible future Soyuz crew and, paired with Zholobov, was assigned to a mission to the Salyut 2 military space station. Unfortunately, Salyut 2 lost attitude control and depressurised within days of launch in April 1973 and no crews ever flew to it. A year later, Volynov and Zholobov served as backups to Gennadi Sarafanov and Lev Dyomin on Soyuz 15, with an expectation that they might fly as the prime crew on a subsequent mission, called 'Soyuz 16A', in October 1974. Had this mission taken place, Volynov and Zholobov would have spent around two months aboard the Salyut 3 military station. Unfortunately, Sarafanov and Dyomin's own rendezvous and docking with the station in August failed and Soyuz 16A quietly disappeared from consideration. On his website, www.astronautix.com, Mark Wade has noted that the old Soyuz 16A – now considered "excess to the programme" – was later flown in an unmanned capacity in the summer of 1975. In spite of the frustration and the long wait, the two men soon found themselves as the prime crew for Soyuz 21, which headed for space at 3:09 pm Moscow Time on 6 July 1976.

Ascent was nominal, but after orbital insertion their rendezvous was complicated when one of the Igla ('Needle') antennas on the spacecraft failed to deploy properly. Interestingly, this same piece of hardware had malfunctioned during the Soyuz 15 mission in August 1974, failing to switch on to its 'final-approach' rendezvous mode and implementing a sequence which should normally have been executed at a much larger distance from its quarry. As a result, Soyuz 15 had accelerated too rapidly towards Salyut and Sarafanov and Dyomin had narrowly missed a collision on not one, but *two*, occasions. Modification of the Igla after this near-miss had been one of the reasons for cancelling Volynov and Zholobov's Soyuz 16A flight, but clearly the underlying problems with the device had still not been eliminated, a full two years later. The Soyuz 21 rendezvous, thankfully, was not as dramatic as its predecessor: visual contact with Salyut 5 was achieved at a distance of 350 m, although the rate of closure suddenly increased to a little over 2 m/sec. Volynov asked for permission to take manual control of the final approach, but was told to wait, as the data indicated nothing out of the ordinary. Shortly thereafter, the Soyuz braking indicator blinked off and although their velocity had decreased back to normal, they were now rapidly losing their line of sight with the station. Seventy metres out, the commander took over from the computer and guided his ship to a perfect docking at 4:40 pm Moscow Time. Barely a day had elapsed since their launch from Tyuratam and Volynov and Zholobov were primed and ready for a mission of 54-66 days. If they were to hit the very top end of that range, they would establish a new Soviet endurance record.

Although Salyut 5 was primarily a military outpost, the experiments conducted within its pressurised confines were devoted to both surveillance and science. In his summary, Phillip Clark divided the work into seven discrete disciplines: science and technology, biological investigations, astronomy, Earth resources observation, 'technical experiments' and medical studies. The station itself had been launched two weeks ahead of Volynov and Zholobov, at 9:04 pm Moscow Time on 22 June, and had been inserted by its Proton booster into an orbit of around 260 km. Since this was at least 70 km lower than the normal operating altitude of the civilian Salyuts, *and* the shortwave transmission frequencies were the same as those used by earlier military stations, it did not take analysts too long to conclude that this was a military flight. "Western observers ... who were expecting an advanced 90-day, six-man mission on Salyut 5 have since reduced their estimates," *Flight International* told its readers on 31 July 1976. "They have concluded that the Salyut 5 station is a direct derivative of the Salyut 3 military reconnaissance craft launched almost two years previously ... "

In terms of heritage, Salyut 5 was part of a family called 'Almaz' ('Diamond'), a code name unknown in the Western world until quite recently. The name derives from the fact that designer Vladimir Chelomei named each of his projects in honour of a precious stone. Its military nature makes it unsurprising that very little information has trickled out from behind the Iron Curtain and, with hindsight, it seems likely that the surveillance capabilities of a manned platform were overestimated and the results did little to justify the expense. For their part, the Americans had long since dispensed with the idea of a manned military station in Earth orbit, since automated reconnaissance satellites had already demonstrated capabilities far superior to even the best-trained human eyes.

As late as 1988, Clark admitted that when trying to describe the features and functions of spacecraft such as Salyut 5, the space analyst was placed "at a great disadvantage", the Soviets having revealed "not a single sketch which hints at the correct design". Virtually nothing was known about their orbital operations and the craft themselves – dubbed 'military Salyuts' by Western observers in the 1970s – had only been described in a very general, almost 'poetic', manner, even comparing their power-producing solar arrays to "a bird in flight". Judging from the dimensions of earlier Almaz-class stations, including the ill-fated Salyut 2 and the successful Salyut 3, it would seem that Salyut 5 measured approximately 14.5 m long and 4.15 m wide and weighed in excess of 19,000 kg. Internally, it had a volume of around 100 m³. It was equipped with two solar arrays and its main habitable compartment was divided into areas for control, work and recreation. The previous Almaz-class station, Salyut 3, had also been fitted with a special 'sofa' for medical experiments, fixed and swinging 'beds', provisions for hot and cold water, a shower and toilet and a small library. One can only speculate that similar facilities may also have been aboard the more advanced Salyut 5.

In fact, one facility that Western observers *did* (incorrectly) suspect Salyut 5 might possess was two docking ports – one at each end of the station – as opposed to the single port aboard each of its predecessors. Volynov and Zholobov conducted a televised tour of their new home a couple of days after arrival … and the background was difficult to discern (probably deliberately so) and their surroundings were significantly darkened. One point about which Clark *was* certain, however, was that the military Salyuts carried a small re-entry capsule, presumably for the return of classified data and photographic canisters. It was several more years before the design of this capsule was revealed in more depth: it apparently consisted of a two-part device, measuring 1.35 m long and 85 cm wide and included a heat shield and a small payload container capable of holding samples or films weighing up to

One of the few television pictures of Volynov (left) and Zholobov during their seven-week occupancy of Salyut 5.

120 kg. Small capsules would have been inserted into a cylindrical airlock, and it was reported that a 'manipulator' on the exterior of the station would have moved samples to the re-entry craft. When fully loaded, the capsule was pneumatically jettisoned at a 60-degree angle towards Earth and against the station's direction of motion. A timer 'spun-up' the craft for stability and a solid-fuelled retrorocket was fired to parachute it to a soft landing on Soviet soil.

One particular instrument which had also flown aboard Salyut 3 was the Agate high-resolution telescope, part of the station's imaging payload, with a focal length of 6.3 m. Clark identified, but did not name, elements of the Earth resources package, but certainly Salyut 3 carried instrumentation for "detailed reconnaissance of Earth's surface in interest of national economy of science", with a focus on meteorological phenomena, observation of breeding grounds, forest and steppe conditions, the extent of change caused by natural disasters ... and, presumably, a significant chunk of time snooping the United States, China and their assets. In fact, the presence of civilian experiments aboard Salyut 5, such as the Kristall ('Crystal') furnace for growing monocrystals in weightlessness, an aquarium and terrarium for observing the development of viviparous guppy fish and turtles, the ITS-5 infrared telescope for observing the Sun, Moon and planets and a broad swathe of medical tasks, did not fool many Westerners that the role of Salyut 5 was primarily one of surveillance. During their time in orbit, for example, Volynov and Zholobov observed the 'Siber' military exercise of Soviet forces, off the eastern coast of Siberia. "Salyut passes over these exercises every local-time afternoon," *Flight International* explained on 31 July, "allowing excellent opportunities for the testing of a variety of optical sensors mounted on operational military equipment." They also undertook a series of propellant-transfer tests, preparatory to plans for future unmanned Progress freighters to routinely refuel future Salyut stations.

But on 17 August, exactly six weeks to the day since their launch, all was not well aboard. Even a televised link-up with schoolchildren that day did not help to alleviate the sense of unease. Life in the orbital environment, with more than a dozen sunrises and sunsets per 24-hour cycle, often made it difficult to judge when one should go to bed and when one should arise to work. Weightlessness itself had required both men to take some time to recondition themselves to the startling realisation that 'up' and 'down' were all but meaningless and this caused some unusual physiological effects: a migration of body fluids from the lower extremities into the abdomen and the chest, a corresponding decrease in thirst, an increased concentration of blood volume and a higher red cell count, an increase in urination to relieve this cardiovascular excess and a drastically reduced appetite. Certainly, rookie Zholobov's adaptation had proven particularly uncomfortable and after the mission his poor physical condition would be partly blamed on lack of proper sleep and partly on not having properly followed the prescribed exercise regime. However, *something else* was conspiring against them; something unseen and potentially deadly. As they were working that August day, the station's alarm suddenly sounded and the lights went out, plunging them into darkness. Several systems also went off-line. Salyut 5 was out of communications range with the ground and the cosmonauts had to endure two unpleasant hours trying to find out what was wrong with their ship. There had been no loss of pressure, but the life-support system had shut down and attitude control had gone. Eventually, Volynov and Zholobov recovered the systems, but they could already discern that something unpleasant was in the air all around them: a curious, acrid odour, which they guessed must have originated from the environmental control systems. Its effect on Zholobov, in particular, was dramatic. In the space of just a few days, Volynov saw his flight engineer's physical and emotional wellbeing deteriorate before his eyes.

It is hard to imagine such a sad state of affairs for a man like Lieutenant-Colonel Vitali Mikhailovich Zholobov, whose experiences during a professional career spent flying and testing advanced military aircraft, pushing them to their structural limits, had been incomparably surpassed by his first and only rocket launch into space. Born on 18 June 1937 in Zburjevka in the southern Ukraine, immediately to the north of the Crimean peninsula, Zholobov studied at the Institute for Petrol and Chemistry in Baku. After graduation, he worked as a chemist and test engineer within the Soviet Air Force and joined the cosmonaut team in January 1963. Two years of intensive training was rapidly followed by various assignments to the ill-fated Soviet lunar landing programme in September 1966 and later to the Almaz effort. By the middle of 1973, he was teamed with Volynov and began to prepare for a long-duration flight to one of the military space stations. His other, rather dubious, claim to fame is that Zholobov happened to be the first spacefarer to ride into orbit wearing a moustache ...

Years later, the question of what happened to Zholobov remains open and medical perspectives remain divided. Certainly, he suffered severe headaches and nausea and there is little doubt that a leakage of nitric acid fumes from Salyut's propellant tanks was to blame. Other reports and rumours even go so far as to employ adjectives such as "emotional" and "psychotic" and suggest that *both* men "developed an unreasonable desire to return to Earth". In his 1998 book *Dragonfly* about the Mir space station, Bryan Burrough recounted conversations with NASA and Russian psychologists which hinted at "interpersonal issues" between the cosmonauts and contributed to the premature end of the mission. More recently, Rex Hall and Dave Shayler have noted that on 24 August Volynov even had to *carry* his weightless flight engineer into the Soyuz spacecraft and strap him into his seat for undocking and re-entry.

On his website, Sven Grahn also made reference to these persistent reports of psychological problems, but admitted that shortwave transcripts of voice signals from Salyut 5, received by himself in Sollentuna, near Stockholm, between 17 July and 13 August, show no signs of distress; quite the opposite, in fact. Zholobov's adaptation to weightlessness does not appear to have affected his work or his enthusiasm, with the last transcript showing him to be in fine spirits. "Good morning, dear viewers," he begins in one light-hearted episode. "Today, we execute one of the series of technological experiments ... for the welding of materials in space. What to say? The conditions of welding, obviously, you immediately think about the images of burning, hot sparks, protection masks, protection filters, but to us this is a little bit different. For the analogue experiments a special installation has been constructed, which includes some reactors, in which the samples ... are enclosed in these packets. There are two materials, in this case. The solder is used, utilised of two difficult combinations. Here are two little pipes and as you can see on your screens ... "

"We can show this," interrupted Volynov.

"Do you see it?" Zholobov continued, addressing his Earthly audience. "Is it good? Visible? In the beginning, we were a little bit anxious about the experiment, for welding is welding and that is why we're a little bit upset, but now all is normal, the results are ... excellent, all pipes are united. The quality can only be determined by experts when we bring back the sample. They do that by the help of laboratory expertise ... "

Periodically, throughout the exchange, laughter could be heard between the two men and, looking at the transcripts of each conversation, across that month-long span from July into August 1976, there does not appear to be the slightest hint of distress in either of them. Nor does there seem to be the remotest trace of any 'interpersonal issues'. Whatever occurred to Zholobov cannot have as its root cause weightlessness or lack of exercise alone – since both of these had also afflicted previous crews in far less dramatic ways – or even to a poor relationship with his commander ... but instead pointed directly at the nitric acid leak from Salyut's environmental control systems, just a few days later, on 17 August. Besides, having worked so closely together on backup and prime crews for more than three years, it would seem unlikely that such interpersonal issues would have arisen in two highly-trained military professionals. Alas, the truth (whatever it is) will probably never be known.

Within a week of first noting the strange smell aboard Salyut 5, the cosmonauts landed back on Earth at 9:33 pm Moscow Time on 24 August. Yet even their return was filled with terrifying moments. At first, as Volynov tried to undock, the latches failed to separate properly, then the mechanism jammed and as they passed out of communications range only one set of emergency instructions were received aboard the spacecraft. Zholobov looked on in mounting horror, but the pair could do nothing except wait a full 90 minutes until the second set of emergency procedures could be transmitted. Eventually, at 6:12 pm, Soyuz 21 undocked successfully, commencing its three-hour descent home. They landed in very gusty conditions, a couple of hundred kilometres south-west of Kokchetav – today's Kokshetau – in northern Kazakhstan and they were described to be in pitiful physical and mental condition. Neither man could stand or walk unaided and Volynov had to help his flight engineer out of the capsule, since Zholobov's helmet had caught on an obstruction. The ultra-secretive Soviets, who always employed a series of code words to describe the health of their crews, said that they were in a "satisfactory" condition. *That* code word actually meant that all was not well and they were both in need of medical attention. Neither man would fly into space again and it is bitterly ironic that Vladimir Shatalov, who was by now the commander of the cosmonaut team, remarked shortly before Soyuz 21 that it would break no new ground, but would be "just another working launch". Midway through the flight, on 31 July, *Flight International* referred to Soyuz 21 as a mission which was "not earning headlines" and whose nature was "unspectacular". *Routine* was the new name of the game, or so it seemed. The reality was quite the opposite. Boris Volynov and Vitali Zholobov had shown through their own misfortune just how unforgiving the space environment could be and how *fallible* their own equipment could be ... for their 49-day mission was *anything* but routine.

DETERIORATING DÉTENTE

It might not have been too difficult for Western space sleuths to ascertain the military nature of Salyut 5, judging from the station's operating altitude, its radio frequencies and the deliberately darkened and grainy television pictures of its interior ... but one mission, which appeared almost out of nowhere in September 1976, caught them by complete surprise. In summing up the eight-day voyage of Soyuz 22, *Flight International* noted that cosmonauts Valeri Bykovsky and Vladimir Aksyonov "broke new ground" by becoming the first Soyuz crew to fly at an orbital inclination other than the standard 51 degrees. In fact, operating in a 64.8-degree path, the two men would fly further north than any previous Soyuz team. Many in the West did not take long to latch onto the fact that a major NATO exercise happened to be going on in Norway at the time – at 64 degrees north. This latitude was totally inaccessible to the imaging equipment aboard Salyut 5 and the assumption was quickly reached that Soyuz 22 had at least some military aims. In retrospect, it is possible that Bykovsky and Aksyonov probably *did* engage in some observations of the NATO forces, but their central focus was undoubtedly scientific research.

At first glance, it seems odd that, barely a year after ASTP and the enormous outpouring of goodwill which it created, relations between East and West were already showing signs of strain. The provisions of the Helsinki Accords, signed in the summer of 1975, had uncovered several irreconcilable differences, notably on the issue of human rights. Jimmy Carter's incoming administration would make no secret of its abhorrence of the Soviet stance in this area, prompting an angry retort from Leonid Brezhnev that the United States should cease meddling in other nations' internal affairs. Besides, spat the Soviets, America was known to actively support regimes with appalling human rights records, including Chile and South Africa, and accusations of hypocrisy and double standards were, to an extent, justified. The Americans also continued to refuse to recognise the Soviet Union's forced hegemony over Estonia, Latvia and Lithuania – a trio of tiny Baltic republics annexed by Joseph Stalin's regime more than three decades earlier – and yet another nail of contention was driven into place. Of course, by embroiling themselves, directly and indirectly, in numerous conflicts around the world, including the Middle East, Africa and Central and South America, neither superpower aided the progress of détente. Four years after ASTP, and despite the provisions of a second round of Strategic Arms Limitation Talks, Soviet troops invaded Afghanistan, kicking off a brutal, decade-long war of attrition against a well-armed and US-supplied mujahideen insurgency. The echoes of that conflict continue to reverberate today, and not only in Afghanistan; for one member of that insurgency – a man who would learn to loathe both the communists who opposed him and the capitalists who once nurtured him – was a young and wealthy Arab named Osama bin Laden ... a man whose name, whose cause and whose perverted vision of the central tenets of Islam would literally reshape the very fabric of the world we live in.

Despite the positive attributes of détente, such as the annual shipments of grain from the United States to make up for shortfalls in the Soviet system of collective farming and promises to reduce missile stocks, the roots of distrust had already begun to grow and fester. In May 1976, the CIA started a notorious exercise, popularly known as 'Team B', to analyse the magnitude of any Soviet threat to the United States. In its report, which leaked out to the press that winter, the team argued that the Soviets had the capacity to build submarine detection systems which did not rely upon sound and were far beyond Western technology. Team B even suggested that the Soviets believed they could win a nuclear war outright and had no intention to adhere to the doctrine of Mutual Assured Destruction. Within the next decade, Team B cautioned, the Soviets' burgeoning gross domestic product would bring about an extraordinary increase in military spending, including the production of aircraft and ballistic missiles. Team B would be proved wildly wrong on virtually *every count* ... but its scary reading unnerved many Americans and hardened many attitudes towards the secretive, closeted Soviets. After all, it *was* only a year since ASTP and cosmonauts were back in space, aboard *military* stations, snooping once again on the West.

Despite the overwhelmingly scientific nature of Soyuz 22, there was a very clear and obvious change in the air. The days of the kindly Konstantin Bushuyev playing with Glynn Lunney's young son and a group of Soviet academicians being warmly greeted by American schoolgirls in a Houston shopping mall were gone. A new and far starker reality had taken over; one that would extend well into the following decade and would progressively nudge both superpowers closer and closer to the nuclear trigger. The increasing view in the West that the Soviets were returning to their old ways cannot have been softened by the menacing radio callsign of the Soyuz 22 commander: *Hawk*.

'Hawk', to be fair, had been chosen years earlier by Valeri Fyodorovich Bykovsky, shortly before he first flew into space in June 1963 aboard a Vostok capsule. Today, his fame primarily rests on the fact that he holds the record for the longest time spent *alone* in space: his solo Vostok 5 mission spent no less than 118 hours off the planet. Bykovsky came from Pavlovsky-Posad, one of Moscow's easternmost towns, famed for its textile industry – including shawl production – and its churches. He was born on 2 August 1934 and was chosen alongside Yuri Gagarin and Alexei Leonov as one of the Soviet Union's first cosmonauts. In fact, Bykovsky had been one of the final six candidates for the first Vostok mission and, although his examination scores were judged as good, it was noted that his natural reticence meant that he made few important or substantial contributions to group discussions. His performance aboard Vostok 5 was outstanding and he was an early contender in April 1964 for command of the first Voskhod, although he was not ultimately selected. Bykovsky subsequently moved to the Soyuz training group and by November 1966 had been formally identified to lead its second manned mission in April of the following year, with an expectation that he would rendezvous, dock and exchange crew members with Soyuz 1 and its lone cosmonaut, Vladimir Komarov. Tragically, Soyuz 1 failed and Komarov was killed during re-entry, when the strands of his primary parachute became entangled and the reserve canopy refused to deploy.

In a sense, Bykovsky's career during this period closely mirrors that of Alexei Leonov, in that both men made spectacular first flights and were then dogged for years afterwards by misfortune. As Leonov lost the opportunity to walk on the lunar surface, so Bykovsky also lost his chance to journey to the Moon. And as Leonov lost his chance for a long-duration space station flight, so Bykovsky seems to have been conspicuously absent from the Salyut crewing roster. At the same time, however, both men had cause to be thankful: but for a few twists and turns of fate, Leonov might have been in command of the ill-fated Soyuz 11, whose crew died during re-entry when their capsule lost pressure, and Bykovsky might have been aboard Soyuz 2, flying a fault-ridden carbon-copy of the ship in which Komarov had lost his life.

Now, in the mid 1970s, it was into a much improved Soyuz that Bykovsky and his flight engineer, Vladimir Viktorovich Aksyonov, clambered on the morning of 15 September 1976. Their launch came at 12:48 pm Moscow Time and Sergei Korolev's venerable old R-7 rocket boosted them perfectly into space. Within the first 24 hours, they would adjust their orbit on two occasions in order to establish themselves at an altitude of around 255 km. Their inclination of almost 65 degrees, wrote Mark Wade on his website, www.astronautix.com, "maximised ground coverage ... especially of the former East Germany". This was particularly important, since one of Soyuz 22's stated goals was "to check and improve the scientific and technical methods and means of studying geological features of Earth's surface in the interests of the national economies of the Soviet Union and the German Democratic Republic". Not surprisingly, a major segment of their research payload originated from East Germany. The 200 kg MKF-6 multi-spectral camera, developed by Carl Zeiss, was mounted in place of the docking mechanism on the front of the orbital module. A larger version would be installed aboard the Salyut 6 station in 1977. Its six lenses covered visible and near-infrared wavelengths and permitted the imaging of over 500,000 square kilometres in less than ten minutes. Typically, Aksyonov would operate the camera from within the confines of the orbital module, whilst his commander handled the orientation controls. Early in the mission, the cosmonauts acquired good images of the route of the Baikal-Amur railway and the vast expanse of Siberia and its myriad environments – steppe, mountain, taiga, tundra, permafrost – which progressed eastwards towards the Sea of Okhotsk. Observations of Central Asia, including Kazakhstan and Azerbaijan, allowed detailed attention to be paid to geological formations and the effects of collectivised agriculture. Shortly before their return to Earth, on the final full day of the mission, Bykovsky and Aksyonov made a detailed survey of East Germany, co-ordinated with cameras aboard an Antonov An-30 aircraft, following their exact flight path. By comparing the two sets of pictures, it would be possible to develop algorithms to 'subtract' the effects of the atmosphere and hence make inferences about the properties of the surface. More than two thousand images would be returned, covering 30 different geographical

Vladimir Aksyonov works aboard the Soyuz 22 orbital module during his eight-day mission with Valeri Bykovsky in September 1976.

areas. Also aboard the Soyuz were a number of biological investigations, including a small centrifuge for plant growth, an aquarium and equipment to observe the effect of cosmic radiation on the eyes.

Bykovsky and Aksyonov returned safely to Earth on 23 September, touching down 150 km north-west of Tselinograd – today's Astana, the capital and second-largest city in Kazakhstan – after a mission lasting a little over a week. For Aksyonov, born in Kazimov Rayon on 1 February 1935, the flight was nothing less than the culmination of a lifetime of work and experience in aviation, engineering and design. He had been picked as a civilian cosmonaut alongside fellow engineers Gennadi Strekalov, Valeri Ryumin and Alexander Ivanchenkov, only three years earlier, and was the first of them to ride into orbit. Fittingly, for a spacecraft design engineer, Aksyonov's two space missions would see the last flight of one generation of Soyuz and the first flight of another. What made Soyuz 22 unique was that it was the backup vehicle for ASTP – an older-specification craft, incompatible with Salyut – and on his next flight, in the early summer of 1980, Aksyonov would fly a new ship called 'Soyuz-T'. Unlike Mr T, it would wear no gold chains or sport a Medinka hairstyle, but would be brimming with its own treasures: advanced solar panels, an Igla which actually *worked*, new attitude thrusters, digital avionics and, perhaps most importantly of all, the ability to fly a three-member crew in full space suits. Team B may have been totally off the mark in their fanciful talk of advanced anti-submarine interceptors and harbingers of nuclear holocaust, but they were right on one point: the Soviets *did* pose a threat on the space scene. The first flight of the Shuttle was several years away and as it suffered landing gear and brake failures, problems with its heat-resistant tiles and maddening explosions of its main rocket engines, the Soviets had *already* staked their claim to the heavens. They intended to dominate space … and for the next half decade, their cosmonauts would do just that.

A WATERY YARN

Scattered across the globe, from Australia to Iceland and Kamchatka to Chile, are almost two thousand unique sites, protected by the provisions of the 40-year-old Ramsar Convention. This intergovernmental treaty, signed in Iran back in February 1971, provided for national and international action to conserve and wisely exploit wetlands and their resources throughout the world. One of the so-called 'Ramsar Sites' is an obscure salt lake called Tengiz, located in north-central Kazakhstan. It has an area of almost 1,400 km^2, but is extremely shallow – even its deepest spots run to only a few metres. It has been identified as the home of around 300 different species of birds, two dozen of which are classified as endangered. Today, it forms part of the Korgalzhyn Nature Reserve, which was nominated a few years ago as Kazakhstan's first natural UNESCO World Heritage site. Over the millennia, Tengiz has provided a haven for many creatures. In October 1976, it even offered a temporary home for two travellers from outer space. They were returning with heavy hearts, disappointment etched on their faces, for it had been their task, just two days earlier, to reinvigorate a home in space which had been abandoned by its previous occupants.

    As summer cooled into autumn, very little comment had emerged from behind the Iron Curtain as to precisely why Boris Volynov and Vitali Zholobov had returned to Earth so abruptly. Even Radio Moscow was caught by surprise when, on the morning of 24 August, they were told that the men would be on the ground in just ten hours! The landing, which took place in darkness, further underlined the need to get them home as quickly as possible. In its summing-up on 4 September, *Flight International* quoted a Russian daily newspaper, which had revealed that psychologists had begun playing music to the crew "to ease the effects of prolonged isolation". At length, it was *Aviation Week & Space Technology* which published the first report in the West that the crew *had* indeed evacuated Salyut 5 following the development of an acrid odour in the station's atmosphere. This suspicion seemed to be vindicated in February 1977, when another crew – that of Soyuz 24 – boarded Salyut, vented and replaced its atmosphere ... and did so with their faces screened behind the safety of oxygen masks.

Sandwiched between Volynov's ill-fated mission and that of Soyuz 24, however, were two other flights; the non-station-related solo science exercise of Bykovsky and Aksyonov and another which failed to dock with Salyut ... but did earn its own place in history as the only time that a Russian cosmonaut team landed in water, rather than on land. Splashing down was, of course, not the intention, but many things occurred during the unlucky voyage of Soyuz 23 which were neither intended, nor anticipated.

In the years to come, Soviet policy towards the composition of each spacegoing crew would change to always include at least one veteran cosmonaut. This was in response to the fact that three all-rookie missions during this period became failures, in each case being forced to return to Earth after just a couple of days. On Soyuz 15 in August 1974, Gennadi Sarafanov and Lev Dyomin had almost *collided* with their station, and on Soyuz 25 in October 1977, Vladimir Kovalyonok and Valeri Ryumin failed in their effort to dock with the new Salyut 6. The third failed mission, Soyuz 23, has always held a special significance for this author, since it took place in the middle of October 1976, a few days before I was born. Crewed by Vyacheslav Dmitryevich Zudov and Valeri Ilyich Rozhdestvensky, it would similarly fail to link up with Salyut 5. In so doing, these two men would miss out on a long-duration flight which many in the West were confidently expecting would exceed Skylab's 84-day record.

Zudov, the commander, was born on 8 January 1942 in the ship-building town of Bor, located on the opposite bank of the mighty River Volga from the city of Gorky. He graduated from the Higher Military Pilots School in Balashov in 1963, having received extensive instruction as both a Soviet Air Force pilot and a parachutist, and entered the elite ranks of the cosmonaut team two years later. Completion of training at the end of 1967 was followed by work on the Almaz military effort. By August 1973, Zudov and Rozhdestvensky – who had been selected in the same cosmonaut class, albeit as an engineer, rather than as a pilot – were working together as a crew, backing up Gennadi Sarafanov and Lev Dyomin for a mission to Salyut 2. When this space station failed soon after launch, they served in a support capacity for the Soyuz 14 and 15 crews and were the backup team for Volynov and Zholobov. A little more than seven weeks after their predecessors returned from a decidedly unpleasant stay aboard Salyut 5, Zudov and Rozhdestvensky may have felt a little trepidation as they prepared to take up the baton themselves.

Rozhdestvensky's naval career prior to becoming a cosmonaut is more than a touch ironic in view of the fact that his one and only space mission would end up by dunking him well and truly in the *drink*. He was born in Leningrad on 13 February 1939, gained an engineering degree and later commanded several deep-sea diving units in the Baltic Sea War Fleet. In October 1965, he was selected as a cosmonaut engineer and from August 1973 until the time of his flight he was teamed with Zudov in support and backup roles. Finally, their careers brought them to the base of the very same launch pad from which Yuri Gagarin had departed on his own historic mission, more than a decade earlier. As they looked up at their Soyuz 23 spacecraft, atop its shimmering rocket, Zudov and Rozhdestvensky knew that it was their responsibility to restore Salyut 5 to operation and they had a few key tools in their kit to do it ... including oxygen masks and air samplers.

On launch morning, 14 October, little seemed to go well. The bus broke down on the way to the pad and then, after an otherwise flawless liftoff at 8:40 pm Moscow Time, high winds aloft caused the rocket to veer off-course and the crew narrowly avoided having to execute an abort. They achieved orbital insertion safely, albeit lower than planned, and ultimately established themselves at an altitude of around 270 km, preparatory to docking late on the evening of the 15th. Seven kilometres from the station, Zudov placed the Soyuz into its automatic rendezvous mode, but shortly thereafter reported "strong lateral fluctuations" in his craft. With less than 1,600 m to go, he was becoming alarmed; for the fluctuations had now increased to the point at which Soyuz 23 was actually *turning away* from Salyut 5 ... even though Zudov's instruments told him the approach was proceeding normally. Five hundred metres out, both men could *see* the target, but instinctively knew that they were still inexplicably turning away from it and were travelling too fast, anyway, for a successful docking to be attempted.

At length, the Soyuz ceased its oscillations and Zudov asked to make a second try. However, the lower-than-normal orbital insertion had already required him to expend a significant proportion of his attitude control propellant and mission controllers knew that sufficient reserves *had* to be maintained for two attempts at retrofire. The cosmonauts protested that they felt confident that a second docking attempt would succeed, but were out of luck. There was nothing left to do but come home, disappointed and disheartened, the following day. One of the greatest surprises was that the Soviets actually *admitted* to the press that the mission had failed, telling journalists in the elaborately roundabout and irritatingly indirect manner for which they had become famous, that the docking of Soyuz 23 with the station "was cancelled because of an unplanned operation of the approach control system of the ship".

Translation? *Zudov and Rozhdestvensky's craft had let them down.*

It was an astonishing admission of failure; for totally honest and forthright admissions from the Soviets had occurred only rarely in their chequered space history, normally in the wake of a disaster or tragedy in which no alternative explanations were possible. In the years to come, as relations between East and West soured once more, such admissions would become yet more infrequent. In fact, not until American astronauts boarded the Mir space station from 1995 onwards would any real concerns about the inadequacy of many aspects of Soviet space technology be brought again into the public arena.

It is at this stage of the mission that Tengiz, the salt lake in Kazakhstan, becomes pertinent. Aboard Soyuz 23, Zudov and Rozhdestvensky performed their retrofire normally and were aware that high winds and blizzards were making conditions in the primary recovery zone dicey. Little could they have realised that those gusts would carry them no less than *one hundred and twenty-one kilometres* from the intended landing point. Moreover, as they descended, their capsule was enveloped by

Zudov and Rozhdestvensky, the unlucky crew of Soyuz 23.

a thick blanket of freezing fog and the only sensation was that of gently swinging beneath the parachute. Zudov told Rozhdestvensky to brace himself for the firing of the solid-fuelled retrorockets at the point of touchdown ... but both men were more than a little surprised when their epic journey ended, not with a hard thud, but with a loud *splash*.

They had come down into the icy waters of Tengiz, about 8 km off its northern shore. As the parachute dragged the capsule, it floated on its side and this prevented the cosmonauts from opening the forward hatch, lest they admit a flood of freezing water. Outside temperatures were -22°C. With only a two-hour supply of oxygen remaining in their suits, Zudov and Rozhdestvensky were relieved to find that the pressure equalisation vent of their capsule was just above the waterline. They duly opened it. They were aware that it would only sustain them for five hours or so, but were not overly concerned, because, surely, the recovery forces would be with them promptly? The two men removed their pressure suits, donned their flight suits and began tucking into their food rations. "To stay in such situation in the suits and do nothing," Zudov later recalled, "we would be frozen and die. That's why, first, we *had* to get out of space suits and to free ourselves of them. We spent an hour and a half to get out ... even used *knives* to cut them, then we managed to wear our plain sport wear." The recovery forces, though, were having their own difficulties simply spotting the floating capsule. Splashdown had occurred at 8:46 pm Moscow Time, in darkness, and the steadily thickening fog had all but obscured the capsule's light beacon from the rescue helicopters.

If it was possible for circumstances to worsen at this point, *they did*. Salt corrosion activated the pyrotechnic charges for Soyuz 23's reserve parachute, causing it to deploy ... and instantly fill with icy water! Tengiz is not a deep lake by any standards, reaching barely 7 m at most, but with the pressure equalisation vent open, the cosmonauts *would* have drowned if their craft had slipped underwater. Zudov and Rozhdestvensky's survival training took over at this stage. In order to conserve their remaining oxygen, they stopped talking and moving, which compounded the rescuers' efforts even further, since they had difficulty establishing communication with the men and had no search signal upon which to focus. At length, a lone helicopter spotted Soyuz 23 by chance and pinpointed its location with a powerful searchlight. Iosif Davydov, a survival training expert and one of the helicopter crew, told the cosmonauts to don their water survival suits. The heat of re-entry had quickly cooled the exterior of the capsule and, within the tiny cabin, the temperature began to fall rapidly and ice started to form on the walls. Outside, the first flurries of snow flickered downwards from an ominous, fog-enshrouded sky. Davydov readied himself to be winched down to the capsule, but the gathering snowstorm prevented the pilot from holding the helicopter in a sufficiently stable position. A second attempt to hover above Soyuz 23 also failed and the would-be rescuers were forced to retreat to the shore.

Progress to reach Zudov and Rozhdestvensky was agonisingly slow and many spectators doubted that the men would still be alive by the time they were extracted from the descent module. Amphibious vehicles arrived at the lakeside, but could not negotiate its many bogs and marshes, and even the deployment of life rafts was hampered by the blocks of ice and sludge on Tengiz' surface. The rescuers had no options but to withdraw until first light the next morning, 17 October. Zudov and Rozhdestvensky would have to wait out the discomfort and the cold. To do so, they first had to switch off all internal power, to preserve the 40-minute limit to its on-board electrical supply, and the cosmonauts spent the night with just a tiny light for company. Within two hours, listeners could hear clear signs of oxygen starvation in the men's voices – they were breathing heavily, then spoke in unusually hoarse tones and at other times were virtually incoherent – and at one stage, as dawn approached, Rozhdestvensky reported that Zudov had lost consciousness. Exhaled carbon dioxide was steadily accumulating. "You could *feel* the CO_2 without *any* instrumentation," the flight engineer explained later of those harrowing hours spent alone, "just *feel* it. When I felt that we could lose consciousness ... then I switched on the regeneration unit. When the mind started to clear, and blue haloes we were starting to see disappeared in the eyes, I switched it off ... *and so on*, all through the night!"

Not everyone was prepared to wait for first light. A handful of intrepid rescuers set out into the icy expanse of Tengiz in rubber boats and one of them, a helicopter pilot named Nikolai Chernavsky, succeeded in reaching the capsule. There was little that he could do, however, for Rozhdestvensky would later liken the surface of the water to "a salted quagmire". As the first glimmers of a wintry dawn shone over the lake, the cosmonauts turned on their exterior lights. With temperatures hovering around -20°C, exterior conditions were still severe, but had improved somewhat during the night. A rescue helicopter finally arrived on the scene, but because it was

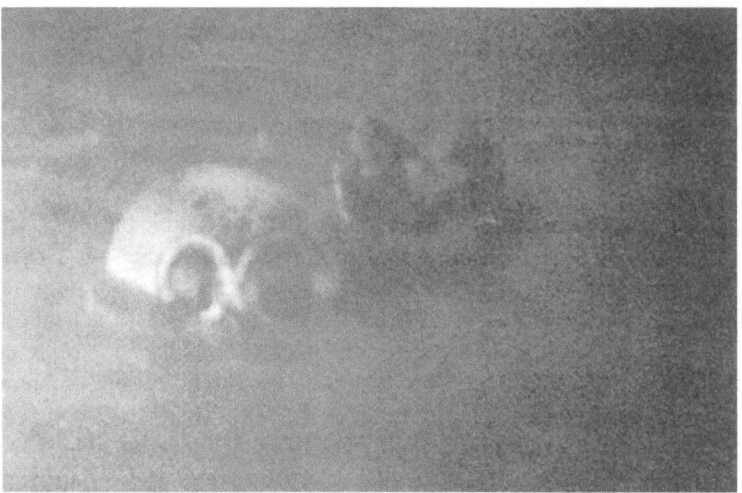

The Soyuz 23 descent module floats in the icy waters of Lake Tengiz in the gloom of an October day in 1976.

impractical to open the hatch, the only option was to tow the capsule to the shore – an untested and hazardous procedure.

"It was a difficult operation," wrote Rex Hall and Dave Shayler, "as the helicopter was nearly dragged down by the reserve parachute when it emerged from the water. The towing caused severe buffeting during a demanding 45-minute trip for both the helicopter crew and for the cosmonauts in the capsule." A few kilometres from the shore, Zudov and Rozhdestvensky came close to suffocating, when Soyuz 23 nearly sank, but at long last, *eleven hours* after hitting the icy waters of Tengiz, they were back on dry land. At seven that morning, Moscow Time, the electrifying news was relayed to mission controllers and to the press: against all the odds, the crew was *alive*!

In the days that followed, the sheer terror of what they had endured became truly apparent. For Valeri Rozhdestvensky – the only cosmonaut to have come from a naval background – the greatest terror did not come from the hours spent inside the cramped confines of the Soyuz ... but from a handful of *photographs* which he saw afterwards. "When I got to see our photos and how they dragged the capsule," he related, "*then* I really was frightened ... the *only* time in my life I was *really* frightened."

The joy of having survived the ordeal was quickly tempered by concern, however, for the cosmonauts had failed in their attempt to dock with and reoccupy Salyut 5. In the days that followed, Zudov and Rozhdestvensky *were* blamed for *not* attempting a second docking and an investigation focused attention on the large oscillations in the signals from the Igla rendezvous device. "It was determined that when the spacecraft had acquired the Salyut," continued Hall and Shayler, "the lateral movement light in the cabin came on, indicating that all such motion had ceased. However, as the *engines* used to stop lateral motion were *not* turned on, the spacecraft continued to swing around." The crew *had* felt the motions, but Zudov's instruments told him that they were stationary. He knew that if he abandoned the final approach, there would be insufficient fuel reserves for a second attempt and therefore waited until the last possible moment before making his decision. In the final report, dated 2 December 1976, it was concluded that the indicator lights on Zudov's panel *had* recorded an accurate approach and it was indeed the Igla which had erred. Armen Mnatsakanyan, Igla's designer, was offered the chance to resign. When he refused to do so, he was fired in January 1977, only weeks before the *next* Soyuz crew – Viktor Gorbatko and Yuri Glazkov – were set to put two bitter doses of bad luck behind them and finally bring some much-needed success to Salyut 5.

SICKLY AND JAUNDICED

Even though it succeeded where its predecessors had failed, the mission of Soyuz 24 was both short and relatively uneventful. It is ironic that some flights are so successful as to be labelled 'boring'. Cosmonauts Gorbatko and Glazkov were launched from Tyuratam at 7:12 pm Moscow Time on 7 February 1977 and docked at the space station late the following evening. Five weeks after the firing of Armen Mnatsakanyan, it was perhaps with little surprise that his troublesome Igla rendezvous device failed yet again, although Gorbatko was able to take manual control and guide the Soyuz into its port. In the three months since Zudov and Rozhdestvensky's ill-fated flight, Salyut 5 had been used to conduct a number of experiments automatically – taking photographs of Earth, for instance, and making remote-controlled observations with its ITS-5 infrared telescope – and its orbital altitude had been regularly adjusted in anticipation of a new crew. When Gorbatko and Glazkov entered the station for the first time, they took no chances. Both men wore breathing apparatus and took pains to carefully sample, and later replace, the atmosphere.

Viktor Vasilyevich Gorbatko, the commander of the mission, was making his second flight into space. He was born in the village of Ventsy-Zarya in the Krasnodar region of southern Russia on 3 December 1934 and, aged 15, completed a course at a stud farm, propitiously named 'Voskhod'. In 1953, he graduated from the Eighth Military Aviation School of Basic Pilot Training in the Ukraine. He subsequently flew as a senior pilot in the Odessa Military District of Moldova and served as a parachute instructor. He precisely fulfilled the requirements of the 1960 cosmonaut intake: he was under 30 years old, less than 1.7 m tall and below 70 kg in weight. In truth, he just *barely* met the criteria, standing almost 1.69 m and weighing 69.5 kg. By the time that Shatalov and Filipchenko were chosen three years later, the criteria had been opened to allow military navigators and engineers to be considered, as well as pilots, and the age limit had risen to 40. Moreover, the second group had to have academic credentials from either a military academy or a civilian university.

Gorbatko's early assignments were within the Vostok project and, as late as September 1963, he confidently expected to fly sometime in the following year. When plans for additional Vostoks were cancelled, he was redirected to train as a possible commander for the Voskhod 2 mission, which featured the world's first spacewalk. Subsequently, he served as Yevgeni Khrunov's backup on the original Soyuz 2 mission and later shadowed his spacewalk training for the Soyuz 4/5 transfer. In the final weeks before the January 1969 launch, however, Nikolai Kamanin noted that Gorbatko made some minor mistakes in their final exams. Nevertheless, he flew aboard Soyuz 7 in October of that same year. Subsequent work involved training on the early Salyut missions and Gorbatko and Glazkov later backed up Zudov and Rozhdestvensky on their flight.

Joining Gorbatko was Yuri Nikolayevich Glazkov, born on 2 October 1939 in Moscow. He attended and graduated from the Kharkov Military Engineering High School and received a candidate of technical sciences degree, before entering the Soviet Air Force as a flight engineer. In October 1965, he joined the cosmonaut team and worked extensively on Almaz, initially supporting the mission of Volynov and Zholobov and then working closely with Gorbatko to back up Soyuz 23. In an update on 19 February 1977, *Flight International* told its readers that Glazkov had earlier written "a PhD-standard thesis on EVA", which lent some credence to Western rumours that a spacewalk involving himself and Gorbatko may have been planned for Soyuz 24. This possibility was also noted by Clark in 1988. The precise objectives have never been identified, although the absence of an EVA compartment on the military Salyuts would have required Glazkov to venture outside through the hatch of the Soyuz orbital module.

Later in his career, he served as deputy chief of the cosmonaut training centre at Star City and worked closely on the joint Shuttle-Mir effort in the 1990s. We are exceptionally fortunate that Glazkov – who died in 2008 – was interviewed by NASA's oral historian and was able to share a few nuggets of anecdotal interest from his flight. Speaking in May 1998, he recalled that he caught a cold aboard Salyut 5 ... but came up with a rather novel idea for finding warmth and effecting a cure. "You just put your nose to the window," he recalled, "and the *solar rays* cure your cold! It's very simple and I suggest to *anybody* who has a cold, try to fly up there!"

One notable event which did take place during their 18 days in space was the replacement of Salyut 5's atmosphere on 21 February, shortly before they returned to Earth. The procedure involved a release of compressed air, which created positive pressure, blowing any contaminated air out through a relief valve. Although the atmosphere was clean, mission controllers decided to perform the experiment to demonstrate that such a total exchange was feasible. It worked admirably and Salyut's atmosphere was replenished from storage tanks in the Soyuz orbital module. "A progress report," it was announced, "described how the cosmonauts, by using the special multi-functional system on-board the space station, could carry out a complete or partial change of the atmosphere. This operation was carried out during the routine communications session, when the cosmonauts said that the atmosphere inside the station was fine and breathing was easy ... " Years later, Glazkov would remark that such atmospheric replacement systems proved beneficial back on Earth, too, and their technology has since helped to reduce fungal growth in vegetable stocks, particularly in regions such as Indonesia.

The Soyuz 24 crew of Viktor Gorbatko (centre) and Yuri Glazkov (right) are briefed by Alexei Leonov in the spacecraft assembly building at Tyuratam.

Over the years, it has been hinted that the replacement of the atmosphere led to the cancellation of a planned EVA. Two manoeuvres were performed: one was a test of the Soyuz spacecraft's propulsion system and the second was executed by the station itself. Gorbatko and Glazkov finally undocked from Salyut 5 on the morning of 25 February and landed safely at 12:38 pm Moscow Time, just to the north-east of Arkalyk in Kazakhstan.

Not until the years of *glasnost* and *perestroika* and their aftermath did it finally become clear that a *fourth* mission to Salyut 5 was planned, but never executed. Its crew would have comprised rookie cosmonauts Anatoli Berezovoi and Mikhail Lisun and, in all likelihood, they would have flown for a couple of weeks in July 1977. An additional Soyuz was in the process of being built, but at a meeting of the State Commission in March it was reported that the completion of the spacecraft was anticipated to take two months, followed by *another* two months to satisfactorily man-rate it for its mission. By the time that Berezovoi and Lisun's

Soyuz was ready to go, the propellant supplies aboard Salyut 5 would have depleted to a level at which another manned visit would be impossible. Some engineers muttered that the 'delay' was actually a ploy to get rid of this last military station and divert attention to the more advanced civilian Salyut 6, whose launch was anticipated later in 1977 and whose cosmonauts were expected to smash the Skylab endurance records once and for all.

Whatever the truth, the Soyuz originally destined to carry Berezovoi and Lisun *was* eventually completed and *was* flown in mid-1978 ... on a mission to Salyut 6. For its predecessor, the final Soviet military space station, however, the end was nigh. Its film canister was released a couple of days before Gorbatko and Glazkov came home and was later sold at Sotheby's in New York. Meanwhile, Salyut 5 ended its days in a shower of burning debris, re-entering the atmosphere to destruction in August 1977.

Even after more than three decades, so little is known – and perhaps no more will *ever* be revealed – about precisely what Volynov and Zholobov and Gorbatko and Glazkov did during their weeks aboard Salyut 5 and what additional work Zudov and Rozhdestvensky and Berezovoi and Lisun might have completed. Undoubtedly, there are documents in the deepest recesses of the Kremlin which will almost certainly never see the light of day. Some spectators have argued convincingly over the years that the only reason the Soviets imposed such secrecy on their entire space programme was as a kind of 'reverse psychology' that went right back to the perceived, albeit erroneous, 'missile gap' with the United States back in the early 1960s.

In essence, the Soviets *knew* that they held no huge technological advantage, they *knew* that America's miniaturised computing technology was ahead of their own and they *knew* that infighting between their own chief designers and politicians made it a constant uphill struggle to sustain an effective and rational space programme. When one considers the psychological impact of this cloak-and-dagger stance on America, it is easily appreciated how the *absolute lack* of reliable information and transparency from behind the Iron Curtain allowed unfounded stories to grow, questions to fester and mysteries and rumours to linger. Simply by revealing nothing, the Soviets were able to play the situation to their advantage. For example, despite all of the mystery, suspense and fear which surrounded their N-1 Moon rocket, it was, in fact, a shadow of its American counterpart, the Saturn V. The much-lauded Salyut stations, too, had barely a fraction of the capacity of Skylab ... and doubtless their living conditions and the standard of their experimental hardware was similarly inferior.

Nevertheless, the lure of the unknown and the insatiable desire *to know* has always appealed to the human spirit. The reality that so few details and hard facts have ever been released surely adds to the fascination of these secrecy-enshrouded missions. The Soviets, as masters of psychological and ideological warfare, clearly played such xenophobic fears and suspicions to their advantage. In *Dragonfly*, Bryan Burrough quoted Yuri Glazkov as having remarked "it's very easy to see your missiles [from space] ... we could look *right down on them*". Clearly, surveillance *was* a primary focus, but the fact that Salyut 5 was the last of its kind underlines for us that the return was simply not worth the enormous investment. If the truth about Almaz and the military Salyut programme is ever revealed, it may come as something of a disappointment. Without even realising it, perhaps the Western rumour mill had done the Soviets' job for them – and done it spectacularly well – by inflating the importance and technological superiority of what is more likely to have been a sickly and jaundiced reply to what American spy satellites had already achieved.

SALYUT 6: THE NEXT GENERATION

On 29 September 1977, as dawn broke over Tyuratam, a new era began. Two full decades had passed since the Soviet Union stunned the world with the launch of its first artificial satellite, Sputnik, then scored one propaganda victory after another, with the exploits of Gagarin and Titov and Tereshkova and Leonov … and then quietly, almost inexplicably, fell into second place behind America in the space race. In rapid succession, the United States took each endurance record, made spacewalks of far greater complexity than Leonov had done, executed intricate rendezvous and docking operations with the utmost precision and delivered the final *coup de grâce* with a manned landing on the Moon. Of course, the amateur student of human space exploration today knows that there was nothing 'inexplicable' about the apparent slowdown in Soviet space ambitions during this period; the death of Sergei Korolev in January 1966 was nothing short of a calamity and the constant infighting between various design bureaux, both before and after his passing, had made the road to a coherent space programme an intensely difficult one. Moreover, regime change from Nikita Khrushchev to Leonid Brezhnev had brought with it a far less sympathetic outlook on the exploration of the new frontier. Yet, in his own way, even Brezhnev respected the space effort as a pillar of Soviet technological prestige and national pride and in late 1969 he decreed that the establishment of Earth-circling stations, crewed by cosmonauts, would form their new road ahead. Each step along that eight-year road had carried the Soviets through a multitude of emotions, from tragedy, when the Soyuz 11 crew died during their return to Earth, to frustration, as several early Salyuts succumbed to launch failures, and finally to triumph, as cosmonauts gradually closed the space endurance gap between themselves and the Americans. Now, two decades after Sputnik, the Russian bear was fully awake. With Salyut 6, a much-lauded 'next-generation' space station, they would not only *close* the gap, but would *eliminate* it totally, ushering in a new era which would see them really *living* off the planet on a near-permanent basis. As a result, today, the endurance record in space is held by a Russian cosmonaut.

The key to this remarkable turnaround in Soviet space fortunes was the Proton. In the autumn of 1977, four years since the final flight of the Saturn V, it now held the enviable reputation of being the world's most powerful operational rocket … as well as one of the most reliable. It had not always been that way and the vehicle had suffered an astonishingly high failure rate during its genesis in 1965-70. The Proton originated as an oversized intercontinental ballistic missile, designed to deliver a huge nuclear warhead across distances of 13,000 km to hardened targets in the West.

It was never actually deployed in this role and was eventually offered by Vladimir Chelomei, its designer, as an alternative to the beleaguered N-1 lunar booster. Had circumstances been different, in 1968 it might have sent a two-man crew around the Moon in a Zond spacecraft, just days ahead of Apollo 8. Today, the Proton's descendants – still built at the same Khrunichev plant in Moscow – can lay claim to a proud heritage and their boast of a 96 percent launch success rate is among the highest in the world.

Standing more than 50 metres tall, with the Salyut 6 space station tucked away inside its bullet-like nose and upper main body, the Proton was also a visually impressive creation. Its three stages were each fed by a combination of unsymmetrical dimethyl hydrazine fuel and an oxidiser of nitrogen tetroxide; both are highly toxic and highly corrosive propellants, but with the advantage that they can be kept at ambient temperatures and are hypergolic, meaning that they burn on coming into contact, eliminating the need for an igniter. As a consequence, the Proton had the ability to sit on the pad for far longer than a cryogenic booster. At 9:50 am Moscow Time, the first stage's cluster of half a dozen single-chambered RD-253 engines thundered to life and the vehicle began its climb for the heavens. Second stage performance was similarly nominal, as was that of the third stage. The payload shroud was jettisoned and Salyut 6 was inserted into orbit.

Physically, the machine which now circled the globe looked similar to the Salyut 4 station, launched several years earlier and occupied for several months by two pairs of cosmonauts in 1975. It weighed 19,824 kg and measured 14.4 m long and 4.15 m across its broadest diameter and it provided a habitable volume of around 90 m^3 for its crew. In his summary of Salyut 6, Phillip Clark was quick to point out that these outward similarities were exceeded by its internal provisions; for it was with *this* new station "that the Soviet space programme reached the peaks of achievement comparable with the American Apollo programme ... a major forward step towards the Soviet goal of permanent manned occupation in space which would be realised a decade later". Shape-wise, Salyut 6 comprised a trio of cylinders – a forward transfer compartment and two working and living areas – and one fundamental difference from its predecessors was the absence of a rear propulsion module, although it did possess a pair of manoeuvring engines for orbital adjustment. In the place of the rear propulsion module was a *second* docking port, intended to permit the arrival of visiting Soyuz crews and a new, unmanned resupply freighter, known as 'Progress'. This additional port traced its origins back to at least December 1973, when Vasili Mishin, then-head of the TsKBEM design bureau (formerly OKB-1 and soon to become Energia), received approval to build a next-generation Salyut and 'borrowed' the idea for a dual-port configuration from an advanced Almaz military station, built by Vladimir Chelomei.

In order to describe the basic layout of the new station, it is perhaps best to imagine oneself aboard a Soyuz ferry which has just docked at the forward port. Upon opening the hatch, a cosmonaut visitor would first enter the cramped forward transfer compartment, a tapering cylinder some 3.5 m in length and 2 m in diameter. Within this compartment was stowage provision for the crew's extravehicular suits, together with bottles of compressed air and the main EVA controls and an inward-opening hatch through which spacewalkers could exit the station. It also housed an astronomical camera, a sextant port and a fixture for an experiment to survey Earth's horizon. Outside was a large, dish-like rendezvous antenna, thermal control panels and EVA handholds. A second axial hatch led into the first work compartment, which measured 3.5 m long and 2.9 m wide and upon which (like the earlier Salyut 4 station) were mounted three steerable solar arrays. In total, these covered an area of 51 m^2 and produced around 4 kW of electrical power. Inside the first work compartment were the main control consoles – together with seats for the cosmonauts – and single drives for the three arrays. Elsewhere were cupboards and other storage facilities, a veloergometer and body mass meter on the 'ceiling', the Rodnik drinking water system and the KATE-140 wide-angle stereographic camera, with which to create topographic maps of Earth's surface. It could take single or strip photographs under cosmonaut or automatic ground command. At a conical 'frustrum', the first work compartment broadened into the second compartment, which measured 2.7 m in length and 4.15 m across; this represented the maximum diameter supportable by the Proton launch vehicle. Within the frustrum was the MKF-6M Earth resources camera, which was a direct development of the device tested by Bykovsky and Aksyonov on Soyuz 22. "It will now work for up to two years," explained *New Scientist*'s Sarah White in September 1978, "it can be handled with just one hand and the image sharpness has been improved." It enabled photographs to be taken in six different spectral bands and had been heavily modified (hence the 'M' suffix) to support operations over several years in space. Its film cassettes were each capable of taking 1,200 exposures and spares would be frequently delivered to the station aboard Progress freighters. Phillip Clark estimated that the MKF-6M's resolution was probably about 20 m or so. The presence of these two powerful topographical instruments had already led the Soviet Ministry of Agriculture to plant a number of specially selected crops at test sites in the Ukraine and the region around Lake Baikal to examine their capabilities.

The main work compartment was dominated by the 650 kg BST-1M telescope, a huge conical structure with a 1.5 m mirror. It was designed to record atmospheric data at infrared, ultraviolet and submillimetre wavelengths, although solar, stellar and planetary observations were also to form part of its research schedule. In time, however, the telescope's power-consumption needs and the limitations of its receivers, cryogenically cooled to around -270°C, would mean that it was actually used sparingly. In fact, it could only be operated whilst Salyut 6 was in the 'night' side of its orbit. Atop the BST-1M was the 'Yelena' gamma ray detector. At other places in this compartment were creature comforts for the crew: a small shower, a treadmill, a set of sleeping bags, a pair of trash airlocks and food lockers. One of the trash airlocks, interestingly, would later be used to house the 'Splav' experimental materials furnace. At the far end of the complex was the second docking port, to be occupied either by Soyuz or Progress visitors. Although Salyut 6 did not possess a large propulsion unit, it *did* carry manoeuvring thrusters: two primary engines, fed by unsymmetrical dimethyl hydrazine and nitrogen tetroxide, each of which Clark believed to have a thrust of some 300 kg.

The provision of a second docking port should be neither underestimated, nor overlooked, for it was perhaps the pivotal feature of the new station which permitted the Soviets to conduct Soyuz changeovers and regular Progress deliveries of supplies and equipment, thereby empowering Salyut 6 to operate on a totally new level over its predecessors. The development of the Progress concept had been in the works since 1973 and the first draft plans had been laid in February of the following year. Early Salyuts had to be launched with almost all of the cosmonauts' supplies aboard and, over time, they would also require periodic re-boosts to maintain their altitude against orbital decay ... re-boosts which demanded many hundreds of kilograms of propellant expenditure *every year*. "To maintain a 250 km orbit," wrote Phillip Clark, "would require [4,300 kg] of propellant. At 350 km, this requirement dropped to about 600 kg. Of course, the Soviets allowed the Salyuts to decay to lower orbits and then raised them again in large manoeuvres; this was more economical in terms of fuel expenditure. All in all, to maintain a fully-functioning Salyut in orbit for two years with a crew permanently on-board required about [18,000 kg] of consumables. It was ... impractical to launch all these supplies with the station – even if a Soviet booster had the necessary lifting power, which none had at the time." Progress was the obvious solution. It was modelled closely on Soyuz, with the orbital and 'descent' modules redesigned to house up to 1,300 kg of foodstuffs, water, experiments – including biological payloads which had to be operated within a short period of time – and up to 1,000 kg of propellant. For ease of removal, the lattice-like framework of cargo racks within the Progress could be unfastened with simple half-turn bolts.

A Soyuz would normally automatically rendezvous and position itself several hundred metres from a station, whereupon the commander would take manual control and complete a docking. The Progress was capable of docking automatically. Its orbital module carried two television cameras to permit flight controllers on Earth to observe Salyut 6 as the procedure was executed. The descent module was replaced by a framework with four large tanks to carry several hundred kilograms of propellants for the station. These would be fed into the storage tanks in Salyut 6's aft compartment using an ingenious system of pipes which mated at the exterior of the docking collar. "The refuelling can be conducted either by the crew," explained Clark, "or automatically under the control of the ground. Once Progress has docked at the back of Salyut, the propellant lines and connections are checked for integrity. This done, a compressor slowly reduces the nitrogen pressure in the propellant tanks. The nitrogen is pumped back into its storage bottles ready for the refuelling operation itself. The fuel and oxidiser are transferred at different times for safety reasons." Firstly, the unsymmetrical dimethyl hydrazine would be fed through a system of pipes into Salyut, after which the nitrogen tetroxide would be transferred. After the completion of this procedure, the connecting propellant lines would be purged with high-pressure nitrogen in order to prevent contamination or spillage. In practical terms, as time would tell, the Soviets would be able to proudly boast that from the first Progress mission in January 1978, the propellant transfer system never once failed them. In their history of Soyuz, Rex Hall and Dave Shayler speculated that the name 'Progress' may have come from the implication that it suggested significant progress in space station operations, although its precise heritage is unclear.

Salyut 6's new level of operation would be demonstrated within months of its arrival in orbit, when, in rapid succession, two teams of cosmonauts would twice demolish the Skylab endurance record. It would not end there. By the conclusion of its life in 1982, this venerable station would have supported no fewer than *five* long-term crews and almost a *dozen* visiting crews ... including representatives of Russia's allies in the Eastern Bloc.

First up would be the Soyuz 25 crew of Vladimir Kovalyonok and Valeri Ryumin, who blasted off from a darkened Tyuratam at 5:40 am Moscow Time on 9 October 1977. The timing of their launch and, indeed, that of Salyut 6 itself were by no means accidental. Rather, they were precisely planned to coincide with the 20th anniversary of Sputnik *and* with a major Communist Party conference; in fact, Kovalyonok rode into orbit with a copy of the newly-published Soviet constitution – the 'Brezhnev Constitution' – in his personal belongings. By the second day of their mission, they had circled the globe 17 times, had completed the rendezvous and were ready to close in and dock. At this point, things began to go wrong. Several attempts to engage the docking latches failed and the contact light on Kovalyonok's control panel refused to illuminate, indicating that a hard mate had not been achieved. It did not take the contact light alone to tell them that; they had instinctively *felt* that the docking was too soft. Perhaps, they thought, the Soyuz probe had failed to engage the mechanism at the tip of the conical drogue. Kovalyonok was therefore instructed to fly in formation with the station, whilst mission managers debated their options.

As with many of these early docking failures, the story of exactly what happened depends on whose opinion the reader prefers. Owing to the previously appalling success record of the Igla, the *modus operandi*, starting with Soyuz 25, was to execute a *manual docking* and Kovalyonok duly took over his ship's controls at a distance of around a hundred metres. However, Boris Chertok, who had served under Sergei Korolev and was at the time of Soyuz 25 the deputy chief designer for control systems, would later note in his memoirs that the cosmonauts found that their longitudinal axis had deviated by a couple of degrees. The maximum allowable deviation was four degrees, meaning Kovalyonok could still have docked successfully, but he and Ryumin were convinced that the situation was not nominal and elected to pull back to a distance of 25 m. In Chertok's mind, the two men had misinterpreted the situation and assumed that a problem existed when in fact it did not. Before launch, simulations had taught the cosmonauts to expect Salyut 6 to be in a particular orientation when the deviation exceeded one degree; *now*, during the actual docking attempt, their view of the station through the Soyuz periscope differed from what they expected to see. According to Chertok, Soyuz 25 made no physical contact with Salyut. However, Flight Director – and former cosmonaut – Alexei Yeliseyev noted in *his* account that Kovalyonok commenced a soft docking and the probe failed to engage with the station's drogue. A second attempt on their 23rd orbit was also frustrated and power and fuel limitations ruled out the option of circling Salyut in order to try the rear port. "Moreover, it had been determined that the fault lay in the Soyuz docking equipment and *not* the Salyut port," wrote Rex Hall and Dave Shayler in their account. "Not wishing to inflict damage on the second port and risk putting the station out of commission before it was manned, it was decided to bring the crew home."

No precise reasons for the docking failures have ever been given – nor, indeed, is it clear how many attempts were made: some sources suggest one, Yeliseyev said two, Chertok said three and the Yorkshire-based Kettering radio group even stated *four*, based on an analysis of communications traffic – although one recollection hinted that Soyuz 25's television system did not function correctly. Yeliseyev would later add that a dangerous situation arose *after* the final docking attempt, when the craft seemingly came close to a collision with Salyut. He opted not to ask the crew to make any burns of their manoeuvring thrusters and instead allowed natural forces to cause the two vehicles to gradually drift apart. Kovalyonok and Ryumin landed safely at 6:25 am Moscow Time on 11 October, a little more than two days after launch. Their descent module hit the ground 195 km north-west of Tselinograd and two very glum cosmonauts emerged shortly thereafter. Their mission – which would have hosted the first Progress, a visiting crew in November *and* would have attempted to beat the Skylab endurance record – had ended before it could even begin.

For the Soviet managers, this caused a serious dilemma. Kovalyonok and Ryumin seem to have shouldered at least some of the blame (they were awarded the Order of Lenin, but *not* the gold star of Hero of the Soviet Union). To be fair to the two men, both were rookies. The implications of the failure of Soyuz 25 to dock would be far-reaching, in that all further crews would include at least one veteran cosmonaut. The original plan, had Kovalyonok and Ryumin succeeded, envisaged a second long-duration crew of Yuri Romanenko and Alexander Ivanchenkov and then a third of Vladimir Lyakhov and Georgi Grechko. With the failure of Soyuz 25, the cards were metaphorically thrown into the air. In the new plan, Grechko, a veteran of an earlier Salyut mission, would join rookie Romanenko to complete the three-month flight, beginning in December 1977.

"Our crew fell victim to a failure," Kovalyonok explained many years later. "We did not manage to dock with the orbital station. The authorities, the Politburo ... *all* were finding fault with us. We *felt* our guilt before many millions of Soviet people. That's why we wanted to find the reasons and not just to defend ourselves. We acted as investigators who knew that the most important thing is that the mistake should not repeat itself. It turned out that it was *not* our fault." In the West, far from *blaming* the cosmonauts, many journalists took a quite different stance. "It seems likely," *Flight International* told its readers on 22 October, "that the Soyuz 15, 23 and 25 problems had the same cause, indicating a *lack* of troubleshooting ability on the part of the Russian design or operations teams."

Whatever the truth behind the refusal to grant Hero of the Soviet Union status to the Soyuz 25 crew, it appears *not* to have significantly impaired their careers ... for, almost immediately, *both* men were placed back into the mix for future flights. Kovalyonok – by now a *veteran*, of course – would lead the second long-duration crew, paired with Ivanchenkov, whilst Ryumin would join Lyakhov on the third. A fourth crew, consisting of rookie commander Leonid Popov and veteran flight engineer Valentin Lebedev, would then wrap up operations aboard Salyut 6.

Vladimir Kovalyonok (left) and Valeri Ryumin before launch. Their failure to dock with Salyut 6 would usher in a new policy, whereby all future crews would include at least one veteran cosmonaut.

Lieutenant-Colonel Vladimir Vasilyevich Kovalyonok was only the second Belarussian spacefarer, having been born in the remote village of Beloye on 3 March 1942. Times were hard in his formative years (he would even recall eating potatoes from snow-covered fields), but the dream of aviation and the lure of the skies attracted him from a young age. As he dug around in the remnants of downed aircraft from the Second World War, he once found a battered helmet with a set of earphones; the discovery earned the young boy the affectionate nickname of 'Pilot'. He was chosen as a cosmonaut in early 1967 and completed more than two years of training. By the summer of 1971, he found himself teamed with military navigator Vladimir Isakov for the 'active' half of a possible dual Soyuz mission to test the 'Kontakt' rendezvous and docking hardware for lunar orbital flights. In the wake of America's triumphant landing on the Moon and several failures of the N-1 rocket, the Kontakt missions disappeared from consideration. Kovalyonok was next paired with engineer Yuri Ponomaryov in 1975 to form the backup crew for Soyuz 18B, commanded by fellow cosmonaut (and fellow Belarussian) Pyotr Klimuk.

Graduation from the Soviet Air Force's prestigious Higher Air Force School in 1976 was followed, a year later, by his first space mission. In completing his two days in orbit, Kovalyonok became the first of his 12-man cosmonaut class to fly. Yet he tried never to neglect the simpler things in life. His own father had abandoned his mother to live with another woman and Kovalyonok found himself the new head of the household, looking after his mother and his younger brother. Touchingly, he recounted his relationship with his own toddler son, one day during training. "I was revising for the flight," he said, "and used to study a lot at the academy. He would come into my office, sit on my knees and get involved with something. In five minutes, not being able to concentrate, I would say to him: 'Son, you're in my way.' Once he opened the door and said, standing in the doorway, 'Daddy, I won't be in your way. I'll just stand here still'." Years later, his son having long since grown to adulthood, Kovalyonok would still well up with tears at the thought of those tender words ...

Valeri Viktorovich Ryumin was a little older than his commander, having been born on 16 August 1939. Not surprisingly, judging from the fact that he grew up in the large industrial city of Komsomolsk-on-Amur, several hundred kilometres north-east of Khabarovsk in the Soviet Far East, Ryumin's education took him into engineering and he graduated from a technical college in Kaliningrad in 1958, specialising in the cold working of metals. At the Department of Electronics and Computing Technology of Moscow's Forestry Engineering Institute, he later worked on spacecraft controls. He also served three years in the army as a tank commander, during which time he heard over the radio that Yuri Gagarin had become the first man in space. Ryumin was electrified by the exciting news. From 1966, however, his career brought him firmly within the sphere of the space programme: he served as a ground electrical test engineer, deputy lead designer for the Salyut orbital stations and ultimately deputy general designer for testing. Since the end of 1969, Ryumin had worked exclusively on space station design. With the possible exception of Konstantin Feoktistov, he was perhaps more intimately involved in the design and definition of each Soviet orbital station – beginning with Salyut 1 – than any other cosmonaut.

He was chosen as a civilian cosmonaut in March 1973, alongside fellow engineers Vladimir Aksyonov, Gennadi Strekalov and Alexander Ivanchenkov, and with Soyuz 25 became the second of his group to journey into space. Much colour has been added to Ryumin's character, however, in the days *after* his cosmonaut career ended. From 1981 until 1989, he served as flight director for the Salyut 7 station and from 1992 he directed the Russian side of the Shuttle-Mir venture with NASA. Bryan Burrough's book *Dragonfly* has received praise and criticism in equal measure over the years for its presentation of the exploits of individual cosmonauts and astronauts. Within its pages, Ryumin as a manager and as a director was painted as a somewhat "blustery" character whom many cosmonauts came to dislike. Among them was Alexander Serebrov, whom Burrough described as once having accused Ryumin of deliberately misleading an investigating commission to divert blame from flight controllers and onto Serebrov's crew when their Soyuz T-8 spacecraft failed to dock with Salyut 7 in early 1983. All this, of course, does nothing to detract from Ryumin's standing as a cosmonaut and, indeed, in June 1998, he would fly again, at the grand age of 58, in the process becoming the oldest Russian ever to journey into space ... a record which he still holds today.

With their return to Earth, of course, Vladimir Kovalyonok and Valeri Ryumin were no longer rookies and their new status as 'veterans' placed them back in the running to fly with *other* rookies on a future mission. In the days after their landing, *Flight International* speculated that the next flight might occur sometime within the following six weeks, perhaps in early December 1977. In the revised planning, Soyuz 26 would now be undertaken by Romanenko and Grechko. Although Grechko had always been pointed at a long-duration mission, the decision to pair him with Romanenko was not accidental; Phillip Clark noted that Grechko was an experienced spacecraft designer in his own right and this experience extended into space ... for he had completed a month-long stay aboard Salyut 4 in the spring of 1975. This experience was acutely needed, for no one knew whether the forward docking port of the new station was somehow damaged or inoperable. If this was the case, the mission of Salyut 6 might be over before it had even begun, for with only *one* working port it would be unable to support Soyuz changeovers and regular Progress deliveries. One of Grechko's tasks was to thoroughly test the station's systems and it was decided that he and Romanenko should perform a tricky EVA to inspect the docking mechanism.

The bubbly, jovial Georgi Mikhailovich Grechko had come from Leningrad – today's St Petersburg – where he was born on 25 May 1931. As a young engineer, he worked for Sergei Korolev's OKB-1 design bureau and became one of only a handful of civilians to pass the preliminary screening for a flight aboard the Voskhod spacecraft in 1964. Two years later, after a slight relaxation of the rules on health requirements, he was picked as a cosmonaut trainee. Later in the decade, Grechko seems to have served as something of a pawn between Korolev's successor, Vasili Mishin, and the powerful Air Force commander of the cosmonaut team, Nikolai Kamanin. In the days preceding the Soyuz 7 mission in October 1969, for example, Mishin pushed for Grechko to replace Vladislav Volkov, but Kamanin nixed the idea. Six months after that, it was *Kamanin* who wanted Grechko to fly Soyuz 9 instead of Vitali Sevastyanov and, this time, *Mishin* strenuously objected. From August 1972, Grechko was teamed with Alexei Gubarev for a long-duration Salyut flight and they began by backing up Vasili Lazarev and Oleg Makarov's Soyuz 12 mission in September of the following year. Grechko and Gubarev flew for 29 days in the spring of 1975. A decade later, Grechko became the then-oldest cosmonaut ever to enter space, travelling to the Salyut 7 station in September 1985 at the age of 54.

In command of Soyuz 26 was Yuri Viktorovich Romanenko, a veteran parachutist with 39 jumps and a Soviet Air Force pilot with a wealth of experience in the Yak-18 and L-29 military trainers and in the MiG-15, MiG-17 and MiG-21 jet fighters. Born on 1 August 1944 in the village of Koltubanovsky, more than a thousand kilometres south-east of Moscow, he was barely 33 years old and already a decorated lieutenant-colonel at the time of his first space mission. A cursory glance at his family hints at a respected tradition in the armed forces: Romanenko's father had been a senior naval commander and his mother a combat medic. As a child, he enjoyed building model aircraft and ships and developed a love of ships, boxing, shooting and underwater fishing. On his second mission in September 1980, he flew with a Cuban cosmonaut named Arnaldo Tamayo-Méndez and several years later he was invited to Cuba at the behest of Fidel Castro ... who not only organised a fishing tour, but participated in the event, freediving with Romanenko to a depth of several metres.

After completing his high school education in Kaliningrad in 1961, Romanenko enrolled in the Chernigov High Air Force School in Ukraine, graduating with honours. He rose to the rank of lieutenant and subsequently served as an instructor. He was selected as one of nine cosmonaut pilots in April 1970, from a pool of more than 400 finalists, passing not only the rigorous selection procedure, but also gaining clearance from the KGB and the Communist Party for admission into the country's spaceflying elite. Progression through the ranks of the Soviet Air Force was swift: Romanenko was promoted to captain in February 1971, major in February 1974, lieutenant-colonel in December 1976 and finally, in March 1978, when he returned from his record-breaking mission with Grechko, he was decorated as a full colonel. When Salyut 6 was launched into orbit, Romanenko watched from the ground as Vladimir Kovalyonok's backup, knowing that his turn would come soon. His expectation that he would fly a long-duration mission in 1978 with engineer Alexander Ivanchenkov, changed rapidly when Soyuz 25 failed and he and Grechko found themselves propelled as prime crew onto the very next flight.

With a radio callsign honouring one of the least known and most unexplored regions of Russia – Taymyr, whose mountainous and lake-studded peninsula is the most northerly point of the Eurasian mainland – the venture of Romanenko and Grechko would also explore new ground, as the Soviets strove to turn the strangest environment yet encountered by human beings into their long-term home. *Flight International* had been right in its expectation that Soyuz 26 would fly sometime in December 1977, for the predictability of launch and landing windows, which cycled every 60 days or so, had allowed Western observers to predict when spacecraft were likely to be launched and recovered. Had the Soyuz 25 mission succeeded, it seems likely that Soyuz 26 (in its 'original' incarnation, carrying Vladimir Dzhanibekov and Pyotr Kolodin) would have been launched sometime in November. Since the Soyuz had an orbital lifetime of only a couple of months, a 'spacecraft switch' would have occurred, with Dzhanibekov and Kolodin returning home in Soyuz 25 and leaving their 'fresh' ferry for the long-duration crew. After the failure of Soyuz 25 to dock, the 60-day cycle meant that a new launch window would open around 8 December and it was therefore unsurprising that Yuri Romanenko and Georgi Grechko blasted off at 4:19 am Moscow Time on the 10th of that month.

Unlike its predecessor, rendezvous and docking with Salyut 6 was triumphantly accomplished, albeit at the *rear* port, some 26 hours after launch. It was only after this successful arrival that the Soviets admitted that the new station was equipped with not one, but *two*, docking ports. "The first is installed on the station's transfer compartment," they announced, "and the second on the opposite side, in the equipment bay. The two docking systems make it possible for two spacecraft to service manned stations." The ambiguous language – making reference to the "opposite side" and an "equipment bay", for example – may have been an attempt to make Salyut 6 appear much more complex than it actually was. After all, Skylab had a pair of docking ports at right angles to one another, whereas the Soviet station possessed them in a much more straightforward fashion, at either end. It has never been proven, but often suspected, that the language used by the Soviets was a deliberate attempt to mislead and imply a greater level of sophistication than was actually the case.

That is not to say, of course, that Salyut 6 was not a sophisticated machine and that its capabilities were not a level or more above its predecessor; they *were*. Still, before the work of this new home in the heavens could commence, any possibility that the forward docking port had been disabled needed to be resolved. Late on 19 December, Romanenko and Grechko donned space suits and secured themselves in the tapering confines of the forward transfer compartment. The 73 kg 'Orlan-D' ('Sea Eagle') suit was the first in a series of extravehicular ensembles whose descendants are still used aboard the International Space Station. The Orlan-D consisted of flexible limbs and a one-piece, rigid body-helmet unit. The cosmonaut entered the suit through a hatch in the rear of the torso and, for Salyut 6 operations, it had a maximum operating time of around three hours. Its integrated design meant that it did not need external oxygen hoses and its operating pressure required a relatively short pre-breathing period of only 30 minutes. Electrical power came from umbilicals connected to the station and the cosmonauts controlled the function of the suit from a chest panel. The distance to which a cosmonaut could venture beyond the hatch was therefore limited by the length of his umbilical.

It would appear that the spacewalk plans called for Grechko to venture outside and for Romanenko to remain inside to provide assistance. At 12:36 am Moscow Time on the 20th, the Salyut hatch opened and the first Soviet spacewalk in almost a decade – and only the third in their entire programme – got underway. Grechko manoeuvred himself with relative ease toward the docking mechanism and quickly verified that it was undamaged; in fact, it had hardly been scratched ... confirming that Soyuz 25 *had* indeed made physical contact on at least one occasion. A few checks to ensure its functionality were conducted and the port was given a clean bill of health. It was with an intense and collective sigh of relief that flight controllers realised that their mission was back on track. In total, the spacewalk lasted 88 minutes from depressurisation to repressurisation and Phillip Clark noted that its routine nature was still filled with a few heart-stopping moments. "It is claimed," he wrote, "that Romanenko fancied a taste of spacewalking experience and slowly left the transfer compartment." This story gained an ominous note when related in the Western press, partly due to Grechko's misunderstood joke that his partner was on

Yuri Romanenko floats inside Salyut 6's transfer compartment before the 20 December EVA with Georgi Grechko. Theirs was the first Soviet spacewalk in almost nine years.

the verge of being 'lost' in space. Romanenko reached a maximum distance of only a few metres and, in any case, his suit's electrical and communications umbilicals would have kept him from drifting away from Salyut. Nevertheless, according to David Portree and Robert Treviño in *Walking to Olympus*, Romanenko "was very angry about the story", since it implied that he had acted irresponsibly. Almost two decades later, in June 1994, this author had the opportunity to meet and talk to Grechko following a lecture at Jodrell Bank in Cheshire. The cosmonaut's eyes twinkled as he told me about the incident. Reacting with split-second timing, he had grabbed hold of Romanenko, just in time ... and also managed to keep hold of his sense of humour. "Yuri," he asked, with mock indignation, "*where* are you going?"

According to reports in *Flight International* in April 1978, a technical problem arose towards the end of the EVA. "While Grechko was waiting in the transfer compartment," the magazine told its readers, "telemetry indicated that a vent valve had stuck in the open position, making pressurisation impossible. Ground controllers ordered Romanenko to begin the flow of air into the transfer compartment to check the telemetry report. The electrical line connected to the vent valve indicator was found to be faulty and pressurisation went ahead without incident. Analysis on the ground later indicated that the fault could have resulted from a short circuit caused by condensation in the wiring and the crew carried out a fruitless search for water in the system."

Despite the euphoria that Salyut's forward docking port was undamaged, there remained another problem. Beginning in 1978, the Soviets were planning to fly an ambitious series of international missions, called 'Intercosmos', whose crews would include cosmonauts from various Eastern Bloc or Communist-aligned nations. The first was to be Vladimir Remek, a Czechoslovakian fighter pilot. The Soviets did not want this mission – a major and high-profile exercise in international relations and socialist propaganda – to risk a docking failure at the forward port. Nor did they want to risk vacating the station, even for a couple of hours, in order to allow Romanenko and Grechko to transfer their craft from one port to another. It was therefore decided to insert *another* short-duration Soyuz into the manifest and launch it in the early part of January 1978. Manned by two Soviet cosmonauts – rookie Vladimir Dzhanibekov in command, joined by veteran flight engineer Oleg Makarov – this would make an attempt on the forward port and after about a week the crew would return to Earth in the old Soyuz 26. They would leave their 'fresh' craft docked at the forward port and free up the rear port, thereby opening the way for Gubarev and Remek to dock there in early March.

The Soyuz 27 crew was an unusual pairing, and neither had been included on the list of long-duration Salyut 6 teams, but at closer inspection both men had much in common with Romanenko and Grechko. In fact, Dzhanibekov and Romanenko had both trained in a backup capacity for the Soviet half of ASTP, whilst Grechko and Makarov had been prime flight engineers assigned to Salyut 4. Clark also noted that, had Soyuz 27 not been inserted at such short notice into the manifest, Dzhanibekov and Makarov were the original prime and backup commanders for a joint Soviet-East German mission, scheduled for later in 1978. When their new assignment was made, Dzhanibekov's place was taken by veteran cosmonaut Valeri Bykovsky and that of Makarov by Viktor Gorbatko.

"A study of the cosmonaut crew assignments," wrote Clark, "shows that there were two different groups of cosmonauts training." One group supported the *expeditsya osnovnoi* (EO, the 'principal expedition'), whose crew would spend several months aloft, as in the case of Romanenko and Grechko, whilst the other was an *expeditsya poseshchenya* (EP, the 'visiting expedition'), which exchanged Soyuz craft at regular intervals and spent about a week docked at Salyut. "One group," continued Clark, "comprised the EO cosmonauts [and] the backup crew for one EO mission would normally be assigned as the prime crew of the next EO mission. The second group were the EP cosmonauts, where the backup commander for one mission would normally be assigned as the prime commander *three EP missions later*." A third group would also be picked in order to introduce the next-generation Soyuz-T spacecraft, later in the lifetime of Salyut 6.

In time, both Makarov and Dzhanibekov would carve out their own records in the annals of Soviet cosmonautics; the former would become the first Russian to chalk up four missions, whilst the latter would be the first to make five. Oleg Grigoryevich Makarov would be making the third of his four missions on Soyuz 27, although the flight would be only his second *orbital* voyage, since he and fellow cosmonaut Vasili Lazarev had endured a harrowing launch abort in early 1975. Known to posterity as 'the 5 April Anomaly', they were assigned a 60-day stay aboard Salyut 4, but their rocket malfunctioned during staging, sending them on a 21-minute suborbital flight, then bringing them back to Earth, with a terrifying descent which finally ended in the snowy Altai Mountains, several hundred kilometres north of the Chinese border. Both men survived, but Lazarev suffered undisclosed internal injuries and never flew into space again. The pair had actually flown together on Soyuz 12 in September 1973 and as Makarov now added Soyuz 27 to his personal tally he became only the third cosmonaut in history to make a third journey into space.

Born on 6 January 1933 in Udomlya, to the north-west of Moscow, Makarov was an outstanding engineering student at the Bauman Higher Technical School. After graduating in 1957, he worked on the design of the Vostok spacecraft at the OKB-1 bureau. "His first job," wrote *Guardian* columnist Pearce Wright in his obituary to Makarov in June 2003, "was to develop the cosmonaut's control panel and instruments for the first manned flights." He was one of only a handful of civilian engineers to pass the first round of preliminary screening for the Voskhod selection in May 1964. A little more than two years later, with OKB-1 now under Vasili Mishin's control, Makarov was selected as a cosmonaut and was almost immediately assigned to the effort to circumnavigate the Moon. By September 1968, he had been teamed with Alexei Leonov for the first lunar-landing mission. In spite of the American success with Apollo 8 and their impending landing on the Moon, candidates were still being selected for Soviet landing missions as late as the following summer and Makarov was one of them. By the middle of 1971 the situation had changed markedly and he found himself teamed with Lazarev, initially for an Earth-orbital flight of the Kontakt system and, after its cancellation that September, as the prime crew for a long-duration Salyut mission and finally Soyuz 12.

Vladimir Alexandrovich Dzhanibekov came from the remote area of Iskandar in the region of Tashkent, today's capital of independent Uzbekistan. He was born on 13 May 1942, the son of a fireman and a nurse. Of his parents, he would later say with admiration that his father was "ready to deploy, 24/7, in any weather conditions and emergency situation to help people" and that his mother was "a dear, taking care of my younger brother and me". His birth surname was 'Krysin', but he appears to have changed it later in life to honour his wife's family, who were descendants of Jani Beg (Russianised to 'Dzhanibek'), a 14th-century khan of the Golden Horde. At the age of 18, he entered Leningrad University to study physics and, whilst there, developed a love of aviation. The following year, he entered a higher military fighter pilot school at Yeisk and on graduating in 1965 he became a Soviet Air Force instructor. In an interview, decades later, he noted that he could never match his friends in sports or gymnastics and feared being considered to be "the weak son of a strong father". One problem for Dzhanibekov in the cosmonaut selection process was his weight; "10 kg too much! I first went to hospital, where therapists helped me reduce. For a week, I consumed only *water* ... and was able to down my weight to 75 kg from 85." He was one of nine pilots picked as cosmonauts in April 1970.

In his early years, Dzhanibekov worked on ASTP, commanding the backup crew for the Soyuz 16 dress rehearsal flight with engineer Boris Andreyev. According to Mark Wade on his website, Dzhanibekov was then assigned to fly with engineer Pyotr Kolodin on the Soyuz 26 visiting mission in November 1977. Their primary objective would have been to perform a spacecraft swap, flying home in Kovalyonok and Ryumin's Soyuz and leaving their own 'fresh' craft behind. Given the decision after the Soyuz 25 mishap to include at least one veteran on each crew, rookie cosmonaut Kolodin was dropped in favour of the veteran Makarov.

Soyuz 27 was launched into space at 3:26 pm Moscow Time on 10 January 1978 and followed the now-standard 26-hour rendezvous profile. Making the first of what would turn into a five-flight cosmonaut career, Dzhanibekov's 'wow' moment was the view of his first orbital sunrise. "There were beautiful stars throughout the sky at night, but *sunrise* ... I saw *more* than the extreme beauty of the colours; something very special and thought-provoking." Decades later, he would struggle to find the right adjectives, but for a fighter pilot who had grown up with an ingrained distrust of the West, he noticed one thing which stood out, sharp and clear: "The *Earth*," he said, "with *no* borders!"

Aware that the docking of Soyuz 27 might compromise the integrity of Salyut 6, it was decided that Romanenko and Grechko must retreat to the safety of Soyuz 26, just in case an emergency should occur. With the exception of a balky hatch, however, nothing untoward occurred and within a few hours of docking, the four men shared laughter and stories together and toasted their success with cold cherry juice. The addition of the new vehicle altered the dynamic characteristics of the station and its inertia and centre of mass had to be recalculated in order for the computer to properly control its attitude. One experiment conducted by Dzhanibekov and Makarov was called 'Resonance' and, at its most basic, involved them running and jumping up and down on an instrumented track to enable sensors to measure the stresses imparted to the station's structure. Some prophets of doom argued that, in worst-case scenarios, such stress could break solar arrays or communications antennas or even transmit vibrations through the complex of such severity as to forcibly uncouple Soyuz 26 from its own port. Thankfully, the Resonance results neatly cleared another thorn from the path to long-duration spaceflight.

In what would become a standard procedure in the coming months and years, Romanenko and Grechko and Dzhanibekov and Makarov removed their custom-moulded seat liners from their original spacecraft and swapped them over. The new arrivals also delivered newspapers and letters ... although one piece of 'news' from Earth was kept quiet, from Grechko at least, for his father had died only a few days earlier. Although Romanenko was told, psychologists felt it was not appropriate to inform Grechko, because he was only a quarter of the way through his three-month mission. One must feel some sympathy for Romanenko, the young commander, who was obliged to keep this to himself for several months *and* take the responsibility to break the news to his crewmate after they had landed. Years later, Grechko would admit that Romanenko's difficult decision was the right one to make.

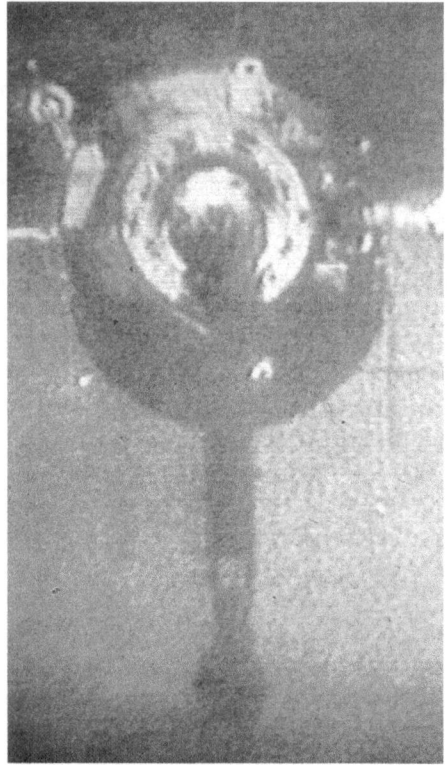

The Progress 1 freighter approaches Salyut 6, as viewed from a camera at the rear of the space station.

Dzhanibekov and Makarov undocked Soyuz 26 from Salyut's rear port on the morning of 16 January and landed at 2:25 pm Moscow Time. It would not be long before Romanenko and Grechko received another visitor. Four days later, the first Progress resupply freighter was launched from Tyuratam. It docked smoothly at the recently vacated port on the afternoon of 22 January. The cosmonauts immediately set to work unpacking the craft in a delicate and somewhat intricate exercise. First, they would prepare cotton gloves, goggles, respirators and waste bags, then don a specialised tool belt to unlock the cargo racks. "Normally," wrote Hall and Shayler, "one crew member works inside the Progress and a second remains in the station side of the access tunnel. To find the items in sequence, a locator diagram is provided in each Progress." This would be the first of many dozens of Progress missions, whose descendants today continue to service the International Space Station. Their arrival has always been an occasion of great joy and anticipation. "We stayed up a few extra minutes as we searched for our crew packages," wrote NASA astronaut John Blaha, who lived aboard the Russian Mir space station between September 1996 and January 1997 and witnessed one late-night Progress arrival. "Once we found our packages, it was like Christmas and your birthday, all rolled together, when you are five years old! We really had a lot of fun reading mail, laughing, opening presents, eating fresh tomatoes [and] cheese. It was an experience I will always remember." For Romanenko and Grechko, experiencing this for the first time, it must have been electrifying to receive such direct contact from home.

Unloading Progress took several days and each new piece of hardware – belts for the seats at Salyut's control panel, carbon dioxide scrubbers, air-purification filters, air circulation fans and so on – had to be carefully positioned in the station. The task of reloading the orbital module with unneeded items, including food waste and useless equipment, was far more complex than simply throwing all of their rubbish into the tiny craft and setting it adrift. Even though the Progress would be destroyed in the atmosphere, it was vital that its centre-of-mass constraints were not violated, in order that it could manoeuvre correctly and properly execute a controlled re-entry. Everything to be discarded had to be carefully recorded and correctly stowed. Early on 1 February, the refuelling procedure was satisfactorily completed and on the 5th the craft undocked. Following a brief test of its backup rendezvous system, it was finally commanded to de-orbit itself a few days later in order to re-enter the atmosphere and burn up. The first Progress mission had lasted 19 days and had been an enormous success, not only refuelling Salyut, but also demonstrating with flying colours that it was practical for an unmanned craft to provide a regular resupply service. The future of Salyut 6 seemed assured. The next step forward would represent a real boon for Soviet propaganda: the first in a series of missions which would feature 'guest' pilots from Communist-aligned nations. NASA had made no secret of the fact that it intended to invite European researchers to work aboard its Shuttle, part of a new effort to foster international participation in what had been a two-horse race between America and Russia. In March 1978, the Soviets beat them to it. The cynic might doubt the long-term sincerity of the Intercosmos missions, for virtually all of them were one-offs, mostly involving nations within the Eastern Bloc, and were conducted purely as an exercise in propaganda to reinforce the image of the 'perfection' of the Communist lifestyle. However, propaganda or not, the fact remained that with Intercosmos the Soviets were able to operate a truly 'international' space station for the first time.

FROM TANKS TO SPACESHIPS

Deep within the central valley of the Vltava River – the longest waterway in the Czech Republic – lies the beautiful city of České Budějovice. Today it is the largest settlement in the South Bohemian Region and has been a significant political and commercial centre for more than seven centuries. Its dramatic, mountain-ringed skyline possesses all the architectural grandeur of the Gothic and of the Renaissance and of the Baroque and hints at a chequered past: its old town hall, with its bronze gargoyles and its decorative murals, its famous 16th-century 'Black Tower', its Dominican Convent and medieval Presentation of the Virgin Mary Church and even one of its more recent additions, the Austro-Hungarian railway station, built in the Belle Epoque style. České Budějovice is particularly well-known for its breweries (in fact, it served the Holy Roman Emperor himself at one stage) and its *Budweiser Bier* eventually led to the American imitation, now known throughout the world. And in this place, on 26 September 1948, was born one Vladimir Remek, who today serves as a Member of the European Parliament for the Czech Republic. More significantly for our story he would also earn world renown, not only by becoming the first Czech cosmonaut, but also as the first person from a nation other than the United States or the Soviet Union to break the surly bonds of Earth and venture beyond.

At first glance, it might seem a little unusual for a Czech to be selected for the first Intercosmos venture ... particularly as Soviet-led Warsaw Pact tanks had rumbled into Prague less than a decade earlier in order to crush the short-lived attempt by liberal reformist Alexander Dubček to institute democratic change. Since April 1969, the Soviet-backed Gustáv Husák had ruled Czechoslovakia with a far stronger hand and his policies of 'normalisation' revoked many of Dubček's reforms, re-established central control over the country's economy and reinstated the powers of political surveillance to the state police. At the same time, Husák's early policies resulted in high economic growth, generally acceptable living conditions and the widespread availability of material goods and helped to appease an otherwise disgruntled population. By the latter part of the decade, however, Czechoslovakia's economy began to stagnate and in the 1980s Husák would be under increasing pressure to institute genuine reforms. It has been speculated over the years that the choice of Vladimir Remek, a pilot in the Czechoslovak Air Force, to fly first in the Intercosmos effort, was designed to build bridges and soften anti-Soviet sentiment in the country. In Remek's mind, however, it was quite the opposite. "If the Soviet leaders had any problems at that time," he told an interviewer, many years later, "it *wasn't* a sense of guilt for entering Czechoslovakia. It could have been partly political, but what was really important was that we were among the strongest partners in the Intercosmos programme and our people were also on the UN space committee. Maybe we weren't the *worst* among those who prepared for the flight!"

Remek was – and still is – a staunch Communist. (In fact, he was born only seven months after the Communist take-over of Czechoslovakia.) At the time of writing, he represents the Communist Party of the Czech Republic in the European Parliament and maintains that his experience as a cosmonaut helped him to develop the skills and credentials to undertake that task. He was the progeny of a Czech maternal line and a Slovak paternal line and his father, in particular, had a strong military background, serving as a lieutenant-general in the army and a key figure in Czechoslovakia's air defence ministry. Not surprisingly, the young Remek also gravitated towards a career in the armed forces, studying mathematics and physics at middle school in Čáslav and at the Higher Aviation School in Koiice, ultimately rising to become a fighter pilot with the Czechoslovak Air Force in 1970. Two years later, he was picked to study at the Yuri Gagarin Air Force Academy in Moscow and earned a master's degree in 1976. Remek returned briefly to his air force unit at České Budějovice and in the early autumn he became one of half a dozen Czech candidates for an Intercosmos mission. On 25 November 1976, this shortlist narrowed still further when Remek was named, along with 33-year-old Czechoslovak Air Force engineer Oldřich

The first group of Intercosmos pilots, selected in late 1976, comprised a pair of Czechs, a pair of Poles and a pair of East Germans. From left to right are Sigmund Jähn, Eberhard Köllner, Vladimir Remek, Miroslaw Hermazewski, Oldrich Pelèzák and Zenon Jankowski.

Pelčzák, to the first Intercosmos group. This group also included two Poles and a pair of East Germans.

Aside from the political propaganda value, these week-long Intercosmos missions also served the important purpose of exchanging Soyuz craft as they approached the end of their two-month operational lifetimes. Yet Intercosmos represented far more than just putting Soviet allies into space. It had been created in 1965 and its sphere encompassed a range of unmanned scientific missions; not until July 1976, in fact, was an agreement finally signed to provide for *manned* flights and in September of that year, Vladimir Shatalov, commander of the cosmonaut team, announced that the training of these non-Soviet pilots would take at least a couple of years, even if there were no language problems. The first candidates would come from Czechoslovakia, Poland and East Germany – "the three most advanced East European countries," according to a US congressional report, published in 1982 – which had already provided almost a decade of talent to unmanned Intercosmos ventures. Following this cosmonaut selection in November, the situation moved rapidly. By May 1977, the first Intercosmos group had completed their basic training and were paired with their Soviet commanders: Remek with Alexei Gubarev and Pelčzák with Yuri Isaulov. These assignments shifted slightly in the wake of the Soyuz 25 failure, when the rule that there must be at least one veteran on each crew was instigated. Isaulov, a rookie, selected as a cosmonaut in 1970, was replaced by Nikolai Rukavishnikov. Since the Soyuz (at the time) could only house two men in pressurised suits, and there was invariably a *civilian* engineer on each *civilian* Salyut crew, this sounded a clear knell for many of those military pilots who were still awaiting their first flights. Among them was Pyotr Kolodin, who had earlier lost his spot to fly to Salyut 1 and would shortly lose his chance to fly to Salyut 6 as well. It cannot have been easy for these men to accept being grounded after so many years of dedicated training.

Nor were the Intercosmos 'pilots' considered much more than passengers by their Soviet commanders. Years later, in a June 1986 article in *Flight International*, space analyst Tim Furniss noted that Remek returned from Soyuz 28 with sore knuckles: "He said it was because every time he went to touch something, he was *smacked* on the hand by the commander!" That commander was Alexei Alexandrovich Gubarev, a colonel in the Soviet Air Force, veteran cosmonaut ... and an old friend of Georgi Grechko; in fact, the pair had flown together for a month to Salyut 4 in the spring of 1975. Gubarev was one of the second generation of older, more experienced and better-qualified cosmonauts picked in January 1963. He was born on 29 March 1931, to a peasant-stock family in the village of Gvardeitsy, on the eastern bank of the Volga River, in the Samara region. After the death of his father, the family moved to the Chashnikovo collective farm, near Moscow, and in 1950 Gubarev graduated from middle school in Kryukovo and enrolled at the Naval Aviation School for Aircraft Mechanics. He later joined the Soviet Air Force and after completing advanced studies was detailed in 1962 to join an aviation unit in the Black Sea as a squadron commander; shortly thereafter, he drew the attention of recruiters for the second group of cosmonauts. Alongside Vladimir Shatalov and Lev Dyomin, Gubarev was one of the bright stars from this class and seems to have worked on the cancelled Soyuz-VI military project, the L1 circumlunar effort and the DOS-2 civilian space station between 1966 and 1971. Together with Vitali Sevastyanov and Anatoli Voronov, he was named in mid-June 1971 to back up Alexei Leonov's scheduled expedition to Salyut 1. However, Leonov's crew was grounded and Gubarev's training was abruptly suspended by the deaths of the Soyuz 11 crew during their return to Earth. Less than four years later, Gubarev finally reached his own space station, with Grechko, and in 1976 he was assigned to the Intercosmos group as commander of its first mission.

Soyuz 28 headed into orbit at 6:28 pm Moscow Time on 2 March 1978, with Czechoslovak Radio's Ilja Jenča likening the faces of the cosmonauts, projected onto a monitor screen in the mission control centre, to "two brothers, or even twins, representing two brother socialist countries ... the symbolism mingles with reality". The very word *propaganda* in Soviet usage was never intended to take on a negative meaning and, whatever one's opinion, *this* propaganda event was being exploited to the limit. Next day, Gubarev expertly guided his craft to a docking at the aft port of Salyut 6. Very little has been published over the years – an occasional anecdote here and there and a few brief references to experiments performed – about the exploits of Remek during his week in space, which is more than a little surprising, owing to the historic nature of his flight. Each Intercosmos nation sponsored its own programme of research, one aspect of which almost always focused upon observations of the home territory of the visitor, often using the MKF-6 camera, and Remek is known to have performed at least four other specific experiments. One was a questionnaire about subjective reactions to the weightless environment, another studied body heat exchange, a third investigated oxygen absorption into human tissue and the final one used the 'Chibis' lower-body negative pressure garment – not dissimilar in function to the device aboard Skylab – to reduce orthostatic stresses and better prepare Remek for his return to Earth. A number of other investigations were conducted, involving Salyut 6's other experimental hardware: a materials processing furnace, known as 'Splav-1' ('Alloy'), was used to perform the Czechoslovakian experiment 'Moravia', whilst another investigated the way in which the light from stars changed as they 'set' below Earth's horizon. This was used as a tool for observations of the composition of the atmosphere.

In fact, research had been ongoing aboard the station since Dzhanibekov and Makarov returned home. As part of their solo programme, Yuri Romanenko and Georgi Grechko had heated Splav-1 to a temperature of 1,000°C to work with various semiconducting alloys, including copper-indium, aluminium-magnesium and indium-antimonide. Since the furnace was mounted inside one of Salyut's trash airlocks, with one 'side' open to the vacuum of space, wrote Phillip Clark, "the excess heat generated by the experiment was easily dissipated". Today, with dozens of Shuttle flights having performed advanced materials processing investigations in Spacelab and Spacehab, it is easy to look back upon this early Soviet work as primitive. Not so, for the work of Romanenko and Grechko and Gubarev and Remek was actually one of the earliest demonstrations of the feasibility of such research ... and opened a new door on the profound applications which it might someday find in the semiconductor industry back on Earth.

Nor, indeed, was Splav-1 itself a primitive creation. "Molybdenum reflectors inside the furnace ensure that the heat always concentrates on the sample," explained *New Scientist*'s Sarah White in September 1978. "This means that the temperature of the furnace wall does not exceed 40°C and does not burden the station's air-conditioning system unnecessarily." The second Progress freighter, later in 1978, would deliver the Kristall furnace, further underlining a Soviet appreciation of the importance of this research.

As the four cosmonauts continued their work aboard Salyut 6, another record was broken on 4 March, when Romanenko and Grechko exceeded the 84-day endurance record set by the last Skylab crew. Six days later, at 6:45 pm Moscow Time, Soyuz 28 touched down on Soviet soil with Gubarev and Remek, concluding a remarkable voyage, lasting just shy of eight days. Indeed, a perusal of each Intercosmos flight in the later 1970s would show that each one lasted for approximately the same duration, give or take an hour or so, in order to ensure that no Eastern Bloc nation could take offence or speculate that the Soviets favoured one above another. Yet the political significance of the mission could not be avoided and nor did Leonid Brezhnev have the slightest intention to neglect exploiting it. During the mission, he lauded "the schooling and courage of the cosmonauts" and "the selfless labour of those who prepared the flights". In the weeks that followed,

The joint crews of Soyuz 26 and 28 gather around the dinner table during their brief time together aboard Salyut 6. From left to right are Vladimir Remek, Alexei Gubarev, Georgi Grechko and Yuri Romanenko.

the solid gold stars of Hero of the Soviet Union and Hero of the Czechoslovak Socialist Republic flowed aplenty and in a particular lavish and poignant affair organised at Prague Castle, awards were presented by Gustáv Husák himself.

Gubarev and Remek's triumphant return to Earth came at the beginning of a two-week 'landing window', during which Romanenko and Grechko were also expected to return. Air sports rules, enshrined by the Fédération Aéronautique Internationale (FAI), only recognised a 'new' record if it broke the 'old' record by at least a ten-percent margin; the landing window for the Soyuz 26 crew fell between 10-21 March, producing a total mission span of between 90 and 101 days. Romanenko and Grechko actually landed at 2:19 pm Moscow Time on 16 March, having spent 96 days aloft ... an increase of almost 15 percent over the achievement of Gerry Carr's crew in February 1974. In general, subsequent Soviet expeditions tended to stick to the FAI's ruling: the next flight would spend four and a half months in orbit (an increase over the Romanenko-Grechko record of more than 40 percent) and the *next* would come close to a full six months (a 25 percent jump). The only 'anomaly', if it can be so called, was the fourth long-duration residency aboard Salyut 6, during which cosmonauts Leonid Popov and Valeri Ryumin broke their predecessor's record ... but only by nine days, a measly five percent increase. They held the absolute endurance record until the end of 1982, but only unofficially, since their achievement was not recognised by the FAI.

Romanenko and Grechko's return to Earth brought mixed emotions: euphoria, undoubtedly, at a job well done, and the Order of Lenin and Hero of the Soviet Union followed, almost by default. For Romanenko, however, the landing brought with it the unenviable task of giving Grechko the devastating news of the death of his father. Some relief must have been afforded to the young commander in other regards, since an excruciating toothache which developed aboard Salyut 6 could finally be treated. During the flight, all that the doctors could do was to suggest that he washed his mouth frequently with warm water ... and by the time he was back on Earth, a *nerve* had been exposed!

For Romanenko and Grechko, however, on that day in mid-March 1978, they were the unquestioned and unrivalled champions of the world in terms of space endurance. Skylab scientist Ed Gibson's hope that his crew's record would soon be broken had come true ... albeit by a team of Soviets, *not* Americans. The Soyuz 26 achievement would be broken by cosmonauts Vladimir Kovalyonok and Alexander Ivanchenkov within months, to be fair, but that misses the point and does nothing to diminish the significance of what Romanenko and Grechko, and their *nation*, had done. *Never again* would the United States reign supreme in long-duration flight and, even today, at the time of writing, in 2011, the current American endurance record-holder, Peggy Whitson – who also happens to be the incumbent chief of NASA's astronaut corps – is a mere *twentieth* in line of most experienced spacefarers in the world. The *nineteen* names ahead of her are all Russian ...

IN FOR THE LONG HAUL

With the triumphant return of Romanenko and Grechko to Earth, the door had been opened on half a decade of Soviet space spectaculars. At 11:17 pm Moscow Time on 15 June 1978, Vladimir Kovalyonok – veteran of the ill-fated attempt to dock with Salyut 6 a few months earlier – and rookie engineer Alexander Ivanchenkov were launched into orbit aboard Soyuz 29, with an expectation in the West that they would break the record set by their predecessors by a margin of at least 30 percent, during a mission of some 120-140 days. In the early hours of 17 June, the two men docked safely at the space station and set to work on what would turn into a four-and-a-half-month marathon. The 'de-mothballing' of Salyut after three months in hibernation got underway with bringing on-line the life-support systems, including air regenerators and thermal regulators and the water recycling unit which condensed vapour. Within days, the station was back in good shape. Internal temperatures, air pressures and humidity levels were normal and the men set to work on routine maintenance and started their scientific programme.

It had been a quite remarkable six months for Kovalyonok, but the Kremlin's refusal to award him a Hero of the Soviet Union accolade after Soyuz 25 seems to have had little, if any, impact on his cosmonaut career. Within weeks, he was back in training, not only for a long-duration Salyut stay, but also as backup to Vladimir Dzhanibekov on the short Soyuz 27 visiting mission. Kovalyonok's flight engineer, Alexander Sergeyevich Ivanchenkov, came from the town of Ivanteyevka, on the Ucha River, a couple of dozen kilometres north-east of Moscow, where he was born on 28 September 1940. A keen sportsman, traveller and guitarist, academia suited Ivanchenkov well and he departed middle school with a gold medal, then entered the Moscow Aviation Institute and graduated in 1964. He spent some years thereafter working for Sergei Korolev's design bureau, "engaged in designing space technology products", according to Mark Wade on his website, then transferred to the test flight division, under Sergei Anokhin, in 1971. Two years later, he was selected as one of four civilian cosmonaut engineers and in 1974 was paired with Yuri Romanenko on the support crew for the ASTP rehearsal mission, Soyuz 16, and also for the joint flight itself, Soyuz 19. He remained with Romanenko to back up Kovalyonok and Ryumin on Soyuz 25. By the logic of the assignment process, these two would have gone on to fly a long-duration Salyut mission together, but this was prevented by the decision at the end of 1977 that every crew must include a veteran cosmonaut. The team was broken and Ivanchenkov found himself reassigned, with Kovalyonok, to back up *both* Soyuz 26 and 27. With so many reserve stints under his belt, it must have been with a sense of euphoria that Ivanchenkov finally set off on a mission of his own.

The mission would be a long one, but the two men would not be alone: not only were *three* Progress resupply freighters scheduled to arrive, carrying a variety of cargoes which included a new materials processing furnace, Kristall, but no fewer than *two* Intercosmos crews were also to visit them in June and August 1978. For Kovalyonok and Ivanchenkov, though, the work would not wait for visitors. For a couple of days, from 24-26 June, they placed Salyut into a gravity-gradient-stabilised attitude in which the long axis was radial to Earth – a mode which avoided thruster firings – to allow them to conduct a delicate materials processing run in the Splav-1 furnace. Shortly afterwards, at 6:27 pm Moscow Time on the 27th, Soyuz 30 was launched, carrying Soviet cosmonaut Pyotr Klimuk and Intercosmos pilot Miroslaw Hermazewski of Poland. They docked at Salyut 6 the following evening and heralded the start of a week of joint activities.

As already discussed, Poland was one of three Soviet-aligned Communist states, along with Czechoslovakia and East Germany, which had provided candidates for the first Intercosmos group in the autumn of 1976 and it is therefore hardly surprising that one of its nationals would fly an early mission. Both Poles – Hermazewski and Zenon Jankowski – were military officers and veteran pilots; the former representing the air force and the latter the army. Hermazewski was born in the small village of Lipniki, in the former Wołyń Voivodeship of eastern Poland, on 15 September 1941. At that time, the country was firmly under the Nazi yoke, but following the overthrow of the Third Reich, the westward march of Soviet Communism produced the People's Republic of Poland, which would endure for more than four decades. Hermazewski's fortunes and the history of his fatherland during this tumultuous period – a period punctuated by social unrest, economic depression and a poor standard of living for many Poles – would be closely intertwined ... and not always for the better. In 1944, Hermazewski's father was killed and he himself narrowly escaped death during the infamous Volhynian Massacres, part of an ethnic cleansing atrocity perpetrated by Ukrainian nationalists, which saw perhaps 80,000 Polish men, women and children slaughtered. *Nineteen* members of Hermazewski's family were murdered during this period ... simply because they were Polish.

After graduating from High Aviation School in Dęblin, he joined the Polish Air Force in 1965 and later served as a fighter pilot. Further academic study followed at the Karol Sverchevski Military Academy and in the autumn of 1976 he was picked from a pool of 500 Polish aviators for the first Intercosmos group, along with 39-year-old army pilot Jankowski. Before focusing on Hermazewski's Soyuz 30 mission, however, it is important to establish some additional context, for his involvement in the fortunes of Poland would stretch much further than simply into space. For several years after his return to Earth, and by now a senior air force officer, Hermazewski found himself mired in one of the darkest periods of Polish history, as a member of the Military Council of National Salvation, a notorious dictatorship established in December 1981 to enforce martial law. Ostensibly, the purpose of the council, headed by General Wojciech Jaruzelski, was "to save the country from collapse", morally, economically and politically, although in reality it was a concerted attempt to crush opposition to the authoritarian regime. Curfews would be imposed, tanks and soldiers would be deployed onto the streets of major cities, telephone lines would be cut, mail censored, national borders sealed and airports closed in one of the most terrifying crackdowns on civil liberties ever seen in the modern era. Freedom of speech and human rights fell by the wayside, three quarters of a million Poles fled their homeland in the 1980s and the crisis produced a flurry of economic sanctions from the West. In a 1997 documentary, entitled 'The Weights of Weightlessness' (details of which are available at www.dqfilms.com/hermazewski_eng.html), Miroslaw Hermazewski distanced himself from the events of those tumultuous years and reflected that his involvement with the military council had rendered him essentially a "slave to the Communist propaganda". In the documentary, he hotly denied charges of providing falsified information, of participating actively in the enforcement of martial law and even of killing a man during a hunting trip.

Whatever the truth, these two infamous episodes in Hermazewski's life – the appalling suffering of his family during the Volhynian Massacres and the equally appalling period of martial law in his homeland – were but sidenotes to his remarkable achievement of becoming Poland's first man in space. Yet the importance of the role taken by Hermazewski in the wake of his Intercosmos flight underlines another reality: that the title of *Cosmonaut* exerted a powerful political pull on the governing elites in those propaganda-hungry days of the late 1970s ... and this continues today. Vladimir Remek, as we have seen, would later enter Czech and pan-European politics, as would other Intercosmos veterans. Bulgaria's Georgi Ivanov would be elected to his country's National Assembly and play a pivotal role in the formation of its post-Soviet constitution in 1990; Hungary's Bertalan Farkas would stand for election as a centre-right candidate for the Hungarian Democratic Forum in 2006; Vietnam's Pham Tuân would hold high office within his nation's defence ministry; and Mongolia's Jügderdemidiin Gürragchaa would serve a spell as his country's defence minister. Others, including Romanian cosmonaut Dumitru-Dorin Prunariu, the youngest of the group to fly, would serve both their country *and* the wider European Union. Militarily, *all* of them would reach the rank of a general officer within their respective armed services.

For many of the tottering Communist regimes in eastern Europe and Asia at this time, the trumpeted space journeys of their cosmonauts provided one of very few glimmers of light during a period which was otherwise marred by the darkness and gloom of economic stagnation, political repression and social unrest. Throughout the 1970s, Poland was under the leadership of Edward Gierek and had borrowed on a massive scale from the West to upgrade its ability to produce export goods and import cars and televisions to reinvigorate its population. At first, it seemed to work and the country experienced a few years of improved living standards, an apparently stable economy and a 40 percent increase in wages. Increased oil prices in the wake of the 1973 Arab-Israeli War, however, triggered recession in the West, sharp increases in the price of imported goods and sharp *decreases* in the demand for Polish exports, such as coal. With the country's foreign debt spiralling into the billions of dollars, borrowing became impossible and when Poland signed the Helsinki Accords in the summer of 1975 its people became more aware than ever before of the basic rights which they were being denied. When Western creditors came knocking at the door, prices increased massively (sugar, in particular, rose by a *hundred percent*) and in June 1976 a wave of strikes and violent demonstrations swept through Poland. Two years later, only two things of international note would occur to excite and unite an otherwise outraged population. The first was the flight of Hermazewski into space and the second, in October 1978, was the election of Kraków's archbishop, Karol Wojtyła, as Pope John Paul II. His repeated calls for calm and talk of creating an "alternative Poland", with mechanisms independent of government, would do little to prevent the calamity which befell the country in the coming decade.

By the end of June 1978, as Hermazewski orbited Earth aboard Salyut 6, he saw a pitiful incarnation of his once-proud nation, on the brink of economic collapse. This knowledge did not diminish the sheer grandeur of Poland itself, for countries viewed from space often show few of the scars which might otherwise be tearing them apart. There are no lines or boundaries. All Hermazewski could see was the beauty of the Baltic Sea coast in the north, extending from the Bay of Pomerania to the Gulf of Gdansk, to the hilly moraine lake districts, the Ice Age river valleys and, in the far south, the mighty ranges of the Świętokrzyskie Mountains and the highest peaks of the Carpathians. Indeed, observation of Earth was one of Hermazewski's key tasks, together with studies of the aurora borealis – the 'Northern Lights' – and a series of cardiovascular and other medical investigations. The Polish 'Siren' experiment put the airlock-mounted Splav-1 furnace to use for 46 hours in studies of the uniformity of alloys and the transfer of mass in a weightless environment. "In each international crew," wrote *New Scientist*'s Sarah White in September 1978, "the research programme carried out during their few days in space has apparently been largely determined by the interests of that particular country." Researchers from the Warsaw Institute of Physics, White continued, had submitted proposals for several investigations for Hermazewski to perform, including mercury-cadmium-telluride, an important semiconductor, which had already been under intense international study for more than a decade, due to its sensitivity to infrared wavelengths.

Salyut 6 was already earning itself a reputation as a truly 'international' space station, as this four-man crew pressed on with its work; in addition to the Polish Hermazewski there was the Moscow-born Ivanchenkov and *two* Soviet cosmonauts from today's Belarus: Vladimir Kovalyonok and the Soyuz 30 commander, Pyotr

Ilyich Klimuk. Space historian Asif Siddiqi has referred to the youthful-looking Klimuk – he was almost 36 years of age – as "something of a child prodigy", but his credentials were impressive: he was a full colonel in the Soviet Air Force, had two previous space missions (one of which set a new endurance record for the Soviets back in the summer of 1975) and he would go on to serve as head of the cosmonaut corps. He was born to a peasant-stock family in the Komarovka district of Brest, near the Polish border, on 10 July 1942. His father had died in the Second World War and, after graduation from middle school in 1959, Klimuk entered Primary Aviation School and later the Leninsky Komsomol Chernigov High Aviation School. Completion of initial flight instruction in 1964 was followed by enrolment in the Soviet Air Force, from whose ranks he would eventually retire as a general. In October 1965, he joined the cosmonaut corps and within two years was assigned to the L1 Moon project. In September 1968, Klimuk was named as a backup crew member for the second circumlunar mission, supporting cosmonauts Valeri Bykovsky and Nikolai Rukavishnikov. Three years later, he was formally paired with military engineer Yuri Artyukhin and a civilian engineer for an early Salyut station visit, before being given the backup command of Soyuz 13 in July 1973. This eight-day 'solo' mission was followed by a two-month period aboard Salyut 4 in 1975, making Klimuk – for a time – the Soviet Union's second most experienced cosmonaut.

If June 1978 was a hectic month for manned missions, with the launches of both Kovalyonok-Ivanchenkov and Klimuk-Hermazewski occurring within two weeks of one another, the pace was not expected to slow down as the year wore into July and thence into August. Soyuz 30 touched down on Soviet soil at 4:30 pm Moscow Time on 5 July, completing a mission of almost eight days, and on the 7th a second Progress resupply freighter blasted off from Tyuratam, bringing with it the Kristall furnace. Kovalyonok and Ivanchenkov installed it into the transfer tunnel at the rear of the station. In its early experiments, wrote *New Scientist*'s Sarah White, it produced pure monocrystals of indium antimonide and gallium arsenide, used today for a variety of applications, including thermal and infrared detectors, microwave circuits and solar cells. Other experiments focused on the semiconducting materials cadmium mercury telluride and niobium aluminium germanium. Progress 2 also carried several hundred kilograms of foodstuffs and propellant and when the fuel and oxidiser were pumped into Salyut's tanks on 19 July it was done under *ground control*, rather than that of the cosmonauts.

Also aboard, it seemed, were medical supplies for the crew, including nitrazepam, a sedative to treat insomnia, which the cosmonauts referred to as 'Eunoctin'. During a voice transmission from the station, which was received and unscrambled by radio analysts Richard Flagg and Sven Grahn and later reported on Grahn's website, Kovalyonok and Ivanchenkov expressed some annoyance either that something was omitted from their medical kit or that the wrong contents were provided. "Although the medical kit on the cargo ship should have been put together, considering our crew," Kovalyonok began, "Sasha [Ivanchenkov] made the same observation. I ask you to log this, for it was authorised with him, pass it up the chain of command and report it … tomorrow, so that he would find who's responsible and that measures

Alexander Ivanchenkov works outside Salyut 6 during the 29 July 1978 EVA. His crewmate, Vladimir Kovalyonok, clutching the camera, can clearly be seen in Ivanchenkov's gold-tinted visor, as can the bulk of Salyut 6 itself.

would be taken to them appropriately ... *up to the penalty*! And inform us about the outcome. Okay?" Clearly the contents of Progress freighters were eagerly awaited by their cosmonaut recipients and poor planning inevitably led to frayed emotions and short tempers.

Betwixt the arrival of the freighter and that of the next Intercosmos crew, another major task for Kovalyonok and Ivanchenkov was a spacewalk – only the second to be conducted outside the station and only the fourth in Soviet history – on 29 July.

They spent a little more than two hours outside, retrieving several pieces of equipment on the exterior which had been exposed to the harsh environment ever since the launch of the station, ten months earlier. According to Phillip Clark, writing a decade later, these included micrometeoroid detectors and a series of polymer and biopolymer cassettes. Ivanchenkov ventured outside at just after 7:00 am Moscow Time, securing himself into a set of 'Yakor' ('Anchor') foot restraints, whilst Kovalyonok brought out the television camera and provided support. The images captured by the commander of his flight engineer, clad in a pure white space suit and gold-tinted visor, set against the iridescent blue of Earth and the unfathomable blackness of space beyond, are among the most awe-inspiring of the entire Soyuz 29 mission. "The two cosmonauts were out of communications range for the first one hour [and] seven minutes," *Flight International* reported on 2 September, "and declined an invitation from the ground to move back inside when they completed the assigned tasks earlier than expected. The crew finally re-entered Salyut 6 as the station passed into darkness over the Pacific." In *Walking to Olympus*, David Portree and Robert Treviño noted that during one period of orbital darkness, Kovalyonok and Ivanchenkov were treated to a brilliant meteor, burning up 'beneath' them in the upper atmosphere, and that the pair photographed areas of the Black Sea, Kazakhstan and China, then watched as Australia and its Great Barrier Reef passed beneath their boots, more than 200 km below.

The Soviet progress in catching up with the Americans may have been enormously successful in terms of *endurance*, but in spacewalking they remained far behind. With the completion of Kovalyonok and Ivanchenkov's excursion, the Soviets had spent a total of only nine and a half hours outside on four EVAs since March 1965, whereas the United States had scored an impressive 103 hours on more than a dozen EVAs *and* a further 162 hours in 15 Moonwalks. By the spring of 2011, when this book was written, a grand total of more than 340 spacewalks have been made by 57 Russians, 123 Americans and a handful of other nationals, including France, Germany, Japan, Switzerland, Canada, Sweden and China. The balance has tipped, but only slightly: it is now a *Russian* cosmonaut – Anatoli Solovyov – who currently holds the record for EVA time … although the *next six places* are all presently held by Americans.

Four days after the completion of Kovalyonok and Ivanchenkov's EVA, the Progress 2 freighter, laden with trash and unneeded equipment, was cast adrift to burn up in the atmosphere, but as its voyage drew to an end, *another* craft was being readied for launch. Progress 3 headed into orbit in the early hours of 8 August and although it did not deliver any more propellant, it *did* bring a few comforts from home … including a guitar for Ivanchenkov. After less than two weeks at the station, the freighter undocked and the rear port of Salyut 6 was free once again to await its next human visitors. These came within a matter of days. At 5:51 pm Moscow Time on 26 August, Soviet cosmonaut Valeri Bykovsky and Intercosmos recruit Sigmund Jähn, a lieutenant-colonel in the East German Air Force, were launched into space aboard Soyuz 31. It was Bykovsky's third flight, having also piloted Vostok 5 and the solo mission with Vladimir Aksyonov in the autumn of 1976, discussed earlier in this chapter. Sigmund Werner Paul Jähn, born on 13

February 1937 in the village of Morgenröthe-Rautenkranz, in the eastern portion of today's Free State of Saxony, became the first German cosmonaut. He received schooling in the local area and trained as a printer for a time, receiving his diploma from the Falkenstein Printing House in Klingenthal. Aviation, though, turned out to be his ultimate calling and Jähn joined the East German Air Force – properly known as the Air Force of the National People's Army of the German Democratic Republic (GDR) – as a cadet in 1955. He graduated three years later and became a fighter pilot. Jähn studied at the Gagarin Air Force Academy in Monino, to the east of Moscow, from 1966 until 1970 and later worked on pilot training and flight safety issues for his national air force.

Like Remek and Hermazewski before them, it would certainly appear that Jähn and his backup, fellow air force pilot Eberhard Köllner, were eager Communists. (In fact, following the 1990 reunification, Köllner, by then a colonel, *refused* to be transferred to the new German Federal Defence Force, the *Bundeswehr*.) In his youth, Jähn had served as secretary of his troop in the socialist organisation, FDJ ('Freie Deutsche Jugend' or 'Free German Youth') and became a member of the Socialist Unity Party in 1955. On his space mission itself, he took with him a portrait of Erich Honecker, a copy of the Communist Manifesto ... and the 'Sandmännchen' puppet, which appeared nightly on East German television – and still does today – to serenade children to sleep. Over the years, the puppet also served a dual role, however, by showcasing socialist technology and success, with references to futuristic cars and flying machines. A mock 'wedding' between the Sandmännchen and a Russian puppet, carried by Bykovsky, would even be broadcast during the mission, underlining the 'marriage' of East German and Soviet technology. Wearily, it seemed, even a children's *puppet* was being used as a propaganda symbol.

This impression of Jähn as an uncompromising Communist, for whom *nothing*, not even a children's *puppet*, would avoid being used to make a political statement, is perhaps a little unfair. In the wake of his flight, he would endear himself to the East German public, on account of his modesty and his desire to remain out of the spotlight, but the political undercurrent continued to flow, nonetheless, with the state-controlled *Neues Deutschland* newspaper referring to Jähn as the "first German in space" and lionising him as an icon of Communism. This was quite remarkable, since the usual custom was to refer *only* to citizens of East or West Germany, but *never* as a united people.

Feelings between the two German nations – the socialist East and the capitalist West – had improved dramatically since the post-war division of their country into Soviet, US, French and British occupation zones. By the 1950s, this administrative arrangement had given rise to the German Democratic Republic (GDR, popularly nicknamed 'East Germany') and the Federal Republic of Germany (FRG, otherwise known as 'West Germany'). Relations between them had reached a nadir in August 1961, when the first steps were implemented to build the Berlin Wall. In September 1973, by then enjoying relatively normal diplomatic relations, they were admitted to the United Nations.

Yet opinion on each side of the Inter-German Border remained sharply polarised and manifested itself most memorably in their humour. West Germans poked fun at the poor quality of cars in the East and their inability to obtain exotic fruits, such as bananas. It was said that a banana placed on the Berlin Wall could actually be used as a *compass* ... because only the side facing *east* would have a bite taken out of it. In response, the East Germans retorted that the very *reason* there were so many bananas in the West was because West Germans were *definitely* descended from apes. Both sides had their own ideas about the motivations of their respective political elites. One notable joke from this period involved a teacher asking a student about the differences between capitalism and socialism. "Capitalism," the student replied, "is the exploitation of *man by man*. Under socialism, it is the *other* way round!"

In West Germany, many perceived the East as illegitimate – a puppet regime, the prostrate slave of a despotic Soviet mastery – whereas East Germans saw themselves as an entirely separate entity, "a socialist nation-state of workers and peasants". The signing of the Basic Treaty between the two sides, which became operative in June 1973, politically recognised both German nations and obliged them to respect each other's sovereignty. Gradually, East-West telephone and postal communications improved, personal ties across the border were restored and the East had more direct access to luxuries from the West, including radios and television sets. Nevertheless, there was a line which the East German government would not cross. As part of the Helsinki Accords, the provision for freedom of movement led to more than a hundred thousand East German applications to emigrate ... *all of which* were rejected. Not for almost two more decades would the dream of a truly 'unified' Germany be realised.

One of Sigmund Jähn's primary tasks during his week aboard Salyut 6 was Earth observation, including the two German nations, and it is unsurprising that this involved the MKF-6M multi-spectral camera, built by the state-run East German firm, Carl Zeiss Jena. This instrument had been touted so often by politicians and the muzzled media that the public, by now bored by the continual propaganda drivel, came to refer to it as a "multi-*spectacle* camera". Nonetheless, it did produce pleasing results and gave detailed imagery in support of agriculture and mining applications. Nor was this the end of Jähn's involvement in the remote sensing of the Home Planet, for in 1983 he would earn a doctorate from the Central Institute for Earth Physics in Potsdam. Other experiments conducted during the Soyuz 31 mission were an audio study to examine sound and noise perception limits, the 'Berolina' investigation, which used Splav-1 to process an ampoule of bismuth and antimonide, and a demonstration of the relative performance of various photographic films.

Notably, this was the first Intercosmos mission to perform a 'Soyuz swap', with Bykovsky and Jähn transferring their custom-moulded seat liners from Soyuz 31 over to Soyuz 29 for their return to Earth. Additionally, more than two dozen containers of experimental data, including a hundred sets of data results and film, were exchanged. Soyuz 29 undocked from Salyut 6 on the morning of 3 September and touched down at 2:40 pm Moscow Time. Years later, it would become clear that Jähn suffered some form of back injury during the rather rocky landing and there is suspicion that this was deliberately hidden from the public. It would seem that

An impressive view of Salyut 6, acquired during the relocation of Soyuz 31 from the rear to the front port of the space station.

whatever happened had little impact on Bykovsky, however, for he would later take the backup command of another Intercosmos mission, which flew in the summer of 1980 as Soyuz 37, before he retired.

Kovalyonok and Ivanchenkov were now alone for the final two months of their own expedition. On 7 September, they boarded Soyuz 31, undocked from Salyut's aft port and after a short period in free flight, redocked with the forward port, thereby opening the way for the arrival of another Progress freighter in October. This 'Soyuz transfer' was the first such exercise and by necessity left the station temporarily unoccupied. With Kovalyonok at the controls, the Soyuz backed away to a distance of a couple of hundred metres, whereupon ground specialists commanded the station to rotate, laterally, by 180 degrees, for a 'half-somersault' which placed the *forward port* directly ahead of them. "The manoeuvre," explained Phillip Clark, "was undertaken during the Salyut landing window, presumably to allow a return to Earth under the nominal flight conditions should the redocking have failed for any reason." Against the eventuality of their having to return home, they had to place the station into a 'hibernation' state prior to undocking and then, upon *redocking*, restore it as if they had just arrived. All of this placed a considerable overhead on the mission.

Two weeks later, on 20 September, amidst a plethora of medical checks, heart measurements and circulatory studies, Kovalyonok and Ivanchenkov quietly broke the 96-day record set by Yuri Romanenko and Georgi Grechko in March ... and prompted a satirical cartoon from David Austin, published in *New Scientist* on the 28th. It is a pity, as Clark later wrote, that these Salyut 6 missions received such little publicity; not just because they were planned and performed behind the secrecy of the Iron Curtain, but also because they were settling into a routine. In the eyes of the

Western media, it was a *boring* routine. Nothing of the calibre of the Moon landings or ASTP was happening, nor was there any chance of picking up juicy tales of on-board mutinies, like Skylab. As this latest record was set, David Austin summed up popular opinion through his cartoon. In it, a pair of men from NASA stand side by side, one holding a newspaper, emblazoned with the headline that the Soviets have just set a new record. One man says to the other: *"Another long, boring shuffle forward for mankind!"*

Boring, maybe, but necessary, nonetheless. For when humans finally journey to Mars, they will do so – as René Descartes once said – by standing on the shoulders of giants. In the final months of 1978 and for the first half of the following year, in terms of endurance, Vladimir Kovalyonok and Alexander Ivanchenkov *were* those giants. A final Progress visit in October brought fresh supplies and propellant for Salyut's tanks and Western observers were already predicting, based on landing window data, that the two men would be back on Earth sometime between Halloween and the middle of November. Their touchdown at 2:05 pm Moscow Time on 2 November concluded a record-breaking mission of more than 139 days ... an increase of just over 40 percent beyond the achievement of the Soyuz 26 crew. The cosmonauts were in good shape – as good, in fact, as their predecessors, Romanenko and Grechko – and within a few days they were able to walk unaided. Medical specialists attributed their remarkable condition to a strict exercise regime, whose intensity had picked up in the final weeks of the mission, and filled the Soviets with confidence that the *next* crew, consisting of Vladimir Lyakhov and Valeri Ryumin, scheduled to launch in the spring of 1979, would push the envelope still further.

HITTING THE STRIDE

Within a year, the Soviets had twice smashed the United States' endurance record in space, extending the time that humans spent off the planet, firstly to three months and then to an impressive four and a half months. What is most surprising is that *both* of these achievements were themselves eclipsed just as rapidly and by the end of 1980 *two more records* would have been established ... *both* involving the *same cosmonaut*, but with only *one* officially recognised by the FAI. For Valeri Ryumin, 1977 had ended with a mixture of disappointment and optimism: disappointment because he and Vladimir Kovalyonok had failed to dock with Salyut 6, thereby losing their chance to be the crew which beat the American record, but optimism in the sense that both men remained in the pipeline for future flights. When Kovalyonok blasted off for his own flight in June 1978, Ryumin and Vladimir Lyakhov formed the backup crew for Soyuz 29. Ryumin knew that his own chance was not too far away, but he could not have suspected that he was in line for not one long-duration mission, but *two*, and in October 1980 would set a new empirical record as the most experienced spacefarer in the world, having spent a grand total of almost a full year in orbit. This was a personal record that he would hold for more than half a decade. Many Western observers would credit the Soviets for their restraint in this period: for there was little of the dreary political crowing that had greeted each new space spectacular a decade earlier. To be fair, however, there didn't *need* to be. Kenneth Gatland of the British Interplanetary Society summed up the general thinking. "Once the record is broken," he noted, "the US has little chance of regaining it for many years, because there is no ongoing space station programme." Competition, at least for a while, was dead. The Shuttle was suffering development problems and was several years away from making its first flight. The Soviets *knew* that they held one key advantage: *human spaceflight was theirs* to dominate.

When Soyuz 32 was launched at 2:53 pm Moscow Time on the gloomy afternoon of 25 February 1979, many in the West confidently anticipated a flight of around six months, exceeding the record set by Kovalyonok and Ivanchenkov by 25 percent. In keeping with now-standard protocol, the rookie commander, Vladimir Afanaseyevich Lyakhov, a lieutenant-colonel in the Soviet Air Force, was accompanied by veteran flight engineer Ryumin. Born on 20 July 1941 in Antratsyt, in what is today south-eastern Ukraine, Lyakhov entered the military pilots' school at Chuguyev in his early twenties and was selected as a cosmonaut candidate in May 1967. Only months into his two-year training programme, he was one of a handful of cosmonauts detailed to the Spiral orbital spaceplane project and after this was cancelled, he was transferred to Salyut. At the end of 1977, he was assigned to the EP training group, for principal expeditions, and anticipated assignment to command the third such crew.

The mission got off to an exceptional start with a picture-perfect launch and a safe arrival at Salyut 6 late on the afternoon of 26 February. 'De-mothballing' the station took a few days and the interior – which, of course, had been uninhabited since the previous November – exhibited an unusual atmosphere when they first came aboard; a little like burned steel, which Ryumin could only describe as the unique odour of space itself. Medical checks filled their first few weeks – with experiments to check their mood, their accomplishments, their food intake, their psychological condition and their cardiovascular systems – and the two men typically exercised for at least a couple of hours per day and consumed 3,100 calories of food and 2.5 litres of water. On 14 March, Progress 5 arrived and Lyakhov and Ryumin set to work unloading its cargo, which included repair tools, a storage battery, a television monitor, a Kristall furnace to replace the old one, which had broken, and a gamma ray telescope. It also carried more than a thousand kilograms of propellant for the station's tanks.

Yet the following days and weeks would bring disappointment. The first hint of trouble came on 16 March, when it was discovered that the fuel tank of the station's propulsion system was leaking into the pressurising nitrogen bellows and it was quickly recognised that critical valves and regulators might be placed at risk. In fact, the existence of a problem had been noted by their predecessors. "At the end of their work aboard the station," the Soviet press reported, "cosmonauts Kovalyonok and Ivanchenkov noted some deviation in the control parameters in the main pneumatic line of the engine installation's supercharge system." Although these deviations did not affect the system's overall functionality, a subsequent analysis revealed that the problem was most likely "damage to the mobile membrane dividing liquid fuel and gaseous nitrogen in one of the three fuel tanks." Following the specified procedures, the crew isolated the tank and switched to the reserve. They then set the station spinning about its transverse axis, such that the centrifugal force would separate the gas from the liquid, and by the 23rd they were able to drain the leaking tank by transferring it into empty tanks aboard Progress 5. At around the same time, the cosmonauts tried out the station's shower for the first time, wearing scuba-type masks to keep water from irritating their eyes and applying shampoo carried aloft aboard Progress 5.

On 24 March, they installed a television monitor to permit a two-way link with ground controllers ... and family members, an essential psychological crutch for long-duration missions. "Setting aside the importance that this would have in the exchange of information between Salyut and the controllers," Phillip Clark noted, "it was important from a *morale* point of view, since the cosmonauts were able to regularly *see* their families and friends, rather than simply hear[ing] disembodied voices." With the new load of propellant having been successfully transferred into Salyut's tanks, a re-boost of their orbit was effected by the engines of both Progress 5 *and* Soyuz 32, in readiness for the expected launch of a new Intercosmos mission, just before the next landing window, in mid-April.

Soyuz 33 headed into orbit in 40 km/h surface winds (the highest yet recorded for a Soviet launch) from a darkened Tyuratam at 8:34 pm Moscow Time on the 10th of that month, with a standard eight-day voyage ahead of it. In command was veteran cosmonaut Nikolai Rukavishnikov and his crewmate was Georgi Ivanov, a major in the Bulgarian People's Air Force. Ivanov had been born in the town of Lovech, in north-central Bulgaria, on 2 July 1940 ... although, curiously, his birth name was 'Kakalov'. With a delightfully respectful use of euphemism, Mark Wade says on his website that this surname "had a *scatological meaning* in Russian" – it has been suggested elsewhere, in a slightly more liberal vernacular, that Kakalov sounded something like 'asshole' – and it was supposedly Leonid Brezhnev himself who suggested a change. 'Ivanov', a well-known Russian patronym, was duly picked. Humour aside, it is perhaps yet another indication of the intense propaganda importance placed on these Intercosmos ventures, for they were a central tenet of Communist pride: and the thought of the names of this newest team of trailblazing heroes – Mr Rukavishnikov and Major Asshole – splashed across the international press simply would not do ...

Ivanov (or Kakalov) followed an appropriately socialist upbringing and education, completing five years of military school in Dolna Mitropolia, near the city of Pleven, before entering the Bulgarian National Army as a fighter pilot. He rose to become an instructor and head of a division and when the call arose for Intercosmos candidates, his name was put forward. It is interesting that the People's Republic of Bulgaria, as it was at the time, had been one of the first socialist states to propose undertaking a manned space mission with the Soviet Union. In August 1964, Russia's minister of defence organised a meeting in Moscow with the Bulgarian military attaché, Zakhari Zakhariev, to discuss such matters. Realistically, it was far too early in the Space Age for such plans to reach fruition, but a decade later, with the establishment of Soyuz and Salyut technology, the time seemed more ripe. When the call for candidates went out in late 1977, it was mandated that all applicants had to be Bulgarian People's Air Force fighter pilots and *had* to be graduates of the Georgi Benkovski Higher People's Air Force School and *must* have completed their studies at some point between 1964 and 1972. The rationale for this requirement was that in 1964 the school had begun to issue science degrees and therefore the first Bulgarian cosmonaut would be a suitably educated scientist. On top of this, the candidate needed at least three years of regular flying experience. An aeromedical commission evaluated several hundred candidates and the most suitable finalists were sent to Sofia for detailed medical examinations. In rank, they varied from squadron leader to the executive officer of an air regiment and eventually fifteen were winnowed down to just six: Ivanov, Alexander Alexandrov, Georgi Yovchev, Ivan Nakov, Clavdar Dzhourov and Kiril Radev. Of these, the latter pair resigned their places, bringing the number down to four when they finally flew to Moscow for training. Yovchev turned out to have a heart problem, which eliminated him from consideration, and the Soviet doctors finally settled on Ivanov as the prime crewman, backed up by Alexandrov. Uniquely, as circumstances would turn out, *both* men would fly into space. The experiments to be conducted on the first mission, according to a booklet released by Bulgaria at the United Nations on 10 April 1979, included physiology and psychology, hygiene, radiobiology and radiation protection.

'Indestructible for eternity' was how a 1969 Soviet stamp presented the friendship between the Bulgarian and Russian nations and, indeed, the People's Republic had been a staunch ally since 1946, when its monarchy was deposed by the Communist Fatherland Front. By the mid-1950s, Todor Zhivkov had assumed absolute power and would control the country for more than three decades; under his regime, Bulgaria would remain a loyal vassal of the Soviet Union – even participating in the invasion of Czechoslovakia in August 1968 and distancing itself diplomatically from China – but he would pursue relatively moderate policies at home, permitting some freedom of expression, ending the persecution of the Church and renewing friendly relations with Yugoslavia and Greece. Despite operating a planned economy, he decreed that surplus produce could be freely sold and in 1965 Bulgaria was the first Communist nation to obtain a Coca Cola licence. Having said this, Zhivkov was intolerant of dissent and only months before Rukavishnikov and Ivanov's mission, the Bulgarian émigré Georgi Markov was assassinated in London by Communist agents. His death – the notorious 'Umbrella Murder', in which a modified umbrella was used to inject a ricin-filled pellet into his leg – cast the People's Republic in a decidedly more unpleasant light and by the early 1980s, the Zhivkov regime was seen as corrupt, autocratic and increasingly erratic in its actions. In 1984, for example, an attempt was made to forcibly assimilate a Turkish minority by forbidding them to speak their own language and adopt Bulgarian names. In a sense, the People's Republic remained a die-hard socialist bastion right to the end and Zhivkov even (barely) survived Mikhail Gorbachev's reforms of the Soviet Union, being finally pushed aside on the basis of age and infirmity in November 1989 after mass reformist demonstrations throughout Bulgaria.

By the time Zhivkov was removed from power and the Communists have given up their stranglehold on the country, not one, but *two*, Bulgarian cosmonauts would have flown into space. The second mission will be left to the next volume in this series. In command of the first Bulgarian mission was the sour-faced Nikolai Rukavishnikov, who had served aboard two earlier flights ... and was the first civilian cosmonaut ever to lead a mission. Having held the flight engineer's mantle aboard Soyuz 10 and 16 and served as backup commander for the Soviet-Czech Soyuz 28, he was one of only three civilian cosmonauts – the others were Valeri Kubasov and Oleg Makarov – to be considered and trained for a command position. Both Rukavishnikov and Kubasov would go on to fly their commands. Makarov was a prospective backup commander of the joint Soviet-East German venture, Soyuz 31, until the new crewing policy of late 1977 denied him the chance to lead his own mission. When he next journeyed into space, it was as a flight engineer under the command of Vladimir Dzhanibekov.

The early phases of Rukavishnikov and Ivanov's mission seemed to go well and on the evening of 10 April 1979, the crew of Salyut 6 spotted Soyuz 33 approaching like a steadily brightening star. On their 18th circuit of Earth, the new arrivals were given a go-ahead to proceed with the final stages of the rendezvous. Approaching to within 3 km of the station, Rukavishnikov reported that all systems aboard his craft were running normally ... but then things began to go wrong. Two days later, the Soviet news agency, Tass, would say that the occurrence of unusual "deviations from the regular mode of operation of the approach-correcting propulsion unit of the Soyuz 33 spacecraft" led to the premature termination of the mission. Rukavishnikov instinctively *knew* that there was something not quite right with his ship's engine – it fired for just three seconds of a planned six, and the burn could only be considered erratic, at best – and both he and Ivanov felt abnormal vibrations which shook them in their couches and rattled the Soyuz itself. Indeed, the vibrations were so violent at one point that Rukavishnikov had to physically *hold* onto the instrument panel. A second engine firing was also fruitless. Lyakhov and Ryumin watched the proceedings with mounting distress from aboard the space station; the latter, indeed, would later write in his diary that he saw the engine flicker to life, after which its exhaust momentarily changed colour and it shut down.

Saddled with the burden of command, Rukavishnikov could do little to diagnose the problem with the engine, for his instrument panel offered only functional displays and no data to investigate the cause. In fact, according to Rex Hall and Dave Shayler, it was flight controllers on the ground who managed the ignition and shutdown of the engine and the cosmonauts' responsibility was merely to monitor the operation with a stopwatch and report its progress. The commander implored the ground to authorise a manual attempt, but his pleas were in vain. If there *was* a serious problem with Soyuz 33's engine, then it might also endanger the Soyuz already docked at Salyut 6. Moreover, an Intercosmos visiting mission, Soyuz 34, was to have been launched on 6 June, crewed by Valeri Kubasov and Hungarian fighter pilot Bertalan Farkas, but *this* would have to be postponed until the problem with the engine could be ascertained and eliminated.

In the meantime, Rukavishnikov and Ivanov had little option but to prepare for an early return to Earth. There *might* have been a remote chance that Salyut 6 itself could have been manoeuvred a thousand metres or so to dock with them, but in actuality the two vehicles were already drifting apart at 28 m every second and *time*

Nikolai Rukavishnikov (left) and Georgi Ivanov are pictured during winter survival training before their ill-fated flight.

would be needed to compute any new manoeuvres. Alas, time was in short supply. Soyuz 33's backup engine, intended only for retrofire, was therefore burned for three and a half minutes – 25 seconds longer than normal – and then *manually* shut down by Rukavishnikov to begin a steep, ballistic descent through the atmosphere. The fall homeward lasted almost nine minutes, but subjected the cosmonauts to deceleration loads of up to *ten times* that of terrestrial gravity and almost *three times higher* than a normal Soyuz re-entry was expected to impose. "Both cosmonauts reported that breathing was difficult for several minutes as the loads pressed against their bodies," related Hall and Shayler, "but despite the difficulty they were able to talk to each other." The view outside, as superheated plasma surrounded the descent module, was described by Rukavishnikov as like flying through the flame of a blowtorch. Owing to the fact that this was an emergency situation, the precise landing point was expected to be somewhere in the far-western part of the scheduled touchdown zone, some 320 km south-east of Dzhezkazgan – today's Jezkazgan, in the Karagandy province of central Kazakhstan.

Two bitterly disappointed, but doubtless relieved cosmonauts scrambled out of the capsule after a mission which had lasted just under an hour shy of two full days. In fact, Rukavishnikov would recall several years later that it had seemed like a month, not least when the startling realisation hit them that their craft's primary *engine* had a major problem, which might leave them stranded in orbit. The backup engine could have been similarly impaired. On the night of 11 April, despite being told to sleep, Rukavishnikov had been unable to rest, wondering if both the primary and backup engines were damaged. His mind drifted to the Gregory Peck movie, *Marooned*, which told the tale of a stranded team of astronauts: if the backup engine did not fire the following morning, he and Ivanov had only five days of oxygen and supplies to sustain them ... but their spacecraft's orbit would not naturally decay for at least *ten* days. The 213-second firing of Soyuz 33's backup engine must have seemed like an eternity, for no one knew if it might suffer the same fate as its primary counterpart. If the firing stopped short of 90 seconds or so, there was a very real risk that the crew would be unable to return home. At one point, Rukavishnikov turned to a nervous Ivanov. "Stare at *this speck of dust*," he said, pointing to something floating in the cabin. "If it goes *down*, we *live!*" Thankfully, as the spacecraft began to experience atmospheric drag as it descended into the sensible atmosphere, the dust, and other free-floating objects in the cabin, did indeed succumb to the welcome tug of gravity.

According to the memoirs of flight director Alexei Yeliseyev, a former cosmonaut, Soyuz 33's engine provided insufficient thrust. Other Western observers have speculated over the years that perhaps its automatic systems tried to compensate by trying to fire for longer than planned. It was the comments of the Salyut 6 crew, Vladimir Lyakhov and Valeri Ryumin, which aided the investigation-board's efforts. "It appeared that a pressure sensor in the rocket's combustion chamber had terminated the burn when it detected 'off-nominal' performance," wrote Hall and Shayler, "and this in turn prevented fuel from being pumped into an engine that was not performing correctly." The board concluded within a month that a gas generator, responsible for feeding the primary engine's turbopump, had failed, despite an impressive history of more than 8,000 tests and 2,000 actual firings since 1967. Repairs and modifications were duly implemented, but there remained a problem, since not only was it entirely possible that Lyakhov and Ryumin's Soyuz 32 might have a similar flaw, but it was clear that *their* craft would reach the end of its 90-day operational lifetime in May. *Another* Soyuz had to be launched ... and *quickly*.

Original plans called for Rukavishnikov and Ivanov to return home in Soyuz 32, leaving their 'fresh' craft for the long-duration crew. After that, the *next* Intercosmos team of Valeri Kubasov and Bertalan Farkas, was not expected to arrive in Soyuz 34 until the second week of June. Speculation was rife in the Western press about what the Soviets planned for their next move. On 28 April, *Flight International* speculated on the dilemma. "It is thought," the magazine explained, "that the Soviet Union has already begun preparations for another Soyuz launch, but the choice of crew poses a political dilemma. One possibility is the backup crew to Soyuz 33, Yuri Romanenko and Alexander Alexandrov. However, *this* would mean that Bulgaria will be able to claim *two* men in space ... to the chagrin of other East European countries. This makes it more likely that the *original* crew, Rukavishnikov and Ivanov, will make a repeat attempt."

The correct answer? *None of the above.*

For reasons which remain unclear to this day, Rukavishnikov and Ivanov were never given another chance to complete the mission which they had been denied. One possibility is that Rukavishnikov was reassigned, later in 1979, as the backup flight engineer for the upcoming Soyuz T-3 mission. *Another* joint Soviet-Bulgarian space voyage *would* take place, but not until the summer of 1988, and it would involve not Ivanov, but his backup, Alexandrov. (In the meantime, Yuri Romanenko, the backup commander for Soyuz 33, would go on to lead a joint Soviet-Cuban mission in the autumn of 1980.) Nor would the Soviet-Hungarian crew of Kubasov and Farkas fly in the summer of 1979. When Soyuz 34 *did* blast off from Tyuratam on the evening of 6 June, it did so in an unmanned capacity, laden with a cargo of biological samples for Lyakhov and Ryumin's programme of experiments. The improvements implemented in the wake of the Soyuz 33 failure clearly worked and the new ship arrived without incident. The crew transferred their specially-contoured couches from the old craft to the new one and on 13 June Soyuz 32 was loaded with 180 kg of photographic film, Splav-1 and Kristall samples and other data and automatically undocked. It made the de-orbit burn without incident and landed shortly thereafter on Soviet territory after a flight of 108 days – nearly a *month* longer than its maximum design life. Meanwhile, Lyakhov and Ryumin undocked Soyuz 34 from the station's rear port and redocked it at the forward port, thereby opening the way for future Progress propellant deliveries.

For a time after the disappointment of Rukavishnikov and Ivanov's mission, little could raise the spirits of the long-duration crew. Lyakhov and Ryumin had responded with grunts to ground controllers' announcement that Soyuz 33 had been terminated and apparently the cosmonauts tersely ended their communications session and went to bed, acutely distressed. It must be borne in mind, of course, that they had been in space, physically alone and isolated, for nearly seven weeks and were clearly expecting and excited by the prospect of receiving other human company. The arrival of the Progress 6 freighter in mid-May went some way to lifting their spirits, but by this time it was evident that their mission would run until August and *neither* of the planned Intercosmos visits would now occur. They actually spent a full six months alone in orbit; the longest period of isolation ever experienced by a spacefaring crew.

In fact, the only 'visitor' received during the long summer of 1979 was Progress 7, launched on 28 June, which brought more than 1,200 kg of supplies, including food, plants, mail and a folded radio telescope, called the KRT-10. The latter consisted of reflectors, focal container and supports, a package of low-frequency radiometers and a control console. The cosmonauts had not actually *seen* the complete system, since it was still undergoing tests at the time when they trained to assemble it. In fact, Lyakhov and Ryumin required two weeks to prepare the equipment for deployment. When the freighter undocked a few weeks later, Phillip Clark wrote, "it was not a routine operation". The telescope, which measured 0.5 m across in its folded configuration, had apparently been attached to the docking tunnel at the rear of the station. "The outer hatch had remained open, while the tunnel hatch inside Salyut was closed," Clark explained. "The folded telescope lay along the tunnel and inside

A lonely six months in orbit finally came to an end in August 1979, when Valeri Ryumin (left) and Vladimir Lyakhov touched down on Soviet soil after a record-breaking 175-day mission. The Soyuz descent module can be seen lying on its side on the barren steppe, whilst the two cosmonauts recline in special couches, tended by recovery personnel. Ryumin could not possibly have known that he would be part of the *very next* long-duration crew to Salyut 6 in the spring of 1980 ... much to his family's distress.

the orbital module of Progress. As the freighter pulled away from the station, the now-exposed wire-mesh parabolic dish unfolded and deployed to its full, 10 m diameter." Televised images of this deployment procedure were transmitted to ground controllers from the departing Progress. Many Western analysts inferred that, since the 200 kg KRT-10 was now blocking Salyut 6's rear port, the station was approaching the end of its operational life.

They were wrong. Not only would it appear that the KRT-10 mission was a short one, but there have been persistent rumours over the years that during deployment it somehow became fouled with either the space station's rear docking target or one of the antennas and as a result its usefulness was drastically impaired. "Certainly," wrote Clark, "the data which were returned were of *lower quality* than one would have expected from a dish of the announced diameter." As July wore into August, the widespread expectation that Lyakhov and Ryumin would shortly return to Earth was seemingly vindicated when the men started the lengthy process of 'mothballing' the station's systems for a lengthy spell of automated operations. Then, on 9 August, a problem arose which demanded their immediate attention on an EVA. The KRT-10 scientific programme ended and the telescope should have detached cleanly from the station's aft port. It did not do so, probably due to its entanglement with the Salyut antennas. The requirement for humans in space was now demonstrated in dramatic style ... but the cosmonauts were concerned. They were tired after almost six months in space and hence were not in the best physical condition to undertake a lengthy EVA. It has even been said that they left *letters* to loved ones in the Soyuz 34 descent module, in case they did not survive the excursion.

A sense of foreboding did not seem ill-placed. Whilst preparing for the excursion on the morning of 15 August, the backup fan in Ryumin's space suit failed, because its controller was exposed to high levels of humidity. Late that afternoon, Ryumin – his heart reaching 130 beats per minute – exited the forward transfer compartment, deployed a folded handrail and worked his way with difficulty along the *entire length* of the station to reach the troublesome telescope. The spacewalk started ten minutes before sunset and Ryumin spent more than half an hour in almost pitch blackness, as Lyakhov steadily paid out the umbilical. Upon inspection, it became apparent that the KRT-10's antenna ribs had torn the station's insulation material and as Ryumin started to cut it away, the huge dish oscillated backwards and forwards, almost hitting him on several occasions. At one point, his heart rate shot up to 146 beats per minute. After freeing the telescope by severing four steel cables, he then used a 1.5 m barbed pole to push it clear. Ryumin rejoined Lyakhov at the transfer compartment and the pair inspected the station's exterior, noting discolouration of the insulation material and collecting micrometeoroid samples, before they retreated inside. The 83-minute excursion was a remarkable success. It also marked the third and final EVA from Salyut 6.

Less than four days later, at 3:30 pm Moscow Time on the 19th, Lyakhov and Ryumin were safely back on Soviet soil, after a triumphant mission of 175 days. Weight proved a problem at first, and the men were unable to even bear the weight of a small bunch of flowers presented to them in celebration – it felt like a giant sheaf of wheat, they said – but within a week their strength returned. Although Ryumin maintained his pre-launch weight, Lyakhov had lost around 5.5 kg and both cosmonauts experienced a 20 percent reduction in lower leg volume. They had eclipsed the record set by Vladimir Kovalyonok and Alexander Ivanchenkov by more than 25 percent. Early plans for Salyut 6 had envisaged three principal expeditions, all of which had now been satisfactorily completed. However, the station was in an exceptionally good condition and it did not take long for the Soviets to redouble their efforts to reoccupy it. In fact, before the end of its life in 1982, Salyut 6 would host another *two* long-duration crews and no fewer than *five* Intercosmos teams and several flights of the updated Soyuz-T spacecraft. Fittingly, the long-duration crews included amongst their number the very same cosmonauts who – but for a quirk of fate and a foiled docking attempt – may have broken the Skylab endurance record with their Soyuz 25 mission: Vladimir Kovalyonok and Valeri Ryumin.

It might seem surprising that Ryumin was granted a second long-duration mission, and so soon, to Salyut 6, particularly as he had only recently returned from his record-breaking 175-day flight. Indeed, no one was more surprised than Ryumin himself ... for the assignment came utterly out of the blue. Before leaving the station with Lyakhov, he had left a note of congratulation to the two men who would arrive next: rookie Soyuz 35 commander Leonid Popov and veteran flight engineer Valentin Lebedev, whose own six-month mission was scheduled to begin in April 1980. Little could he have known that he would open and read his *own* letter, deadpanning that he was *not* in the habit of writing letters to himself . . .

"OPEN-MOUTHED"

As already explained, détente between the Soviet Union and the United States veered sharply off course in the late 1970s, but this did not prevent Leonid Brezhnev and Jimmy Carter from adding their signatures to the agreements of a second round of Strategic Arms Limitation Talks in Vienna in June 1979. Six months later, though, in an event which would leave Carter "open-mouthed", Soviet forces swept across the border into Afghanistan and landed in the capital, Kabul, on Christmas Day. Quickly, they took government and media buildings and assaulted the Tajbeg Palace, where they killed the unpopular president, Hafizullah Amin. By February 1980, more than 100,000 Soviet troops had occupied Afghanistan, bringing with them thousands of tanks and amphibious armoured vehicles. However, far from creating stability and imposing a Soviet-style socialist state on a population which was ruled by tradition and religion, the invasion actually kicked off a vicious, decade-long war against a well-armed (and US-supplied) mujahideen insurgency.

The roots of the conflict can be traced to July 1973, when a former Afghan prime minister, Mohammad Daoud Khan, seized power in a military coup, having accused the autocratic government of King Mohammad Zahir Shah of corruption and blaming it for poor economic conditions. Daoud's regime proved extremely unpopular and he was executed in April 1978, at which point a pro-Soviet politician, Nur Muhammad Taraki, assumed control of the new 'Democratic Republic of Afghanistan'. His effort to implement socialist reforms, including changing marriage customs and land rights, were received with anger by many Afghans, whose traditions were deeply rooted in Islam. By August 1979, civil strife had spread throughout the country and Taraki was brutally suppressing opposition and relying upon Soviet military support to prop up his government. At around the same time, in the United States, Jimmy Carter authorised the CIA to begin covert propaganda operations against Taraki and made available funds for anti-Communist guerrilla factions in Afghanistan. A few weeks later, Deputy Prime Minister Hafizullah Amin seized power after a palace shoot-out in which Taraki was murdered. He remained in control for barely two months, for the Soviets already knew that he had participated in secret meetings with the Americans and there was suspicion that Amin was in the employ of the CIA. His removal was essential to the 'stabilisation' of the region. Yet the death of Amin and its aftermath did nothing to smooth the way towards a new socialist state, centred on Kabul. Three dozen Islamic countries demanded the immediate withdrawal of Soviet troops and Jimmy Carter prohibited American athletes from participating in the 1980 Summer Olympics in Moscow. His boycott was endorsed by other nations, including China, Japan and West Germany.

Interestingly, the only Warsaw Pact country to condemn the Soviet invasion was Romania ... which would send its first cosmonaut into space aboard a Soyuz craft a little more than a year later.

By the time the final Soviet troops withdrew from Afghanistan in February 1989, their military superiority had fared badly against the well-armed guerrilla insurgents. They had *not* enforced socialism on the mountainous, land-locked country. A decade of bitter fighting had actually inspired *nationalistic* feelings in the mujahideen ... but not peace. By then, more than a million men, women and children were dead and an astonishing *half* of the world's refugees were Afghans. Irrigation systems, crucial to the country's agriculture, were in ruins as a result of Soviet or government bombings, populations in the cities were decimated and the Geneva Accords, signed in May 1988, left post-war Afghanistan in ruins, with no viable support for future governance. It was perhaps more than a little ironic that the first Afghan cosmonaut rode a Soyuz to the Mir space station a few months later, but would gain few of the honours that the other Intercosmos fliers received upon their return. Vicious inter-tribal warfare in the early 1990s eventually forced him to leave his homeland and he became another tragic statistic on the world's list of Afghan refugees. Today, he lives in Germany.

Meanwhile, in the spring of 1980, weeks after the invasion, another attack was being planned: a new attack on the heavens, with a new crew slated to occupy the Salyut 6 orbital station. The two men destined to fly for perhaps six months or more were Leonid Ivanovich Popov and Valentin Vitalyevich Lebedev. The former came from Oleksandria in the central portion of today's Ukraine, where he was born on 31 August 1945. Popov graduated from the Chernigov Higher Air Force School with an electrical engineering degree in 1968 and was selected as one of nine cosmonauts in March 1970. After obtaining further credentials from the Gagarin Military Academy, he served on the support crew for Valeri Bykovsky's Soyuz 22 mission, before being paired with Valentin Lebedev on the backup team for Soyuz 32. In a sense, career-wise, Popov can be compared quite favourably to fellow cosmonaut Pyotr Klimuk, since both flew on three occasions in relatively quick succession and at a relatively young age. Indeed, Popov would have completed each of his missions, and spent a grand total of some 200 days in space, before his 37th birthday.

Joining Popov for the long-duration Soyuz 35 mission, which was expected to host at least one visiting Intercosmos crew, was Valentin Lebedev, who also entered the cosmonaut corps at a young age and rapidly rose from a wet-behind-the-ears selectee to a hardened veteran spacefarer. In fact, he had been picked as a civilian engineer in March 1972 ... and flew his first mission, Soyuz 13, barely 21 months later! Born on 14 April 1942 in Moscow, Lebedev completed high school in Naro-Fominsk, then studied for a year at the Higher Air Force Navigators School, near Orenburg in the Volga District. Unfortunately, due to a reduction in the numbers of personnel in the Soviet armed forces, he was soon discharged. He switched to the Moscow Aviation Institute, from which he graduated in 1966. After working for a time as an aircraft designer, specialising in structures and materials, he joined the cosmonaut team a few weeks before his 30th birthday. However, his involvement in space-related matters had actually begun a few years earlier, when he served aboard

Valeri Ryumin (left) and Leonid Popov display their Sokol pressure suits whilst aboard Salyut 6.

an expedition of the Eighth Naval Squadron to the Indian Ocean and was later based in Bombay (today's Mumbai) to support rescue operations for two unmanned Zond missions. Following his eight-day Soyuz 13 mission, Lebedev began Salyut training and was expected to participate in a six-month mission with Popov. Then, just a few weeks before launch, he injured his knee in a trampoline accident. Devastated, he was pulled from the crew.

This placed the Soviets in an awkward position with regard to the policy, which required that each mission include at least one veteran cosmonaut to be aboard. The backup crew for Soyuz 35 consisted of a veteran commander, Vyacheslav Zudov, but the flight engineer, Boris Andreyev, was a rookie. Writing in 1988, Phillip Clark also hinted that cosmonauts Valeri Illiarianov, Vladimir Titov and Gennadi Strekalov provided support for the mission ... but *all three* were rookies at the time and therefore none were qualified to fill Lebedev's veteran shoes. The decision to draft Valeri Ryumin into the mix seems to have been based on two overarching factors: firstly, he was an experienced Salyut crewman, and secondly, he was one of the few cosmonauts *not* training at the time on the new Soyuz-T spacecraft. This made him an ideal partner for rookie Popov.

If Jimmy Carter had been left open-mouthed by the Soviet invasion of Afghanistan, then certainly Ryumin's family were left in a similar position when they learned the news that he was heading into orbit *again* – for *another* six-month flight – less than a year after returning to Earth. Judging from reports in the Western press over the years, it would appear that his family were far from happy with the decision. Nonetheless, he and Popov were launched from Tyuratam at 4:38 pm Moscow Time on 9 April 1980 and docked at Salyut 6's forward port the following evening. On entering the station, Ryumin quickly opened the letter that he had written and read its contents with a wry smile. There had been a few changes since he was last here; the two windows in the forward transfer compartment had lost their transparency and were chipped as a result of impacting micrometeoroids and orbital debris. It had been 244 days since Ryumin left Salyut 6, but the station had not been ignored. A Progress freighter had arrived a few weeks before Soyuz 35 and the cosmonauts spent several days unpacking its cargo: replacement parts for the attitude-control and life support systems, a new caution and warning device, a replacement battery and other supplies. However, one visitor which Popov and Ryumin did *not* witness was the unmanned maiden voyage of a new, upgraded version of the manned Soyuz craft. Known as 'Soyuz-T', it had been launched on 16 December 1979 and spent more than three months at the station, successfully using its own propulsion system to re-boost the orbit on one occasion. It would appear that this inaugural flight was a test of the craft's systems, in advance of a manned mission in the summer of 1980. Fundamentally, as shall be seen later in this chapter, the Soyuz-T was capable of supporting crews of up to *three* cosmonauts in pressure suits.

Another Progress craft arrived at the end of April, bringing with it more than two thousand kilograms of additional supplies and equipment and – for the first time – *water* was transferred into the space station's storage tanks. Only weeks later, with a new landing window looming from 31 May until 11 June, an Intercosmos mission was anticipated. Soyuz 36 duly blasted off at 9:21 pm Moscow Time on 26 May. Not surprisingly, the crew was the Soviet-Hungarian team of Valeri Kubasov and Bertalan Farkas, who would have flown Soyuz 34 the previous summer, had the misadventure of Rukavishnikov and Ivanov not put their mission on hold. Interestingly, when the Press Office at the Hungarian Embassy in London issued the cosmonauts' biographies in May 1980, they mistakenly failed to adjust the dates … which *still* stipulated 'June 1979'. This confirmed what most Western observers already knew: that the Hungarian mission *had* originally been scheduled to occur during the Soyuz 32 residency and had been delayed by almost a full year.

The man who became Hungary's first spacefarer, Bertalan Farkas – the surname means *wolf* in the Magyar tongue – came from Gyuláháza in the Northern Great Plain region, in the far east of the country, where he was born on 2 August 1949. Today, he is perhaps less well-known as the first spacefarer to be fluent in Esperanto, the most widely-used international auxiliary language in the world. "When I was a child," he recalled to an interviewer in 1998, "I had no thought of being an astronaut. I seem to remember I wanted to be a footballer; I was a big fan of Ferencváros. My parents considered football too dangerous so I chose a less risky career and decided to be a pilot!" After graduating from the George Kilián Aeronautical College in Szolnok in 1969, he entered the Krasnodar Military Aviation Institute in the Soviet Union and completed pilot training in 1972. By the time that he applied as a candidate for the Intercosmos group in March 1978, he had logged more than a thousand hours as a fighter pilot in the Hungarian Air Force.

Hungary had gradually entered the Soviet orbit in the wake of the Second World War and in the following decade it suffered one of the worst periods of persecution of its *intelligentsia* and bourgeois classes, with thousands executed, tens of thousands imprisoned and *hundreds* of thousands 'purged', often violently, from the country. Declining living standards and the iron-fisted Stalinist government of Mátyás Rákosi prompted the Hungarian Uprising in October 1956, which was eventually crushed by 150,000 Soviet troops and 2,000 tanks. Twenty thousand Hungarians were left dead in the calamity. In the aftermath of the uprising, János Kádár rose to power, currying Soviet favour by denouncing the 'revolutionaries' and the 'mavericks' ... yet by the end of the 1960s, his form of Communism had introduced a more liberal economy, higher standards of living and less restrictions on the press. It came to be nicknamed 'Goulash Communism', since, just like the contents of the traditional Hungarian dish, it comprised a mixture of dissimilar ideological ingredients. Given the conditions in which the peoples of the *other* Intercosmos nations – Czechoslovakia, Poland, East Germany, Bulgaria – existed, in the Soviet era, Hungary was perhaps the most liberal corner of the Eastern Bloc.

During their week aboard Salyut 6, the experiment load for Kubasov and Farkas was not only a large one ... but a heavy one on their time. In fact, it has been said in the Western space literature that they averaged only three hours of sleep each night! A sizeable portion of their time involved photography of geomorphological objects, geological formations, oceans, wave motions and plankton and meteorological phenomena. The second theme was life sciences, with Farkas spending many hours monitoring changes to his physiological state, his protein metabolism, the calcium and potassium content of his hair, the effect of microgravity on the production of interferon in his lymphatic system and the mental loads on his capacity to work. In fact, the Hungarian press later remarked that Farkas had adapted more quickly to weightlessness than did his veteran commander, Kubasov.

Shortly before leaving the station, the visitors used the engines of their craft to boost Salyut's orbit, after which they swapped seat liners, space suits and personal items and undocked aboard the old Soyuz 35 on 3 June. They landed at 6:07 pm Moscow Time, completing a mission of just over a week, although an altimeter glitch caused the solid-fuelled soft-landing rockets to fail, giving them a harsh thump back onto *terra firma*. Fortunately, neither man was injured.

The following day, Popov and Ryumin undocked 'their' new vehicle, Soyuz 36, from the station's aft port and redocked at the front port, leaving the former – and its plumbing for fuel and water transfer – free for subsequent Progress visitors. But the *next* visitor would not be another Progress ... but *another Soyuz*. "This rapid switch of the ferry vehicle," wrote Phillip Clark, "coupled with the Soyuz 36 launch coming at virtually the earliest date to allow a crew recovery in the nominal landing window, raised the question of a second ferry mission being planned to Salyut during the June landing opportunity." Speculation was rife that *another* Intercosmos crew was about to be launched – this time with a cosmonaut from Vietnam, or possibly Cuba – but there was considerable surprise when Soyuz T-2 headed into orbit at 5:19 pm Moscow Time on 5 June.

Physically, the new spacecraft looked much the same as its predecessors, with a spheroidal orbital module, bell-shaped descent module and instrument module, although its capabilities were much improved. It could survive, independently, for up

Valeri Kubasov (left) and Bertalan Farkas exchange their custom-moulded seat liners between the two Soyuz craft, prior to their return to Earth.

to 14 days, *without* the full powering-down of its systems, and could be kept in 'orbital storage' for six full months. Its orbital module could also be left accessible during long-duration missions to provide a few extra cubic metres of storage space, *and* could be jettisoned *prior* to the de-orbit burn, thereby allowing a ten-percent reduction in propellant to around 250 kg at the end of each mission. This allowed for the inclusion of a third (fully-suited) crew member or two cosmonauts and a hundred kilograms of cargo. As for the descent module itself, this component included improved window covers, capable of being jettisoned after re-entry to allow better views of parachute deployment, and the cosmonauts themselves would benefit from a new 'Sokol' ('Falcon') space suit, weighing only eight kilograms, which was lighter and considerably more flexible than earlier models. The escape tower atop the R-7 booster was also improved. It could be jettisoned 123 seconds after liftoff, rather than the previous 160 seconds, and the upgraded solid-fuel rocket meant that in an abort situation the orbital and descent modules would be pulled to a higher altitude, thereby enabling the *main* parachute – rather than the less reliable backup canopy – to deploy and bring the craft safely down to the ground. Six soft-landing rockets in the base of the Soyuz-T descent module replaced four in the previous model and, internally, the 'Chaika' ('Seagull') control systems incorporated a digital computer called 'Argon', cathode ray tube displays and lightweight circuitry. The new instrument module had been designed in a similar manner to that of Progress, with smaller attitude-control thrusters incorporated into the main propulsion system, so that both could draw their nitrogen tetroxide and unsymmetrical dimethyl hydrazine propellants from the same supply. Finally, two wing-like solar arrays spanned 10.6 m (slightly smaller than those of the original Soyuz) and generated a little over half a kilowatt of power.

The group of men who would first fly this new ship had been assembled several years earlier. In January 1974, Soviet Air Force pilots Leonid Kizim, Vladimir

Lyakhov, Yuri Malyshev and Leonid Popov were selected for command positions, teamed with a mixture of military and civilian flight engineers: Vladimir Aksyonov, Anatoli Voronov, Gennadi Strekalov and Mikhail Burdayev. Two years later, Malyshev, Popov, Aksyonov and Strekalov were reassigned to support the scientific Soyuz 22 mission and, shortly thereafter, Voronov and Burdayev were dropped from consideration after failing their exams. Hall and Shayler cite claims that Voronov and Burdayev, who were both Soviet Air Force officers, were *deliberately* failed because the civilian NPO Energia organisation did not want military flight engineers aboard its new craft. We have already seen that Vladimir Lyakhov transferred to training for a long-duration Salyut 6 mission, as did Leonid Popov, and this state of affairs left only Malyshev, Aksyonov, Kizim and Strekalov. Veterans Vasili Lazarev and Oleg Makarov also joined the group. By 1978, this revised group produced three early Soyuz-T pairs: Malyshev-Aksyonov, Kizim-Makarov and Lazarev-Strekalov. Original plans called for a 'solo' manned test flight, but this was dropped in favour of a docking flight to Salyut 6. Soyuz-T was actually under development for the next station, Salyut 7, whose launch was slipped because Salyut 6 was still usable. It is reasonable to assume that original plans for this solo manned test would have seen it fly sometime after the abandonment of one station and before the launch of the next. With Salyut 6 still available, however, it made sense to send the new mission to it.

In command of Soyuz T-2 was Lieutenant-Colonel Yuri Vasilyevich Malyshev, who came from the village of Nikolayevsk, near the city of Volgograd, on the eastern shore of the Volga River, where he was born on 27 August 1941. He completed high school in Taganrog, a port on the Sea of Azov, at the age of 18 and promptly joined the military, graduating from the Higher Air Force Academy at Chuguyev in 1963. Four years later, he was selected – with Burdayev and Lyakhov – as a cosmonaut candidate and was initially assigned to follow the development of the Spiral winged spaceplane. Selection to the Soyuz-T training group in January 1974 was followed in January 1976 by assignment as backup commander of Soyuz 22, a mission discussed earlier in this chapter. In 1979, he was given command of the first manned Soyuz-T, teamed with veteran cosmonaut Vladimir Aksyonov, who had previously served as Valeri Bykovsky's flight engineer on the Soyuz 22 mission. Backing them up would be Kizim and Makarov. The prime crew for the next flight would then be Lazarev, Strekalov and a physician named Valeri Polyakov.

On the evening of 6 June 1980, Malyshev and Aksyonov guided their ship safely to a docking with Salyut 6. Having launched two crews within only *ten days* of each other, the Soviets had broken by one day the record established by America with the Gemini VII and Gemini VI-A missions in December 1965. "The approach" to the space station, wrote Phillip Clark, "was completed automatically, while the final 180 m and the docking itself were accomplished manually." In their summary, Hall and Shayler have explained the important role played by the new Argon computer, which enabled Malyshev to select the best possible approach to the station from a series of options, based on real-time data. They pointed out that, at 180 m, the commander was "unsure" of the chosen approach – a fact documented by his heart rate, which rose to 130 beats per minute – and chose to override the Argon. "Later analysis," Hall and Shayler concluded, "indicated that the computer program had selected a flight path that *would* have achieved a successful docking and that Malyshev was perhaps a little eager or over-cautious."

Having said this, there have been persistent rumours over the years that the Argon did *not* perform well. Some reports have hinted that it left Soyuz T-2 *perpendicular* to the station and in 1988 Clark expressed his belief that it was "probable" that at least a partial failure of the new computer occurred. Also surprising to many in the West was the short duration of the Soyuz T-2 mission, which ran for only four days, three of which were spent aboard Salyut, unloading and then performing biomedical experiments. It has been suggested that the need to carry these experiments into orbit was one factor which dictated only a two-man crew on the first manned flight of a new spacecraft which could support three cosmonauts. Malyshev and Aksyonov undocked on 9 June and touched down at 3:40 pm Moscow Time, several hundred kilometres south-east of Jezkazgan.

For the resident station crew, Popov and Ryumin, their time alone would be short. Another Progress freighter – the tenth in a little over two years – arrived on the first day of July, bringing with it spare equipment and supplies, including intensifiers for the BST-1M telescope, and additional propellant. It departed on 17 July, opening the way for a second Intercosmos crew, which many in the West expected to involve a Cuban cosmonaut. They were disappointed when Soyuz 37 was launched at night at 9:33 pm Moscow Time on 23 July with Viktor Gorbatko and a *Vietnamese* pilot by the name of Pham Tuân. Nor was Pham just *any* Vietnamese fighter pilot: he would be honoured in the Soviet press as being the *only* North Vietnamese aviator to have successfully engaged and shot down a US Air Force B-52 Stratofortress in air-to-air combat. The incident occurred on 26 December 1972, during Operation Linebacker II, a concerted 11-day campaign in which the US Air Force and Navy bombed targets in North Vietnam. According to the official record, *two* B-52s were lost that day, but American military sources assert that the aircraft claimed by Pham was actually downed by a surface-to-air missile. Pham came from Quoc Tuan in North Vietnam's Thái Bình coastal province, a hundred kilometres or so from Hanoi. He was born on 14 February 1947 and joined the North Vietnamese Air Force – properly the 'Vietnam People's Air Force' – at the age of 18, rising through the ranks to become a commissioned combat officer. During the conflict with the United States, he flew MiG-21 fighters.

With the fall of Saigon in April 1975, the US-backed puppet regime collapsed and South Vietnam came under the control of a Provisional Revolutionary Government and in the summer of the following year both halves of the fragmented country were formally merged into a new Socialist Republic, based on fundamental Communist principles, whose central tenets were reasserted in 1992. Soviet collectivisation of farms and factories was implemented, prompting catastrophic economic collapse and triple-digit inflation, together with an agonising slowness in reconstruction after the war and unresolved humanitarian problems. The result was a mass exodus, with millions fleeing the country in makeshift boats. In 1978, Vietnam invaded Cambodia and successfully removed Pol Pot's despotic regime ... but in the process drew itself into conflict with China. The new leader of the People's Republic, Deng Xiaoping, argued that the small country's ethnic Chinese minority were being mistreated and

In a remarkable renewal of diplomatic relations with the People's Republic of China, Chris Kraft (front right) briefs Deng Xiaoping (front centre) and his wife during a visit to the Johnson Space Center in February 1979. Two weeks after this photograph was taken, a quarter of a million Chinese troops swarmed across the border into northern Vietnam. Along with the Soviet invasion of Afghanistan, the United States' resumption of talks with the People's Republic polarised opinions on either side of the Iron Curtain. The Soviets would demonstrate their ideological support of Vietnam by flying cosmonaut Pham Tuân into space, whilst America would counter by arming an anti-Communist mujahideen insurgency in the mountains of Afghanistan. Four years after the glory of Apollo-Soyuz, two former partners had become bitter enemies once more.

that Vietnam was in illegal possession of the Spratly Islands, to which China laid claim. Later that same year, the Soviet Union signed a 25-year mutual defence treaty with Vietnam, prompting an enraged Deng to tell Jimmy Carter in January 1979 that "children who don't listen have to be spanked". Nor was Deng prepared to stop there. In tough language, he warned the Brezhnev regime that China *was* ready for a full-scale war against the Soviet Union. Words were quickly followed by actions, when more than half a million Chinese troops moved into positions along the border with Russia.

Leonid Brezhnev knew that there was no realistic way for him to support Vietnam against China – the distances involved were simply too great – and consequently took steps to restart an already bubbling border conflict in the north of the country. On 17 February 1979, a few days after their 30-year alliance of friendship with the Soviets expired, 200,000 Chinese troops and 400 tanks from the People's Republic swarmed across the border into northern Vietnam. Within three weeks, the short, bloody war was over: the Chinese claimed to have made significant 'punitive' gains, declaring that they had opened the gates to Hanoi, whilst the Vietnamese retorted that the invaders had mostly fought against mere border militias. In a sense, *both* claimed military victory and *both* were able to declare that they had taught their opponent a harsh lesson. In truth, the destructive departure of the Chinese had left much local infrastructure in northern Vietnam in tatters and had paralysed its already fragile economy. It can certainly have been no coincidence that, on 1 April 1979, just three weeks after the Chinese withdrawal, the game of power politics turned to space and the Soviets selected Pham and another fighter pilot named Bùi Thanh Liêm for their Intercosmos programme.

A year later, in the summer of 1980, the *timing* of Soyuz 37's mission was also no coincidence, since the Summer Olympics in Moscow – the first to be held in Eastern Europe – had opened just a few days earlier, on 19 July. In March, three months after the Soviet invasion of Afghanistan, Jimmy Carter had publicly boycotted the Games and his decision had been quickly followed by 65 other nations, leaving only 80 to participate; the fewest in almost three decades. Even the opening ceremonies caused fraught emotions, with more than a dozen countries – including Denmark, France and the United Kingdom – choosing to march under the Olympic flag, rather than national flags. Elsewhere, in Philadelphia, many of the boycotting nations participated in a so-called 'Olympic Boycott Games'. Consequently, the 1980 Summer Olympics did not offer the Soviets the platform from which to crow about their socialist achievements that they had sought.

To be fair, the mission of Soyuz 37 *was* governed by the Salyut landing window, which ran from 1-13 August, but the presence of the Western media in Moscow was simply too tempting to ignore and the Vietnamese mission would demonstrate Soviet co-operation and assistance of a smaller ally against the larger foe, China. On the 19th, Popov and Ryumin sent greetings to the athletes in a live communication link with the opening ceremony in Moscow's Central Lenin Stadium. Their faces appeared on the stadium's scoreboard and their voices were transmitted over loudspeakers. Unfortunately, the impact of having a Vietnamese cosmonaut in space created little of the anticipated impact and many saw it for what it was: a propaganda stunt, pure and simple. In its summing-up of the mission on 2 August, even *Flight International* had little to say, apart from the footnote that Pham was "the first person from a Third World country to fly in space".

In command of Soyuz 37 was Viktor Gorbatko, who was making his third journey into space. Following the unique 'troika' flight in 1969, he led a two-week voyage to the Salyut 5 military space station and went on to serve as backup commander of the joint Soviet-East German Intercosmos mission, paired with Eberhard Köllner. A day after launch, Gorbatko docked successfully at the space station and Pham set to work on his week-long programme of 30 experiments, which included photographing and mapping his homeland, a number of biological experiments – including one with Vietnamese azolla water ferns – and a series of materials science investigations, which used the Kristall furnace to melt a series of siliceous mineral samples. After the now-customary eight days, Gorbako and Pham returned to Earth aboard Soyuz 36, landing at 6:15 pm Moscow Time on 31 July, having left their newer craft behind for Popov and Ryumin.

The following afternoon, the long-duration crew boarded Soyuz 37 and moved it to the front port. Many in the West anticipated *another* manned launch, but this did not happen. Nor, indeed, was another Progress visitation imminent. In fact, Progress 10, launched in July, had apparently carried sufficient supplies to sustain Popov and Ryumin until the scheduled end of their mission in October. The landing window fell between 2-15 October and a final Intercosmos team – the Soyuz 38 crew of veteran Yuri Romanenko and Cuban fighter pilot Arnaldo Tamayo-Méndez – were expected to rise from Earth on about 24 September. This would bring their eight-day voyage back to Earth at the start of the landing window. It was with some surprise, therefore, that Romanenko and Tamayo-Méndez set off a week *earlier*, at 10:11 pm Moscow Time on 18 September.

The early launch posed something of a mystery ... until the Cuban connection was made. "It was required that when an international cosmonaut was on Salyut, it should be possible to see Salyut in the night sky from the country which had supported the guest cosmonaut," wrote Phillip Clark. "For most of the Intercosmos nations this was not a major consideration, since orbital mechanics meant [that] for countries close to the eastern Soviet Union [the station] would be seen at night automatically when an international mission was launched with a landing targeted for the nominal window." In the cases of Vietnam and Cuba, however, a slightly earlier launch was required to enable their peoples to witness history in the making. In the case of Pham's mission, Clark continued, the difference was barely a day and it had gone unremarked by the press, but for Tamayo-Méndez – whose home was in the Western Hemisphere – it amounted to almost a full week.

Although Cuba shared many of the Communist ideals of the Soviet Union, it was in a quite different area of the world and operated within a totally different sphere of influence. Today, in the second decade of the 21st century, it is the only surviving bastion of single-party socialist government in the Western Hemisphere, a little more than a hundred kilometres south-east of the Florida Keys ... and in the 1960s and 1970s was perceived as a constant threat by successive US presidents. For five decades, Fidel Castro and, more recently, his younger brother, Raúl, have controlled the Republic of Cuba with an iron fist and through their 'Revolutionary Government' have implemented a strict, Soviet-style planned economy on the Caribbean island nation. Literacy rates in Cuba are amongst the highest in the world, medical care and higher education are free and life expectancy is higher than in the United States ... but political dissent, to this day, is mercilessly persecuted and the rights of freedom of speech, freedom of association, freedom of movement and the freedom of the press are severely curtailed.

Numerous attempts have been made over the years to topple Castro, but the wily old leader has outlived them all. A flashpoint arose in April 1961, just a couple of years after his revolution, when a group of Cuban exiles, around 1,500 in total, were trained and equipped by the CIA to invade the island and overthrow this new Communist dictatorship on America's doorstep. The attempt failed catastrophically and caused President John Kennedy and his administration profound embarrass-ment. Eighteen months later, in October 1962, reconnaissance photographs clearly showed the construction of ballistic missile bases in Cuba and, since June, Soviet ships had been delivering nuclear-tipped warheads, bombers, MiG fighter jets and mechanised infantry units. The United States acted decisively, rapidly blockading the island to prevent the arrival of further Soviet military support. Kennedy knew that stationing such ballistic missiles in Cuba would reduce the warning of a nuclear attack on the United States to just a few minutes. This was insufficient for the deterrence doctrine of Mutual Assured Destruction. Castro insisted that the bases were for self-defence against US aggression, which was not outlandish given the previous year's attempted invasion, but *would* be used if necessary. A tense war of words erupted between Kennedy and Nikita Khrushchev and pushed the world right to the brink of nuclear conflict, before the situation was defused. Other attempts on Castro's life have been made over the years and the dictator has himself supported Soviet wars in Africa, sending Cuban troops to Angola and Ethiopia in the 1970s, but discontent at home was rife. By 1975, failed economic policies forced Castro to implement reforms. A few months before the mission of Arnaldo Tamayo-Méndez, ten thousand Cubans stormed the Peruvian embassy in Havana, demanding political asylum, and many thousands more were granted permission to emigrate and between April and June 1980 they sailed in makeshift boats to Costa Rica and Miami.

If many of the past Intercosmos spacefarers could boast their socialist principles – from the staunchly Communist Vladimir Remek to the 'marriage' of Sigmund Jähn's Sandmännchen doll with Valeri Bykovsky's Russian puppet – then the presence of Arnaldo Tamayo-Méndez on Soyuz 38 highlighted something a little different: for he had arisen from the kind of background that Nikita Khrushchev and Leonid Brezhnev and Fidel Castro would have loved. Within months of his birth in Guantánamo on 29 January 1942, both of his parents died. He was adopted soon after his first birthday. By the age of 13, he worked as a shoeshine boy and later as a carpentry assistant. In the wake of Castro's revolution in 1959 he entered the Cuban Association of Young Rebels. Just a few weeks after the abortive invasion at the Bay of Pigs, he was sent to the Soviet Union to begin MiG-15 training. By October 1962, during the Cuban Missile Crisis, he was a fully-fledged fighter pilot and performed two dozen reconnaissance missions. Five years later, and by now a squadron leader in the Cuban Air Force, he joined the Communist Party. Tamayo-Méndez finished his studies at the Máximo Gómez Military Academy in 1971 and went on to serve as a head of staff of a fighter brigade. He rapidly rose through the ranks, becoming a major in 1975 and a lieutenant-colonel in 1976. Alongside fellow Cuban fighter pilot Captain José Armando López Falcón, he was selected for Intercosmos training in March 1978. At length, the two men were paired with veteran Soviet commanders: Tamayo-Méndez with Yuri Romanenko and López Falcón with Yevgeni Khrunov.

During his week aboard Salyut 6, Tamayo-Méndez – who became the first black person, the first non-US citizen from the Western Hemisphere *and* the first person of African heritage to venture into space – undertook more than two dozen experiments on materials processing and life sciences. He monitored the circulation of his blood, the patterns of electrical activity of his brain, the changes in his metabolism, motor co-ordination in weightlessness and the size and structure of his muscles and bones. Other studies, explained *Flight International* on 11 October 1980, "covered changes in the structure and function of the human foot arch, intra-cellular investigation of rapid-growth yeasts, growing layers of gallium arsenide and psychomotor co-ordination between left and right hands ... In addition, some organic monocrystals were grown for the first time".

When Romanenko and Tamayo-Méndez returned to Earth at 6:54 pm Moscow Time on 26 September, touching down in darkness just a couple of kilometres off-target, some 175 km south-east of Jezkazgan, some Westerners were surprised that they came home in their *own* craft, Soyuz 38. "It is not clear," *Flight International* editorialised, "why [they] returned to Earth in Soyuz 38 ... in the past, crews visiting the orbiting laboratory, Salyut 6, have often re-entered in an earlier craft, leaving a fresh capsule for the long-stay occupants." It was suspected in the West that Soyuz had an orbital lifetime of around three months; a suspicion already demonstrated by the safe recovery, albeit unmanned, of Soyuz 32. Since Popov and Ryumin's Soyuz 37 spacecraft had been aloft since July, the logic was now inescapable: they *must* be preparing to return to Earth during the Salyut landing window in October 1980. Yet *Flight International*'s confusion is understandable, because the FAI would officially fail to recognise a new endurance record if it did not surpass its predecessor by at least a ten-percent margin. Popov and Ryumin exceeded the previous 175-day record on the first day of October and to establish a new record they would need to land no earlier than the 19th, the 193rd day of their mission. However, the landing window *closed* on the 15th and it was perhaps with a little dismay that the Soyuz 37 descent module, carrying Popov and Ryumin, hit the ground at 12:50 pm Moscow Time on the 11th. They *were* the new empirical world record-holders, with almost 185 days in orbit ... but in the eyes of the FAI, their achievement failed to count. Not for another two years, until December 1982, would a 211-day mission be accomplished to finally push the endurance limit well beyond six months.

As for Popov and Ryumin themselves, they were in fine shape and could walk for half an hour, unaided, within a day of landing and play tennis together a week or so later. Both had gained a couple of kilograms, which Soviet doctors attributed to the strictness of their exercise regime – despite the running track having broken for a few days back in June – and Ryumin had now spent almost a full year (352 days) in orbit, spread across three space missions.

Even after four long-duration missions, Salyut 6 was still not ready to be retired and consigned to a fiery destruction in the atmosphere ... but the situation aboard the space station was not good, either. It would appear that Popov and Ryumin spent a considerable portion of their time in their final weeks evaluating its capacity to support further crews and a key obstacle was a problem with the thermal control system, as well as number of other pieces of equipment. Another flight of the upgraded Soyuz-T spacecraft had long been scheduled to occur during the Salyut landing window of 4-15 December 1980, but the *content* of that mission changed a great deal from the first assignment of cosmonaut crewmen in October 1979 to its launch, a little more than a year later. In its first incarnation, veteran commander Vasili Lazarev and rookie flight engineer Gennadi Strekalov were named as the prime crew, with rookie commander Yuri Isaulov and veteran flight engineer Nikolai Rukavishnikov as their backups. The objective appears to have been merely a short-duration manned test. By December 1979, the plan had changed, to place significant emphasis on medical research, although its precise foci remain unclear. For that reason, a civilian medical doctor was attached to both crews, with Valeri Polyakov joining Lazarev's team and Mikhail Potapov joining Isaulov's team. (In fact, the prime crew actually comprised *two* medical doctors; for Vasili Lazarev held the impressive credentials of *both* a physician *and* a Soviet Air Force pilot.) The presence of Polyakov and Potapov on the prime and backup crews led fellow cosmonauts to dub them, rather tongue-in-cheek, as 'The Pol-Pot Team'. It can be assumed that if this medical-oriented mission had taken place, it would probably have lasted in the order of a couple of weeks.

By the late summer of 1980, the first two flights of Soyuz-T – an unmanned mission in December of the previous year and the crewed voyage of Malyshev and Aksyonov in June – had both been satisfactorily accomplished ... but the problems aboard Salyut 6 needed urgent attention and the exclusively medical research flight was cancelled. In late June, having baselined Soyuz T-3's new repair objectives, a 'revised' prime crew was named as Leonid Kizim, Oleg Makarov and Konstantin Feoktistov, with Lazarev, Strekalov and Polyakov backing them up and Isaulov, Rukavishnikov and Valentin Lebedev attached to a 'third' support team. Feoktistov is a notable addition, since he had flown aboard Voskhod 1 in October 1964 and his key role and expertise in the *design* of Salyut 6 must have prompted his assignment to this repair mission. However, born in 1926, he was now 54 years old and even at the time of his Voskhod flight, his health had been questionable; he had been criticised for his poor medical condition, suffering from ulcers, near-sightedness, deformation of the spine, gastritis and even missing fingers on his left hand. In October 1980, just weeks before the scheduled launch of Soyuz T-3, Feoktistov was replaced by Strekalov, whose *own* place on the backup crew was taken by rookie cosmonaut Viktor Savinykh. Ironically, despite these health issues, Feoktistov would actually live far longer than many of his contemporaries within the Soviet cosmonaut team, dying in November 2009 at the grand age of 83. He would outlive the younger Lazarev, Makarov, Rukavishnikov *and* Strekalov ... and even Kizim – more than 15 years his junior – would die only a few months after Feoktistov.

In a private correspondence with this author in February 2011, Soviet space historian Bart Hendrickx speculated that the focus of the Pol-Pot 'medical research programme' on the 'original' Soyuz T-3 may have been to investigate the adaptation of older cosmonauts to the microgravity environment. During ASTP, Deke Slayton had flown into orbit at the record-breaking age of 51 and, had Vasili Lazarev flown Soyuz T-3 in the autumn of 1980 he would have been 52 years old. Hendrickx felt that it might not be totally outlandish to suspect that the Soviets were planning to break the record for the oldest man in space and that Polyakov's addition to the crew in December 1979 might have been engineered in part to study the adaptation of Lazarev to weightlessness. In reinforcing this line of thought, Hendrickx noted that it would have made more sense to fly the Soyuz T-2 backup crew of Leonid Kizim and Oleg Makarov on T-3, but as it turned out the *second* backup team of Lazarev and Strekalov were picked instead. "It is also puzzling," Hendrickx explained, "that when Lazarev-Strekalov were moved to backup position after T-3 became a repair mission, Polyakov *remained* part of the backup crew, even though the medical programme had reportedly been cancelled. Officially, he backed up Feoktistov, who was picked to fly the mission because of his engineering experience." Years after the event, Hendrickx wondered if the plan was designed so that whichever crew (prime or backup) ended up flying, *one* of its members (Feoktistov or Lazarev) would have a stab at gaining the oldest-man-in-space record. Ironically, *both* men were disqualified...on *medical* grounds; Feoktistov in October 1980 and Lazarev in the spring of 1981.

Of course, today, all this is now academic. Soyuz T-2 backups Kizim and Makarov – and Gennadi Strekalov, too – ended up flying the mission and at 5:18 pm Moscow Time on 27 November 1980, the Soviet Union's first three-man cosmonaut crew in almost a decade took to the skies. In the centre (commander's) seat aboard Soyuz T-3 was Lieutenant-Colonel Leonid Denisovich Kizim, whose chance of being in this exalted position might have seemed impossible to him a couple of decades earlier. He was born in Krasnyi Lyman, a Cossack-founded town in eastern Ukraine, on 5 August 1941, and like many of his fellows in the Soviet spaceflying corps his dream of aviation started at a young age. However, his shortness of stature caused him to be turned down for flying school; only Kizim's tenacity finally enabled him to achieve his goal and in 1963 he graduated from the Chernigov Lenin Komsomol Higher Air Force School and joined the Soviet Air Force, rising to become an accomplished test pilot and parachutist. In October 1965, he was one of 22 cosmonaut candidates, both pilots and engineers, to be chosen for an anticipated flurry of Soyuz and space station missions. Of those 22 selectees, only *six* ended up actually flying into the heavens ... and of those six, arguably Kizim would achieve the most: three space voyages in total, one of which spent a world-record-breaking *eight months* in orbit, and command of the first (and so far only) mission to visit *two* space stations in a single flight. The wait for his first flight, though, was long and arduous. Completion of initial cosmonaut training at the end of 1967 was followed by assignment to the Spiral spaceplane effort and in January 1974 he was picked as one of the inaugural members of the Soyuz-T team. A year later, he graduated from Higher Air Force School.

The new Soyuz-T spacecraft in orbit. This photograph was taken during the approach of Yuri Malyshev and Vladimir Aksyonov towards Salyut 6 in June 1980.

Seated to Kizim's left side as Soyuz T-3 rose from its Tyuratam launch pad that gloomy November evening in 1980 was one of the Soviet Union's most experienced cosmonauts, Oleg Makarov, who became the first person from behind the Iron Curtain to perform a fourth space mission. As has already been seen, Soyuz T-3 would only be Makarov's third *orbital* trek, since an earlier voyage with Vasili Lazarev had suffered a booster malfunction and was effectively suborbital, but the credentials of this frail-looking flight engineer were unquestioned. In fact, Makarov had been in line for a backup command slot on one of the early Intercosmos missions – probably the East German venture – and this made him one of only three civilians to be offered this opportunity. In a sense, *all three* men aboard Soyuz T-3 would carve out their own impressive niches in the annals of space history, for the third crewmember, Gennadi Mikhailovich Strekalov – a unique 'research cosmonaut', seated to Kizim's right side – would end up flying no fewer than *five* times.

Strekalov was born in Mytishchi, a major industrial hub, situated to the north-east of Moscow, on 26 October 1940, and judging from the nature of his birthplace it is perhaps not surprising that he forged a career for himself in science and engineering. His father was killed in 1945, during the Red Army's liberation of Poland, only weeks before the end of the bloody conflict with Nazi Germany. The young Strekalov completed his schooling and became an apprentice coppersmith, before enrolling at the prestigious Bauman Moscow Higher Technical School. He received his engineer's diploma in 1965 and moved directly to work for the organisation which evolved under Sergei Korolev into TsKBEM and eventually Energia, helping with the design of Soyuz. Strekalov was chosen as a civilian cosmonaut in March 1973 and within months began formal training. His first crew assignment was as backup flight engineer to Valeri Bykovsky's Soyuz 22 mission. Two years after that flight took place, in October 1978, he recommenced flight training and at the end of 1979 was teamed with Vasili Lazarev for the 'original' Soyuz T-3. Six months later, following changes to the mission, he was reassigned as backup flight engineer to Konstantin Feoktistov. When Feoktistov was grounded in October 1980, Strekalov found himself back on the prime crew. In this way, he became the only member of the 'original' Soyuz T-3 to actually fly the mission.

Late on the afternoon of 28 November, Kizim guided his spacecraft to a smooth docking with Salyut 6 and a lengthy programme of repairs to the station got underway. Many observers in the Western world were unsure what to make of the mission: would it be a long-duration voyage, perhaps eclipsing that of Popov and Ryumin, or a shorter one? The state-run Soviet press revealed little. "Tass has not stated," *New Scientist* told its readers on 4 December, "whether this mission will be a long one. However, it seems unlikely that the Soviets will keep all three cosmonauts on-board for any extended period, because the Salyut 6-Soyuz-T complex is not that large." What was *not* known, of course, was that the purpose of Kizim, Makarov and Strekalov's mission was an extensive series of repairs in order to support a possible future crew. Of crucial importance was work on the thermal control system and the cosmonauts fitted a new hydraulic unit with four pumps and conducted the 'Mikroklimat' ('Microclimate') experiment to assess the living conditions aboard Salyut, declaring them to be satisfactory. Other repairs included work on electrical faults, the replacement of a timing device in the station's control system and fitting a new power supply unit for the compressor on the Progress refuelling mechanism. As they progressed, they received regular advice from Popov and Ryumin, who were both stationed in the mission control centre in Kaliningrad and knew the condition of the station better than anyone. Kizim, Makarov and Strekalov also unloaded the Progress 11 freighter, which had docked a week or so before Popov and Ryumin came home and which had earlier performed the first automated refuelling of Salyut. In spite of the repair work, they also had the opportunity to conduct a few scientific experiments, loading new samples into the Splav-1 and Kristall furnaces and studying biological processes in the 'Oazis' ('Oasis') mini-greenhouse. Additionally, explained *Flight International*, they made the first hologram in orbit. "The cosmonauts used a portable helium-neon laser," the magazine reported on 20 December, "to make a hologram of a crystal being dissolved, as part of a Soviet-Cuban experiment." Based on the *nature* of these experiments – particularly the biological ones – it was suspected that a fifth long-duration period of occupancy was scheduled to take place, sometime in the spring of the following year. In the meantime, less than two weeks since their launch from Soviet Central Asia, the Soyuz T-3 cosmonauts were back on Earth, landing in Kazakhstan at 12:26 pm Moscow Time on 10 December. "One report," noted *Flight International*, "says that Cossack horsemen were the first on the scene." It has been speculated over the years that the orbital modules for Soyuz T-3 and the *next* mission, Soyuz T-4, were left attached to Salyut 6 after undocking, although such reports have never been confirmed by the Russians. "Some of the cosmonauts involved in those missions have been asked ... if this was really true," wrote Rex Hall and Dave Shayler, "but they denied it. A possible explanation for these reports is that

NORAD tracked the [orbital module] after the return of the [descent module] and though that they had separated from Salyut 6, whereas they had actually been separated from the [remainder of the Soyuz itself] before retrofire. Indeed, the orbital module of Soyuz T-3 did not re-enter the atmosphere to destruction until late January 1981." In any case, it would have been pointless to leave an orbital module attached to the station, because its 'end' would be open to space and therefore inaccessible to anyone aboard Salyut.

Flight International and others were right in their suspicion that another long-duration mission was on the cards ... though the flight of Soyuz T-4 would not come close to ensnaring any new records. In fact, it would fall ten days short of the 84-day Skylab achievement. When one looks at the steadily increasing lengths of Soviet missions in the 1970s – from one to two months, then four and a half months and finally six months – the 74-day voyage of Vladimir Kovalyonok and Viktor Savinykh, which began in March 1981, looks like a historical anomaly. It was not. In fact, its very occurrence and its very duration were dictated primarily by the need to satisfy two more Intercosmos nations and conclude the programme. By rescheduling the Hungarian mission from June 1979 until May 1980 and making the short-notice decision to add a Vietnamese Intercosmos venture, the Soviets now found themselves with Mongolian and Romanian cosmonauts still waiting to fly. That the Vietnamese mission was based purely on political convenience (and the need to make a timely and bold statement to the People's Republic of China) is further underlined when one considers that the Mongolian and Romanian Intercosmos teams had been selected in March 1978 ... more than a year *earlier* than Pham and Bùi.

At face value, this seems to speak of double standards, particularly since the Soviets had sought to ensure as much 'fairness' as possible with each Intercosmos mission; even the *duration* of each flight was timed, as closely as possible, to a couple of hours shy of eight days, thereby avoiding charges of favouritism. The People's Republic of Mongolia, for example, had long proven itself to be a loyal Soviet ally and a staunch Communist state from its formation in 1924 through to its collapse in 1991. Attempts by China's Qing Dynasty to annex Outer Mongolia in the early years of the 20th century had drawn the Mongolian nobility into the orbit of the Russian Empire and, after renewed Chinese aggression in the wake of the Bolshevik Revolution, into the Soviet sphere of influence. Gradually, the nomadic lifestyles of the Mongols changed into a strict Communist system of collective herding and the expropriation of the nobility and the Buddhist monasteries. By the end of the Great Patriotic War, the provisions of the Yalta Conference obliged China to recognise Outer Mongolia's independence and in the 1950s the relationship between the two neighbours improved, with the People's Republic providing much-needed economic aid and restructuring. This changed abruptly in 1963, when Mongolia sided with Russia against China in the Sino-Soviet ideological dispute. A few years later, Mongolia and the Soviets signed a mutual assistance treaty and in February 1967 Leonid Brezhnev authorised the emplacement of Red Army troops in the country. Attempts by Mongolia to enter the United Nations were repeatedly thwarted by China and even the United States would not establish diplomatic relations until the end of the Cold War.

Within the People's Republic of Mongolia – a region perched on the fence between centuries of traditional nomadic herding and the new demands of a Communist system based on shared ownership – was born, on 5 December 1947, Jügderdemidiin Gürragchaa, who would one day become the first of his countrymen to venture into space. He came from Gurvanbulag, a district in the Bulgan province of northern Mongolia, whose diverse geography offers a series of dramatic contrasts, from alpine forest to arid steppe and from the meandering rivers of the Orkhon and the Selenge to surprisingly fertile areas of arable farming. Little is known about his youth, except that he studied aerospace engineering in the capital, Ulaanbaatar, and joined the Mongolian People's Army, reaching the rank of captain by the time of his space mission. "If [he] had been a pioneer in the Gagarin mould," noted space historian Tim Furniss in 1986, "his near-unpronounceable name would have caused havoc in many a television newsroom". (In fact, had Gürragchaa not flown, even the name of his backup – Maidarzhavyn Ganzorig – proved only marginally more merciful on the Western tongue. To the Soviet cosmonauts with whom he trained, Jügderdemidiin Gürragchaa received a simple nickname: 'Gurr'.)

Clan names within Mongolia had been banned during the Communist era, primarily on the insistence of the Soviets, who sought to eradicate the heritage of a region and the memories of a people historically associated with the dynasty of Genghis Khan, which, from the 12th century onwards, had brought wave after wave of havoc and destruction across Central Asia and into the fringes of Eastern Europe. Members of the Mongolian nobility who claimed descent from the Great Khan were systematically arrested and executed and in 1925 the use of clan names was formally abolished. Even in the 1980s, no Mongolian museum was permitted to display imagery – or even the *name* – of its most famous son. As Stalinist purges swept across the country, killing perhaps a tenth of a million-strong population and breaking down ancient loyalties to the nobility, clan and family names were quite simply *forgotten* within two or three generations of the 20th century. It is perhaps the greatest tragedy to befall Mongolia in modern times.

In 1997, having by now overthrown the Communist yoke, the ban was removed and steps were taken to allow Mongolians to recreate their sense of national identity. At one stage, in the summer of 2004, more than 60 percent of the population had chosen Genghis Khan's clan name – 'Borjigin' ('Master of the Blue Wolf') – as their own. Like many of his fellow countrymen in the post-Communist era, Jügderdemidiin Gürragchaa was unable to identify his own clan name. Some Mongolians chose the name of their profession as a 'new' clan name – Mr Writer, Mr Hunter, even Mr Policeman – and for Gürragchaa the final decision, as his nation's first cosmonaut, was straightforward. He chose as his new clan name the word 'Sansar', meaning 'Cosmos' or 'Space'. Today, perhaps Tim Furniss and countless other journalists across the world might be relieved to learn that the space explorer with the unpronounceable name, previously known as Mr Gürragchaa, is no more. In his place is 'Mr Space'. Prince, the American singer-songwriter, who changed his own name so frequently in the 1990s, would be proud ...

Today, Mr Space's achievements and extraterrestrial exploits are commemorated not only through his new name, but also on a stylised panel, high in the hills to the

Vladimir Dzhanibekov (background) and Jügderdemidiin Gürragchaa are pictured during water survival training. Rumour would abound for many years that the Mongolian cosmonaut was either ill or incapacitated during his mission ... a rumour which has never been conclusively confirmed or denied.

south of Mongolia's capital, Ulaanbaatar, on the so-called Zaisan Memorial. This enormous concrete structure was built to honour Soviet and Mongolian relations. In addition to its colossal statue of a soldier, the memorial boasts an astonishing circular mural, bearing colourful images of heroic events from Mongolia's past: Soviet support independence from China, the defeat of Japan's Kwantung Army in 1939, victory over Nazi Germany ... and a vibrant, stylised image of Cosmonaut Jügderdemidiin Gürragchaa – our very own Mr Space himself – encircled by doves, clutching his space helmet with one hand and the palm of a particularly attractive female Russian scientist with the other.

It has now been three decades since a rocket bearing Gürragchaa and his Soviet crewmate, Soyuz 27 veteran Vladimir Dzhanibekov, lit up the Tyuratam sky at 5:59 pm Moscow Time on 22 March 1981 and carried the dreams of another Communist nation into the heavens. As with each of its predecessors, sadly, we have few firm facts from the eight-day voyage itself, although there have been persistent rumours over the years that Gürragchaa was ill or even incapacitated during the mission. Certainly, only a handful of images from Soyuz 39 have ever been released. The experiments performed by Mongolia's first cosmonaut were also fairly humdrum in nature: multi-spectral observations of his homeland, measurements with a cosmic ray detector, a specially instrumented collar to restrict head movements as part of a space sickness study and the 'Gologramma' ('Hologram') investigation which transmitted images to and from ground stations and created images of crystals growing in the microgravity environment. The cosmonauts returned to Earth on 30 March, touching down in drizzle and fog, some 170 km south-east of Jezkazgan at

2:42 pm Moscow Time. Today, Gürragchaa's Sokol space suit gloves and headset reside in a dedicated room within Ulaanbaatar's Museum of Mongolian History, together with a copy of his 1984 autobiography ... one of the rarest in the world.

When Dzhanibekov and Gürragchaa arrived at Salyut 6, a new long-duration crew was in residency; a crew which offered a glimpse back to the past and a wistful glance towards the future. Commander Vladimir Kovalyonok might, but for a twist of fate, have been Salyut 6's first skipper, back in October 1977, whilst his flight engineer aboard Soyuz T-4, Viktor Savinykh, would go down in the annals of space history as the one hundredth spacefarer since Yuri Gagarin. When Kovalyonok and Savinykh blasted off from Tyuratam at 10:00 pm Moscow Time on 12 March 1981, it was exactly a month shy of the 20th anniversary of Gagarin's flight ... and the Western space press had wondered for months *who* would seize the coveted title of Spacefarer No. 100. When Soyuz 38 returned to Earth in September 1980, *Flight International* made reference to the fact that Cuban pilot Arnaldo Tamayo-Méndez was the 97th spacefarer and pondered the question. "With the US Space Shuttle roughly six months from flight," the magazine noted on 11 October, "the Soviet Union is almost certain to achieve the 100th [person] in orbit. Russia currently has Mongolian, Romanian and Cuban cosmonauts in training ... but there has been no indication on who the 100th man will be." Of course, no one in the West anticipated a *three-man* Soyuz T-3 mission at the end of 1980 and many could therefore have been forgiven for wondering if the title might fall to either the Mongolian or Romanian cosmonaut.

In fact, when Kovalyonok and Savinykh reached Salyut 6, shortly before midnight, Moscow Time, on 13 March, there were some observers who speculated that the *next* Intercosmos crew might contain *both* Mongolian *and* Romanian crew members, together with a Soviet commander, now that the three-man capability of Soyuz-T had been demonstrated. Yet the assumption that each mission would henceforth launch with three crew members as standard was a mistaken one. The Soviets must have enjoyed the confusion when Soyuz T-4 launched with only Kovalyonok and Savinykh and *Flight International* admitted that it was "not clear how long the cosmonauts will spend in space, nor why there are only two crew in the ... capsule". Would they attempt to break the six-month records of the Lyakhov and Popov missions? No one knew. As for Kovalyonok and Savinykh themselves, they were paired up as a crew in the latter part of 1980, "meaning," wrote Phillip Clark, "that they would not be ready for a launch until just before the March 1981 landing window, rather than the earlier February window". Since the Soyuz T-4 mission had to host *two* Intercosmos missions, the Soviets were presented with the opportunity to fly the new-generation craft for a period spanning one landing window to the next, a period of some two months. "Since Salyut had to be operating with a resident crew before the Mongolian mission was launched and the crew had to remain on the station until after the recovery of the Romanian mission," continued Clark, "the residency would have to last for longer than the nominal 60 days interval between landing windows."

In their first few days aboard Salyut, Kovalyonok and Savinykh unloaded the Progress 12 craft, which had docked automatically in January, and conducted

As the last flight of the 'standard' Soyuz ended, the traditions continued, including this one: cosmonauts signing the exterior of their descent module. Here, Dumitru-Dorin Prunariu scrawls his name, whilst Leonid Popov looks on.

general repairs on the station itself. Viktor Petrovich Savinykh came from Berezkiny in Russia's Kirov Oblast, where he was born on 7 March 1940. He studied at the Moscow State University of Geodesy and Cartography, graduating in 1969, and was one of seven civilian cosmonauts selected by Energia in December 1978. During the reshuffling of crew assignments for Soyuz T-3 in the summer and autumn of 1980, Savinykh found himself on the backup crew for a time, then commenced his formal training with Kovalyonok for Soyuz T-4 before the year's end. In becoming the world's hundredth spacefarer, Savinykh was also the 50th Soviet cosmonaut; the remainder of the hundred was made up by 43 Americans and one person each from Czechoslovakia, Poland, East Germany, Bulgaria, Hungary, Vietnam and Cuba.

The hundredth person in space was quickly followed by one of the first few members of the *next hundred*, a Romanian pilot named Dumitru-Dorin Prunariu, who launched into orbit aboard Soyuz 40 at 8:17 pm Moscow Time on 14 May. The youthful-looking Prunariu, who was just 28 years old and therefore the youngest Intercosmos flier so far, was joined by veteran commander Leonid Popov, who had himself returned from Salyut 6 only a few months earlier. It would appear that most cosmonauts had by now shifted over to train on the new Soyuz-T vehicle and Rex Hall and Dave Shayler remarked that it was not possible for either the Dzhanibekov or Popov crews to perform the customary spacecraft swap, "as neither had trained on Soyuz-T". It is probably also reasonable to assume that Popov's recent experience with the 'older' design, having flown Soyuz 35 the previous year, explained the rapid turnaround between his first and second missions. Only *seven months* had elapsed since he and Ryumin hit the Kazakh steppe at the end of their 185-day mission; a remarkably short interval of time. It did not establish a new *world* record, for American astronaut Tom Stafford had flown twice in less than six months, more than a decade earlier, but it *was* a new record for a Soviet cosmonaut. Moreover, when one considers that Popov's backup commander, Yuri Romanenko, had *also* flown relatively recently, less than eight months earlier, it seems to reinforce the likelihood that cosmonauts with recent experience were having their skills tapped for the final few flights of the old-style Soyuz.

Indeed, with the new craft, its systems and its capabilities to support three-man crews *and* lengthy space station missions more or less proven, Soyuz 40 was intended to be the final flight of the old-style craft. In fact, after landing on the evening of 22 May, Popov stated that his mission "was the last flight of ships of the Soyuz series" and praised the venerable old workhorse's designers for the magnificent job they had done. This caused some confusion in the West, with a few journalists wondering if Soyuz itself was being retired. "Does this mean that a new spacecraft is being planned?" asked *Flight International* on 6 June, "or does it perhaps mark a pause in Soviet spaceflight?" Neither, it seemed, for a new Soviet space station, called Salyut 7, was primed and ready to fly in 1982 and its crews would arrive aboard the now-fully-operational Soyuz-T.

Like his Intercosmos predecessors, Dumitru-Dorin Prunariu conducted a series of experiments aboard Salyut 6, including measurements of heart parameters and blood circulation, growing crystals by capillary action in the Splav-1 and Kristall furnaces and Earth observations. Naturally, he was a military pilot in the Romanian Air Force, although academically he was an aerospace engineer. Born on 27 September 1952 in Braşov, he studied physics and mathematics at high school, before enrolling at the Polytechnic University of Bucharest and completing his degree in 1976; in later years he would go on to achieve a PhD. He subsequently worked as a diplomatic engineer for Industria Aeronautică Română (IAR), the Romanian Aeronautic Industry, based near Braşov, which, since 1925, has built and upgraded hundreds of helicopters, gliders and other aircraft for commercial and military users. Prunariu joined the reserve officers' school of the air force in Bacău for six months of inaugural military training from March to September 1977. Despite having a website devoted to his mission, www.prunariu.org, he offers little detail about either the training or the execution of Soyuz 40 and even facts about his upbringing and education are scant. Whilst working in Bacău, Prunariu noticed a call for Intercosmos candidates and in February 1978 he and electrical engineer Dumitru Dediu of the Romanian Army were selected. A few weeks later, on 13 March, Prunariu became an active officer, receiving the rank of lieutenant-major, roughly equivalent to a first lieutenant in most Western militaries. He finally won his seat on Soyuz 40 after scoring the maximum awardable number of marks in his final exam.

If Mongolia had remained a steadfast ally of Russia since the 1920s, then Romania was one of the few Eastern Bloc states to have drifted somewhat from the Soviet orbit ... and it has been suggested that, despite being selected *ahead* of Vietnam, this may have been a factor in why its cosmonaut flew *last* in the series. Unlike many of the socialist satellites, Romania entered the Warsaw Pact as a co-belligerent with Nazi Germany, having actively participated in the 'Operation Barbarossa' invasion of the Soviet Union and for some years was treated as a conquered territory. Then, in the early 1960s, the Romanian People's Republic (later renamed the Socialist Republic) began to assert a degree of independence. In August 1968, the newly-appointed President, Nicolae Ceauşescu, openly criticised the Soviet invasion of Czechoslovakia; coupled with his moderate stance against internal dissent, he was seen in a positive light, both at home and abroad. However, he was not a liberal reformer and a period of economic growth, bankrolled by foreign creditors, was quickly replaced by an era of austerity and repression. By the beginning of the 1980s, this had led to widespread rationing – bread, milk, butter, meat and even potatoes in some areas – and the food which *was* available tended to be poor-quality export rejects. Despite the country's vast refining capability, petrol was restricted, gas heating was frequently turned off, electricity supplies were rationed and it has been reported that *one in three* Romanians was an informant for Ceauşescu's feared secret police, the 'Securitate'.

If its western neighbour, Hungary, was one of the most liberal corners of the Eastern Bloc, and if its southern neighbour, Bulgaria, was staunchly pro-Soviet, then Romania in the latter part of the Communist era pursued a completely opposite tack. The appalling treatment meted out to the Romanian people by Ceauşescu during this period was reciprocated by the violence with which his regime was finally deposed; the old dictator and his wife were executed by firing squad on Christmas Day 1989. In that year of revolutions, Romania would be the only Communist nation to forcibly overthrow its government and exact violent retribution against its former master.

A difficult decade lay ahead for Romania when Dumitru-Dorin Prunariu returned from his space mission at 4:58 pm Moscow Time on 22 May 1981, although the awards and medals came aplenty – the Supreme Title of Hero of the Socialist Republic of Romania, the gold star of a Hero of the Soviet Union and the German Rocket Society's Hermann Oberth Gold Medal, to name just a handful – and 'friendship' with the Soviet Union was overtly proclaimed. Yet the end of Ceauşescu's despotic two decades in power was within sight and, in orbit, so too was the end of Salyut 6's monopoly on space operations. When Kovalyonok and Savinykh finally undocked their Soyuz T-4 craft from the old station on the afternoon of 26 May and touched down at 3:38 pm Moscow Time, they brought a remarkable period to a close. No fewer than *five* teams of cosmonauts had extended the endurance record to more than *double* that of the final Skylab crew. Endurance, though, was only part of the achievement, for Progress had enabled such missions to take place and the capability to refuel and restock Salyut marked a great increase in operational capability compared to Skylab. Moreover, for all the American plans to fly international scientists on the Space Shuttle, the Soviets had *already* established strong international partnerships and had executed no fewer than *nine* Intercosmos missions, eight of which had successfully docked with the station. Yet Salyut still had a further role to play. Three weeks after Kovalyonok and Savinykh returned to Earth, the unmanned Cosmos 1267, described by the Soviets as being more than twice as massive as a Soyuz, was launched atop a Proton rocket and docked automatically with the old station. In July 1982 the complex was de-orbited, its burning debris splashing harmlessly into a sparse expanse of the Pacific Ocean.

In his survey of the early Soviet space programme, published in 1988, Phillip Clark noted that this triumphant half-decade of Soviet primacy in the heavens was balanced, ironically, by a profound lack of information, as each long-duration mission and each Intercosmos flight turned steadily 'vanilla' in colour and attracted precious little interest in the Western press. It is this author's fervent hope that, someday, the archives to these missions will be fully opened to shed something more than just a glint of light onto what they were *really* all about. What *really* happened to Nikolai Rukavishnikov's craft during his ill-fated approach to Salyut 6? Was Jügderdemidiin Gürragchaa *really* ill or incapacitated during his mission? Was the decision to fly a Vietnamese cosmonaut at such short notice *really* a cynical political ploy to score points over the People's Republic of China? Was Soyuz T-3 *really* planned to seize the oldest-man-in-space crown from Deke Slayton? Maybe such questions will never be answered, but the paucity of information from behind the Iron Curtain during this era has another, more positive side. By *becoming* 'vanilla', and seemingly more routine, the missions underlined that space *was* becoming more accessible to different people and different nations. The Soviets did little to trumpet their achievements in smashing the Skylab record, over and over and over again, because there was no longer any *need* to do so ... for they were *already* firmly ahead. When the Shuttle made its maiden voyage in April 1981, its creators justified its existence on the basis that it could fly people into space reliably, cheaply and more frequently than ever before.

Yet 'frequently' does not necessarily mean the same as 'routine'. *Routine*, as a course of standardised actions followed with some repetition, was a state of mind with which the Americans allowed themselves to become bogged down in the early part of the 1980s. For them, launching the Shuttle every few weeks marked it out as being 'routine' ... but its engineers and managers quietly ignored a multitude of unacknowledged problems which lurked behind the scenes and which, at length, would catch out an unsuspecting crew one January morning in 1986, with catastrophic consequences. The Soviets, for their part, would not launch their missions with the same frequency as their counterparts in the United States, and their hardware was unquestionably inferior to that of the Shuttle, but they would nevertheless send their cosmonauts into space reliably, relatively often and would do one thing that the Americans *could not* do: they would not just *visit* space, but they would *live* there; creating and sustaining a real 'home', off the planet, for an extended period of time. America and its international partners – Germany, France, Japan, Canada, Italy and a handful of others – are only today *beginning* to match what Soviet cosmonauts achieved, *routinely*, almost three decades ago.

By the time Salyut 6 re-entered the atmosphere to destruction, one day in late July 1982, *another* station – the more capable Salyut 7, to be discussed in the next volume – had already been in orbit for several months and its first crew was partway through a long-duration mission which was expected to break the 185-day record of Popov and Ryumin. The 1980s would allow the Soviets to take longer strides. Other nations, and not just Soviet-aligned ones, would be active participants in their space missions, including France, India and Syria, and they would fly their first female cosmonaut in almost two decades. By the latter half of the decade, an altogether different complex, 'Mir' ('Peace'), would be in orbit ... and *its* achievements would truly bridge the gap between the relatively primitive Salyuts and the astonishing engineering endeavour that is today's International Space Station.

5

Dreams and nightmares

"THE ENTERPRISE IS SET!"

She was not returning from an ambitious science mission, trailing double sonic booms in her wake as she swept into California's desolate Edwards Air Force Base or the marshy expanse of the Kennedy Space Center in Florida. Nor was she roaring into orbit under the combined thrust of two immense Solid Rocket Boosters and three main rocket engines to deploy an important satellite, upgrade a world-class observatory or haul hardware aloft for the International Space Station. Rather, she was hurtling a couple of kilometres over Birmingham, right in the industrial heart of England, securely fixed to the top of a Boeing 747 Shuttle Carrier Aircraft (SCA). Her name was Enterprise. It was June 1983. It was sports day at my infant school in Tyseley, a small town which even today continues to be dominated by a long heritage of industry and railways. I was six years old. As I stumbled clumsily along, desperately trying to keep the egg fixed, equally securely, to the spoon in my hand, the heads of everyone were suddenly jarred upwards as Enterprise flew overhead, midway through her tour of Germany, Italy, England and ultimately Canada and the United States. She had already stolen that year's Paris Air Show. Today, she sits in the McDonnell Space Hangar at the Smithsonian's Steven F. Udvar-Hazy Center, near Dulles International Airport, in Washington, DC. The museum, into which she was placed in November 2003, now hosts a myriad of other rockets and spacecraft. Even three decades after she was built and flight-tested over the Californian desert, she remains in good shape, albeit somewhat dusty and with parts missing. Enterprise's career has been distinctly overshadowed by those of her spacegoing sisters, the Shuttle orbiters Columbia, Challenger, Discovery, Atlantis and Endeavour. Yet in some ways she is the most important of them all, for she removed much of the uncertainty in the 1970s over the handling, approach and landing characteristics of this new, delta-winged spaceplane, which would be blasted in orbit, attached, bullet-like, to the side of a huge External Tank and a pair of Solid Rocket Boosters, and which would then plunge, brick-like, through the atmosphere at journey's end to perform a pinpoint touchdown on a conventional runway. Removing this uncertainty would lead directly to the triumphant homecoming of Columbia on 14 April 1981: the first manned orbital spacecraft to land like a conventional airliner.

Enterprise, sadly, was never given the opportunity to travel into space. Instead, since her completion in March 1976 and rollout six months later, she has served as a 'hangar queen', a Second World War reference to an aircraft which, though non-flying itself, provides parts for those which do fly and has been instrumental in testing safety upgrades across the Shuttle fleet, which, at the time of writing, in the spring of 2011, has completed more than a hundred and thirty orbital missions and is now on the verge of retirement. Most recently, in the summer of 2003, only months before being transferred to the Udvar-Hazy Center, NASA 'borrowed' fibreglass panels from Enterprise's wings to aid the Columbia Accident Investigation Board in its inquiry. During the course of that inquiry, which included firing chunks of insulating foam at Shuttle wing panels, the sections of fibreglass from Enterprise, though not broken, suffered permanent damage to their seals. Since the protective panels on Columbia's wings had been two and a half times *weaker* than Enterprise's fibreglass, the contribution made by the hangar queen to the investigation helped to validate the theory that flyaway foam and a breached thermal-protection system were indeed responsible for this second Shuttle tragedy.

Her contribution, however, goes still deeper. Although best known for a series of captive and free flights performed with the Boeing 747 SCA, she was brought out of retirement in the wake of the first Shuttle disaster – that of Challenger in January 1986 – to evaluate a new net-like 'barricade' system for capturing an orbiter whose brakes might fail during rollout. She also supported practice runs for new crew-escape techniques, 'lent out' her nose landing gear and elevon flipper doors for testing and even demonstrated procedures for transferring propellant between the Shuttle's nose-mounted forward thrusters and tail-mounted orbital manoeuvring pods. Over the years, NASA has had lots of different uses for Enterprise, according to structural engineer Julie Kramer White, particularly as the agency explored its option for follow-on concepts. "As the Shuttle has evolved and as we considered a second-generation orbiter," she explained, "we have used Enterprise as a testbed for subsystems development." Remarkably, therefore, even three decades after her completion, Enterprise continued to play a key role in the Shuttle's evolution. It is a pity that she was never upgraded to make her spaceworthy; an irony not lost on *Star Trek* moviemakers and moviegoers, whose own Enterprise carried a commemorative mural in its ready room showing Space Shuttle Enterprise docked at the International Space Station. In fact, Gene Roddenberry's fictional series is an appropriate place to begin exploring her history, for it was responsible for changing the name on her payload bay doors from NASA's preferred 'Constitution' to that which she bears today.

Original plans, dating back to July 1972, when NASA awarded a contract to North American Rockwell for her construction, called for just two spacefaring Shuttles: 'Constitution' (Orbiter Vehicle 101), which would perform a number of Approach and Landing Tests (ALT) in the low atmosphere before being upgraded for operational missions, and 'Columbia' (Orbiter Vehicle 102). However, in the first of many changes that would be applied to Constitution, following a mass influx of more than 100,000 letters to then-President Gerald Ford from 'Trekkies', NASA bowed to pressure that she should be renamed 'Enterprise'. The agency, apparently, did not approve of the name, preferring 'Constitution', as it honoured the upcoming 1977 bicentennial of the United States' declaration of independence and its aftermath. Nonetheless, when Enterprise finally rolled out of Rockwell's Air Force Plant 42 in Palmdale, California, on 17 September 1976, the Trekkies' wish was very visibly granted. She had been structurally 'complete' since March, but had spent several months undergoing a series of Horizontal Ground Vibration Tests to validate her integrity and ability to withstand the stresses of several simulated 'launch' and 'landing' phases.

Although Enterprise was not of an identical configuration to sister ship Columbia, Rockwell engineers understood their differences well enough to incorporate them into mathematical testing models. Unlike Columbia, Enterprise was not equipped with provisions for attaching 'real' manoeuvring pods in her aft fuselage (she carried only 'boilerplate' replicas); nor did she have the same integrally-machined vertical stabiliser tailfin as her spacegoing sibling. Four months after her very public rollout, on 31 January 1977 she was towed 58 km overland to Edwards Air Force Base, deep in California's inhospitable Mojave Desert, in readiness for her atmospheric flight tests. Following arrival in the gigantic Mate-Demate Device, Enterprise was hoisted atop the Boeing SCA on 7 February in readiness for three 'taxi tests' along the runway on the 15th. Even after three decades, the sight of the heavily-modified jumbo with a Shuttle riding piggyback is among the most iconic and awe-inspiring images to emerge from the space programme and continues to cause jaws to drop, as my June 1983 eyewitness perspective of Enterprise amply demonstrated. The original airliner was purchased by NASA from American Airlines in June 1974, partly because it was the largest available to accommodate the DC-3-sized Space Shuttle orbiter. Even so, it required the removal of virtually all of its interior equipment, including passenger seating and galley and extensive modifications to its air-conditioning ducts, electrical wiring and plumbing, together with the installation of higher-thrust engines. Its upper fuselage was beefed-up with internal support structures and, following wind-tunnel tests, endplate-style vertical stabilisers were fitted to its horizontal tail. Like Enterprise, it was also provided with an emergency ejection capability for its four-man crew. It completed its maiden flight in December 1976, preparatory to the Approach and Landing Tests with Enterprise the following summer.

All three taxiing tests were performed in the early morning to reduce heating on the 747's tyres and brakes, which, in addition to carrying the 181,000 kg load of the airliner itself, also had the 68,000 kg Enterprise on top, and were performed without incident. The combo's speed along the runway was steadily increased from 143 to 225 and finally 250 km/h, after which procedures for an aborted takeoff at high speed were simulated, together with tests of full braking, thrust reversers and speedbrakes. The success of the taxiing runs cleared the way for Enterprise's first captive flights. During a two-week period from 18 February, the Boeing's crew of pilots Fitz Fulton and Tom

McMurtry and flight engineers Vic Horton and Louis 'Skip' Guidry performed five airborne runs with the still-unmanned Enterprise to assess the combined vehicles' structural integrity and handling characteristics under flight conditions. So successful were these so-called 'captive-inert' tests that a planned sixth run was dropped. In fact, both Fulton and McMurtry found that the presence of the Shuttle above them had little adverse effect on their ability to control the aircraft, partly due to a large aerodynamic tailcone installed onto Enterprise's aft fuselage.

Moreover, the presence of the orbiter's delta-shaped wings actually turned out to generate more 'lift' than expected, prompting some aerospace journalists after the flight to comment that they had just seen the debut of the world's largest biplane. During the last two captive-inert tests, Fulton and McMurtry perfectly executed a so-called 'short-field' landing in readiness for an anticipated touchdown on the 2,300 m runway at NASA's Marshall Space Flight Center in Huntsville, Alabama. The combo demonstrated that it could slow to a complete halt in less than 1,800 m. When the last captive-inert flight ended on 2 March 1977, it was at last time to modify Enterprise to accept her first crew. Three 'captive-active' tests were planned for June and July, in which she would remain attached to the Boeing, followed by a series of up to eight 'free flights', beginning in August, in which she would separate and glide to an unpowered, 'deadstick' touchdown at Edwards.

Aboard Enterprise for her first captive-active flight were astronauts Fred Haise, a veteran of the ill-fated Apollo 13 lunar mission, and rookie Gordon Fullerton. Haise's selection to lead this inaugural exercise was no accident, for his experience with the Shuttle extended back to April 1973, when he was appointed as technical assistant to Aaron Cohen, head of the Orbiter Project Office in Houston. However, according to fellow astronaut Joe Allen, Haise's very chance of making it back into the cockpit – in fact, *any* cockpit – again was fortuitous. Four months after his assignment to Cohen's office, he was almost killed in the crash of a Second World War training aircraft; he suffered second-degree burns over half of his body and third-degree burns to his legs. Allen, who worked closely with Haise during this period, remembered being "thrilled" that he was back in the pilot's seat once more for the Enterprise tests. In February 1976, along with Fullerton and fellow astronauts Joe Engle and Dick Truly, Haise was assigned to lead the ALT programme. Eight months later, following the delivery by Grumman of a pair of modified Gulfstream II Shuttle Training Aircraft, the foursome began their formal flight preparations.

"The initial Approach and Landing Tests on the orbiter were, in fact, just that," recalled Engle in a May 2004 oral history for NASA, "and that was to place the vehicle in aerodynamic flight by itself and exercise all of the systems that we could: hydraulic systems, electronic systems, the flight control systems and landing gear, in a *real* flight environment, and to gather as much flight test information as far as stability and control parameters and performance parameters and do it partly in an ideal environment. In other words, not have to worry about coming in to land and the wind coming up and giving you a big cross wind, or low clouds or things like that. You could take off and an hour later, drop, and you knew what the weather was going to be. Of course, at Edwards, it was normally pretty good anyway. But you could set yourself up ideally over the lakebed, too, so you didn't have the navigation concerns that you do coming back from orbit. Plus, the vehicle itself was ready to go *before* the rocket engines [were] ready, so that gave NASA an opportunity to get a look at the orbiter vehicle, its basic configuration, its flight control system, and make sure that it had an airplane or vehicle that could fly the pattern, the approach, the flare, and the landing, which was a very, very small part of its mission, but a very, very critical part of its mission, and gain confidence in that prior to committing to launch into orbit. The Approach and Landing Tests [were] initially designed for about 11 flights . . . and as we flew more and more flights, there was indeed pressure from the other end to hurry up and finish, so that those engineers could be assigned to the orbital flight test vehicle and the *orbital* flight looked like it was going to take resources and shorten up the initial approach and landing tests. The approach and landing tests were going very efficiently, too, so we were getting a lot of data and were able to condense the program from eleven flights down to what ended up to be five."

Far from simply being 'operators' of the vehicle, the men were heavily involved in the design and development of its software and procedures both for the ALT series and for operational missions. "If I'd never flown the Enterprise," Fullerton would say later, "doing the training was challenging and intriguing in its own right. People say 'How do you train?', thinking, 'Well, you go to a school and somebody tells you how to do it'. It's not like that at all. Somebody's got to write the checklist, so *you* end up writing the checklist, working with each subsystems person and trying to come up with a pre-launch checklist for the Approach and Landing Tests. So you're doing the work [and] the learning comes from doing jobs that needed to be done. We worried about doing this deadstick landing, so we had to train for that. I built a gadget to work on the T-38 [training jets] that would allow you with any given weight to set the power with the speedbrakes down to simulate what the data said the orbiter would fly it at, so that we could go fly the pattern we intended to fly in T-38s, making steep descents, flaring and touching down. The Shuttle training airplane, a Gulfstream II, was built as an airborne trainer, and so the four of us assigned to ALT served as the Shuttle pilots along with a Gulfstream pilot to do many dives at the ground to get the aircraft built and working right."

Haise and Fullerton's first captive-active flight was a day late in coming when one of Enterprise's computers exhibited a fault and had to be replaced on 17 June 1977. All was ready by the following morning, however, and the Boeing lumbered off the runway with its oversized cargo at 8:06 am Pacific Standard Time. The peculiar sensation of sitting aboard the Shuttle at such a great height came as a surprise to both astronauts. "When we first rode on top," Haise recalled, "you couldn't see the 747, no matter how [much] you'd try to lean over and try to look out the side windows. Not even a wingtip! It was kind of like a magic carpet ride. You're just moving along the ground and you take off; and something below you [was carrying you]; you knew it was there, but you couldn't see what was taking you aloft. It was also deceptive sitting up that high. Things always looked like it was going slower than it was, for your taxiing and particularly the first takeoff. I really thought Fitz had rotated too early. It didn't look like we were going fast enough."

Clad in their dark flight suits, Gordon Fullerton (left) and Fred Haise listen to a briefing before the first captive-active test flight aboard Enterprise in June 1977.

The feeling of nervous excitement and anticipation was equally intense for the crew of the Boeing. "Fitz was a great leader," Tom McMurtry recalled years later, citing the quiet moment, seconds before taxiing for takeoff on 18 June, when Fulton turned and shook hands with him, Horton and Guidry. "I thought that was a nice gesture. I think he just wanted to do that as a friend." During this flight, Haise and Fullerton were able to briefly test Enterprise's aerosurfaces, rudder and speedbrake. Ten days later, their colleagues Engle and Truly set off for a second run, conducting low-speed tests of her control system and simulating the separation manoeuvre from the Boeing that they would follow on the free flights. To accomplish these tasks, Fulton climbed to 6,700 m, before pushing-over and descending at around 900 m per minute, allowing the astronauts to position Enterprise's elevons in their ready-to-separate orientation. Haise and Fullerton completed the captive-active roster on 26 July to finalise avionics and control surface checks before the green light was given for the free flights to commence in August. The only minor problem was a faulty sensor in one of Enterprise's auxiliary power units, which triggered a caution-and-warning alarm in the cockpit and obliged Fullerton to shut it down. After landing on concrete Runway 22 at Edwards, Haise was able to deploy the Shuttle's landing gear, whilst still atop the Boeing, in readiness for the free flights.

Naturally, those free flights required close co-ordination between Enterprise's crew and that of the Boeing 747. "Our main concern," explained Joe Engle, "was to develop a separation manoeuvre with Fitz Fulton to optimise the separation between the two vehicles, both vertically and laterally. Fitz would put the combination in a slight dive to get the right airspeed in level flight, with the tailcone off. He would 'dive' the airplane and when he got on speed, he would call. We would separate and at *that* time the orbiter was sitting with a 15-degree angle of incidence; in other words, 15 degrees angle of attack and was trying to fly off the 747 at that time. Fitz would dump lift on the 747, throttle back to idle, so that as we came off, we didn't slide back and take his tail off; then the two of us would turn in different directions as well, so that as soon as we lost energy and started to come back down, we didn't come back down on top of him. In looking at the videos, there was lots of room, lots of separation, but initially we weren't sure, so we optimised everything we could. We didn't compromise anything by doing that, but we did have plenty of room for separation, and it was a co-ordinated manoeuvre. We would pull off to the right; he'd dive off to the left, then we'd go wings level and go right into the data-gathering manoeuvres, because we had very little time to get data. We had about a minute and a half to get data and then, well, the rest of the time was flare and land, to get the gear down and touch down."

Sixty-five thousand people, including 900 accredited members of the press and some 2,000 special guests, had assembled at Edwards in the pre-dawn hours of 12 August 1977 to see Enterprise perform her independent landing. Among those special guests were NASA Administrator Jim Fletcher, scientist-astronaut Joe Allen – then serving as the agency's head of legislative affairs – and a handful of Congressmen and senators. With Haise and Fullerton once again at the controls, Fulton duly ran the Boeing's engines up to full power at precisely 8:00 am and within seconds the combo swept, "remarkably quietly," onlookers would later comment, off Runway 22 and into the steadily-brightening desert sky, accompanied by five NASA T-38 jets. Higher-than-normal air temperatures at altitude delayed the original 8:30 am release time, but at 8:48 am, flying at 500 km/h, Fulton nosed the aircraft into a seven-degree dive. "The Enterprise is set," Haise radioed. "Thanks for the lift." Without further ado, he pushed the separation button on his instrument panel and seven explosive bolts fired, 'popping' the Shuttle away from the Boeing. As planned, Fulton immediately put his aircraft into a descending left turn, while Haise placed Enterprise into a right-hand turn and pitched upwards to increase the separation distance.

Fullerton, meanwhile, was busily scrambling to remove circuit breakers and reset switches within seconds of leaving the top of the Boeing. "The instant we pushed the button to blow the bolts and hop off the 747," he said years later, "the shock of that actually dislodged a little solder ball and a transistor on one of the computers and we had the caution tone go off and the red light. We had three [cathode-ray tube monitors] and one of those essentially went to halt. This is pretty fundamental. All your control of the airplane is through fly-by-wire and these computers. I had a cue card with a procedure if that happened, that we'd practiced in the simulator, and I had to turn around and pull some circuit breakers and throw a couple of switches to reduce your susceptibility to the next failure. I did that and, by the time I looked around, I realised, hey, this is flying pretty good, because I was really distracted from the fundamental evaluation of the airplane at first." As Fullerton worked, Haise held two degrees of pitch for three seconds, before banking 20 degrees to the right and heading for dry lakebed Runway 17. Maintaining a nine-degree, nose-down attitude, Enterprise executed a pair of 90-degree turns and Haise aimed her for the runway centreline and opened the speedbrake. Mission Control in Houston erroneously radioed that he had a lower lift-to-drag ratio than predicted in wind-tunnel tests, to which Haise responded by flying his final approach at higher speed, conserving energy to extend the glide. In fact, Enterprise's lift-to-drag ratio was exactly as expected.

Realising that the Shuttle was actually 'high and hot', and that he would land 'long', Haise opened the speedbrake from 30 to 50 percent to slow down and began the landing flare at an altitude of 275 m. As Enterprise levelled out, he deployed her landing gear and touched down 900 m 'long' at 340 km/h. As expected, in view of her low lift-to-drag ratio, the Shuttle had remained independently airborne for almost five and a half minutes. One unusual aspect of the touchdown was provided by the shortness of the nose landing gear, which caused the wings to tip downwards in a 'negative' angle of attack as soon as all six wheels were on the ground. "It felt almost funny," Haise recalled, "the first time you de-rotate, or try to the put the nose down: for a little bit, you almost think you don't have a nose gear, because it goes down so far! It does present a problem today, where the vehicle's heavier, with [the need to follow] a ritual on de-rotating to get the nose gear on. I've never been on an airplane that you had to actually worry about a sequence to do that effectively, because if you do de-rotate too fast, too early, while you're still at high speed, the effect of that negative lift, putting pressure down on the tyres, can conceivably blow [them]. You have to go down to a point in pitch to hold and wait till you get below a certain speed to then continue the de-rotation to effectively get the nose gear on the ground. At the same time, you can't hold it off too long, while it's still too high, or you'll lose the ability to arrest the fall [through], and so, if you kept it up too long, it would fall through and damage the nose gear from the standpoint of hitting down too hard. You've kind of got to work in between [with] a scheme of getting the nose gear on the runway."

For Jim Fletcher, the first independent landing of Enterprise turned into something which left him more than a little red-faced. As the vehicle dropped closer and closer to the runway, he began to panic that the Shuttle was going to crash and shouted "Landing gear! Landing gear!" at the top of his voice. The congressmen and senators around him remained silent. Seconds later, the landing gear duly popped out. "The flying procedure," explained Joe Allen, "is you can't put [the gear down] until the airplane's going a certain speed, lest you rip [it] off. *That* speed is very close to the ground, so it's unnerving to watch. No question." Later that day, Fletcher told Allen of his embarrassment at having shouted and having not known that the gear would be deployed at virtually the last second …

Engle and Truly took Enterprise aloft for her second free flight on 13 September for a more extensive series of manoeuvres and a perfect landing. In fact, the only problem of note was a radar failure at Edwards, which could have led to an abort had it not been promptly brought back online. Ten days later, Haise and Fullerton evaluated the 'autoland' system that NASA hoped would ultimately be capable of landing the vehicle without crew interaction. The astronauts allowed the computers to guide them down to the 270 m flare point, before taking over and hard-braking Enterprise to a perfect halt.

Original plans called for four flights with the tailcone in place, followed by two final attempts to land with the dummy main engines and OMS pods exposed, in order to assess the 'real' aerodynamic obstacles a Shuttle might encounter whilst returning from space. However, managers felt that the first three flights were so successful that Enterprise's fourth landing could be done without her tailcone and a trio of dummy main engines were installed. It had been expected that the Boeing would encounter increased aerodynamic 'buffeting' as the vehicle separated, although Fulton and his crew reported conditions as moderate, but acceptable. Enterprise did, however, descend far more steeply and rapidly than on her three previous flights and, despite some problems with the tactical air navigation system, Engle and Truly landed perfectly in just over two minutes.

"The orbiter flew pretty benignly with the tailcone on," Engle reflected, "a relatively shallow glide slope. You could get to a higher altitude for launch, because there was less drag, and the flight duration was well over five minutes. In fact, it was maybe up to *seven minutes* with the tailcone on, but that was *not* the configuration that we needed to really have confidence in, in order to commit for an orbital launch, because that re-entry and landing would be made with the engines *exposed* and required a much steeper glide slope, much more demanding profile, much more condensed time period from flare to touchdown, because the air speed would bleed off much faster with the additional drag. Although we were able to get a lot of really good systems data and time on the hydraulic systems and electrical systems and computers and flight control system ... with the tailcone on, from a performance standpoint and a piloting task standpoint, we really didn't have what we needed until we flew it *tailcone off*, and those flights were only about two and a half minutes long."

The final test, on 26 October 1977, with Haise and Fullerton at the helm for the last time, involved a precision touchdown on concrete Runway 22. Although the flight itself was successful, with both pilots reporting that the Shuttle performed better than predicted, trouble cropped up during their final approach to the landing site. As he emerged from the pre-flare manoeuvre, Haise found that Enterprise was dropping at 540 km/h, considerably faster than planned. In an effort to slow the orbiter, he opened her speedbrake early, but instead of slowing down, her speed increased. In response, he deployed the landing gear and pitched her nose down to achieve the desired touchdown point on the runway. The wings dipped and Haise struggled to correct the problem as Enterprise's main landing gear hit the concrete hard, before suddenly taking off again and 'bouncing' several metres into the air.

Eventually, after several long seconds, the vehicle stabilised herself for the remainder of the rollout and her nose finally dropped down onto the runway surface. The problem was ultimately traced to a 270-millisecond 'time lag' problem in the flight control software: the delay between Haise's inputs on the stick and Enterprise's response had led him to overcorrect, resulting in the 'pilot-induced oscillation', a porpoising motion and a bouncy landing. Although very unwelcome for Haise

Enterprise sits serenely on the runway at Edwards Air Force Base after the fourth free flight, piloted by Joe Engle and Dick Truly. Hurtling overhead is the Boeing SCA and a chase plane.

himself, the touchdown and rollout provided necessary data for engineers monitoring the landing gear. "The landing gear folks were quite chagrined through most of that test programme," he said later, "because we were not landing hard enough to get them good data for the instrumentation they had on the landing gear struts; although *I* solved their problem on the fifth landing flight, where I landed and *bounced* the vehicle! *That* gave them the data [they needed] and they were very happy with that, although I *wasn't*!"

Despite the somewhat problematic landing, NASA decided that the ALT series had achieved its objectives: to guide the approach and landing phases of a Shuttle mission, test the orbiter's autoland capability and demonstrate her subsonic airworthiness and the performance of her systems. Haise would also comment that verifying the performance of Enterprise's brakes and nosewheel steering had to be accomplished in *reverse*, compared to normal flight-test projects. "In an airplane," he explained, "you have a jet engine and normally approach flight test by first of all doing some taxi tests around the ramp and then some runs down the runway, progressively getting faster and faster. We had no way of doing taxi tests because Enterprise had no engines, so we were going to have to face taxi tests from the upper end of the speed spectrum, *backwards*! In other words, after we landed at 190 knots or so [about 350 km/h], somewhere down that rollout we were going to do taxi tests and we did it by each flight: [the] first flight started it at very low speed [and we] didn't touch anything 'til we got down slow. Then each flight stepped backwards up the speed spectrum to check out braking and nosewheel at progressively higher speeds. That's the way taxi tests were done with Enterprise."

Overall, the astronauts were happy with the performance of the Shuttle. "It handled better, in a piloting sense, than we had seen in any simulation," said Haise, "either our mission simulators or the Shuttle Training Aircraft. The term I use is: it was *tighter*. [It was] crisper in terms of control inputs and selecting a new attitude in any axis and being able to hold that attitude; it was just a better-handling vehicle than we'd seen in the simulation." The cause of Haise's bouncy landing, the time lag in the flight software, was subsequently corrected in time for the Shuttle's first orbital mission in April 1981. Although Enterprise had completed her approach and landing test series with flying colours, she still had several more tests to complete before, it was hoped, she would be extensively upgraded to make her capable of travelling into space. In March 1978, she was moved to NASA's Marshall Space Flight Center and spent the following year undergoing vertical tests, attached to an empty External Tank and two inert Solid Rocket Boosters. The results of this work ultimately led to several design changes to the system, particularly the boosters. It was after the completion of these tests that Enterprise was scheduled to be modified for her first orbital voyage.

As history has shown us, it never happened.

Lessons learned during her construction were subsequently incorporated into the design of the first spaceworthy craft, Columbia, and it became evident in 1978 that Enterprise weighed too much to transport a full payload into orbit. She would need a new set of plans, quite different from those of her sister ship, to make her spaceworthy. Moreover, she contained no propulsion system, plumbing, fuel lines or tankage of any kind, her main engines were dummies, her payload bay had no mounting hardware for cargo, its doors had no opening-and-closing mechanisms or cooling radiators and her 'thermal-protection system' was nothing more than black-and-white polyurethane foam and fibreglass. Inside Enterprise's crew cabin, her instrument suite was sparsely populated with switches and dials compared to later orbiters: she had no guidance equipment, such as star trackers or heads-up displays and no indicators of External Tank or Solid Rocket Booster systems. Elsewhere, she had no aft flight deck or overhead windows, no airlock, no middeck lockers, no galley, no shower, no other crew-related items and her simple fuel cells were fed by high-pressure tanks, rather than cryogenic dewars. Her landing gear operated by explosive bolts and gravity, with no hydraulic mechanisms or manual backup systems available. She did, however, contain Lockheed-built ejection seats for her pilots, which would have fired them through two blow-out aluminium panels in the ceiling in the event of an emergency. Modifying her for space missions, therefore, was envisaged to be a long, complex and expensive task. Additionally, the new design specifications called for much stronger wings and mid-fuselage than were installed aboard Enterprise and several aluminium components had to be changed to titanium to save weight. Transportation and modification costs to accomplish these changes were simply unavailable and, as 1978 wore on, NASA was already considering a high-fidelity Shuttle structural test article known as STA-099 as a cheaper option to upgrade for orbital service. On 29 January 1979, it was official: STA-099 would become the second space-capable orbiter, later renamed 'Challenger' and NASA contracted with Rockwell to build two more vehicles.

Several months after the contracts to modify STA-099 and build Discovery and Atlantis were signed, Enterprise was transported to the Kennedy Space Center in Florida, mated to an empty External Tank and two inert Solid Rocket Boosters and rolled out to Pad 39A. The disappointment at seemingly having had the opportunity to be made spaceworthy taken from her was sweetened somewhat, in that she became the first structurally 'complete' Shuttle stack ever to sit on the launch pad. Her task for the next 11 weeks was to test the crew-escape systems and ensure that all of the work platforms were in the right places to access different parts of the orbiter, tank and boosters. Returned to California in the autumn of 1979, she was juggled between Edwards Air Force Base and Rockwell's Palmdale plant for two years, before undertaking her European tour in the early summer of 1983, offering me my chance to see her, albeit from afar, whilst returning to the United States to Louisiana as an exhibit in New Orleans for the 1984 World Fair. More tests were afoot in the winter of that year, when she was moved to Vandenberg Air Force Base in California for six months of fit checks at the Department of Defense's Shuttle launch complex. Ironically, this also made her the only complete Shuttle stack to sit on the launch pad at 'Slick Six', for the site was abandoned, at least for manned missions, in the wake of the Challenger disaster. A brief spell sitting next to the Saturn V exhibit at the Kennedy Space Center was followed by installation at Dulles Airport in November 1985, becoming the formal property of the Smithsonian. Following the loss of Challenger, the option of modifying her for space missions to return the fleet to four operational vehicles was briefly considered, but it was deemed cheaper to use an already-fabricated set of structural spares as the basis for a new orbiter, which was duly named 'Endeavour'.

Her last, remotely viable, chance of being upgraded into a spacegoing vehicle came more than a decade ago, in February 1996, when a team from the Johnson Space Center in Houston, Texas, assessed her structural condition for possible refurbishment as an 'additional' Space Shuttle. In general, even after two decades of operations, she was in remarkably good condition, with no significant corrosion around her aft body flap or elevons and her wings were found to be in acceptable condition. More serious corrosion, probably caused by the effects of standing water, was detected on the floor of her aft fuselage and around her nose landing gear wheel well. Ultimately, NASA derived greater benefit for the remaining members of its Shuttle fleet by using Enterprise as a testbed for continuing improvements to the system and helping to pinpoint the cause of the February 2003 Columbia tragedy. Today, as she sits in pride of place in the Smithsonian's Udvar-Hazy Center, it is fortunate, in a way, that she was never made spaceworthy. Unlike Columbia, whose loss, astronaut Jay Buckey once lamented, had robbed him of the chance to someday take his grandchildren to see the ship he once flew, the happy ending for Enterprise, at least, is that she is still with us.

WINGS AT THE EDGE OF SPACE

Happiness was a precious commodity during much of the Shuttle's genesis, for the notion of changing the United States' spacegoing philosophy from small capsules, lofted atop the descendants of ballistic missiles, to something the size and shape of a conventional airliner, ferried into orbit attached to a large fuel tank and a pair of rocket boosters, was a vast effort and posed monumental, and frequently maddening, obstacles. Naturally, the idea for the Shuttle did not appear from nowhere; it could trace its development across several decades and many of its systems had been pioneered and proven by a series of aerospace projects organised by NASA and its forerunner, the National Advisory Committee for Aeronautics (NACA). As early as the 1930s, the German aerospace engineer Eugen Sänger developed early concepts for rocket-propelled aircraft, some of which were capable of velocities as high as ten times the speed of sound and could reach altitudes of up to 70 km. In his later work, Sänger showed how the addition of wings could extend the potential of a spacegoing rocket: although its initial range would be quite modest, following an arc-like trajectory, much akin to an artillery shell, the *lift* generated by wings during re-entry would carry it *upward*, allowing it to 'skip' off the atmosphere and opening up possibilities of a craft which could circle the globe and return to its launch site. Within a few years, such dreams rose from the drawing boards and into hardware: in 1947, Chuck Yeager flew the rocket-propelled Bell X-1 aircraft through the sound barrier for the first time and in 1951 the Douglas Skyrocket and its pilot, Scott Crossfield, reached an impressive Mach 2.

At these velocities, issues of aerodynamic overheating did not yet prove a significant obstacle, but as the US military focused more attention to the prospects of delivering heavy nuclear warheads to Moscow, speed and stability became more important. In January 1952, NACA was urged to pursue a manned research aircraft, capable of exceeding Mach 5, with early reports even suggesting a *commercial* hypersonic vehicle. If such a plan was to be realised, it needed to overcome issues of aerodynamic overheating and stability. The latter problem was resolved by Charles McLellan, an aerodynamicist at NACA's Langley Aeronautical Laboratory in Hampton, Virginia, who proposed that small, wedge-shaped vertical fins and horizontal stabilisers would be much more effective than conventional thin surfaces. Overheating proved a harder nut to crack. If the aircraft re-entered the atmosphere with its nose pointing in the direction of flight, its streamlined shape would subject it to disastrous overheating and destructive aerodynamic stresses. A re-entry with the nose positioned at a higher angle, with the flat undersurface presented to the hypersonic airflow, would be a more manageable approach, permitting the craft to lose speed in the upper atmosphere, ease overheating and lower aerodynamic loads.

This realisation that overheating would be lessened did not detract from the reality that such vehicles would be subjected to re-entry temperatures of significantly higher severity than had previously been encountered. Bell Aircraft had already begun to explore temperature-resistant materials, including a chrome-nickel alloy called 'Inconel-X' and stainless steel 'shingles', capable of radiating heat away from the airframe, coupled with techniques for water-cooling areas along the leading edges of the wings. In October 1954, the Aircraft Panel of the Scientific Advisory Board expressed its belief that the next decade's research and development goal should focus on the field of hypersonic flows. "This is one of the fields in which an ingenious and clever application of the existing laws of mechanics is probably not adequate," the panel reported to Air Force Chief of Staff Nathan Twining. "It is one in which much of the necessary physical knowledge still remains unknown at present and must be developed before we arrive at a true understanding and competence." Summing up, the panel considered the time ripe for the construction of a new aircraft to surpass Mach 5 and reach altitudes as high as 150 km. The project received overwhelming support from within the Air Force and NACA and in the spring of 1955 it also received a name: the X-15. In time, it would become the fastest and highest-flying aircraft before the arrival of the Shuttle. In August 1963, with NASA test pilot Joe Walker at the controls, the X-15 reached a record altitude of 107 km, and a few years later, the Air Force's William 'Pete' Knight achieved a maximum speed of Mach 6.72 (some 7,270 km/h). The Air Force decided to award its own 'astronaut wings' to pilots who flew higher than 80 km, but the Fédération Aéronautique Internationale (FAI) decreed that only flights which exceeded 100 km officially reached the threshold of space. The FAI's ruling was based on the so-called 'Kármán Line', named after Theodore von Kármán, the Hungarian-American engineer who first described it as an altitude at which the atmosphere is too thin for aeronautical purposes, since vehicles would need to travel faster than orbital velocity to achieve sufficient lift. Following these two rulings, between July 1962 and August 1968, *thirteen* X-15 missions by eight men exceeded the Air Force's limit, but only *two* of those missions (both flown by NASA test pilot Joe Walker) successfully passed the Kármán Line and were considered official 'spaceflights' by the FAI.

Years later, however, NASA honoured the 80 km Air Force limit and in August 2005 awarded 'astronaut wings' to its three civilian X-15 pilots: posthumously in the cases of Walker and Jack McKay and in person to Bill Dana. (The five Air Force X-15 pilots had long since been awarded such wings by their service.) Tim Furniss has referred to this baker's dozen of X-15 voyages as 'astro-flights', thereby differentiating them from actual orbital or suborbital 'spaceflights'.

Prime contractor for the X-15's airframe was North American Aviation, which later merged with Pittsburgh-based Rockwell Standard to form North American Rockwell in 1967 and, perhaps fittingly, in later years the corporation would build the Shuttle. Three X-15s were built – numbered, unsurprisingly, No. 1, No. 2 and No. 3 – although only two survive today: one resides in the Smithsonian, the other in the National Museum of the Air Force at Wright-Patterson Air Force Base in Dayton, Ohio. The third was lost in a November 1967 crash which killed its pilot, Mike Adams. The X-15 was a one-man aircraft and took the form of a cylindrical fuselage, 15.5 m long and 4.1 m high, with a total span across its stubby wings of 6.8 m and a total weight, when fully loaded, of 15,420 kg. It possessed thick dorsal and ventral wedge-like stabilisers, a nose-wheel carriage and a pair of rear skis. Meanwhile, the powerplant for the X-15 was built by Reaction Motors. The initial pair of XLR-11 rocket engines was later replaced by a single, *throttleable* XLR-99, fuelled by ammonia and liquid oxygen and capable of 26,000 kg of thrust. The aircraft was carried aloft under the wing of a modified B-52 bomber and was typically released at an altitude of about 13 km and a speed of some 800 km/h, after which the engine would ignite to commence its climb to the desired maximum altitude.

"Many felt that the throttleable feature might have been a needless luxury that complicated and delayed the development of the XLR-99," wrote Dennis Jenkins in 2000, "but in the mid-1960s these attributes were considered vital to the development of a rocket engine to power the Space Shuttle. At the time, the Shuttle was to consist of two totally reusable stages; essentially a large hypersonic aircraft that carries a smaller winged spacecraft, much like the B-52s carried the X-15s. The same basic engine was going to power both stages; the pilots therefore needed to be able to control its thrust output. At some points in the early Shuttle concept development phases, the same engines would also be used *on-orbit* to effect changes in the orbital plane, so the original concept for the Space Shuttle Main Engines (SSME) included the ability to operate at ten percent of their rated thrust, and to be restarted multiple times during flight. In the end, the production SSMEs are throttleable within much the same range as the XLR-99 (65 to 109 percent, in one percent increments). In actuality, about the only routine use of this ability is to throttle down as the vehicle reaches the point of maximum dynamic pressure during ascent, easing stresses on the vehicle for a few seconds on each flight. Nevertheless, the development pains experienced by Reaction Motors provided insight for Pratt & Whitney and Rocketdyne (the two main SSME competitors) during the design and development of the SSMEs."

Scott Crossfield, the first man to fly at twice the speed of sound, was the X-15's maiden pilot, taking the craft on an unpowered sortie in June 1959, followed by the first powered mission in September *and* the first XLR-99 test in November 1960. Of key importance for the purposes of space travel were those astro-flights which reached 80 km or higher between July 1962 and August 1968, featuring Air Force pilots Bob White, Bob Rushworth, Joe Engle (on *three* occasions), William 'Pete' Knight and Mike Adams and civilians Jack McKay, Bill Dana (twice) and Joe Walker (thrice). Of Walker's missions, two of them – on 19 July and 22 August 1963 – reached 105.9 and 107.8 km, respectively, making him the only one of the eight pilots to have been officially recognised as an astronaut by the FAI.

During his July 1963 flight, Walker and his B-52 ferry set off from Smith Ranch Dry Lake in Nevada and performed experiments ranging from using a balloon instrumented to measure air densities, a horizon scanner and infrared and ultraviolet photometers. The entire mission lasted about 12 minutes, with the XLR-99 burning for 85 seconds, subjecting Walker to four times the force of terrestrial gravity. Less than five weeks later, he exceeded his previous height by a little under 2 km and established a world aviation altitude record which would not be surpassed until Brian Binnie flew SpaceShipOne to an altitude of 112 km in October 2004. A year after Binnie's flight, Walker – who, as a civilian, had previously been barred from receiving astronaut wings – was posthumously awarded the honour by NASA. With this recognition, he can now be credited not only as America's first civilian astronaut, preceding Neil Armstrong by almost three years, but also can lay claim to having been the first man to make *two* spaceflights.

Joseph Albert Walker was born in Washington, Pennsylvania, on 20 February 1921 and developed a keen mechanical mind, once taking components from an old farm motor and buggy and building a scooter ... *without brakes*. He came fourth in his class at Trinity High School and enrolled at Washington and Jefferson College to study physics. In his final year, he enrolled in the Civilian Pilot Training Program and was told by an instructor that he was too cautious to succeed, but Walker toughed it out and achieved the highest score of any cadet in the Pittsburgh area. Graduation in 1942 was followed by admission into the US Army Air Force, during which time he flew the Lockheed P-38 Lightning fighter and the F-5A weather reconnaissance aircraft during the Second World War, receiving the Distinguished Flying Cross and an Air Medal with seven oak leaf clusters.

He later resigned from active military duty, with the rank of captain, to work for NACA, first as a civilian physicist and later as a research pilot conducting analyses of icing processes. He transferred to the agency's High Speed Flight Station at Edwards Air Force Base in 1951 and rose rapidly to become the chief research pilot. Walker flew the Bell X-1, the Skyrocket and served as chief project pilot for Douglas' X-3 Stiletto – the latter, he would remark, was the *worst* aircraft he had ever known. In June 1958, Walker was one of nine pilots picked by the Air Force for 'Man In Space Soonest' (MISS), a short-lived attempt to launch the world's first astronaut. After the cancellation of MISS, he joined the X-15 project, flying the rocket aircraft two dozen times, "touched the skirts of space", according to *Time* magazine, and supported the Apollo Moon effort by piloting NASA's notoriously unpredictable Lunar Landing Research Vehicle.

Awarded an honorary doctorate by his *alma mater* in 1961 and recognised as Pilot of the Year by the National Pilots Association in 1963, Walker was one of the most famous names in American aviation when he met his untimely death on 8 June 1966. He was flying an F-104 Starfighter in tight formation with the XB-70 Valkyrie, an experimental version of a planned supersonic bomber, during a publicity event in Barstow, California. Without warning, his aircraft drifted into contact with the Valkyrie's wing tip, flipped over, struck the bomber's vertical stabilisers and left wing and exploded. Walker was killed instantly and although one of the Valkyrie's two pilots safely ejected, the other tragically also died. Had the accident not occurred, Walker was scheduled to make his first flight at the XB-70's controls ... *the very next day*. In its summing-up of the disaster, a week later, *Time* noted that Walker's "shattered corpse" was found inside the wreckage of the F-104, "minus his helmet ... leading to speculation that he might have been killed as the [Valkyrie] sheared through his canopy ... "

The X-15 project concluded a little more than two years after Walker's death, completing a grand total of 199 missions (a 200th *was* planned to be flown by Pete Knight, but later cancelled due to bad weather and technical problems). However, as the X-15 pushed altitude and velocity to new levels and demonstrated a throttleable rocket engine, other aerodynamicists had made great strides in the development of manned 'lifting bodies' and Air Force Projects ASSET (Aerothermodynamic/elastic Structural Systems Environment Test) and PRIME (Precision Recovery Including Manoeuvrable Entry) flew a series of small hypersonic gliders at speeds of up to 25,000 km/h and achieved cross-ranges of more than 1,000 km. Although the ASSET and PRIME prototypes did not have the capacity to land on runways – they used parachutes for their final descent to Earth – they did nevertheless demonstrate their manoeuvrability and their ability to survive the furnace heat of re-entry. These early concepts offered a proving ground for a series of manned variants, flown at Edwards Air Force Base from 1963 until 1975, which evaluated the ability of a pilot to manoeuvre and land safely in a wingless vehicle, whose *shape* provided them with the aerodynamic lift normally afforded by the wings. The concept of lifting bodies originated with a team led by NACA aerodynamicist Al Eggers in 1957 and their design ultimately became the M2-F1, nicknamed 'the flying bathtub'. Subsequent developments included the M2-F2 (later rebuilt as the M2-F3 after a near-fatal crash in 1967), the HL-10 and the Air Force's X-24, all of which were launched from the wing of a modified B-52 and utilised XLR-11 rocket engines to accelerate to speeds of almost 2,000 km/h.

Of course, America's space ambitions in the 1960s were bound up with the drive to land a man on the Moon, but in the opening months of Richard Nixon's administration a Space Task Group was established, chaired by Vice-President Ted Agnew, with the mandate to chart possible courses for the immediate and long-term future. Sadly, Agnew's lack of real clout in the White House meant that only one of his group's four recommendations – the Shuttle – was actually endorsed by Nixon. Plans for advanced Moon bases, a manned voyage to Mars and an expanded space station fell on deaf ears. The president liked the 'heroic' image of astronauts, but his enthusiasm and excitement for space travel itself was low. Unlike John Kennedy and, to a lesser extent, Lyndon Johnson, Nixon saw little political, scientific or technological value in exploring the heavens. The American Institute of Aeronautics and Astronautics (AIAA) also endorsed the Shuttle above the other three options, noting that a low-cost manned spacecraft for delivering medium to large payloads into orbit would enable it to "effectively compete with present expendable boosters" and collectively felt that "commitment to an entirely new space station is less urgent than commitment to a new logistics system". The President's Science Advisory Committee (PSAC) went still further, suggesting that the Shuttle – "a reusable space transportation system with an early goal of replacing *all* existing launch vehicles" – would not only allow for the deployment and recovery of satellites *and* the orbital assembly of a variety of different structures, but would also lead to a "radical reduction in unit cost of space transportation". In general, the Shuttle was growing in popularity in the opening months of 1969, supported by the AIAA, the PSAC, NASA itself and the US Air Force, with the latter expected to be one of its major customers and a key player in the definition of its final design.

"During the year that followed the landing of Apollo 11 in the Sea of Tranquillity," space advocate Thomas Heppenheimer wrote in *The Space Shuttle Decision*, "NASA received a cold bath in the Sea of Reality." A pessimistic remark, perhaps, but appropriate, nonetheless, since 1970 was a time of retreat from lofty goals of permanent lunar bases and expeditions to Mars and a refocusing of national priorities: ending the war in Vietnam, beginning the war against poverty, improving the health care system ... and sharply cutting back the space budget. Although Congress *would* ultimately agree that the Shuttle *was* the most important 'next step', and would appropriate $110 million to start the work in Fiscal Year 1971, the chances of this reusable winged spaceplane ever getting off the ground seemed slim. Its initial funding had but barely survived in the House and had only won by four votes in the Senate; clearly even this low-cost approach to future space exploration was vulnerable to even the slightest change in anti-space sentiment.

The physical appearance of what would become the Shuttle was also steadily evolving into a delta-winged spacecraft, because this would produce considerably more 'lift' at hypersonic speeds and permit a substantial 'cross-range' capability – flying large distances to the left or right of an initial direction of flight – so as to attract the US Air Force, whose plans for large photographic reconnaissance and imaging satellites required the kind of payload capacity and capabilities that the Shuttle could offer. In 1967 and 1968, their attempts to acquire detailed satellite imagery of troop movements during the Six-Day War in the Middle East and the Soviet invasion of Czechoslovakia had been thwarted due to the limitations of the system and the amount of time required for photographic films to be returned and developed. 'Real-time' reconnaissance required not only new technologies, such as charge-coupled imaging devices, but also the ability of the Shuttle to fly into orbit, perform a single orbit and return to base with film exposed less than an *hour* earlier. Other plans included using the craft to quickly snare Soviet spy satellites from orbit. For the Shuttle to receive Congressional approval, NASA *needed* the Air Force on its side, because such important missions of national security would benefit from the capabilities of a delta-winged, high-cross-range vehicle. "These Air Force leaders knew that they held the upper hand," wrote Heppenheimer. "They were well aware that NASA needed a Shuttle programme and therefore needed both the Air Force's payloads *and* its political support. The payloads represented a tempting prize, for that service was launching over *two hundred* reconnaissance missions between 1959 and 1970."

Unfortunately, the Air Force did *not* need NASA to launch their satellites; they could rely perfectly well on expendable boosters, such as the Titan III. Their support for the civilian space agency would come at a cost: *only* if NASA gave the Shuttle a large enough payload bay to hold big satellites *and* a 1,700 km cross-range capability – more than twice as large as NASA had planned – which would enable them to land back at Vandenberg Air Force Base in California after single-orbit, 90-minute-long reconnaissance or satellite-snaring missions. NASA's Max Faget, who wanted a straight-winged Shuttle optimised for subsonic performance, was disappointed when the delta wing shape was accepted, but it turned out to be a small price to pay for the Air Force's support. In order to achieve the larger cross-range capability, the Shuttle would glide hypersonically, thereby increasing both the *rate* of aerodynamic heating and the *duration* of that heating ... which, in turn, would require the additional weight (and cost) of a beefed-up thermal protection system.

Linked to the requirements of *both* NASA and the Air Force was the size of the Shuttle's payload bay, which then-Administrator Tom Paine described as being around 5 m in diameter and 20 m long; this was significantly larger than earlier concepts, which had bays about 7.5 m in diameter and 10 m long. "The Air Force needed length," explained Heppenheimer, "for its reconnaissance satellites amounted to orbiting telescopes and these had to be long to yield the sharpest images. Moreover, such satellites were growing markedly in length. The Corona spacecraft of the 1960s, each with an attached Agena upper stage, had *started* at [5.5 m] and quickly grew to [8 m]." Already, the CIA was preparing for new generations of satellites, known as 'Big Bird' and 'Kennan', and the latter was only in the preliminary design stage, but was expected to have a focal length of around 20 m. Accordingly, the Air Force told NASA that they would require the Shuttle's payload bay to accommodate a 20 m satellite. Any reduction in size would be simply unacceptable: even a 14 m bay would mean that less than *half* of the Air Force's satellites could fly on the Shuttle. "This requirement," the Air Force told NASA, "is *still* considered valid. Should you elect to develop the Shuttle with a [smaller] payload compartment, it will preclude our full use of the potential capability and operational flexibility ... Also, if a portion of the present expendable launch vehicle stable must be retained to satisfy some mission requirements, then the potential economic attractiveness and the utility of the Shuttle ... is severely diminished." For the civilian space agency, the bay would enable the Shuttle to carry components for a modular space station, to be launched, piecemeal, in the early 1980s ... but that was a mere sidenote, compared to the needs of the Air Force and the Department of Defense. In terms of weight, the Shuttle *had* to be capable of delivering an 18,000 kg payload into low polar orbit; Big Bird was known to be around 13,000 kg and future satellites would almost certainly exceed this. With few quibbles from NASA, the military got its way. One anecdote from this period concerns an attempt made by Max Faget, one of the world's foremost aerospace engineers, designer of the

Mercury spacecraft and then-head of engineering and development at the Manned Spacecraft Center in Houston, to slightly reduce the *diameter* of the payload bay (a dimension in which he thought the military had little interest) from 5 m to 4 m. "It took the Air Force only three days to put him in his place," wrote Heppenheimer, "with a reply that read: *The USAF fully supports and stands firm on the present Level 1 requirement for a payload diameter of 5 m and a length of 20 m.* This reply came from one Patrick Crotty, whose rank was no higher than Major." NASA can have been under no illusions that the Air Force was in control of virtually every detail of the Shuttle.

This control was underlined, once and for all, in January 1971, when NASA formally presented the requirements of the Shuttle to all study contractors at a meeting in Williamsburg, Virginia ... and gave the Air Force all it demanded: the cross-range capability, the payload bay length *and* diameter, the delivery capacity to polar orbit; *everything*. "One sometimes hears that when two parties are in a relationship, the one that wants it *more* is the weaker," remarked Heppenheimer. "NASA certainly had been pursuing support for the Shuttle with unmaidenly eagerness and the Williamsburg rules were the result. The agency now was promising to build a bigger and heavier Shuttle than it had *wanted* for its own uses, with considerably more thermal protection. It also was prepared to treat the Shuttle as a *national asset*, which meant the Air Force would not pay for its development or production and yet would receive the equivalent of exclusive use of one or more of these vehicles, entirely *gratis*." The Air Force would, to be fair, be responsible for the development of their own Shuttle launch facility at Vandenberg Air Force Base in California, but setting that to one side Heppenheimer is not alone in comparing the relationship to a treaty between a superpower and a banana republic ...

As the Shuttle danced to the Air Force's tune, other changes were ongoing in the design of the rocket booster. Potential contractors had considered expendable or only partially-reusable elements, but from August 1969, NASA turned its gaze to a fully-reusable booster. "Partially-reusable designs had represented an effort to meet economic goals by seeking a Shuttle that would cost less to develop than a fully-reusable system," wrote Heppenheimer, "even while imposing higher costs per flight. This approach had held promise prior to the spring of 1969, when the Shuttle had been considered largely as a means of providing space station logistics. Now, its intended uses were broadening to include launches of automated spacecraft, which meant it might fly far more often. The low cost per flight of a fully-reusable [system] now made it attractive and encouraged NASA to accept its higher development cost."

Flying more often, it was hoped, would gradually help to reduce overall launch costs and NASA anticipated it might also cut the cost of the *payloads* themselves. As more payloads entered orbit, frequently and reliably, they might stimulate new uses for space, thereby encouraging contractors to build further satellites, with the Shuttle serving much the same role as commercial aviation had done, achieving enormous growth by cutting the prices of its passengers' tickets. Its payload bay offered a vast volume, which eased restrictions on weight and size, and the satellites to be placed into orbit could be fault-checked by the Shuttle crew, *in orbit*, thus lessening the mass of paperwork on the ground. Subsystems aboard these satellites could be standardised and when they started to fail in orbit after a few years of service, the Shuttle had the capability to retrieve, refurbish and deploy them for a fraction of the cost of a full replacement.

Processing of the Shuttle and its booster was also required to contribute to substantial cost savings. In the early part of 1971, the booster was expected to be longer than a Boeing 747 and considerably fatter in the fuselage, and it would be swung into an upright position and mated with the orbiter. An enormous, diesel-powered 'crawler' would then transport it from the Vehicle Assembly Building (VAB) to the pad. Launch would occur under the combined thrust of maybe a dozen main engines – the contract for which would ultimately go to Rocketdyne, a division of North American Rockwell – and the booster and Shuttle would part company in the high atmosphere, with the former returning to a conventional runway landing, under the power of up to a dozen jet engines. The Shuttle itself would return to a similar style of landing after completing its mission in space. Processing of the booster and orbiter between flights was expected to require two weeks. Mathematical and economic studies from this period suggest that NASA was confident that each mission could be launched for around $4.6 million – a million dollars less than for an expendable Delta rocket, one of the smallest rockets then in use – and a full order of magnitude lower than the $55 million cost of the Saturn IB. Early development cost outlays, to be fair, *would* be significant, but within a few years the Shuttle would begin to show massive savings over expendable boosters, perhaps as high as a billion dollars per annum. This dramatic reduction in launch costs was then expected to spark the new revolution in spacecraft design, but it would not come to pass; for in Thomas Heppenheimer's words, "the *first* statement was a speculation [and] the *second* then amounted to a speculation that *rested* on a speculation". Few could have foreseen at that time the difficult development process which would precede its maiden launch.

The final approval of the Shuttle came in April 1972 and was announced, by an interesting and coincidental quirk, whilst the man who would subsequently be chosen to command its first flight, astronaut John Young, was on the Moon, halfway through his Apollo 16 mission. NASA had drummed up a list of possible names for the new craft – Pegasus, Hermes, Astroplane, Skylark, to name a few – but it was Richard Nixon himself who ultimately made the final choice, by referring to it simply as 'the Space Shuttle'. It would thus break with previous practice of actually *naming* the project, although, as time would tell, individual orbiters *would* receive their own unique names. Of course, Nixon's acceptance of the Shuttle was rooted securely in down-to-Earth politics and his motivation was to strengthen employment within the aerospace industry as he entered election year for his second term in office. NASA Administrator Jim Fletcher had already told the White House in November 1971 that an early start on the Shuttle "would lead to a direct employment of 8,800 by the end of 1972 and 24,000 by the end of 1973". In terms of employment and the need to pacify America's so-called 'battleground states' – those pivotal areas in control of large blocks of electoral votes – NASA and the space programme was recognised to have an importance which was out of proportion to its budget.

In its final form, the Shuttle looked quite unlike any previous manned spacecraft; an asymmetrical vehicle with the winged orbiter attached to a bulbous External Tank and a pair of solid-fuelled rocket boosters. Here, Columbia rolls towards Pad 39A in December 1980, preparatory to her maiden mission.

On 26 July 1972, out of a pool of proposals which also included Lockheed, McDonnell Douglas and Grumman, the Nixon administration selected North American Rockwell, based in *California*, with its 55-vote monopoly in the Electoral College, to receive the $2.6 billion contract to build the Shuttle. This prompted Jean Westwood, chair of the Democratic National Committee, to harshly criticise Nixon's "calculated use of the American taxpayers' dollars for his own pre-election purposes". Westwood queried the 'help' that five of the corporation's directors provided to Nixon's 1968 election campaign and requested "a full airing" of the background to the contract award. Nevertheless, there *were* solid engineering reasons for not selecting the others. Lockheed's craft, for example, was heavy, "unnecessarily complex", according to Jim Fletcher, and it left a minute-long gap during ascent when there was no emergency abort provision. McDonnell Douglas' proposal was technically deficient and weak, whilst Grumman – which had earlier built the Apollo lunar lander for NASA – came second to North American Rockwell. Grumman's presentation was impressive, identifying fundamental problems and offering good solutions, but fell down in terms of costs and overall project management. Even North American Rockwell itself, whose cost estimate was the *lowest* of the four, had a number of weaknesses in its design, not least a crew cabin which would be difficult to build. It is an interesting footnote that the corporation also gained an edge through its approach to the hiring of *minorities* into its workforce; by 1972 it had more blacks, Hispanics and Asians that the other three competitors. Notably, the Equal Employment Opportunity Act was passed that same year. "We're *not* crusaders for civil rights," Richard McCurdy, NASA's then-associate administrator of organisation and management, noted, "but the fact that North American moved forward on this front tells us something about how the company is thinking ahead."

High scores and low costs, therefore, were the reason behind the decision to pick North American Rockwell to build the Shuttle. In fact, when NASA Deputy Administrator George Low asked the three losing bidders to comment on the fairness of the competition, all three felt that it was the best and fairest competition that they had ever participated in. The situation was quite the opposite for the leading contenders to build the Shuttle's main engines, with Rocketdyne – a subdivision of North American Rockwell – having been chosen for the $450 million contract over Pratt & Whitney. The latter had confidently *expected* to win, even taking out adverts in major aerospace magazines and declaring themselves ready to get the project started. Pratt & Whitney had even lodged a complaint against NASA, alleging that the space agency had performed unfair acts in selecting Rocketdyne, but in March 1972 the Comptroller-General, Elmer Staats, upheld the decision. He slammed Pratt & Whitney, telling their lawyers that it was *also* unfair on NASA, which had helped them "to bring [an] original, *inadequate* proposal up to the level of other, *adequate* proposals, by pointing out those weaknesses which were the result of [Pratt & Whitney's] *own* lack of diligence, competence or inventiveness".

As the contracts rolled out, North American Rockwell – which became 'Rockwell International' in March 1973 – proved itself more than honourable, by offering important subcontracts to its rivals. Grumman would take charge of building the orbiter's delta-shaped wings *and* the Gulfstream II Shuttle Training Aircraft, for example, and an unsuccessful attempt had earlier been made to interest Pratt & Whitney themselves in sharing in the main engine development programme. McDonnell Douglas would take responsibility for the Shuttle's tail-mounted Orbital Manoeuvring System (OMS), whilst in August 1973 NASA chose Martin Marietta to build a large External Tank to carry cryogenic propellants for the main engines and in June 1974 picked Morton Thiokol for a pair of reusable, solid-fuelled rocket boosters, which would provide 80 percent of the thrust needed for the orbiter to reach space.

By this time, of course, the shape, design and definition of the Shuttle system had evolved into the form that we can recognise today. The orbiter itself is similar in dimensions to the DC-9 airliner, roughly 36 m long with its wings spanning 24 m from tip to tip. The first set of wings, built by Grumman, were among the earliest sections to be completed and were delivered to Rockwell's Palmdale plant in California in April 1975. The habitable area of the Shuttle, meanwhile, consisted of a two-tiered cockpit – a 'flight deck' for operations and a 'middeck' for experimental work, eating and sleeping – which backed onto the 20 m long payload bay and an aft compartment to house the three main engines and two OMS pods and support a vertical stabiliser fin. Forty-four tiny Reaction Control System (RCS) thrusters in the Shuttle's nose and tail would provide attitude control and additional manoeuvrability whilst in space. The graphite-epoxy payload bay doors were, at the time, the largest aerospace structures yet built from composite material and *had* to be opened within a couple of hours of reaching orbit, to enable the radiators lining their interior faces to shed excess heat from electrical systems into space. The five-piece doors were hinged at either side of the mid-fuselage, mechanically latched at the forward and aft bulkheads and thermally sealed at the centreline. Ordinarily, they are driven 'open' and 'closed' by electromechanical power, but if the astronauts are unable to open them, the vehicle must return to Earth at the earliest opportunity. Conversely, if the doors do not *close* properly at mission's end, two crew members are trained to go outside on an EVA to operate the mechanism manually.

In the years between the initial 1972 contract awards and its maiden voyage in April 1981, the Shuttle's designers faced setback after setback: frustrating problems with the patchwork of heat-resistant tiles and thermal blankets to shield it during its hypersonic descent through the atmosphere and maddening explosions of the main engines on the test stand. Throughout the 1970s, powerful Congressional opponents questioned the need for a reusable manned spacecraft. The unusual appearance of the Shuttle 'stack' on the launch pad has been described by Story Musgrave – the only astronaut to have flown aboard all five orbiters – as "like bolting a butterfly onto a bullet". The *bullet* was the External Tank, which resembled an enormous aluminium zeppelin, standing on end, and measured 46.6 m tall. It comprised two tanks for liquid oxygen and liquid hydrogen, separated by an 'inter-tank' for instrumentation and umbilicals. The oxygen tank at the top housed up to 542,640 litres of oxidiser and the hydrogen tank held some 1.4 million litres of fuel, both of which were fed through a pair of 43 cm diameter lines into disconnect valves in the Shuttle's aft compartment and thence into the combustion chambers of the three main engines.

"The main engine is very high performance," said Henry Pohl, NASA's former head of engineering and development, "with a very high chamber pressure for that day and time and very lightweight for the thrust that they were producing. I would say that we came out with that at the only time when it would have been successful. If we had waited another two years before starting development on the Shuttle, we probably would not have been able to do it, because the people that designed the main engine were the same that designed previous rocket engines. That group had designed and built seven different engines before they started the Shuttle development. A lot of them retired and so if we'd waited another two years, those people would all have been gone and we would have had to learn all over again on the engine development."

At the start of each Shuttle mission, the main engines burn for about eight minutes and are shut down a few seconds before the External Tank is jettisoned, right on the edge of space. Each engine measures 4.2 m long, weighs 3,400 kg and (like the XLR-99 on the X-15 aircraft) is 'throttleable' at one-percent incremental steps from 65 percent to 104 percent rated thrust. "How one can run an engine at *more* than 100 percent never made much sense to *me*, either," wrote Shuttle astronaut Jerry Linenger in his autobiography, *Off the Planet*, "but the bottom line is that the Shuttle's main engines turned out to be more powerful than the designers thought they would be. Consequently, we *can* actually run the main engines at four percent *greater* thrust than what was originally thought to be full speed ahead." The throttle was controlled by the Shuttle's five General Purpose Computers (GPCs) and throttling back reduced stress on the vehicle during periods of maximum aerodynamic turbulence and also served to limit the G-loads in the final phase of ascent.

Development of the main engines was undertaken at the National Space Technology Laboratory (NSTL) in Mississippi, a NASA site previously used for testing of the Saturn booster, and got underway in 1974. A full-thrust chamber test of an integrated subsystem demonstrator was conducted in the summer of the following year and in March 1976 the engine was fired successfully for 42 seconds at 65 percent rated performance. It uncovered a number of potentially serious problems, including failures of its high-pressure fuel and oxidiser turbopumps, which prompted modifications and the addition of more monitoring instrumentation. Twenty-five further tests got underway in April 1977, with no serious difficulties, although turbopump problems *would* continue to plague the effort throughout its development. Two years later, in July 1979, the main fuel valve experienced a major fracture during a test firing which allowed hydrogen to leak into a mockup aft compartment; the system was commanded to shut down, but not before causing serious structural damage. Four months later, the high-pressure oxidiser turbopump failed again, less than ten seconds into a full-flight-duration, 510-second firing of a three-engine cluster. Triumph followed hard on the heels of failures in the following months: a perfect static test in December 1979, then a premature shutdown in April 1980, then another success ... *and another* shutdown. For astronaut Terry Hart, who was tasked with following these engine developments, it was an embarrassing time, for *he* had to stand up in front of his peers, each Monday morning, to explain the problems as each new obstacle arose.

Despite the immense power generated by each engine and the colossal amount of propellant needed to run them for such a short length of time, they in fact provide the orbiter with only about 20 percent of the muscle to reach space. The remainder comes from the two 45.4 m tall Solid Rocket Boosters (SRBs), the first solid-fuelled rockets ever to be used in conjunction with a manned spacecraft. Loaded with a powdery aluminium fuel and an oxidiser of ammonium perchlorate, the boosters, built by Thiokol (now ATK Thiokol) of Utah, are mounted like a pair of big Roman candles on either side of the External Tank. When the decision to use solid-fuelled rockets was made in March 1972, Jim Fletcher explained that they would help to reduce the Shuttle's development costs from $5.5 billion to around $5.15 billion. Typically, during pre-flight preparations, the SRBs are paired in matching sets and filled with propellant ingredients from identical 'batches' to minimise the risk of thrust imbalances during ascent. This unusual combination, nicknamed 'the stack', is not wholly reusable and originated from financial and technical compromises back in the early 1970s. Each orbiter was designed to make a hundred missions before major refurbishment would become necessary, although of the surviving vehicles even frequent-flying Discovery had made fewer than 40 space voyages when it was finally retired in 2011. The SRBs were to be capable of flying 25 times apiece (although they would require to be stripped down and reassembled in between each mission), but the External Tank was to be discarded and burn up in the atmosphere. It would have proven more costly to recover and modify the tank than to simply build a new one for each mission. "The Shuttle is an asymmetric vehicle," said Neil Hutchinson, one of the flight directors for the first mission. "It doesn't look like it ought to launch right, because it's *not* a pencil. Some of us, in the early days, wondered how that was going to work. In fact, it's [still] a very tricky vehicle to launch. It has to be pointed carefully in the right direction at certain times or you'll tear the wings off or tear it off the External Tank. It's *not* a casual launch process."

The boosters represented relatively new technology for NASA and the first tests of an empty casing took place in September 1977 at Morton Thiokol's facility in Utah to verify structural integrity and fracture mechanics and to permit analysis of the growth and development of cracks. Four static firings of the motors were conducted between July 1977 and February 1979, followed by a trio of qualification tests from June 1979 to February 1980, which validated the design. Parachute drops were conducted in El Centro, California, and Jenkins has noted that the development and verification of the SRBs over a total of just seven test-firings was dwarfed by more than *seven hundred* firings required to qualify the Shuttle's main engines.

Preparing for each Shuttle flight requires several years, but the actual bringing together of the components begins with setting up the boosters on a Mobile Launch Platform (MLP) in the Vehicle Assembly Building at the Kennedy Space Center. Each booster comprises six blocks, called 'segments', each of which is positioned by overhead cranes with pinpoint grace, one atop the next and joined by a ring of bolts. To prevent a leakage of searing gases while operating, a series of rubberised O-rings seal the joints between the segments. After propelling the Shuttle and External Tank to an altitude of about 45.7 km, pyrotechnics separate the boosters, explosive rockets at their nose and tail push them away and parachutes in the nose compartment are deployed to lower them to a gentle splashdown in the Atlantic Ocean. They are then recovered, stripped down, refurbished and reused. When the assembly of the SRBs is complete, the External Tank is moved into position between them and connected by a series of spindly, but strong, attachment struts. After checks to verify its mechanical and electrical compatibility, the orbiter is moved from the nearby Orbiter Processing Facility (OPF), tilted by crane onto its tail and mated to the tank. The transfer of the 1.8 million kg stack from the VAB to one of the two pads at launch complex 39 – a distance of 5.6 km – takes six hours, with the aptly named 'crawler' inching the precious, $2.2 billion national asset along a track made from specially imported Mississippi river gravels. Once the stack is 'hard down' on the pad surface, further checks are conducted, payloads installed and the crew participates in a Terminal Countdown Demonstration Test (TCDT). This is essentially a full dress rehearsal of the final part of the countdown, after which there is a simulated main engine failure and emergency evacuation exercise. As for the *return* of the orbiter from space, in October 1974 NASA decided that the early development missions would land at Edwards Air Force Base in California; according to Jim Fletcher, this offered "the added safety margins and good weather conditions" needed on those early flights. These safety margins were particularly evident in the two primary Edwards landing strips: Runway 17, which crosses a salt flat and measures 13 km long, and the all-concrete Runway 22, whose 4.5 km length was extended by a pair of 500 m asphalt overruns at each end. However, when the vehicle became fully operational, it was expected to routinely use a specially-built 4.5 km runway at the Kennedy Space Center. Indeed, when Shuttle construction work got underway in Florida in 1974, this runway was one of the first structures to be built.

"WOULD YOU LIKE TO FLY THE FIRST ONE?"

For a few days in March 1978, Enterprise came to Houston. She was on her way from Edwards in California to NASA's Marshall Space Flight Center in Huntsville, Alabama. Almost six months had passed since the last of her ALT flights and she was now destined to undergo fit checks with a mockup External Tank and twin Solid Rocket Boosters. The first launch of her spaceworthy sister, Columbia, was at least a year away, but many within the astronaut corps were becoming itchy that a flurry of crew assignments might be just around the corner. Nearly three years had passed since Apollo-Soyuz and many of the pilots had neither the will, nor the inclination, to wait for the better part of a decade in order to fly again. In the wake of the heady days of the Apollo lunar landings, the middle of the decade was a quiet time in American human spaceflight ... yet in the summer of 1976, a strange paradox occurred. The astronaut corps, which numbered around 30 men at the time, faced a crisis: no missions were available in the foreseeable future, yet *more* astronauts were urgently needed. By the time the Shuttle entered operational service in 1980, NASA optimistically hoped that missions would be flying as often as once every two weeks.

In other words, more crews would rocket into the heavens during its first couple

Prime and backup crewmen for STS-1 pose in front of Enterprise, which was then undergoing launch pad fit checks at the Kennedy Space Center in June 1979. From right to left are Joe Engle, Bob Crippen, John Young and Dick Truly.

of years than had previously ridden *every* American spacecraft since 1961. A corps of less than three dozen could not support such an ambitious flight rate, obliging NASA, in July 1976, to announce plans to hire "at least 15 pilot candidates and 15 mission specialist candidates" for the Shuttle effort. Crucially, and totally at odds with previous selections, this group would specifically include both *minorities* and *women*. Pilots, the agency declared, would need at least a degree in engineering, biological or physical science or mathematics, with advanced qualifications desirable. Moreover, they needed to have accrued at least a thousand hours of pilot-in-command time in high-performance jet aircraft, with test piloting expertise preferable. For the mission specialists, similar academic credentials – plus three years of related professional experience or advanced degrees – were demanded, although flight experience was not mandatory. More than 8,000 applications were submitted by the end of June 1977, including over a thousand from women. Later that same year, in groups of less than 20, the promising few were summoned to the Johnson Space Center in Houston for screening. At length, on 16 January 1978, NASA selected 35 new astronauts, its first such selection in nearly nine years ... and a selection unlike any that had preceded it. Deke Slayton, who sat on the selection panel, admitted that the original selection had 20 pilots and 15 mission specialists ... and only *one* woman who made the cut. "Five pilots were dropped from the list," Slayton wrote, "and replaced with five women mission specialists." (Interestingly, those five 'dropped' pilots were selected in the *next* astronaut intake, two years later.) Among the 35 selectees in January 1978, meanwhile, were *six* women, two blacks and an Asian-American and more besides. When they got together at the Johnson Space Center for their first press conference on 1 February, the diversity of the group was clear. NASA's earnest, heartfelt attempt to open up the astronaut selection process to minorities and women had, it seemed, turned into something of a circus and the pendulum of equal opportunity had swung in the *opposite* direction. "There was a mother of three, two astronauts of the Jewish faith and one Buddhist," wrote 1978 selectee Mike Mullane in his autobiography, *Riding Rockets*. "There were Catholics and Protestants, atheists and fundamentalists. Every press camera was focused on this rainbow coalition, particularly the females. I could have *mooned* the press corps and I would not have been noticed. The *white* males were invisible!"

By July 1978, the new arrivals had effectively more than doubled NASA's existing astronaut corps. However, unlike previous selections, they were positively welcomed by the 'old heads' from the Gemini and Apollo eras. In the two years before their selection, no fewer than four veteran pilots – Stu Roosa, Gene Cernan, Ron Evans and Gerry Carr – had retired from the space agency to pursue other interests. The remainder knew that they would have to get the 35 new astronauts up to speed with Shuttle development as quickly as possible. "We got a year of training," reflected selectee Rick Hauck, "that involved, for virtually eight hours a day, lecturers about the systems or observations from space or visiting one of the NASA centres. Then, eventually, we got 'in', so we were assigned on-the-job training, assigned specifically to one of the old guys and I was assigned to Dick Truly. Everyone was very hospitable to us, bending over backwards to make us comfortable and telling us how much they needed us." Training co-ordinators from those heady days would recall that the classes for the new astronauts were more like briefings on ascent and re-entry aerodynamics, space physics, tracking techniques and physiology, followed by practical experience in various simulators. Kathy Sullivan, one of the six women in the group, admitted that there were few formal tests, but each astronaut was keenly aware that they would someday need everything they had learned, quite possibly to keep them alive.

"The first few months were spent in more of an 'observer' mode," said Skylab veteran Ed Gibson, who co-ordinated the mission specialists' training schedules. "After that, they'd be assuming responsibility the same as anybody else in the office." Some candidates, like Fred Gregory, were detailed to work on enhancing the Shuttle's cockpit instrument suite, while others, including George 'Pinky' Nelson, Anna Fisher and Jim Buchli, worked on procedures for donning and doffing spacesuits. Elsewhere, Dan Brandenstein described the training as an incredibly intense learning experience. "A common joke was that training as an astronaut was like drinking water out of a fire hose," he said, "because it just kept coming and coming and coming! Probably the good point was you weren't given written tests, so they could heap as much on you as possible and you captured what you could."

By the time they completed initial training in August 1979, they began working hand in glove with the 'Devil's Advocates' – the team of instructors who dropped fault after fault and failure upon failure into each mission simulation, testing their knowledge to the limit and proving themselves as the astronauts' worst enemies and closest friends, all rolled into one. Years later, many would look back warmly on the Devil's Advocates as having prepared them to be able to respond to almost any emergency during a 'real' mission. "My first big project for Dick Truly," recalled Hauck, "was to develop the emergency procedures for flying the Shuttle. I was supposed to be a co-ordinator for the flight crew in how [the procedures] would be formatted, how they would read and what kind of book they would be in. This project was to put in one document all the procedures that would have to be acted on quickly, either during launch or re-entry. Many of them would be on cue cards Velcroed to the panels around the cockpit, but there wasn't enough space to Velcro all the procedures of the Shuttle. It was much more complicated in terms of crew interaction than any of the previous vehicles, so I looked at the existing T-38 jet trainer emergency checklist and proposed a certain format and certain flip pages, how they'd be tabbed, how it would be organised and what it would look like. Dick and I tried several versions and that's what became the emergency procedures checklist. Then my job was to work with the flight controllers, who would develop the specific reactions to emergencies, put them in words, try to format them in a way that could be used – in not too much detail – by the flight crew and that was a massive effort. Eventually, we got to the point where we had an ascent and re-entry pocket checklist because, depending on the environment, you had different reactions to the same problem."

Despite the hard work, the newcomers bonded exceptionally well; so well, in fact, that two astronaut marriages resulted. One was between Sally Ride and Steve Hawley, another between Robert 'Hoot' Gibson and Rhea Seddon. Years later, Gibson remembered that the group was so large that it had to be split into two halves, both of which frequently entered into friendly competition through 'red' and 'blue' football matches. They organised happy hours on Friday nights, Christmas parties and New Year celebrations and turned, said Gibson, into an extended family as much as a spacefaring flight squadron.

To highlight the distinction between themselves and the Grizzled Veteran Astronauts already in Houston since the 1960s, they gave themselves the nickname 'Thirty Five New Guys', designing 'TFNG' patches and T-shirts to foster closer camaraderie. Mike Mullane has noted that military pilots also knew of an obscene double entendre with the same acronym – 'The F***ing New Guys' – but that, as far as the outside world was concerned, TFNG reflected solely the number of candidates in the class. Judy Resnik, who died aboard Challenger in January 1986, came up with a design for the group's T-shirt: a forward facing view of the orbiter, literally overflowing with 35 astronauts crammed, sardine-like, into every available orifice, and proudly displaying the Shuttle's 'We Deliver' motto that would later become world-famous.

In March 1978, therefore, one question dominated every astronaut mind: *Who* would get to ride the first Shuttle into orbit? The TFNGs would not complete their inaugural training for more than a year and many of the scientist-astronauts, selected more than a decade earlier, were under no illusions that one of those coveted seats might be theirs. In fact, as Enterprise sat on the Ellington runway, the man who would one day sit in the pilot's seat aboard Columbia for her maiden voyage happened to be walking around the craft with George Abbey, head of NASA's Flight Operations Directorate. All at once, Abbey stopped in his tracks, looked Captain Bob Crippen of the US Navy straight in the eye and spoke:

"Hey, Crip, would you like to fly the first one?"

Years later, Crippen would remember the moment lucidly. It was the best news of his career and left him metaphorically *floating* off the Ellington tarmac. "I had anticipated that I would get to fly on *one* of the Shuttle flights, early on," he told a NASA oral historian in May 2006, "because there weren't that many of us in the astronaut office during that period of time", but it did not make Abbey's offer any less surprising. There was only one answer to the question, of course, and a few days later, on 16 March, it was official: Crippen would join John Young, the chief of the astronaut corps, on the first flight of Columbia. Other names were not quite so surprising: ALT pilots Fred Haise, Gordon Fullerton, Joe Engle and Dick Truly were also on the list for follow-on flights, together with Skylab veteran Jack Lousma and ASTP's Vance Brand. However, when Haise resigned from NASA in June 1979, he was replaced in command by Jack Lousma, who in turn was replaced by Gordon Fullerton. Brand, meanwhile, was reassigned a new pilot, Bob Overmyer. At the time, NASA's plan called for up to six Orbital Flight Tests (OFTs) of the Shuttle to evaluate its performance during ascent, in orbit, during re-entry and with a variety of different payloads and pieces of equipment, including a Canadian-built mechanical arm. Young and Crippen would form the 'A' crew, Engle and Truly the 'B' crew, Lousma and Fullerton the 'C' crew and Brand and Overmyer the 'D' crew. *If* the fifth and sixth OFT missions were needed – and NASA was keen to get the series out of the way as soon as possible, to get the Shuttle operational – they would be flown, respectively, by the Young-Crippen and Engle-Truly pairs. According to one spaceflight historian in Internet chatter on www.nasaspaceflight.com in March 2011, the draft flight requirements for the fifth and sixth OFTs "always were speculative" and "never appeared in anything other than draft". By late 1979, the number of OFTs had indeed been reduced to four and several of the key objectives of the fifth and sixth flights – performing an EVA, landing at the Kennedy Space Center in Florida and working with fully-functional payloads – were shifted onto the early operational missions. When the four OFT missions had been satisfactorily concluded, the winged spacecraft would be declared fully operational, sometime in 1980.

NASA's 'keenness' to get the OFT missions done is underlined by Dennis Jenkins in his concise textbook on the Shuttle, in which he noted 1979 recommendations from the Office of Management and Budget to cancel the programme entirely, on the basis of its delays. "Fortunately," wrote Jenkins, "the Department of Defense again came to the rescue and convinced the Carter administration to allow the programme to continue." Meetings between Carter and NASA Administrator Robert Frosch revealed a low likelihood that even the first OFT mission would be underway before the end of 1980, let alone an operational flight, and problems with the final flight certification of the main engines pushed this prediction further to the right. At length, the pathfinding voyage of Young and Crippen settled on a date, sometime in the early spring of 1981.

Robert Laurel Crippen was born in Beaumont, Texas, part of the Gulf Coast's Golden Triangle of industry, on 11 September 1937. He completed his education at New Caney High School, before moving to the University of Texas at Austin to study for a degree in aerospace engineering, which he received in 1960. Crippen joined the US Navy immediately after graduation and underwent aviator training in Pensacola, Florida, then Whiting Field and finally to Beeville, back in his home state of Texas, to gain his wings. For more than two years, from June 1962 until November 1964, he served as an attack pilot aboard the aircraft carrier USS *Independence*, flying the A-4 Skyhawk. This was prior to the outbreak of the Vietnam conflict and Crippen found himself involved in operations to support President Kennedy's firm stance in Berlin and during the Cuban Missile Crisis, as well as a series of NATO exercises in the Mediterranean and Adriatic Seas. His career within the Navy progressed rapidly and he was sent to the Aerospace Research Pilot School at Edwards Air Force Base in California, from where he graduated as a test pilot and was selected for the Air Force's Manned Orbiting Laboratory (MOL) military space station in October 1966. "The opportunity came to apply *both* to NASA *and* the military," he told the oral historian. "I applied for both. At some point in the process, I had to decide one way or the other and ended up picking MOL, because I thought NASA had more astronauts than they knew what to do with and the [Apollo] programme ... was already starting to have some of the flights cancelled. I ended up being selected for MOL and, sure enough, after a couple of years on that programme, it got cancelled!" MOL guzzled $2.2 billion of taxpayers' money without making a single flight and was formally cancelled by President Nixon in June 1969. The Air Force pressured NASA into hiring the ex-

MOL pilots as astronauts and Deke Slayton, against his better judgement, it seems, agreed to accept those men who were under the maximum allowable age limit of 35. When the new class was announced in September 1969, Crippen was among them. So too was Karol 'Bo' Bobko, who later joined Crippen for the 56-day SMEAT test in the summer of 1972 *and* on the support crew for ASTP. Following the conclusion of the final Apollo mission, the focus of everyone moved onto the Shuttle and Crippen was assigned to work on the development of its computers, software and cockpit displays. Years later, he would wonder if his computer knowledge played a part in Abbey's decision to offer him the pilot's seat on the first mission.

Seated alongside Crippen was John Watts Young Jr, already a four-flight space veteran, having completed two Gemini and two Apollo missions *and* walked on the Moon. Since April 1974, he had served as chief of NASA's astronaut corps, a post he would hold for 13 years, longer than any other person. Even today, spacefarers continue to refer to him as 'The Astronauts' Astronaut' and even those who did not get along with him or disliked his management style are unfailing in their praise of his piloting and engineering skills. Over the years, Young has been regarded as something of a mystery; his misleading "aw, shucks" demeanour and country-boy drawl cleverly concealed a sharp, analytical and talented mind. When Young was worried about an engineering problem, Crippen once said, "then *I* should be worried about it as well!" He came from San Francisco, where he was born on 24 September 1930, and aged just three years old his family moved to Cartersville, Georgia, then settled permanently in Florida, in the city of Orlando. It was at around this time that Young began to build model aircraft, a hobby that would remain with him throughout high school, together with rockets, which he chose for a speech to his classmates in the 11th grade.

In 1952, Young earned his degree in aeronautical engineering from Georgia Institute of Technology with highest honours, receiving coveted membership of the institute's prestigious Anak Society. Among his earliest assignments after joining the US Navy in June of that year, he served as fire control officer aboard the destroyer USS *Laws*. During this time, he completed a tour in Korea and a former shipmate would remember his coolness under duress. Joseph LaMantia (quoted on the website www.johnwyoung.com) recalled: "Though only an ensign at the time, he was the most respected officer on the ship. When we sustained counter-battery fire and enemy rounds were striking the ship, it was John Young's leadership which kept us all cool and focused on returning that enemy fire ... which won the day." On returning home, he entered flight school at Naval Basic Air Training Command in Pensacola, Florida, where he learned to fly props, jets and helicopters. Later, he undertook a six-month course at the Navy's Advanced Training School in Corpus Christi, Texas. With receipt of his wings came four years of service as a pilot in Fighter Squadron 103, flying F-9 Cougars from the USS *Coral Sea* aircraft carrier and F-8 Crusaders from the USS *Forrestal* supercarrier. During these years, colleagues would describe him as "the epitome of swashbuckling aviators ... he exuded confidence coupled with uncommon ability".

This ability, indeed, would ultimately guide him into the hallowed ranks of NASA's spacefaring corps. But not yet. The selection process for the Mercury Seven

Bob Crippen (left) and John Young during pre-flight training in January 1980.

began early in 1959, at which time Young was just starting Naval Test Pilot School at Patuxent River, Maryland, and test-flying credentials were a prerequisite for astronaut training. After graduation, he served as a project test pilot and programme manager for the F-4H Phantom II weapons system at the Naval Air Test Center in Maryland, evaluating armaments, radar and bombing fire controls for both the Crusader and Phantom fighters. During one air-to-air missile test, he and another pilot approached each other's aircraft at closing speeds of more than three times the speed of sound. "I got a telegram from the chief of naval operations," Young later quipped, "asking me *not* to do this anymore!"

In early 1962, he set two time-to-climb world records. By now a lieutenant-commander, Young's experience with the 'Phabulous' Phantom had made him the obvious choice to set the records as part of Project High Jump. The first, on 21 February, saw him climb to 3,000 m above Naval Air Station Brunswick in Maine in 34.5 seconds; and six weeks later he made another attempt from Point Mugu in California and achieved 25,000 m in 230.4 seconds. In September of that year, after leaving active naval duties as a maintenance officer in Phantom Fighter Squadron 143, he got a phone call from Deke Slayton which marked the start of his astronaut career. The training was arduous. "You had to learn a lot of stuff," he said later. "You probably only needed to know one percent of all the stuff you had to learn ... but you didn't know which one percent it was!" In the decade which followed, Young would journey into space four times: twice aboard Gemini and twice aboard Apollo, becoming the first man to fly solo in lunar orbit and the ninth to leave his bootprints on the Moon.

If Crippen and Young were polar opposites in terms of experience, it did not show. "When you're a rookie going on a test flight like this, you want to go with an old pro," Crippen told the NASA oral historian, "and John was *our* old pro. He's got a dry wit that a lot of people don't appreciate fully at first, but he has got so many one-liners!" As any son-in-law will understand, arguably the very best was uttered not before the launch of Columbia, but on the eve of Young's very first mission, way back in March 1965. A journalist asked him if he any qualms about flying into space with his somewhat fiery crewmate, Gus Grissom. In a split-second response, timed to perfection, Young deadpanned: "Are you *kidding*? I'd have gone with my mother-in-law!"

BIRTHING PAINS

With less than eight weeks to go before her first orbital flight, Space Shuttle Columbia finally got the chance to flex her muscles on 20 February 1981. By now, the winged vehicle was running three years behind its advertised schedule; its first launch was originally targeted for 1978 and its highest-profile mission – a delicate orbital ballet to re-boost Skylab and possibly prepare it for reoccupation – had been missed. Billions of dollars had been invested in the Shuttle, which NASA paraded as the world's first reusable manned spacecraft, capable of flying once every fortnight and trucking commercial satellites, a scientific laboratory called 'Spacelab', astronomical instruments and – for the first time – *ordinary civilians* into orbit. Plans were already advanced to send teachers, journalists and foreign nationals into space, with up to seven seats available on each mission. The Shuttle, it seemed, was aptly named: it would whisk people into orbit frequently, reliably, relatively cheaply and in conditions a world apart from the cramped, one-use-only capsules of the 1960s.

However, before it could be declared 'operational', it had to be extensively tested. Much of this work had taken place during and after Columbia's construction, but ground runs were no substitute for actually flying into space. Original plans had called for six OFT missions, each carrying a pair of astronauts – a commander and a pilot – to demonstrate its capabilities and manoeuvrability. As we have already seen, by 1980 that number had long since been reduced to four, all of which were to be flown by Columbia, the first space-rated vehicle. Assuming that all four went to plan, the so-called 'Space Transportation System' (STS), as the Shuttle was formally known, would be ready to fly commercially on its fifth flight.

The name 'Columbia' had arisen following a NASA decision in the mid-1970s to give *names*, rather than *numbers*, to each of the orbiter vehicles. The public desire to change one of the earliest proposals, 'Constitution', to 'Enterprise', has already been mentioned, and in May 1978 John Yardley, the agency's associate administrator for Space Transportation Systems, sent a list of options to the Public Affairs Office. Each of these names – which included 'America', 'Independence', 'Liberty' and 'Freedom' – was required to bear some significant importance to the heritage of the United States. Later that year, Arnold Frutkin chaired another orbiter-naming committee, which broadened the list of options to include sailing vessels and even Native American tribes. In December 1978, three categories of names were selected: (1) Explorers' Vessels, which included 'Enterprise', 'Endeavour' and 'Discovery', (2) American Tradition and Spirit, which included 'Columbia', 'Constitution' and 'Republic', and (3) Stars and Constellations, which included 'Orion', 'Arcturus' and 'Pegasus'. At length, the committee settled on Explorers' Vessels and even the Star Trek-themed 'Enterprise' had a proud naval heritage, having lent its name to the world's first nuclear-powered aircraft carrier. 'Columbia', meanwhile, owed her name to a privately-owned sailing frigate, built in 1773, which was commanded by Robert Gray in a series of expeditions to the Pacific north-west coast of America in support of the maritime fur trade. Later, in 1790, Columbia (whose full name was 'Columbia Rediviva', Latin for 'Columbia Revived') became the first US vessel to circumnavigate the globe. Significantly, as well as being the female personification of the United States, Columbia also served as the name of Apollo 11's command module.

Columbia was physically identical to Enterprise, at least at first glance, but unlike her sister she would fly much further than just the last few minutes from the low atmosphere to the runway. She would, for the first time, undertake the violent climb into space under the combined thrust of her three main engines and two behemoth SRBs, withstand the swinging extremes of heat and cold in Earth orbit and bear the full brunt of a hypersonic descent back through the atmosphere. Moreover, in spite of carrying thousands of parts whose failure could doom the Shuttle and kill the crew, Columbia's audacious first launch would be done ... *with astronauts aboard*!

Never in the history of the American space programme had a crew been aboard for the maiden flight of a new spacecraft. The Soviets, too, had always first tested their spacecraft in an unmanned capacity. Yet there were sound reasons for the decision. Had Fred Haise not resigned from NASA in June 1979, he would have commanded one of its early missions. He saw an *unmanned* first flight as potentially trickier than a manned one. "It would have been *very* difficult," he told the oral historian, "to have devised a scheme to have flown unmanned. I guess you could've used a [communications] link and really had a pilot on a stick on the ground, like they have flown some other programmes, but to totally mechanically programme it to do that – and *inherent* within the vehicle – would have been very difficult. There *was* initially a planned unmanned flight, [but] it was of great complexity and handling the myriad of potential systems problems [made it hard to automate]. With a crew on-board, able to handle the multitude of things that you could work around, inherently made the success potential of a flight much greater." Others, including Henry Pohl, who was then NASA's head of engineering and development, were more sceptical. "I didn't see *any* need in risking humans and I didn't think humans would be as proficient as automated equipment," he explained. "By that time [the late 1970s], we had the know-how and we could build robots or the automated equipment that can detect things long before a human can detect it and I thought the vehicle was going to be so difficult to land that we really ought to land it with automated equipment."

"We were tickled silly," said Joe Engle of the very *idea* that the Shuttle would fly unmanned or that a hardened, experienced test pilot would conceivably hand over control to the computers. "We were very vain and thought: you've *got* to have a pilot there to land it. Fortunately, the certification of the autopilot all the way down to landing would have required a whole lot more cost and development time, delay in launch, and I think the rationale that we put forward to discourage the idea of developing the autoland was that you can leave it engaged down to a certain altitude, but you *always* have to be ready to assume that you're going to have an anomaly and the pilot has to take over and land. The pilot is in a much better position to affect the final landing, having become familiar and acclimated to the responses of the vehicle, after being in orbit and knowing what kinds of displacements give him certain types of responses."

Aside from the astronauts' simple piloting *desire* to be aboard, the Shuttle Training Aircraft – Grumman's Gulfstream II – had also been instrumental in wringing out many of the unknowns in how the vehicle would fly in the subsonic flight regime; John Young would later remark that he felt it had paved the way for how Columbia would behave at lower speeds. During the more dynamic stages of the flight regime, both hypersonic and transonic, more than 22,000 hours of wind tunnel tests had been conducted to understand and predict the Shuttle's performance as closely as possible. Years later, STS-4 pilot Hank Hartsfield remarked that the sheer expense and complexity had also prevented NASA from implementing an unmanned mission. "It would have [required] major modifications to the current orbiter design," he told the NASA oral historian, "because you had to think about the software conversions: we had separate software modules that we had to load once on orbit [and] into the entry. All that had to be loaded off storage devices or carried in core, because the Shuttle computer was only a 65 K machine. It's *small*. When you think about the complexity of the vehicle, with full computers *voting* in tight sync and flying orbital dynamic flight phase and [giving] the crew displays ... and we're doing that with 65 K of memory ... it's *mind-boggling*. Take the little computer you carry in your purse and it's got more memory than that!"

Whatever one's opinion, each of these views does nothing to detract from a keen awareness that STS-1 was *dangerous* and Young and Crippen were under no illusions that they might not survive Columbia's first mission into space. Shortly before the astronauts headed to Cape Canaveral for the final time, Joe Allen bought Young lunch in the Johnson Space Center cafeteria. "John hadn't had any money," he recalled, "so I'd bought his lunch for him. We got in the car and John reached in his flight suit and he took out the money and he gave it to me. I said 'Come on, John'. He said: 'No, you *don't* go fly these things when you got *debts*!' He paid me. He was correct ... and *I* was correct to accept it, so he had no debts. 'All my debts are *paid*,'" he said."

By 20 February 1981, almost all preparations for STS-1 had been completed; in fact, attached to her External Tank and Solid Rocket Boosters, Columbia had sat majestically on Pad 39A since 29 December of the previous year. She had been at the Cape even longer. Her construction took almost five years from the start of work to build her cockpit in June 1974 to rollout of the finished article in March 1979. A week later, to the amazement of motorists in the sweltering California heat, Columbia was towed *overland* from Rockwell International's Palmdale plant to Edwards Air Force Base. She was then flown 'piggyback', atop the modified Shuttle Carrier Aircraft, to the Kennedy Space Center and ensconced in one of two bays in the Orbiter Processing Facility. The latter, which is positively dwarfed by the immensity of the Vehicle Assembly Building, is far more than 'just' a spacecraft hangar and the volatility of the propellants being handled in the OPF require it to be fitted with detectors of such sensitivity that visitors are forbidden to use *camera flashes* when taking photographs.

Columbia's time in the OPF, preparatory to STS-1, was almost two years. Although she was structurally 'complete', she was far from ready to fly. She had no main engines, her thermal protection system needed attention and her External Tank and SRB segments were not destined to arrive in Florida until the summer of 1979. By the end of the *following* year, however, significant progress had been made and in November she rolled into the VAB for stacking and from thence to the launch pad.

The thermal protection system – particularly its thousands of tiles, each of which was *individually* designed and *not* interchangeable – had been the biggest headache, due to the sheer novelty of the design. "When it took off on the back of the 747 from Palmdale, a whole bunch of the tiles came off as they went down the runway," former Shuttle manager Arnold Aldrich wryly recalled. "That led to the requirement to have a better understanding of how the tiles were attached and how to know they were well-attached and *that* problem took two years to solve." As its name implies, the system was designed to protect the Shuttle during its searing hypersonic descent through the atmosphere and comprised several key components: Reinforced Carbon

Columbia rolls into the Orbiter Processing Facility in March 1979. More than two years and a tremendous amount of work awaited her before she could be launched into space.

Carbon (RCC) for the leading edges of the wings and nose cap, capable of withstanding re-entry extremes of up to 1,260°C, black High-Temperature Reusable Surface Insulation (HRSI) tiles for the belly and parts of the OMS pods and fuselage, white Low-Temperature Reusable Surface Insulation (LRSI) tiles for the vertical stabiliser and main fuselage; and a series of durable, lightweight 'blankets' – called 'Advanced Flexible Reusable Surface Insulation' (AFRSI) – for lower-temperature areas.

After the resolution of the thermal protection system problems, one of the next major exercises was a Wet Countdown Demonstration Test (WCDT), which lasted six days and culminated on 20 February 1981 in a 20-second firing of Columbia's three main engines. This Flight Readiness Firing (FRF) was needed to demonstrate their ability to throttle between 94 and 100 percent thrust and 'gimbal' just as they would be expected to do during flight. Similar 'wet', or fully-fuelled, tests had been performed prior to the Saturn V launches, although on those occasions the engines had not actually been fired. Preparations for the FRF proceeded in a manner not dissimilar to a real countdown: launch controllers started the clock at T-53 hours, when they powered-up the SRBs, ground-support equipment and the systems aboard Columbia herself. Four seconds before the simulated liftoff, the Shuttle's engines roared to life at 120-millisecond intervals, reaching 90 percent of their rated performance within three seconds and hitting 100 percent precisely at T-zero. Three seconds later, engineers simulated the retraction of the External Tank umbilical and the SRBs' hold-down posts; and after a further 15 seconds of stable thrust, shutdown commands were issued to all three engines. The test was a great success and another significant milestone had been cleared.

According to the STS-1 press kit, released around this time, the launch was provisionally booked for "no earlier than" 17 March, but a number of technical issues and a human tragedy conspired to delay Columbia's first flight by several weeks. Following the FRF, engineers had to repair a section of super-light ablator insulation, which had become debonded from the External Tank during a test of its cryogenic propellants back in January. This pushed the target date for launch out to 5 April. A strike against Boeing by machinists and aerospace workers then postponed it further to the 10th. Throughout March, the attention of the world's media was riveted on Young and Crippen as they maintained their proficiency, participated in a TCDT and practiced how to escape from the Shuttle in the event of a main engine failure just before launch. The biggest fear in such a scenario was the presence of invisible hydrogen flames, through which the pressure-suited astronauts would have to run to reach a slidewire escape basket that would whisk them from the 58 m level of the launch pad down to the ground and a waiting armoured personnel carrier.

It was only a few days after Young and Crippen had returned to Houston, following their TCDT, when the Shuttle claimed its first two lives. Several technicians working inside Columbia's aft compartment were rendered unconscious by a dangerous accumulation of nitrogen gas; although they were quickly extracted, one man died that same day and another succumbed a couple of weeks later. The cause was traced to a breakdown in communications: a warning sign had been mistakenly removed and a supervisor called away. Bob Crippen would later pay a heartfelt tribute to the two dead men, John Bjomstad and Forrest Cole, whilst in orbit.

Meanwhile, in spite of the setbacks, Columbia was now firmly on schedule to liftoff on 10 April 1981, which, as it happened, was just two days shy of the 20th anniversary of Yuri Gagarin's pioneering flight. The six-and-a-half-hour 'window', which opened at 6:50 am Eastern Standard Time, was dictated by a need for adequate lighting conditions to satisfactorily photograph the Shuttle's ascent for engineering analysis and to preserve the option for a daylight landing opportunity at White Sands Missile Range, should an abort oblige Young and Crippen to make an emergency landing after one orbit. Since March 1979, White Sands – an enormous blotch of hard-packed salt and gypsum in New Mexico's inhospitable Tularosa Basin, to the west of Alamogordo – had been held in reserve by NASA as a backup landing strip. Although it has only been used once, by STS-3 in March 1982, White Sands, with its year-round stable weather and excellent visibility, has played a crucial role in the Shuttle programme ever since. Indeed, it almost saw service a second time in December 2006, when the returning STS-116 crew were temporarily thwarted in their efforts to land by poor weather conditions at both the Kennedy Space Center and Edwards Air Force Base.

Shortly before four in the morning of 10 April, Young and Crippen boarded Columbia for what turned out to be a relatively uneventful countdown – at least, that is, until its final stages. Then, with just nine minutes to go, during a pre-programmed hold, a problem cropped up in one of the General Purpose Computers. It was described by NASA as "a timing skew"; in effect, the backup flight software was unable to synchronise itself with the primary set. Unlike earlier manned spacecraft, the Shuttle was totally reliant upon its computers to run the main engines, move the elevons, control its heading and operate the manoeuvring thrusters, to name just a handful of many thousands of different functions. The units were so critical that five GPCs were carried: four primaries, which ran the same software and 'voted' before issuing commands, and a backup. If one of the primaries disagreed with the others, it was 'outvoted' and considered faulty. The backup computer contained a different set of flight software, so that if *all four* primaries became corrupted, it could take over.

The problem faced by Columbia on the morning of 10 April was essentially that the four primary GPCs were not communicating with each other correctly. Taking advantage of the lengthy launch window, the liftoff was rescheduled for 10:20 am as computer engineers wrestled with the software, but when a solution could not be found it was considered prudent to stand down until 12 April. A disappointed Young and Crippen clambered out of the orbiter. Late that evening, the GPC problem was isolated and the countdown resumed on the 11th. "The software," recalled Gordon Fullerton, who flew as Columbia's pilot on STS-3, "became the *biggest* stumbling block. The software in these computers not only control where you fly and the flight path, but almost *every other* subsystem! Getting the software wrung out and simulators writing the checklists … we didn't really have it nailed down by STS-1. There were a *lot* of unknowns, [but] you just finally have to set a launch date and say 'We're going to go'. You *cannot* be 100 percent sure of everything."

Young and Crippen address the media after arriving in Florida aboard their T-38 jet on 8 April 1981.

After spending the 11th maintaining their proficiency, Young and Crippen again departed the crew quarters in the early hours of 12 April and took their seats aboard Columbia's flight deck. Both were clad in bulky US Air Force high-altitude pressure suits, which afforded them full-body protection and would be worn by all of the OFT crews. Since these were deemed 'test flights' and were also equipped with ejection seats, the full-pressure garments were mandatory. On later 'operational' missions, when restrictions were relaxed somewhat and it became impractical to give *everyone* an ejection seat, it was planned for the astronauts to fly in lighter, overall-type suits and helmets. If an emergency had demanded their use, the rocket-propelled ejection seats would have boosted Young and Crippen through a pair of overhead panels ... but they could only be used up to an altitude of 30 km, meaning that realistically they were useless in the ascent phase and could only work at selected intervals during re-entry. Astronaut Jack Lousma, who commanded STS-3, would later remark that his Shuttle launch was far riskier than his Skylab ascent a decade earlier; the opportunities to escape in an emergency were much reduced.

Whereas previous manned craft had taken the form of ballistic capsules atop expendable boosters and an escape rocket could have lifted them several thousand metres into the air and parachute them a couple of kilometres out to sea, the Shuttle offered no such option. An on-the-pad emergency would have precluded the use of the ejection seats, because the astronauts would have hit the ground *before* their parachutes could open. Furthermore, any attempt to eject during the first two minutes of ascent would have flung them into the SRBs' exhaust plumes. Consequently, they could only have been used at very specific stages of ascent and re-entry and, even then, the astronauts' chances of survival were extremely slim.

Aboard Columbia for the second time, Young and Crippen must have tried hard to cast such thoughts to the back of their minds ... for another problem was afoot: they lowered their visors and found that they could not *breathe* properly! It turned out that a quick-disconnect fitting for the oxygen system, located beneath the instrument panel, had been set incorrectly. After this glitch had been rectified, the countdown proceeded normally.

RIDE OF A LIFETIME

The history of jokes and pranks between astronauts and the ground crews who strap them into the spacecraft before launch has become the stuff of legend, ever since the pioneering Mercury missions of the 1960s. "John Young made a big deal about the size of the American flag on his suit," recalled technician Jean Alexander. "It came in with kind of a small version and they got several sizes before he was satisfied and it was kind of a joke, so on launch morning there was a motel that we stayed at Cocoa Beach and they had this huge flag on a pole [outside] a real-estate office next door. One of the suit techs that was down there for launch talked the real-estate people into letting him take that flag down and he took it to the suit room for suit-up morning and had it actually *cover* one whole wall! When John walked in, he said 'John, is *that* big enough?'" The episode lightened the mood sufficiently for what was about to come.

After three years spent training together for the most complex flying and engineering challenge of their careers, the smoothness of the countdown on only their second launch attempt was a great surprise to Crippen. As the clock was paused, as planned, at the T-9 minute point, Launch Director George Page told the men that he would extend the 'hold' slightly, to ensure that the team were sufficiently calm and focused for the events ahead. "It was for a few minutes," Crippen noted, "to get relaxed." Five minutes to go: right on cue, the pilot started Columbia's Auxiliary Power Units (APUs) and verified that they were up and running normally. All was well. As the clock reached the final minute, their excitement began to build – they were *really* going to do this, *today* – but despite Young's wealth of experience and four previous missions, *both* men were rookies as far as the Shuttle was concerned and *neither* knew fully what to expect. Neither could the public do anything but listen, as the launch commentator ticked off the final events in the countdown with the kind of technobabble which only intensified the excitement: "The firing system for the sound suppression water system will be armed just a couple of seconds from now ... T-45 seconds and counting ... the Development Flight Instrumentation recorders are on ... T-35 seconds, we're just a few seconds away from switching to the redundant set sequencer ... T-27 seconds, we've gone for redundant set sequencer start ... T-20 seconds and counting ... T-15 ... 14 ... 13 ... 12...T-10, nine, eight, seven, six, five, four ... we've gone for main engine start ... " as a shower of sparks from the hydrogen burn igniters gave way to a sudden, low-pitched rumble and a sheet of translucent orange flame, " ... we *have* main engine start ... "

Six seconds before 7:00 am, anything slumbering in the vicinity of Cape Canaveral was jarred instantly awake as the rumble of Columbia's engines intensified into a mighty crescendo. Almost as quickly as it had appeared, the orange sheet of flame was gone, to be replaced by a trio of white shock diamonds, as supersonic exhaust gases surged from the engine bells. A vast cloud of smoke quickly obscured the entire vehicle. The commentator's next few words were drowned out by the ear-splitting staccato crackle and brilliant fireshow of the SRBs, which ignited precisely at T-zero, and *precisely* on the hour. From the press site, 5 km away, Columbia seemed to punch its way upwards from the smoke, accompanied by the shouts, whoops and cheers of three and a half thousand spectators and doubtless hundreds of thousands more who were watching on television. "We have *liftoff* of America's first Space Shuttle ... and the Shuttle has cleared the tower ... "

From their seats, the astronauts would later recall that the Shuttle rocked, perceptibly, backwards and forwards, accompanied by a sharp increase of noise inside the cabin. Crippen would later remark that, although the roar of the main engines definitely got their attention, it was the punch-in-the-back ignition of the SRBs which convinced and assured them that they were *really* going somewhere. Veteran astronaut Jerry Ross, who flew seven times on the Shuttle between 1985 and 2002, described the instant of booster ignition as "somebody taking a baseball bat and swinging it pretty smartly and hitting the back of your seat; a real *bam*." As for the vibration, Charlie Walker, who flew three Shuttle missions in 1984 and 1985, would later comment on his relief that the *helmets* were so well-insulated. "The

Spectacular liftoff of STS-1.

acoustic level is [huge] in the crew compartment," he said. "We would readily be *deafened* if we didn't have the insulation of the helmets around our ears." For the first few seconds, as Columbia cleared the tower and soared into the clear Florida sky atop two dazzling columns of flame, the cockpit instruments were blurred by the vibrations, though not unreadable.

"*Roll Program!*" radioed Young as Columbia performed an axial rotation to orient itself onto the proper flight azimuth, seemingly 'rolling' onto its back.

"Roger, roll," replied Capcom Dan Brandenstein.

By the time that the Shuttle rolled over onto her back, ten seconds into the flight, and established herself on the correct heading for a 40.3-degree-inclination orbit, the two men reported that the vibrations had lessened to a point that allowed them to read their instruments without problems.

"When you get the vehicle going uphill and you're still in the 'sensible' atmosphere, there are tremendous aerodynamic pressures on it," recalled Neil Hutchinson, "and you have to get the angle at which it is going through the airstream *exactly correct* ... it has a *very* narrow performance corridor. In order to get the proper inclination, when the Shuttle takes off, it 'rolls'. What it's doing is getting itself oriented, so it goes into orbit on its back; it goes 'upside down'. You've got to get that roll out of the way and get that whole thing set up long before you get the maximum [dynamic pressure], when the amount of atmosphere combined the direction the vehicle's going and the velocity is the worst."

At the post-flight debriefings, Young would tell engineers that Columbia's ascent was considerably more rapid than he had experienced during his two Saturn V launches to the Moon. Analysis also showed that STS-1 had caused significant damage to Pad 39A, which could have been catastrophic: the shockwaves produced by the main engines and the SRBs had buckled a strut linking Columbia with the External Tank. Had the strut *failed*, it was subsequently determined, the result could have been the loss of the whole vehicle and the crew and steps were taken to strengthen the struts in readiness for future missions.

Climbing through the low atmosphere, the wind noise outside gradually intensified into something which could only be likened to a *screaming*. A minute into the flight, as Columbia approached an altitude of 15 km, she passed through a period of maximum aerodynamic turbulence, which required the GPCs to throttle the engines back to just under two-thirds of their rated thrust. The passage through this period, nicknamed 'Max Q', was accompanied by an increase in the noise and vibration of the engines, although their performance remained within structural expectations. Shortly thereafter, the three engines were throttled back up to full power.

"Columbia, you're Go at throttle up," radioed Brandenstein.

"Roger, go at throttle up," acknowledged Young.

The sound from the boosters, meanwhile, remained sporadic and decreased to virtually nothing as the time approached, two minutes and 12 seconds into the ascent, for their separation. Shortly before the SRBs burned out, Brandenstein, told the crew that they were now "negative seats", meaning that Columbia was too high to use the ejection seats, questionable though their usefulness would have been.

Fortunately, the vehicle was performing beautifully. In his autobiography, Mike Mullane recalled listening with relief as each abort-boundary call was passed up by Brandenstein; each call signalled "the sweet song of nominal flight". The boosters actually generated *more* lift than anticipated and they separated at an altitude several kilometres 'higher' than thought. When the separation motors fired and the SRBs fell away, Young and Crippen reported a bright, orange-yellow 'flash' which appeared to stream up in front of the Shuttle's nose and back above the front windows. By this point, with the boosters tailing off, the sensation was almost that they had *stopped accelerating*, but Columbia was already above much of the sensible atmosphere, travelling at close to four times the speed of sound ... and 35 km high. Separation of the boosters was accompanied by a harsh, grating sound, although both performed nominally, parachuting down into the Atlantic Ocean about 250 km downrange for recovery and reuse. With them gone, the astronauts now found it much easier to flip switches in the cockpit. At this point, the so-called 'T-fail-pitchover' manoeuvre was executed, placing the horizon in their direct field of view for the first time and Young and Crippen saw penny-sized to fist-sized particles flooding past the windows.

"What a view! What a view!" radioed Crippen, jubilantly.

"Glad you're enjoying it," replied Dan Brandenstein. It was not an idle comment; Brandenstein was just as excited, sitting at his console in Houston, and although he would end up flying four Shuttle missions of his own, he would describe his stint in Mission Control on STS-1 as the most exciting episode of his astronaut career. His main concern had been the clearances between the launch pad and Columbia and he had listened carefully for the first few seconds, finally breathing a sigh of relief when the Shuttle cleared the tower safely. This combination of worried concern and exhilaration showed in the tone of Brandenstein's voice as he maintained a series of calls to Young throughout ascent. "I still go back and listen to the tapes," he recalled in an interview for the Smithsonian in the summer of 2000. "The *first* call or two, I was almost *yelling* at them! I was kinda pumped up, I guess ... "

Young and Crippen flew on for six more minutes after SRB separation, reaching Mach 19, close to 23,340 km/h, at which point the engines were throttled back to maintain around three times the force of terrestrial gravity in order not to over-stress the vehicle. Throughout the entire ascent, the commander's heart rate rose no higher than 90 beats per minute, whereas that of rookie Crippen peaked at nearly 130. After the mission, Young would quip that he was so old that his heart would not *beat* any faster ... but Neil Hutchinson had another, tongue-in-cheek, explanation: the cool, calm commander *must* have been asleep the whole time!

By eight minutes after launch, with three times the force of terrestrial gravity bearing down upon them, it felt that the two men had someone heavy sitting on their chests. Breathing became difficult and talking was reduced to a series of guttural grunts. "This period is when the orbiter's three main start reducing their power output," explained Jerry Ross, "so that you don't exceed the structural limit of 3 G, and so for that last minute the Shuttle's main engines are coming back. You're getting lighter and lighter. You're accelerating at 100 [feet] per second, *per second*, which is basically like going from zero to 70 miles per hour, *every second*. It's pretty good. Then, at the time the computers sense the proper conditions, the main engines basically go from around 70 percent power on a 3 G acceleration, shut off … and you're in *zero-G*." At 12:08 pm, some eight minutes and 32 seconds since leaving Pad 39A to the cheers and prayers of thousands of spectators, the main engines of America's first Space Shuttle were shut down and more than 2,000 kg of residual propellant was dumped through their nozzles. During this procedure to 'stow' the engines for orbital operations, Columbia's nose unexpectedly pitched 'upwards' about five degrees. Nineteen seconds after the engines went out, the External Tank was jettisoned to follow a ballistic, suborbital re-entry and burned up over a sparsely inhabited stretch of the Indian Ocean. Young and Crippen pulsed the Reaction Control System thrusters to push themselves away from the now-unneeded tank; they later reported no noise associated with the separation and that, in fact, the only indication that they had of what was happening was when the red main engine lights on the instrument panel suddenly blinked out. They were in space … and although still tightly strapped into their seats, the first traces of orbital flight were readily apparent, as bits of debris – washers, filings, screws and wire – began floating freely around the cabin.

The next task was the first of two firings of the big Orbital Manoeuvring System engines, which began a little over ten minutes after launch and was described by Flight Director Jay Greene as "normal". The astronauts agreed: "We're lookin' good!" A second OMS burn at apogee, about 35 minutes later, circularised Columbia's orbit at around 145 km. Although both firings were satisfactory, the instruments providing quantity readings for the two pods turned out to be erroneous; they were showing sporadic propellant quantities throughout the flight, often staying constant for some seconds, then changing at faster-than-expected rates. None of this had been seen in ground simulations. Still, all evaluations of the systems – using both thrusters in unison and singly – were performed without incident. To test emergency procedures, Young and Crippen also fired the right-hand OMS engine using a cross-over to draw propellant from the left-hand tank. Columbia passed her tests with flying colours.

With a stable, circular orbit having been established, the men now turned their attention to opening the Shuttle's clamshell doors and exposing the cavernous payload bay to space for the first time. The bay itself was bare for this first mission, but on later flights it was expected to be crammed with commercial satellites, scientific instruments and laboratories – the most important of which was the European Spacelab – and major astronomical observatories. For STS-1, it was equipped with a series of Development Flight Instrumentation (DFI) sensors to record Columbia's perfor- mance and the stresses and strains that she endured at key stages in the mission. It was essential that the payload bay doors were opened within the first few hours after reaching orbit, so that the radiators lining their inner surfaces could begin dumping excess heat into space. If, for whatever reason, the astronauts had been unable to open the doors, flight rules dictated that they should return to Earth, at the very latest, by the end of their fifth orbit. Although extra systems *were* carried to dissipate heat, these could be used for maybe a day at most. At the other extreme, if problems arose whilst *closing* the doors at mission's end, Bob Crippen had trained to make a spacewalk to secure them manually. Early plans had actually called for STS-1 to fly with the doors closed throughout the mission, relying upon the flash evaporator system, rather than the radiators, but it was recognised that opening them was critical to dissipating the heat load. Had a contingency EVA been needed, the entire cabin pressure would have been reduced from the normal 101.3 kPa to 62 kPa and after nine hours of 'pre-breathing' Crippen would have suited up and entered the payload bay. By lowering the atmospheric pressure in this manner, the crew's work day would have been shortened by two hours and allowed Young to also be properly 'pre-breathed', in case he too needed to go outside to assist his pilot.

Fortunately, Crippen opened the doors perfectly at the end of Columbia's second orbit. He gingerly unlatched the starboard door first – "Here comes the right door and, boy, that is *really* beautiful out there" – and immediately closed it again to verify the performance of its seal. "All the latches work just fine," he told Mission Control, "and the door looks like she's doing her thing." Both doors would be opened and closed several times during the next two days to evaluate not only the seals, but also their latches and actuators. Throughout each procedure, the astronauts worked at the rear of the cockpit, facing a pair of small square windows, overlooking the payload bay; they reported that they could work there quite comfortably without the need for any kind of foot restraints. Looking into the pristine, white, insulation-enshrouded bay, there was not much to see. There were no payloads to deploy and not much in the way of experiments to perform. The biggest 'experiment', Columbia herself, was already underway ... and she was running perfectly. In fact, her only real 'payloads' on STS-1 were two small, unassuming, yet vital, boxes of measuring devices and sensors. Known as Development Flight Instrumentation, these were mainly to be used during the early Shuttle test flights, but several components would remain aboard Columbia for later missions. Weighing 9,290 kg, it offered the first 'real' measurements of the Shuttle's performance and the stresses endured during launch, ascent, in space and throughout re-entry and landing. Previous data had only been available through computer simulations and the DFI data was expected to provide the first hard details. It stored this information on three magnetic tape recorders, to be analysed after the flight. Other devices on the DFI pallet were microphones to acquire acoustic data and an array of six different materials – including Teflon and gold – to assess their level of degradation in the harsh environment of low-Earth orbit.

Two more experiments were also carried, one of which included actual hardware and the other took advantage of Columbia's re-entry flight path. The Aerodynamic Coefficient Identification Package (ACIP) complemented the DFI by collecting data during all flight phases, but particularly during the hypersonic, supersonic and transonic regimes of re-entry, in order to validate wind tunnel predictions. As well as helping to advance engineers' understanding of the thermal and structural dynamics of the Shuttle during its glide back to Earth, ACIP measured the positions of each flight surface and gathered around four hours' worth of data in total. The second experiment was a wholly passive one, the Infrared Imagery of Shuttle (IRIS). It featured no on-board equipment, but involved NASA's Kuiper Airborne Observatory taking high-resolution infrared pictures of Columbia's belly and fuselage as she re-entered the atmosphere.

Young and Crippen had little direct involvement in any of these experiments, which operated autonomously. In any case, they had their hands full with the many engineering tasks planned. It was during the first day of the flight, while Crippen was evaluating the payload bay doors, that the men noticed that several thermal protection tiles were missing from one of the OMS pods.

"Okay," Young told Mission Control, "what camera are y'all looking at now?"

"We're looking out the forward camera," replied the duty capcom.

"Okay. We want to tell y'all here we *do* have a few tiles missing off the starboard pod. Basically, it's got what appears to be three tiles and some smaller pieces; and off the port pod, looks like ... I can see *one full square* and looks like a few little triangular shapes that are missing and we are trying to put that on the TV right now." Young's observation highlighted, for the first time, a problem which would become commonplace on Shuttle missions: tiles being shed from certain areas of the airframe during the violent climb into orbit. He also remarked that, after visual inspection, *no* tiles seemed to be missing from the Shuttle's wings, vertical stabiliser or nose. However, it was impossible to determine if any had gone from her belly, which, by design, would bear the full brunt of atmospheric heating during re-entry. Back on Earth, NASA managers watched the transmissions, but decided that none of the lost tiles was in a critical area. The most that could happen, they said, was that after landing a patch of aluminium skin underneath some tiles might need to be replaced. Yet even at this early stage, on the vehicle's very first flight, a problem which, further down the line, would turn out to be a catastrophic flaw, had *already* reared its head.

It also raised a new awareness that NASA had the capacity to tap into its links with the military and their state-of-the-art imaging equipment to photograph the Shuttle to ensure that it was in a suitable condition to survive re-entry. When the missing tiles were first noticed, Hart recalled, it caused concern in Mission Control. "Was there something *underneath* missing, too?" he wondered. "I think they found some pieces of tile in the *flame trench* after launch, as well, so there was kind of a tone of concern at the time, not knowing what condition the bottom of the Shuttle was in. We had *no* way to do an inspection, so we were all wringing our hands after the first shift or two. All of a sudden, the word started buzzing around Mission Control that we don't have to worry anymore!" Gene Kranz, the legendary former Apollo flight director, entered the control room with a sheaf of images in his hand. The images clearly showed the underside of Columbia, presumably taken using high-powered (and classified) equipment on the ground.

"How *did* you get those?" was the astonished reaction.

"I can't tell you," grinned Kranz.

Clearly, the images had been acquired from what Hart euphemistically described as "some of our national technical assets". For the controllers in Houston, it came as a moment of relief, since they now knew that Columbia's belly was undamaged. A little more than two decades later, another returning crew, aboard the very same orbiter, would not be so lucky.

For Young and Crippen, though, the future was getting their ship checked out and returned safely to Earth. Three and a half hours after launch, the men doffed their bulky pressure suits and stowed them in Columbia's middeck. With the exception of a practice on 13 April, they would not need the suits until a few hours before re-entry. For the rest of the mission, the Shuttle flew with her topside and open payload bay doors facing 'down', towards Earth. In general, the astronauts found that the pristine new craft was performing with very few problems – "like a champ", Young would say – and that life in space was comfortable and positively roomy. *Roomy*, maybe, but *spartan*, nonetheless. Young and Crippen slept in their seats on the flight deck, although plans were already advanced to include sleeping bags, bunks and even phonebox-sized 'sleep stations' on future missions. These would be particularly useful on Spacelab flights, whose crews would work in round-the-clock, 12-hour shifts.

The 'middeck' was situated directly beneath the flight deck and, in space, the astronauts reached it by means of a small, 66 × 71 cm opening; there were actually *two* openings, one on each side of the vehicle, but normally only one was used. Essentially, the middeck provided a living area for the crew, including stowage lockers for experiments and equipment, sleep stations, a galley, toilet and the huge airlock, which provided entry to the payload bay. Before launch and after touchdown, the crew entered and departed the Shuttle through a circular hatch on the middeck's port-side wall. Above the middeck, the ten-windowed 'flight deck' – once described as "our favourite place" by Kalpana Chawla, a member of Columbia's last crew – was the location for controlling the vehicle during ascent, re-entry and orbital activities. Its forward portion contained fixed seats for the commander and pilot, although on subsequent flight a pair of collapsible seats for mission specialists could be mounted directly behind them, and a bewildering array of displays, dials and switches. Six windows wrapped, airliner-like, around the front of the flight deck, with another pair in the 'ceiling' and two more overlooking the interior of the payload bay.

Young and Crippen experienced a rather cold first night in space, thanks to a problem with a temperature controller. Conditions improved on the second night. They prepared their meals with a food warmer, although a larger and more elaborate galley was planned for inclusion on subsequent flights. A few minor problems were encountered with a suction hose on the multi-million-dollar toilet, which stubbornly refused to work properly. When the time came to return home, the two men – rather ignominiously for space explorers – were obliged to stuff paper towels into the hose to prevent it from overflowing. They were also forced to use the urine-collection devices in their pressure suits.

After a whirlwind two days, the time came to put Columbia to the ultimate test: knifing through the atmosphere at *twenty-five* times the speed of sound, subjecting some of the tiles to immense thermal stress and performing an unpowered, 'deadstick' touchdown at Edwards Air Force Base in California. Although the last few minutes, from passing subsonic in the low atmosphere to the runway, had been exhaustively rehearsed, the 45 minutes from the 'de-orbit' burn of the OMS engines, through the searing furnace of re-entry and the complex series of aerodynamic turns needed to 'bleed off' the craft's speed and align her for touchdown, were largely unknown. To play things safe, NASA opted to use the wide expanse of dry lakebed at Edwards for the first four test flights. This would offer Young and Crippen a more forgiving runway and greater margins for error, although it was anticipated that when the Shuttle became fully operational and its aerodynamic performance was better understood, precision landings on a narrower concrete runway at the Kennedy Space Center would become the norm. Four hours before landing, at around 9:00 am Eastern Standard Time on 14 April, the crew closed and latched the payload bay doors for the final time.

Twenty minutes before the de-orbit burn, which would drop Columbia out of space and into the upper reaches of the atmosphere, the astronauts oriented their craft tail-first and switched on two of the three Auxiliary Power Units (APUs). These were responsible for controlling the Shuttle's flight surfaces and hydraulics throughout re-entry. Fifty-three hours and 28 minutes after launch, passing over the Indian Ocean, the OMS engines ignited in the vacuum, slowing Columbia sufficiently to begin her perilous, high-speed glide to a landing strip on the opposite side of the planet. The two-and-a-half-minute burn was reported with typical coolness by Young: "Burn went nominal."

"Nice and easy does it, John," replied Capcom Joe Allen. "We are *all* riding with you."

His words echoed the prayers of not only the men and women of NASA, but also of countless observers around the world.

Minutes later, Columbia was turned around and her nose pitched 'upwards' at a 39-degree angle. Young and Crippen removed the safety pins from their ejection seats and the overhead escape panels, then switched on the third APU. As the spacecraft entered the denser portion of the atmosphere, the tracking station on the island of Guam in the Central Pacific noted bursts of Columbia's pulsing thrusters. Travelling at close to 25,750 km/h, they hurtled onwards and onwards, as the colour of ionised atmospheric gases outside steadily morphed from a pale pink into a deeper pinkish-red, then reddish-orange, creating a blast-furnace-like scene which would not have been out of place in Dante's *Inferno*.

As a tense world waited, the NASA public affairs commentator reeled off a steady stream of updates. "We will be out of communication with Columbia for approximately 21 minutes," he noted, making reference to the lengthier-than-normal period of radio blackout, caused by the accumulation of a plasma 'sheath' around the re-entering spacecraft. "No tracking stations before the West Coast … and there is a period of about 16 minutes of aerodynamic re-entry heating that communications are impossible … " During this time, the Kuiper Airborne Observatory, flying almost directly beneath Columbia's path, acquired its infrared images. They showed the orbiter coming back to Earth like a meteor. The aircraft had earlier taken off from Hickam Air Force Base in Hawaii and established itself at an altitude of 13.7 km, about an hour before the spacecraft reached 'entry interface'.

Descending lower now, the astronauts were, at length, able to receive UHF radio calls, crackling between Mission Control and one of the T-38 chase aircraft which would accompany the Shuttle down to the runway. "Hello, Houston," Young called, "Columbia's here! We're doing Mach 10.3 at 188 [thousand feet]." For the majority of this period, except for the so-called 'roll reversals' – a series of S-shaped curves to reduce speed – the computers were primarily responsible for flying the vehicle. Shortly after the orbiter crossed the California coastline, near Big Sur, Young took manual control. Long-range tracking cameras on Anderson Peak captured the first ground-based images of Columbia, flying at an altitude of more than 35 km.

"What a way to come to California!" exulted Crippen.

Still travelling at well over four times the speed of sound, the Shuttle passed over Bakersfield, Lake Isabella and Mojave Airport, enabling the astronauts to verify by glancing through their windows that the ground track was "right on the money". Young then executed a sweeping, 225-degree turn to align his ship with the lakebed Runway 23 at Edwards. Dropping to below 12 km, he took Columbia's stick and would later remark that control was crisp and precise.

Watching the arrival of America's first Space Shuttle from orbit were tens of thousands of people, including Larry Eichel of the *Philadelphia Inquirer*. His testimony perfectly encapsulated the anxiety and nervous excitement of everybody awaiting this historic event. "The Shuttle appeared far above the north-east horizon," he explained, "a white dot against a cloudless blue sky. That dot was dropping so fast that to an eye accustomed to watching the more gradual descent of commercial jets, it seemed inevitable that the Shuttle would crash to the desert floor." As Columbia drew closer, her speedbrake was gradually retracted and was fully closed by the time the vehicle was 600 m above the runway. Falling precipitously, *seven times steeper* than a commercial airliner, and almost *twice* the speed, the reaction of Eichel that a crash was about to occur can, perhaps, be forgiven. It was at *this* point, however, that John Young pulled back on the stick, lifted the nose and transformed his ship, in a split second, from a falling brick into a graceful flying machine.

"The Shuttle has a very, very steep glide slope," explained Neil Hutchinson, "about eight or nine degrees. That doesn't sound very steep, but if you were in an airliner doing that, you'd think you were headed for sure death!"

Weather conditions in the California desert were near-perfect and surface winds were calm. At 10:20:35 am Pacific Standard Time (or 1:20 pm in Florida and 12:20 pm in Houston), Bob Crippen deployed the landing gear and all six wheels were down and locked into position within the mandatory ten-second time limit. Columbia touched down perfectly, 22 seconds later, at a speed of 342 km/h, and rolled for almost 3 km before coming to a smooth halt. The speedbrake was opened and full-down elevons were applied, giving the astronauts an impression of considerable deceleration. "As it touched down," recalled Eichel, "at a speed 80-90 miles an hour faster than a commercial airliner does, the rear wheels nestled into the hard-packed sand, kicking a rooster-tail high into the air." The countdown to landing was echoed by both the public affairs spokesmen at Edwards and by the crew of one of the T-38s, who were first to welcome Young and Crippen back home with a resounding "Beautiful! Beautiful!"

Rookie astronaut John Creighton was aboard a US Army helicopter at Edwards that day and he later described the remarkable efforts of some spectators to get a close-up view of Columbia's first return from orbit. "All kinds of people had camped out there for several days," he explained. "There was a fence and there'd been a

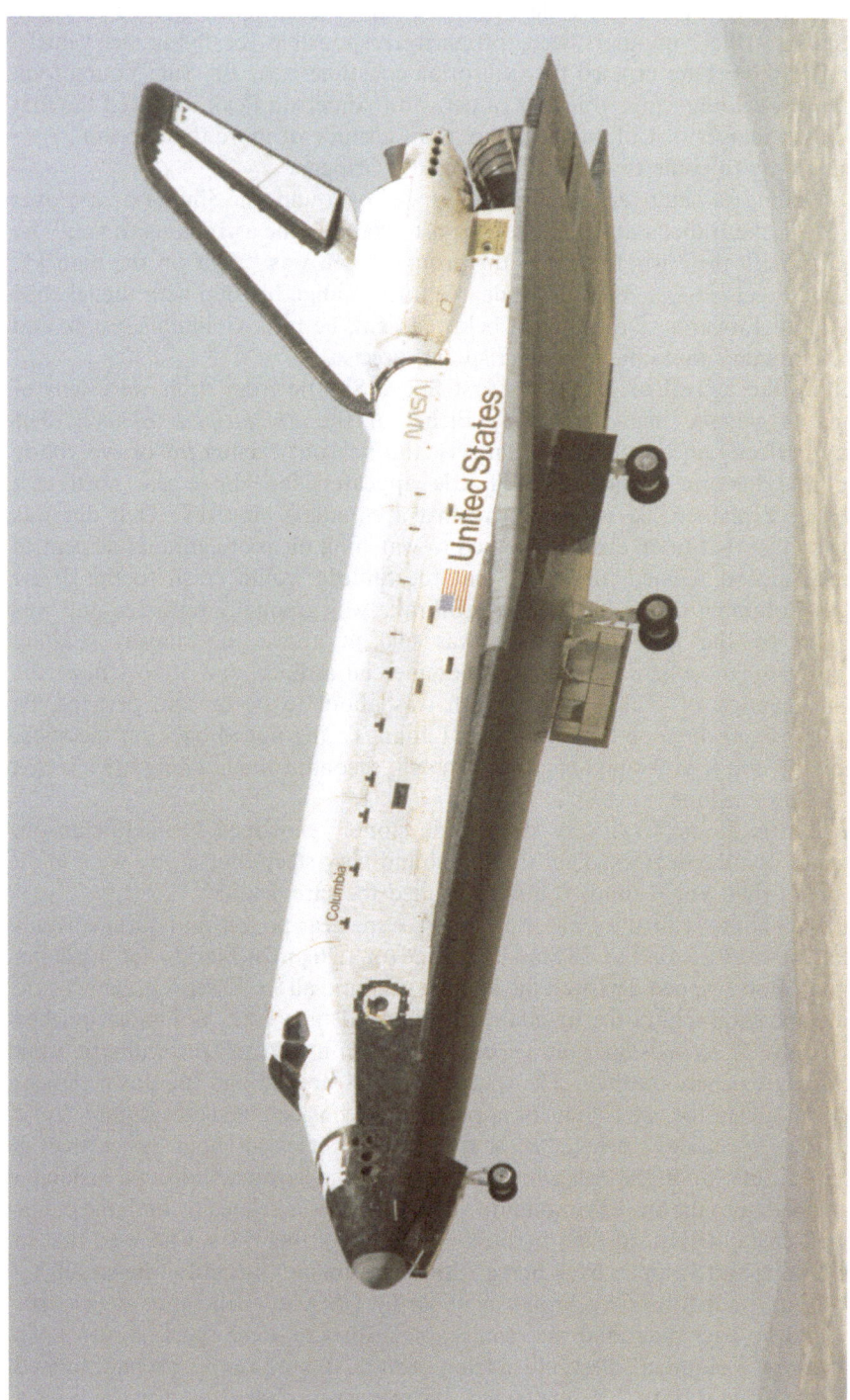

Columbia touches down at Edwards Air Force Base on 14 April 1981.

patrol to keep people back there. As soon as the Shuttle rolled to a stop, these people charged forward, [this] fence went down and they got motorcycles and cars that went out racing. This was about five miles from where the Shuttle actually landed and the only way you could see [it] was with binoculars, but, boy, *they* wanted to get an *up-front* view! The security folks didn't know what to do, so they told the helicopters to try to get this crowd under control, so these helicopters would swoop down in front of the on-charging group of cars. The helicopter pilots *loved it*. They were having a great time trying to head off all of these people!"

Post-landing analysis revealed that Columbia's right-hand inboard brakes suffered higher-than-anticipated pressure, which caused a slight tug to the right, just before the wheels stopped. Young compensated for this by balancing the total braking to either side of the Shuttle, maintaining a near-perfect course straight down the runway centreline, stopping at the intersection of Runways 23 and 15. One notable surprise was the sheer *amount* of lakebed debris – pebbles and grains of sand – kicked up by the wheels.

"Do I have to take it to the hangar, Joe?" asked Young.

"We're gonna dust it off first," retorted Joe Allen with a chuckle.

Immediately after wheelstop, the astronauts unstrapped and began safing the RCS and OMS switches before the arrival of the ground crew. When the latter arrived, they first hooked up sensitive 'sniffer' devices to verify the absence of toxic or explosive gases and attached coolant and purging lines to Columbia's aft compartment to air-condition her systems and payload bay and dissipate residual fumes. Whilst this procedure was underway, the ground teams worked in protective suits, then moved an airport-type stairway over to the hatch. (Years later, Joe Allen would find it amusing to watch Young and Crippen, who looked like ordinary people as they came down the steps ... surrounded by the ground team, whose cumbersome protective suits made *them* look like the astronauts!) John Young, who had remained totally cool throughout re-entry, *now* let his excitement get the better of him. As soon as he got outside, about an hour after touchdown, he bounded down the steps, checked out the tyres and landing gear and jabbed the air triumphantly with both fists. He even *kicked* the tyres ... which scared the life out of Henry Pohl. "I was really worried about *that*, because those tyres have got 375 psi pressure in them ... and I *knew* the brakes got hot," he explained. "I was afraid the tyres were going to explode. It would have been a shame to do all that flying and a terrific landing and then have a tyre blow up because you went over and *kicked* it!"

Young, of course, could be forgiven. He was over-excited ... and it showed. "I've often claimed that John calmed down" by the time he got outside, Crippen said later, but noted with a twinkle in his eye: "You should've seen him when he was *inside the cockpit*!"

A RACE AGAINST THE CLOCK

"Okay, the arm is out and it works *beautifully*!" STS-2 pilot Dick Truly told Mission Control excitedly on the afternoon of 13 November 1981. On only his first flight into space, the 44-year-old naval officer had been given the enviable task of putting Columbia's Canadian-built Remote Manipulator System (RMS) – a gigantic, $100 million robot arm – through its paces for the first time. It was Canada's contribution to the Shuttle programme – a contribution which dated back to 1974, when Spar Space Robotics Corporation was contracted by the country's National Research Council to build a mechanical manipulator for deploying and retrieving satellites and, someday, building a permanent space station. The decision to involve Canada was formally announced by NASA in a May 1975 memorandum of understanding, which was formalised in 1979. By the early summer of the following year, astronauts John Fabian, Sally Ride, Judy Resnik, Story Musgrave and Norm Thagard were actively testing its systems, rotational and translational hand controllers and associated work stations in the Shuttle mission simulators. Ride, in fact, would virtually *write* the procedures for operating the arm in each of its automatic and manual control modes. Not surprisingly, she played a key role in STS-2, as one of the on-orbit capcoms.

The challenges involved in building an arm capable of such complex and dexterous tasks were enormous: it needed to operate automatically *and* under manual control and meet strict weight and safety requirements. Moreover, nothing quite like it had been built or used in space before, which made Spar's task still more difficult. Although a horizontal floor rig had been set up to test its joints in 1977, the first real demonstration would not come until the RMS was physically uncradled in orbit. The first space-rated arm was delivered to the Kennedy Space Center in April 1981 and was installed into Columbia's payload bay on 20 June. It measured 15.2 m long – enabling it to reach the far end of the payload bay – and consisted, like a human arm, of shoulder, elbow and wrist joints, linked by a pair of graphite-epoxy booms. Other components were made from titanium and stainless steel. To protect it from thermal extremes in space, the RMS was covered in white insulation and fitted with heaters to maintain its temperature within required limits. Without a payload attached, its tip could move at up to 60 cm per minute, but only a tenth of that rate when fully loaded. Ingeniously, the means by which the arm could 'pick up' and 'put down' objects was achieved by the 'end effector': essentially a *hand* with a kind of wire snare to capture a prong-like grapple fixture on the payload. Already, one of NASA's most important upcoming projects, the Hubble Space Telescope, then scheduled for launch in the mid-1980s, had an in-built grapple fixture which would allow it not only to be *deployed*, but also *retrieved* and *repaired* in space, by future Shuttle crews.

During operational flights, astronauts used a pair of television cameras on the arm's wrist and elbow to guide the end effector over a target's grapple fixture, before commanding three wire snares to close around it at just the right moment. When this was done, it would impart a force of 500 kg onto the grapple fixture and allow the RMS to move the target into or out of the payload bay. When Dick Truly uncradled the RMS that November afternoon in 1981, he had no payloads to deploy, but on the *next* flight, STS-3, a desk-sized Induced Environmental Contamination Monitor (IECM) would be carried to flex its muscles. The excitement at being the first person to wring out the mechanical arm in orbit was, however, tempered by disappointment. The previous evening, only hours after reaching orbit, Truly and his commander, fellow ALT veteran Joe Engle, had lost more than half of their mission.

Captain Richard Harrison Truly of the US Navy became the world's first spacefarer to blast off on his first spaceflight on his *birthday*, for he had been born on 12 November 1937 in Fayette, Mississippi. "I *was* interested in flying" as a child, he told the NASA oral historian, "but I *never* really intended to be a pilot. It just never occurred to me that *that* would be a possibility." He attended local schools and studied aeronautical engineering at Georgia Institute of Technology, from which he received his degree in 1959. A military career, by now, was firmly on his radar, since he held a scholarship with the Reserve Officers Training Corps, but even at this late stage becoming a pilot was not really part of the plan. "It wasn't that I *didn't* want to be," he explained, "it just never really occurred to me. I was going to be an *engineer*." A handful of training flights changed his mind and Truly entered flight school and was designated as a naval aviator in October 1960. His first tour was with Fighter Squadron 33, in which he flew F-8 Crusaders from the USS *Intrepid* and (propitiously) *Enterprise*, performing more than 300 carrier landings.

Whilst aboard the *Enterprise*, Truly's squadron leader, Commander Larry Ned Smith, advised him to consider test pilot school and for two years, from 1963 until 1965, he worked at Edwards Air Force Base, first as a student and later as an instructor in the Aerospace Research Pilot School. During this period, he was the youngest of nine candidates put forward for the Air Force's MOL project; additionally, he and another aviator, Bob Crippen, were the only Navy members of the group. When MOL was cancelled in June 1969, Truly and Crippen were among the handful of pilots essentially forced down NASA's throat by the Pentagon. The unbridled joy at having entered the ranks of the world's most elite flying fraternity had its ups and downs, though, and one of the notable 'downs' was the lack of seats on space missions. Dave Scott asked him to serve on the Apollo 15 support crew, but Truly felt his chances of a flight were better in Skylab and elected to take duties there instead. "I foolishly thought that maybe I would actually get to fly on Skylab," he said, "but I really didn't account for the fact that ... you were *in a line*. There were a *lot* of people ahead of me." Capcom duties in Skylab were followed by a support role on ASTP and finally, in the spring of 1976, assignment to ALT and the Shuttle.

Five and a half years later, in space, Truly's job, along with Engle, was to somehow cram five days' worth of scientific research and complex engineering objectives into just over 54 hours. It had all seemed so much brighter on the morning of 12 November 1981 and a perfect birthday gift for Truly when they rocketed into orbit from Pad 39A, becoming the first team of astronauts to fly a 'second-hand' spacecraft. Although they had been in training since 1978, the formal NASA announcement of their assignment to fly STS-2 did not come until 23 April 1981, prompting Young and Crippen to design and present them with a ceremonial cardboard 'key' to the orbiter. "I think the hope was that would be a traditional handover of the vehicle to the next crew," Engle recalled in a NASA oral history. "In fact, it was done at a pilots' meeting. John and Crip handed Dick and I the key and I think there were so many comments about buying a used car from Crip and John, that it became more of a joke thing than a serious traditional thing and I don't recall that it really lasted very long. I think it turned from cardboard into plywood and I don't recall that it was done very long after that. Once we got the next vehicles on line, Discovery and Challenger, it lost some significance. Besides, you weren't really sure *which* vehicle you were going to fly after that, so you didn't know *who* to give the key to!"

Despite the stupendous achievement of sending Columbia into space the previous April, bringing her home like an airliner and turning her around to fly again, NASA's promise of a Shuttle launch every two weeks was a long way from reality and would *never* be achieved. To be fair, all four OFT missions were considered 'test flights' and were not under great schedule pressure, but the space agency had hoped to have Columbia in flight-ready condition for STS-2 in less than seven months. Delays had set in, partly due to problems experienced on her maiden voyage, and partly from the fact that NASA had simply underestimated the amount of attention that the vehicle would require between flights. It was *not* an airliner and it was doubtful if it could *ever* be operated like one.

Still, the preparation for STS-2 *was* faster than for STS-1. When Columbia arrived in Florida in March 1979, she spent almost two years undergoing flight preparations; the work to ready her for her *second* mission required only a quarter of that time. She returned to the Kennedy Space Center, atop the Boeing SCA, on 29 April 1981, but she was in need of a great deal of attention. The most pressing issue was tiles: 350 required replacement, 818 demanded removal and repair and a further two thousand would be serviced, in place, whilst in the Orbiter Processing Facility. NASA's hopes of launching Columbia on 30 September quickly evaporated and a revised target of 9 October was chosen for a mission which would represent an order of magnitude over STS-1 in terms of complexity. For the first time, the Shuttle would transport a fully-fledged scientific research platform – developed by NASA's Office of Space and Terrestrial Applications and therefore dubbed 'OSTA-1' – and to make room for it in the payload bay, technicians needed to move the DFI pallet further aft.

The history of OSTA-1 can be traced back to 1976, when a five-day Earth-observation mission was first sketched out. Shortly thereafter, NASA picked six experiments from 32 proposals; a seventh – a demonstration of a study destined for Spacelab – was added later. Initial analyses indicated that Columbia's payload bay could be pointed Earthward for up to 88 hours of a five-day voyage, which would demonstrate her ability to carry major scientific instruments in a stabilised manner and acquire valuable data.

The OSTA-1 experiments weighed 1,190 kg in total and were mounted on an engineering version of the U-shaped Spacelab 'pallet', located at the midpoint of the payload bay. The pallet was 3 m long and 3.9 m wide and was virtually identical in appearance to those scheduled to be flown aboard dedicated Spacelab missions, although for STS-2 it was not fully equipped. 'Operational' flights would also feature a cylindrical, temperature-controlled 'igloo' to provide the pallets with cooling, power and data-management facilities. Pallets, it was hoped, would be used to carry

The large Shuttle Imaging Radar (SIR) is clearly visible in this image of the OSTA-1 pallet being installed into Columbia's payload bay in June 1981.

instruments which demanded unobstructed fields of view, such as a terrain-mapping radar and astronomical telescopes. Five of the OSTA-1 experiments were affixed to the pallet, of which the largest, by far, was the Shuttle Imaging Radar (SIR), designed to assess Columbia's performance as a research platform by furthering geologists' understanding of the radar signatures of various terrestrial features for mineral and petroleum exploration. Measuring 9.4 m long and 2.2 m wide and weighing 180 kg, SIR was a 'synthetic-aperture' radar and filled almost half of the bay. Assembled from spares left over from NASA's 1978 Seasat mission, it was mounted on its own truss structure, which, in turn, was fixed to the Spacelab pallet, providing a 'side-looking' viewing angle 47 degrees from nadir. As a result, the Shuttle would need to orient itself to direct SIR at ground targets. The radar operated at two frequencies in the L-band to provide information to construct two-dimensional radar images of the surface. It worked by transmitting microwave signals and receiving reflected 'echoes', recording data onto computer tapes for post-flight analysis. During STS-2, it would prove hugely successful and would acquire some intriguing data and lead to some tantalising results.

Somewhat less visible on the pallet, yet still capable of holding their own in terms of scientific data, were the Shuttle Multispectral Infrared Radiometer (SMIRR), the Feature Identification and Location Experiment (FILE), the Ocean Colour Experiment (OCE) and the Measurement of Air Pollution by Satellite (MAPS). Integration of the five experiments onto the Spacelab pallet was completed in the Operations and Checkout Building in the early months of 1981 and in June the full payload was loaded aboard Columbia. A series of tests verified its compatibility with the orbiter and the ability of the astronauts to operate it from their stations on the aft flight deck. On 10 August, the Shuttle was transferred to the Vehicle Assembly Building to be stacked onto her External Tank and SRBs. Rollout to the launch pad, though, did not seem to bring the mission much closer. The original 9 October target was scrubbed and postponed by nearly a month, following an accidental spillage of nitrogen tetroxide. Technicians were busy loading the highly toxic oxidiser into Columbia's forward RCS unit, when five and a half litres spilled onto the nose; some 379 delicate tiles had to be removed, painstakingly cleaned and replaced. A revised launch date of 4 November was then set and all seemed on track until, with two days to go, engineers set to work loading the oxygen tanks for the orbiter's fuel cells ... and noticed some unusual readings: one tank was *losing pressure*. A change in the loading procedure seemed to do the trick and the countdown continued. On the morning of the 4th, Joe Engle and Dick Truly arrived at the pad and were strapped into their seats aboard Columbia. Then, with just *nine minutes* remaining, managers called a halt when the oxygen tanks showed up lower-than-allowable pressures. *This* glitch, too, was quickly resolved, but others followed. When Truly turned on the Auxiliary Power Units at T-5 minutes, they showed higher-than-normal oil pressures. Nonetheless, the count proceeded, in the hope that a solution could be found. Thirty-one seconds before launch, just before the ground launch sequencer prepared to hand over control to the Shuttle's computers, the effort was called off. Oxygen tank pressures were too low, the

APUs were shut down and the countdown clock was recycled back to T-9 minutes ... then T-20 minutes ... and then, when the weather started to close in, NASA decided to call off the attempt for the day.

The weather hammered the final nail into the coffin for a launch; the APU oil pressures were too high to allow Columbia to fly – in fact, they were around 690 kPal, instead of the allowable limit of 414 kPal – and it was the Mission Management Team which made the final decision. Subsequent analysis revealed that the APUs' oil filters had become clogged by pentaerythritol, a crystal which formed when hydrazine penetrated their gearboxes. This had caused the unexpected rise in temperature. Both gearboxes were flushed, their filters were replaced and the launch was rescheduled for 12 November. During their lengthy wait on the pad, Engle and Truly also reported that their visibility was only marginal, so the forward flight deck windows were cleaned in readiness for the next attempt.

Eight days later, the men returned to the pad for a second try. There had already been minor problems, again, during the loading of the fuel cells' oxygen tanks, which meant that the No. 3 tank had to be loaded and pressurised separately from the others. Another problem occurred late on 11 November, when one of four Multiplexer-Demultiplexers (MDMs) failed. These provided instrumentation measurements, commands and data to the Shuttle's cockpit displays. A spare was fitted, but it too was faulty, requiring *another* unit to be flown from California in the early hours of the 12th. The new MDM, interestingly, came from the second space-rated orbiter, Challenger, which was then undergoing final checkout at Rockwell International's Palmdale plant in readiness for transportation to the East Coast. It was the first of many occasions in which parts were 'cannibalised' from one orbiter to enable another to fly ... and in a few years' time, NASA would find itself heavily criticised for the practice ...

HALVED MISSION

"I would still call it a successful mission," Dick Truly said, early on 12 November 1981, "even if we had to come home early." Presumably, *he* was referring to the achievement of simply *getting* Columbia safely into space for the second time ... but he could not have known how prophetic his words would prove to be. STS-2 was supposed to last more than twice as long as Young and Crippen's mission and to complete it the orbiter was fitted with a number of modifications, the most significant of which was an extra set of cryogenic hydrogen and oxygen tanks for the fuel cells underneath her payload bay floor. The mission got off to a good start with a perfect liftoff at 10:09 am Eastern Standard Time, reaching space within nine minutes. The launch was *much* less damaging than that of STS-1: the modification to the External Tank struts proved effective and the shattering effect of the SRBs, which had previously caused excessive pressures, were alleviated by improvements to the water sound-suppression system. Indeed, sensors affixed to Columbia's base during STS-2 revealed that the pressures had fallen to barely a *tenth* of those previously encountered.

Moreover, thanks to these improvements, the DFI recorders operated without interruption throughout the ascent and *no* tiles were lost. Post-mission analysis of STS-1 seemed to imply that the missing tiles identified by Young and Crippen had most likely become 'liberated' and were shaken loose as a result of the SRB over-pressure problem. Columbia's second launch, on the other hand, suffered only from the need to shut down one of the APUs slightly earlier than expected, due to high oil temperatures. After dumping residual main engine propellant into space – a process terminated 16 seconds early, due to the APU problem – Engle and Truly performed a pair of OMS firings to circularise their orbit at approximately 193 × 201 km, inclined 38 degrees to the equator. The low orbit permitted OSTA-1 to gather its data at the required resolution. The astronauts also opened the payload bay doors to expose the scientific platform to space for the first time.

Although only the first two OMS burns were necessary to establish Columbia in her correct orbit, two *additional* firings were performed to raise the altitude to 220 km. The third burn was split into two halves in order to satisfy one of STS-2's flight test objectives: an ability to turn off an OMS engine and restart it a few minutes later in the vacuum of space, with no ill-effects. The fourth firing then demonstrated the ability of the OMS to feed the right-hand engine with the left-hand pod and vice versa. All of the tests proved successful. What did *not* turn out to operate well, only a couple of hours into the mission, was one of the electricity-generating fuel cells. This had damaging implications for the accomplishment of the scheduled mission, during which Engle and Truly were to conduct extensive tests of the RMS in both manual and automatic modes and try out the new Shuttle EVA space suit in the middeck. All that began to change dramatically late on the afternoon of the 12th, when ground controllers spotted a high pH indication on the No. 1 cell. Its overall performance, though, at least at this stage, remained more or less normal.

The situation deteriorated rapidly and within two hours a sharp voltage drop was recorded on the cell, indicating the probable failure of one or more 'stacks' inside it. If that was the case, it meant that the cell's capability to generate electricity for the orbiter and, as a byproduct, also drinking water for the astronauts, might be compromised. With the likelihood of a contaminated water supply, the No. 1 cell was switched off later that evening and in response to worries that the water was being electrolysed – thus forming a potentially explosive mixture – it was also depressurised. Under prescriptive mission rules, laid down long before STS-1, all three fuel cells *had* to be fully operational in order for a mission to continue. (The problem was subsequently traced to a deposit of aluminium hydroxide in an aspirator – perhaps as little as just a speck, in fact – which prevented the proper removal of water from the cell.) It was with disappointment that, early on 13 November, Capcom Sally Ride told Engle and Truly that they were going to come home the next day. "That's not so good," was all a dejected Truly could say. He and Ride had spent a significant amount of time together, working on the RMS development tasks – in fact, one of the reasons that Ride would be selected to become the first American woman in space in 1983 was due to her expertise with the Canadian-built arm – and it seemed that their efforts were now in vain.

The External Tank drifts away from Columbia, eight and a half minutes after the STS-2 launch.

Also dejected was Engle himself, whose long NASA career had quite literally been a rollercoaster, with equally as many triumphs and successes as disappointments and setbacks. It is no accident that Colonel Joe Henry Engle of the US Air Force became only the fourth NASA astronaut to command his very first space mission ... and also the *last*, for no American spacefarer has since taken charge of a crew on his or her rookie flight. Whether or not Engle *was* a 'true' first-timer when he rode Columbia into the heavens on 12 November 1981 is very much open to debate: for his parent service, the US Air Force, had already awarded him astronaut's wings back in 1965 for passing an altitude of 80 km on no fewer than *three* occasions in the X-15 rocket-propelled aircraft. To follow this line of thought has drawn some observers to acknowledge Engle as the first man to fly into space three times. Others stick to the FAI ruling and opt to believe that only missions which fly above the 100 km Kármán Line may be considered 'spaceflights'. By the end of his astronaut career, Engle had completed three X-15 missions and two Space Shuttle voyages ... leaving the way open for some to credit him as a five-flight veteran and others to acknowledge only two flights. Whichever you prefer, the real tragedy in the story is that this outstanding pilot came within a whisker of becoming one of the few humans to set foot on the Moon.

Engle came from the small city of Chapman, in Dickinson County, Kansas, where he was born on 26 August 1932, with aviation in his blood. "I don't know, honestly, when I did *not* want to fly airplanes," he told the NASA oral historian. "My mom used to say the same thing to me, that she couldn't remember me seriously wanting to do *anything* but fly airplanes. Of course, I went through the fireman and the cowboy games, but my core desires and my core toys were always airplanes and flying." As a young boy, he recalled his older sister cutting a small aircraft from an old tin can; since it was sharp-edged, Mrs Engle forbade her son to play with it. Years later, the young boy who would grow into a test pilot, astronaut, 'almost' a Moonwalker and ultimately a major-general in the Air Force would wonder if this 'forbidden fruit' had actually spurred him on to choose aviation as his life's work.

For Engle's hometown, Kansas State University was the place to go, since it was primarily an agricultural college and Chapman was a farming community. However, Kansas State did not offer an aeronautical engineering degree and so he went to the University of Kansas, instead, graduating in 1955. Shortly thereafter, Engle received his Air Force commission through the Reserve Officers Training Program and entered flying school in 1957. "Chapman didn't have a runway," he told the oral historian, "so my only exposure to flying was at the annual Labour Day celebration, [when] all the farmers would bring in their goods and display them. There was a guy who landed on an alfalfa field and was giving rides." Engle and a friend bought a ride. During the summers at the University of Kansas, he worked as a draughtsman for Cessna Aircraft and received flying lessons from his supervisor, Henry Dittmer, a man whom Engle greatly respected, not only for showing him the rudiments of aviation, but also the *responsibilities* demanded of a pilot. Dittmer even got him a job sweeping the hangars at Cessna, in exchange for hours of flying instruction.

The Air Force was calling, however, and when Engle completed fighter gunnery school he transferred over to the F-100 Super Sabre to begin training as a fighter pilot at George Air Force Base in Victorville, California. The F-100 was a supersonic jet, one of the hottest in the sky at the time, and he recalled with glee his experiences, practicing air-to-air combat and dogfighting operations, high above the inhospitable terrain of Death Valley and Stovepipe Wells. On one particular occasion, he would be brought – literally – into a head-on confrontation with Chuck Yeager, perhaps the most famous test pilot on the planet and commandant of the Air Force's Test Pilot School. "I had a flight of four up and *that* was where you always found somebody to engage in a mock dogfight," Engle recalled. "I noticed two airplanes coming back from the north-east, heading back toward George, so I called the flight and we set up for an attack. I was salivating, because everything was ideal and rolling in. It turned out that those two airplanes were *Chuck Yeager* and our operations officer, who had been up to Nellis [Air Force Base in Nevada] to check out some advanced gunnery school classes that we were going to do. They were undoubtedly the two best fighter pilots at George and I found that out *real* quick! They *completely* tore up my flight and just scattered us to the winds and I learned then to be a little more cautious when making attacks and not get too over-confident!" This did not prevent Yeager from *recommending* Engle in 1960 for admission into Test Pilot School and only the second class of the new Aerospace Research Test Pilot School – in fact, it may well have *aided* it, for the commandant's philosophy had always been simple: the *best* fliers are the ones who have the *most* experience (Engle accrued more than 1,500 hours), who worked the *hardest* and who took it the most *seriously*.

After graduation, Engle moved over to fighter test operations at Edwards, which he described as "a pilot's heaven", and in 1963 he and fellow pilot Mike Collins applied for admission into NASA's third group of astronauts. Collins was accepted, but Engle was asked to withdraw his application by his commanding officer, Major-General Irving 'Twig' Branch, head of the Air Force's Flight Test Center. Branch had other ideas for Engle: in June 1963, he recommended him to replace Bob White on the X-15 project. "*That* just thrilled me to death," Engle remembered, "because it was a chance to get into *space* . . . and to do it with a *winged* airplane, with a stick and rudder." Intuitively, he *knew* that he was still young enough to make another application to NASA at a later date, but flying the X-15 was *not* something a pilot *applied* for: it was a gift from the gods, the kind of assignment that one could only wait and hope would someday come their way. Engle made his first flight on 7 October of that same year and during his career would pilot the X-15 no fewer than 16 times. Perhaps his greatest accomplishments were exceeding the Air Force's – though *not* the FAI's – requirement for a spaceflight, surpassing an altitude of 80 km, on *three* occasions in June, August and October 1965.

Years later, Engle would explain that passing the magical Air Force limit *was* planned, and was a "threshold accomplishment", but it was *more* important to him to get as close as possible to each mission's planned altitude. "You *could* make sure you got into space," he said, "by letting the engine run another second or two . . . or just 'squeaking' a little bit more on the pitch attitude and by the time the ground could see it on radar, it was too late to do anything about it. A lot of times, people would err a little bit on the conservative side to make sure they got high enough." Engle's parents came to Edwards on 29 June 1965 to see their son perform his first mission above 80 km and a few weeks later he joined NASA's Jim McDivitt and Ed White – newly back from their Gemini IV mission – at the Pentagon to receive his astronaut's wings.

"I was thrilled," said Engle, "and a little hesitant to rain on their parade of getting their astronaut wings, but what happened was Air Force Secretary [Eugene] Zuckert had some kind of schedule conflict, but wanted us all there at the same time. Initially, there was a lot of good-natured and some maybe serious rivalry going back and forth between the Edwards pilots and the NASA astronauts, but I didn't sense any of that at all. I think it was just three Air Force officers had qualified for their astronaut wings and, at that time, you got to go to the Pentagon to get your astronaut wings."

Engle instinctively knew that his time at Edwards was not open-ended and that he would shortly be reassigned, so he decided to reapply to NASA in the spring of 1966 and was selected as a member of the agency's fifth astronaut group in April. It brought mixed emotions, but he was leaving the best flying job in the world and was hopeful that he might someday be considered for a seat on a mission to the Moon. As circumstances transpired, *he was*. In the summer of 1969, Engle was named as a member of the Apollo 14 backup crew and might have rotated into the prime slot as lunar pilot on Apollo 17 ... but was ultimately bumped from *that* mission when NASA, under pressure from the science community, decided to fly a geologist, Jack Schmitt, instead. Engle was bitterly disappointed at the outcome and once remarked that the hardest part was explaining to his children that he *wouldn't* be going to the Moon ... but persevered and supported Schmitt in his preparations for what turned out to be one of the most remarkable missions of lunar exploration ever undertaken. In the light of this admirable attitude, Deke Slayton offered to support Engle in whatever he wanted to do next. Slayton could not *promise* him a seat on Skylab or ASTP or the Shuttle, but he would do what he could. Joe Engle, forever the pilot, opted for the Shuttle and Slayton duly obliged him.

Now, ten years after Engle might have walked on the Moon and more than four years since he had flown Enterprise, he and Truly were faced with having to 'front-load' as many of their mission objectives as possible. Fortunately, this had *already* been timetabled into each of the four OFT missions, in anticipation of just such an eventuality. The OSTA-1 payload had been switched on by Truly within hours of reaching orbit and was already gathering what would prove to be spectacular data. In fact, by the time STS-2 returned to Earth, late on 14 November, the astronauts would have ticked off almost 90 percent of their programmed tasks. The fuel cell problem did have other impacts, however, not least of which was the slower-than-normal dispensation of drinking water.

Years later, Engle would credit their training with having enabled them to get so much done in so little time. "We had trained enough to know, precisely, what had to be done," he said, "and we prioritised things as much as we could. We only had the ground stations, so we didn't have continuous voice communication with Mission Control and [they] didn't have continuous data downlink from the vehicle either, only when we'd fly over the ground stations; so when our sleep cycle was approaching, we did, in fact, power down some of the systems and we did tell Mission Control goodnight, but as soon as we went LOS, loss of signal, from the ground station, *then* we got busy and scrambled and cranked up the [RMS] and ran through the sequence of tests for the arm, ran through as much of the other data that we could, got as much done as we could during the night. We didn't sleep that night; we stayed up all night. Then the next morning, when the wakeup call came from the ground, why, we tried to pretend like we were sleepy and just waking up!" Burning the midnight oil did not go unnoticed, however. After the flight, veteran flight director Don Puddy told Engle that Mission Control *knew* they were awake, "because we could see you were drawing more power than you should've been if you were asleep".

As OSTA-1 gathered data, the astronauts focused their attention on the RMS. "Its movements," Truly told Sally Ride, "are *much* more flexible than they appeared during the training simulations." Although it was controlled by Columbia's General Purpose Computers, the arm's actual movements were commanded by the pilot himself, using a joystick in the aft flight deck. Under Truly's control, the shoulder and elbow joints were moved up and down and left and right and the wrist was pitched, yawed and rolled. As he issued each instruction, the GPCs examined it and determined which joints had to be moved, in which direction and at what particular speed or angle. Meanwhile, the computers looked at each joint at 80-millisecond intervals and, in the event of a failure, could automatically apply a series of brakes and notify the crew. As Truly worked, a continuous flow of data on joint rates and speeds appeared on monitors in the flight deck. An hour after it had been uncradled from the port-side sill of the payload bay, Ride asked Columbia's crew to establish the television feed.

"Transmitting," radioed Engle. "You people seeing anything down there yet?"

After a moment of silence, Ride confirmed that, yes, the first pictures of the Canadian arm in an inverted 'V' shape had appeared on Mission Control's screens. The images drew spontaneous applause from the Canadian delegation. For the next four hours, Truly and Engle took turns operating the arm and tested it in all five control modes, using both the primary and backup software. Four days of tests were now restricted to 24 hours. As Truly 'flew' the RMS, Engle fired a series of reaction control thruster bursts to assess its performance under duress. The arm's television camera were also evaluated and, on one priceless occasion, showed a grinning Truly peering through the aft flight deck windows with the inevitable sign, clutched in his hand: *HI MOM*.

To say that the RMS was a proud achievement for Canada is something of an understatement; and they made absolutely certain that an enormous national flag adorned the arm's insulation. At this, Engle's sense of fun kicked into high gear. He and Truly arranged for a large American flag to be sewn onto the aft bulkhead of Columbia's payload bay, "so when the *cameras* come on, we'll have [them] pointed toward [the US] flag and *that's* the first thing that will be downlinked to the ground!" This one-upmanship in revealing the Stars and Stripes *before* the Maple Leaf was, however, short-lived, for the Canadians had packed a few 'surprises' of their own ... amongst the astronauts' *clothing*. "All of our clothes were packed in storage

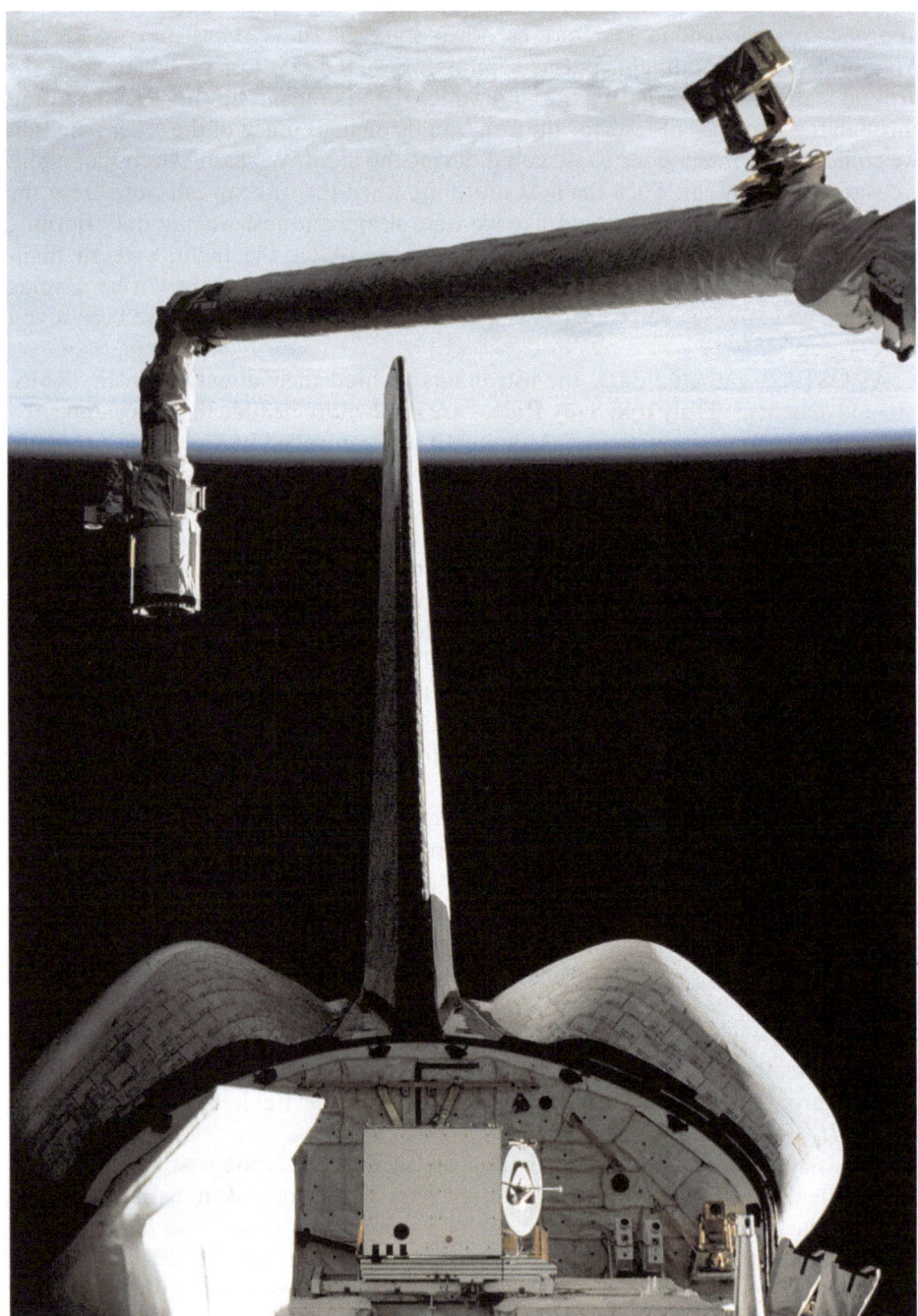

The Canadian-built Remote Manipulator System (RMS) is put through its paces by Joe Engle and Dick Truly.

containers," Engle said, "and you'd pull them out on Day One and get clean clothes out, clean socks and underwear and all. They had *modified* some jockey shorts and replaced *mine* with these Canadian jockey shorts, in which the side panels were *red* and the centre section was *white*, with a big *maple leaf* on the centre!" The Canadians may have lost out to the Stars and Stripes, but when it came to clothing a particularly vital area of Engle's anatomy with the Maple Leaf, they *definitely* got the last laugh.

Only a couple of minor problems were encountered with the mechanical arm itself. The first was a failure in the primary control mode, caused by a broken wire, which the astronauts managed to bypass using the backup electronics. Additionally, the 'elbow' camera – one of six cameras attached to the RMS – suffered a short circuit and failed near the end of the mission. Otherwise, this first demonstration of the arm in orbit was a spectacular success and it performed within expectations under a wide range of temperature extremes.

Although it would not be manoeuvred by the RMS on STS-2, the Induced Environmental Contamination Monitor was carried in the payload bay, attached to the DFI pallet. Its task was to help scientists to better understand the effect of the Shuttle on its local environment, and in particular to assess the 'cleanliness' of the payload bay, before committing sensitive telescopes and detectors to future flights. The IECM was carried on the last three OFT missions *and* as part of the first pair of Spacelab payloads and was either fixed in the bay or was picked up and moved around by the RMS. Its data from STS-2 revealed that contamination levels were within expected limits. Promisingly, it confirmed that exhaust byproducts from the main engines had *not* leached into the bay and that more than 90 percent of all particulate contaminants tended to 'boil off' into space within 36 hours of launch. Generally speaking, the environment was characterised as remarkably debris-free. It did, however, show that RCS thruster firings caused small, short-lived 'clouds' of particles to gather around the bay.

In the meantime, the OSTA-1 experiments were performing extremely well and continued to return their own bounty of scientific results. SIR, for one, was in the process of taping its eight hours of radar data, acquiring images with resolutions of just 40 m of North America, southern Asia and Europe, Australia and Oceania, North Africa and the northern portion of South America. It permitted geologists to determine surface 'roughness' and picked out faults, drainage patterns and evidence of stratification, as well as making a truly unexpected find: ancient *watercourses*, deep beneath the arid sands of Egypt, just as they might have appeared thousands of years ago. They were also, said Engle, able to gather images of the Nazi battleship, *Admiral Graf Spee*, scuttled by her commander just off Montevideo after sustaining damage from British and New Zealand cruisers in December 1939. Such data offered a tantalising preview of what might be achieved on *operational* research missions. The other instruments also more than proved their worth. As Engle and Truly took photographs through Columbia's flight deck windows, SMIRR determined the best spectral resolution necessary to identify and map rock or mineral deposits, whilst FILE evaluated novel techniques to automatically classify surface features, including water, vegetation, bare land and snow, clouds or ice, to better prioritise the timing of future Earth-resources missions. It complemented SIR and SMIRR by offering a means for them to be activated *only* when conditions were right for data collection. When perfected, this capability would enable the operation of instruments carried on satellites to be optimised, thereby reducing the amount of 'unwanted' data to be downlinked.

Another experiment, MAPS, provided the first clear indication of how severe the levels of atmospheric carbon monoxide really were. It surveyed the lower atmosphere – from the surface to an altitude of some 18 km, covering the region known as the 'troposphere', in which the 'weather system' is active – and began a series of missions which identified this uncomfortable trend. MAPS would be flown again in October 1984 and on a pair of Space Radar Laboratory mission in 1994, revealing worrying pollution levels, particularly in the tropics, caused by seasonal burning of biomass. Lastly, OCE evaluated a method for mapping the colour patterns of plankton and chlorophyll, as part of efforts to better identify schools of fish.

The final pair of OSTA-1 experiments were housed not on the pallet, but inside Columbia's crew cabin. These were the Heflex Bioengineering Test (HBT) and the Night-Day Optical Survey of Lightning (NOSL). The former investigated the effects of weightlessness and soil composition on the growth of the dwarf sunflower – *Helianthus annuus* – to better understand the relationship between height and moisture content. It was a precursor of an experiment slated for the first Spacelab mission in 1983, which would analyse the sunflowers in greater detail. The STS-2 test, unfortunately, achieved only partial success, due to the shortened length of the mission. NOSL, too, was similarly affected. It required Engle and Truly to take photographs of lightning flashes over land and water, during orbital 'daytime' and 'nighttime', in the hope that it could stimulate the development of new systems to give early warnings of particularly severe storms. The astronauts removed the NOSL hardware from a middeck locker shortly after reaching orbit, assembled it and successfully gathered a series of nighttime images *and* motion-picture sequences of six large daytime thunderstorms. In recognition of the 'lost' time, it was subsequently decided by NASA to refly both the sunflower and lightning experiment tests on Columbia's next mission, STS-3.

After a jam-packed two days in space, on 14 November, the astronauts began to prepare their ship for the return to Earth. No fewer than *twenty-nine* manoeuvres were planned during the period from Mach 24 to subsonic and this meant that Engle would become the *only* astronaut to manually fly the Shuttle throughout re-entry to landing. "The rationale behind the manoeuvres was [that] we were very anxious to see how much margin the Shuttle had in the way of stability and control authority," he related, "how much *muscle* the surfaces had at different Mach numbers and angles of attack. Also, in the event that a de-orbit had to be made on an orbit that had excessive cross range to the landing site, in order to get more cross range rather than S-turn back and forth to deplete energy, the technique was to just leave the vehicle in the bank in one direction and keep flying toward the landing site, off your straight ground track toward your landing site. You could increase that cross-range ability by actually decreasing the angle of attack. It allowed the leading edge of the wing to heat up a bit more and would cut down on the total number of missions that a [particular] Shuttle could fly, but it would allow you to get that extra performance, that extra range, to make it to the landing site.

"How much the leading edge would heat up and just how much more lift-to-drag that would give you – turning ability, cross-range ability – *was* theoretically known and had some wind-tunnel test data, but the wind tunnels are very susceptible to a lot of variables, so you really want to know for sure what you have in the way of capabilities if you ever have to use them, and that's what our purpose was. During the entry, I would pulse the vehicle in all three axes to see what the effectiveness of the surfaces were during entry and how quickly the vehicle would damp out after being disturbed. Getting that data to verify and confirm the capabilities of the vehicle was something that we wanted very much to do and, quite honestly, not everyone at NASA thought it was all that important. There was an element in the engineering community that felt that we could always fly it with the variables and the unknowns, just as they were from wind tunnel data, and always come down the chute. Then there was the other school (which I will readily admit that I was one of) that felt you just don't know when you may have a payload you weren't able to deploy, so you have maybe the [centre of mass] not in the optimum place and you can't do anything about it, and just how much manoeuvring will you be able to do with that vehicle in that condition? How much control authority is really out there on the elevons and how much crossrange do you really have if you need to come down on an orbit that is not the one that you really intended to come down on? It was something that ... there were good, healthy discussions on and ultimately the data showed that, yes, it *was* really worthwhile to get and, therefore, those manoeuvres that we did on STS-2 were programmed into the automatic flight control system, into the entry flight control system so that subsequent to that, those manoeuvres *continued* to be made and data continued to be gotten, but it was done automatically by the computer."

The Shuttle's re-entry profile as she hurtled through the atmosphere, bound for Edwards Air Force Base, was also markedly different to even these ambitious plans, thanks to the shortened mission. The RCS thrusters housed in the aft fuselage were commanded to fire over 1,000 times – in the process consuming more than 800 kg of propellant, far more than planned – because the predicted fuel-consumption rate after two days differed from pre-mission estimates, which had been based on the demands of a five-day mission. Shortly before hitting the uppermost reaches of the atmosphere, a large quantity of propellant was dumped out of the RCS unit in Columbia's nose, to permit more precise control of the ship's centre of mass during descent. A series of flight tests were also conducted, the most important of which was a 'push-over/pull-up' exercise, performed by Dick Truly. As the orbiter plummeted Earthward, he pushed the nose 'down' from a 40-degree angle of attack to 35 degrees, then lifted it to 45 degrees and returned it to the original 40 degrees. This provided additional, 'real-world' data on the vehicle's aerodynamic performance throughout the re-entry phase. It enabled them to evaluate, at 35 degrees, how much more cross-range capability it afforded the Shuttle, and at 45 degrees, how they could pull up to a higher angle of attack if the demand arose to lower the heat on the leading edges of the wings.

Joe Engle (left) and Dick Truly (right) in joyous spirits as they discuss their mission with George Abbey, then-head of Flight Operations.

Knowing that their mission would last barely two days, the astronauts minimised their sleep, in order to achieve as many of their objectives as possible. "The fact that we were up all night," Engle said later, "may not have been a good plan, in retrospect, in getting ready to do the de-orbit [burn] and landing." The problem with Columbia's water supply had not helped matters, either, when hydrogen from a burst fuel cell membrane caused their drinks to bubble. Inevitably, this forced them to belch to rid themselves of the unpleasant water ... so they decided *not* to drink any more water. As re-entry neared, they were tired and dehydrated; hardly an appropriate combination for a *manual*, hypersonic glide through the atmosphere, *twenty-nine* manual manoeuvres and an unpowered, one-chance-only 'deadstick' landing on a desert runway.

As we have seen, Joe Engle had a long and chequered past with Edwards Air Force Base and, in fact, he and Truly had spent many weekends there during their training, flying simulated Shuttle landing approaches. On one occasion, half-jokingly, the base's control tower operator told Engle to give him a call on Columbia's final approach, "and I'll *clear you*" to land. "Of course," said Engle, "that was *not* a normal thing to do, because we were talking to a capcom here at Houston, throughout the flight." However, for fun, "and just kind of an *instinctive* thing", he made the call:

"Eddy Tower, it's Columbia, rolling out on high final. I'll call the gear on the flare!"

The controller came back with split-second timing. "Roger, Columbia, you're cleared No. 1. Call your gear!" In Houston's Mission Control, this mysterious third voice came as a surprise, but the team at Edwards loved it. It was a gracious acknowledgement from Engle of their important role in the mission.

As the runway neared, the commander pointed his craft directly into a 37 km/h crosswind and Truly deployed the landing gear. Eighteen seconds later, at 1:23 pm Pacific Standard Time, the world's first 'used' manned spacecraft touched down on Runway 23, wrapping up a voyage which had lasted only seven minutes short of STS-1. Nose-wheel steering was not yet available and Engle found himself having to apply differential braking to maintain a straight course along the centreline. He would tell the post-flight debriefing that a fluctuating indicator on his instrument panel also made it difficult to maintain a constant deceleration rate. For Engle, returning to the hallowed ground of Edwards, where he had spent so much time as a test pilot, X-15 veteran and research flier, almost two decades earlier, was a moving experience. "*That's* where I felt the most comfortable," he told the NASA oral historian. "the most *at home*, going back to Edwards. At the end of the flight, when we rolled out on final approach, going into the dry lakebed, that turned out to *really* be the case. It was a demanding mission and there were a *lot* of strange things that went on during our flight, but when we got back into the landing pattern, it just felt like I was back at Edwards again ... ready to land another airplane!"

PATHFINDERS

A Space Shuttle with a difference headed into dreary, overcast Florida skies at precisely 11:00 am Eastern Standard Time on 22 March 1982. For the first time, Columbia flew into orbit attached to a rust-coloured External Tank, the result of deleting a white coat of fire-retardant latex primer, thereby saving more than $15,000 and increasing the payload capacity by 270 kg. Although this was only the third mission, and STS-1 and STS-2 had already demonstrated the reusability of the system and its viability as a scientific research platform, there was still much to prove and its powerful Congressional enemies continued to call for the Shuttle to be cancelled. In addition to STS-3, one more test flight was scheduled for July 1982, after which there would be the first 'operational' mission, also using Columbia, in September or October.

Fortunately, with the exception of a few minor technical issues – eagerly seized upon by the press, but, in reality, insignificant in terms of their effect on the mission – Columbia's third orbital voyage was a spectacular success. Not only did its pilots, Jack Lousma and Gordon Fullerton, almost *quadruple* the two-day limit of their predecessors, but they also conducted the first tests of the mettle of the RMS by hauling 'real' payloads and carried another prototype of the Spacelab pallet; this time outfitted for a series of experiments sponsored by NASA's Office of Space Science and called 'OSS-1'. This payload had originally been scheduled for the *fourth*

OFT slot, *Flight International* revealed in March 1981, but was advanced to Lousma and Fullerton's mission when the Department of Defense exercised its option to carry its own classified military payloads on STS-4.

In fact, by the time OSS-1 was launched, the office itself had assumed a new name: the Office of Space Sciences and Applications (OSTA). Original plans called for OSS-1 to be the first in a series of flights to ferry sophisticated astronomical instruments and plasma detectors into orbit, but before this was possible a better knowledge of the impact of 'outgassing', waste water dumps and thruster firings on experiments in the Shuttle's payload bay was acutely needed. Such waste products were known to deposit thin 'films' of debris, capable of causing the optics of very sensitive instruments to become degraded. One of the cornerstone missions under consideration at this time was a trio of ultraviolet telescopes for detailed observations of the Universe. The data gathered as a result of STS-3 helped to get this plan off the drawing board and – after a *long* wait – into space in December 1990. Known as 'ASTRO-1', it was one of the most complex, yet brilliant, astronomical missions even attempted. Results from the OSS-1 investigations also enabled NASA to build advanced solar and atmospheric physics investigations, which would fly aboard other Shuttle payloads in the 1980s and 1990s.

No fewer than *eight* of the nine OSS-1 experiments were attached to the Spacelab pallet and were devoted to examining the near-Earth environment and measuring levels of contamination produced by Columbia herself. One of the most intriguing was the University of Iowa's Plasma Diagnostics Package (PDP), a small cylindrical canister of electromagnetic and particle sensors to 'sniff out' the environment around the Shuttle. Its results would prove hugely important in allowing NASA to commit highly sensitive experiments to future missions. In order to support PDP, Lousma and Fullerton were extensively trained to use the RMS to lift the payload from the OSS-1 pallet and manoeuvre it to various positions in the bay. According to their pre-mission press kit, they would employ the arm not only for the PDP, but also to hoist the desk-sized Induced Environmental Contamination Monitor, previously carried on STS-2. It was in eager anticipation of one of the most demanding missions ever attempted that the fifth and sixth men to ride the Shuttle headed into orbit on that murky March day in 1982.

Lousma, by now a full colonel in the US Marine Corps, had advanced through both success and disappointment since his Skylab mission, almost a decade earlier. He had served as Deke Slayton's backup on ASTP and might have accompanied Fred Haise as pilot of the Shuttle flight to re-boost Skylab. When unexpectedly fierce solar activity brought the aging station down earlier than intended, Haise resigned from NASA in June 1979 and Lousma received 'promotion' to the commander's seat of what became the third mission of Columbia. Rookie astronaut Gordon Fullerton, previously assigned to fly with Vance Brand on STS-4, was moved up to become Lousma's pilot. For Fullerton, himself a full colonel in the US Air Force, gaining a seat on one of the Shuttle's test flights was the culmination of an aviation career which had begun long before – and would continue long after – his astronaut career.

Charles Gordon Fullerton was born in Rochester, New York, on 11 October 1936 and an interest in aviation came as naturally to him as it had to Joe Engle. However, whereas Engle had parents who were supportive, but not aviators, Fullerton had a father in the Army Air Corps, forerunner of the US Air Force, and from a very young age he could remember receiving an 'educational' instrument panel, cardboard rudder pedals and control stick – "toy is *not* the word," he told the NASA oral historian – one Christmas. Although Mr Fullerton Senior left the military after the Second World War, he *did* take his son for a flight in a small, two-seater Aeronca aircraft. As the young Fullerton matured, his interests were primarily in mathematics and science and he attended California Institute of Technology, receiving bachelor's and master's credentials in mechanical engineering in 1957 and 1958. Flying remained a driving ambition, though. As a member of the Reserve Officers Training Program, he had the opportunity, from time to time, to fly aboard the T-33 Shooting Star at Norton Air Force Base in California, after which he formally entered the Air Force and was admitted into aviation school.

Fighters were the most desirable aircraft, Fullerton recalled, and *that* was what pushed him to choose training on the F-86 Sabre over the F-100 Super Sabre; pilots of the latter, he felt, were very often shunted into flying heavy bombers later in their careers. As circumstances worked out, ironically, quite the *opposite* happened in Fullerton's case and initial instruction in the F-86 led to an assignment to become a B-47 Stratojet bomber pilot at Davis-Monthan Air Force Base in Arizona. "My *whole class*," he said, "out of F-86s were sent to bombers and transporters! In looking back, though, while it seemed like a terrible thing at the time – some of my classmates almost wanted to *slit their wrists* – I went on with it and decided to be the *best* B-47 pilot the Air Force had ... and, in the long term, it paid off." Fullerton felt that the assignment gave him both fighter experience and heavy aircraft experience and possibly proved pivotal in his selection to fly the Approach and Landing Tests aboard Enterprise, more than a decade later. Fullerton remained at Davis-Monthan for four years and his air wing was on full alert at virtually all times. "It wasn't like you were on the verge of World War Three every minute," he explained, although the Cuban Missile Crisis in October 1962 *certainly* caught everyone's attention and Fullerton was dispersed to Hill Air Force Base in Utah for a time to ensure that a strike on the home base would not catch all the aircraft.

Selection to attend the Aerospace Research Pilot School at Edwards in 1964 was followed by assignment as a bomber test pilot at Wright-Patterson Air Force Base in Ohio. At this point in his career, the space programme came knocking at his door, when he was selected as one of the pilots for the Manned Orbiting Laboratory ... a project which would later be cancelled in the summer of 1969. At 32 years old, Fullerton was one of only a handful of men to be selected by NASA, in September of that year, for astronaut training. After a spell working in a support capacity for the final four Apollo lunar missions, he began actively pursuing the Shuttle and was heavily involved in the design of its cockpit displays and controls. At length, Fullerton was paired with Fred Haise for the ALT flights, then with Vance Brand for the fourth OFT mission and finally, in the summer of 1979, with Jack Lousma for STS-3.

When the time for his first launch into space finally arrived, one of the most unexpected surprises for Fullerton was conditioning his body and brain to the fact that Columbia was now sitting *vertically* on the pad and everything had been *rotated* by 90 degrees. "When I first went to the Cape and crawled in the hatch," he said, "I was just *flabbergasted* how … it becomes an entirely different outlook. I was lost. Wait a minute. Where's upstairs? Upstairs is *this* way; so it's a huge psychological [and] physiological difference when you get on the pad and that whole part of it. You get over it, of course. You find yourself [realising] 'Wait a minute. I'm standing on an instrument panel. I'm *not* supposed to be standing on it', but that's the way it is. We knew we were going to do that. We built the switches recessed so you could stand on it, but that's a whole different thing."

Also totally different was the experience of launch itself and the ethereal feeling of looking 'down' on Earth from orbit. Although it was only his first flight, Fullerton was placed in charge of the RMS for its multitude of tasks on STS-3. His training had taken place in several different venues, including the robotic arm's home in Canada and in Houston. "I went a couple of times up to Toronto to see how [it] worked," he said. "Then we had a full-size mockup at Houston, with a 1-G-capable arm, driven by hydraulics. We had an electronic version of the arm, looking at screens in the windows and the simulator. There were a lot of tools to get the hang of working the arm. *That* was pretty cool."

"A real barn-burner," was how Lousma described Columbia's rousing launch; Fullerton was inclined to agree, although it did not go quite as intended. Firstly, it set off an hour late, after the failure of a heater on a ground-based nitrogen gas line. Next, an Auxiliary Power Unit overheated four and a half minutes into the ascent, triggering a caution and warning alarm in the cockpit and obliging the astronauts to shut it down early. This left one of Columbia's three main engines running at only 82 percent performance later in the climb to orbit. The vehicle's overall performance, though, was unaffected, and the minor scare proved more than worth it: but for one thing – space sickness. As the reader will recall, Lousma had suffered badly from this malaise a decade earlier, aboard Skylab, and *this* time, on STS-3, it showed little mercy on either himself or Fullerton. Both men took Dexedrine and Scopolamine and consumed their required amount of calories each day. Thankfully, the sickness did not impair their ability to work and post-landing medical checks would show them to both be in excellent health.

"Of course, *everybody* has their acclimation problems," Fullerton told the NASA oral historian. "That's pretty consistent through the population. It takes about 24 hours to get to feel normal, at varying levels of discomfort. Most can hang in there and do their stuff, even though they don't feel good." Clearly, *all* astronauts, upon entering a new and totally alien environment, need time to gain their 'space legs' and this had posed particular problems for Joe Engle and Dick Truly, neither of whom had flown into orbit before. "They had not had time, in the two-and-a-half-day flight [to acclimatise, when] they were cut short," Fullerton continued. "By the time *they* got on-orbit and traced down the problem and the decision was made to come back early, they were getting ready to come back, so they had *no time* other than to kind of *respond*. They had some dizziness and orientation problems on entry that Jack and I worried about a lot." In fact, with only a hint of humour, Joe Engle

The 'barn-burning' launch of STS-3.

recalled that Dick Truly had dosed himself up on anti-motion-sickness medication before the STS-2 re-entry ... and *still* felt queasy. He performed well, but it was *not* an ideal situation for pilots who were expected to fly a hypersonic vehicle halfway across the world to a pinpoint, unpowered landing with only one chance to succeed.

"One thing that we had was a 'G-suit', like they wear in the F-18 [Hornet], except that, for entry, you could pump up the suit and keep it that way," added Fullerton, "and so that helped keep your blood flow up near your head. The other thing about the motion sickness is we're not sure there's a direct correlation to flying airplanes. I know if you go up and do a *lot* of aerobatics, day after day, you get to be *much* more tolerant. Jack and I flew literally *hundreds* of aileron rolls. If I did roll after roll, I could make myself sick and I got to the point where it took *hundreds* of them to make me sick! For the first day or so [in space], I didn't ever throw up or anything; I never got disorientated, but I felt kinda fifty-fifty. You're pretty happy to just float around and relax, rather than keeping on charging ... and into the second day, this is really fun and great and you feel 100 percent. Whether the aileron rolls helped or not, I'm not sure, but it *was* relatively easy."

Luckily, it was on the second day of the STS-3 mission, when Fullerton again felt "great", that he uncradled the RMS for the first of what would turn out to be a marathon 48 hours of tests. Late on the morning of 23 March, he flexed the arm's robotic muscles, then returned it to its berth on the left-side sill of the payload bay after four hours. One problem that cropped up during these tests was the failure of the wrist-mounted television camera, which was deemed essential to enabling Fullerton to view the grapple fixtures on the IECM package and pick it up properly. These tests were vital, owing to the unit's relatively large size and 385 kg mass. It was important to demonstrate the arm's handling and manoeuvring characteristics, before the Shuttle could confidently be assigned ambitious satellite deployments and retrievals. Already, a risky recovery of the space agency's Solar Max observatory – which would involve intensive use of the arm *and* a pair of lengthy EVAs – had been provisionally booked into the Shuttle's schedule for 1984.

With the wrist camera out of action, fears grew that faults in *other* cameras might occur and fail to provide Lousma and Fullerton with acceptable views of the IECM on their monitors in the aft flight deck. This in turn might then have prevented them from successfully reberthing it onto the DFI pallet, in which case they would have been obliged to discard the valuable payload into space. Nervous NASA managers decided, therefore, to defer the IECM deployment until the fourth and final Shuttle test flight, STS-4 in July 1982. In its stead, the package was substituted by the smaller PDP, although the handling characteristics of the arm would be significantly different because this was less than half of the intended mass. In the event, the tests were satisfactory. Original plans called for Fullerton to unberth the PDP for a series of eight-hour tests on 24 and 26 March, allowing it to examine the electromagnetic and particle environment within a range of about 14 m from Columbia. In the event, the astronauts completed *three* PDP deployments and acquired more than a day and a half of additional information whilst it was attached to the OSS-1 pallet. The results provided, for the first time, insights into the strange plasma 'wake' generated as the Shuttle passed, boat-like, through the ionosphere at low orbital altitude. This wake might, it was theorised, complicate the measurements made by very sensitive scientific detectors and the STS-3 data was useful in planning and developing state-of-the-art space plasma instruments for the Spacelab-2 mission.

The PDP was also used in conjunction with several other OSS-1 experiments, including the Vehicle Charging and Potential (VCAP) study, provided by Utah State University, whose purpose was to investigate the Shuttle's electrical characteristics and effect on surrounding ionospheric plasma. The experiment consisted of a fast-pulse 'gun', which fired 100-volt bursts of electrons for durations ranging from 500 nanoseconds to several minutes. It studied the extent to which electrical charges accumulated on the orbiter's insulated surfaces and how 'return currents' could be established through a limited area of surface-conducting materials to neutralise active electron emissions. Investigators hoped that the VCAP data might provide practical experience of using electron accelerators on later missions, particularly Spacelab-1. Plans were already advanced to build a revolutionary 'tethered satellite' with Italy; a satellite which would be trawled through the upper atmospheric plasma on the end of a 20 km conducting cable. Such a tether, researchers argued, could provide a future spacecraft with a steady supply of electrical power.

Several other OSS-1 investigations were also intended as forerunners of more advanced experiments planned for later Spacelab missions. Two instruments – the US Naval Laboratory's Solar Ultraviolet and Spectral Irradiance Monitor (SUSIM) and Columbia University's Solar Flare X-ray Photometer (SFXP) – were devoted to observations of radiation emitted from the Sun, to better understand the processes responsible for their origin and their impact on our planet. To support their needs, the STS-3 flight plan called for Columbia to face her payload bay directly at the Sun for several protracted periods of time.

In fact, positioning Columbia in a series of different attitudes also satisfied another in a long list of tasks which had to be demonstrated before the vehicle could be declared fully operational. During their eight days in space, Lousma and Fullerton oriented her in four 'inertial' attitudes to subject different parts of her airframe to maximum solar heating. She spent 30 hours with her tail facing the Sun, 80 hours with her nose aimed at the Sun and 36 hours with her open payload bay facing the Sun. The men also executed a series of 'barbecue rolls' for passive thermal-conditioning. With the tail directed toward the Sun, the payload bay was exposed to its *coldest* environment. Temperatures in the bay at one point were so low that 'outgassed' condensation formed on the aft flight deck windows! With these tests complete, the radiators were stowed and latched and Columbia's port-side payload bay door was closed. In general, the doors performed nominally under intensely frigid conditions, with the exception of a minor problem when a 'latched' indication was not received for one of the aft bulkhead latches. A spell of passive thermal conditioning quickly rectified this.

The week aloft enabled Lousma and Fullerton to indulge in taking photographs of Earth. According to oceanographer Bob Stevenson, this mission had given them an opportunity to photograph a virtually cloud-free China and one of their shots almost got them into diplomatic hot water after landing. "Jack and Gordon were invited to China to speak to a huge audience," Stevenson later recalled, "and they showed a picture of a lake. It was such a beautiful picture that they had it enlarged and matted and framed and they signed it off to the Premier of China as a gift. When they got to this picture, there was *silence*. When the talk was over, there was subdued clapping and they didn't know what to think, so they turned to the US ambassador and said 'We want to give the picture to the premier.'" The ambassador looked the two men straight in the eye, grabbed the picture and told them not to do that. It turned out that Lousma and Fullerton had inadvertently photographed a top-secret Chinese nuclear base. "Jack wasn't sure he was going to come home *alive*," Stevenson recalled. Naturally, the incident demanded some good-natured ribbing. After the flight, Stevenson and his colleague, Paul Scully-Power, arranged with a Chinese friend to put some important-looking comments and signatures in Mandarin on a blown-up copy of the photograph and presented it to Lousma and Fullerton at one of the astronauts' Monday morning meetings.

The faked inscription read: 'If you damn Yankees *ever* come over China again ...'

MONSTER STORM

In March 1979, two years before Columbia made her first flight, NASA took the seemingly "illogical" step of selecting a great white blotch of compacted salt and gypsum in New Mexico's Tularosa Valley as a potential Shuttle landing strip. Although it had long been planned for the primary end-of-mission landing sites to be Edwards Air Force Base in California and, later, the Kennedy Space Center in Florida, the area known as 'White Sands' offered something more: near-perfect weather conditions, *all year round*, and an enormous runway which provided the margins of safety necessary for the Shuttle. Additionally, its vast size and colour make it easily visible and virtually unobstructed. It can even be seen from space. It lies in a mountain-ringed area called 'Alkali Flats' and was first used by Northrop Aviation in the 1940s to test a series of military target drones. Quickly, it acquired the nickname 'Northrop Strip' and, later, following a typo on a press release, '*Northrup* Strip'. The new name stuck. By 1952, it had become part of White Sands Missile Range and soon gained a pair of 10.6 km runways, crossing each other in an 'X' shape. During the first two Shuttle missions, it would have been used in the event that an emergency demanded an immediate return to Earth after the first orbit. "Should the orbiter not be in a safe orbit," noted a NASA news release, "the spacecraft would be slowed down by a de-orbit burn, high over the South Pacific, east of Samoa. The flight path would cross Baja California and the Mexican state of Sonora, until the spacecraft was in the denser atmosphere and the crew would fly it, 'dead-stick', into Northrup Strip." The only mission *ever* to do this was STS-3. However, even up to the very end of the Shuttle programme, White Sands remained on NASA's list of contingency landing sites and the space agency's pilot astronauts continued to hone their flying skills here.

Landing would have certainly played on Lousma and Fullerton's minds as their mission wore on, but they could hardly have expected it to take the turn that it did. Besides, they remained far too busy with their multitude of engineering and scientific tasks. The OSS-1 payload was performing beautifully. In addition to the instruments already mentioned, the pallet carried the Space Shuttle Induced Atmosphere (SSIA), the Thermal Control Experiment (TCE), the Contamination Monitor Package (CMP) and – a boon for the United Kingdom's space ambitions – the University of Kent's Microabrasion Foil Experiment (MFE). The latter marked the first experiment built *outside* of the United States to fly aboard the Shuttle. In effect, it was a square section of about 50 layers of tin foil. During STS-3, it 'operated' in an entirely passive mode, measuring the numbers, chemical compositions and densities of tiny micrometeoroids in low-Earth orbit. Back on Earth, the foil was removed from its place atop the cube-shaped TCE and laboratory analysis enabled scientists to determine not only the depths to which micrometeoroids had penetrated it, but also their *impact velocities*. Heavier particles punched *right through* the foil and very often left debris, whilst lighter, icier ones frequently left shallow craters.

Meanwhile, the TCE investigation had been built by NASA's Goddard Space Flight Center in Greenbelt, Maryland, and was designed to evaluate a new method of protecting scientific experiments from the swinging extremes of heat and cold – a *three-hundred-degree* range, extending from 200°C to -100°C – in low-Earth orbit. It employed a series of 'heat pipes', which maintained several 'dummy' instruments at specific temperatures under various thermal loads and radiated waste heat into space. The canister actually performed *better* in orbit than it had done in ground tests and would later be used in the electronics module on the ASTRO-1 payload. It also provided useful data for an ambitious experiment, scheduled for Spacelab-2, which sought to better comprehend the physical properties of a peculiar substance known as 'superfluid helium' – the coldest-known liquid – and demonstrate its potential as a cryogenic coolant for future astronomical telescopes. The Spacelab-2 investigation would build on the STS-3 results by evaluating the behaviour of this strange liquid and testing a prototype containment vessel. Within NASA, the OSS-1 payload as a whole came to be known as the agency's 'Pathfinder' mission. In many ways, its experiments would find applications on missions during the 1990s and beyond. The remaining pallet-mounted experiments (the induced atmosphere study and the contamination monitor) assessed the impact of plumes of waste particles, ejected from the Shuttle, on delicate instruments. SSIA measured the brightness of these particles, whilst CMP employed a pair of aluminium mirrors, coated with magnesium fluoride, commonly used in ultraviolet sensors, whose sensitivity had been carefully determined before launch.

Scientific activity was also pursued inside Columbia's cabin, with several experiments housed in middeck lockers. These were tended by the astronauts throughout the mission. One experiment utilised a new, filing-cabinet-sized container known as the Plant Growth Unit (PGU), which, in fact, was so large that an entire middeck *locker* had to be removed in order to make room for it. The unit carried all the equipment necessary – growth lamps, temperature sensors, batteries, fans, data-storage – to grow almost a hundred plants in the weightless environment. One of the key objectives was to test whether or not 'lignification' was a response to gravity or a genetically-determined process with little environmental influence. Lignin is a structured polymer, which allows plants to maintain a vertical posture, despite the

"Unreal what it had done." Columbia becomes the only orbiter to touch down at White Sands in New Mexico after a space mission . . . but would pay a price: even many years later, gypsum contamination from the desert landing site was still evident.

effects of gravity, and is thus very important for the plant's ability to grow properly. The experiments tended by Lousma and Fullerton sought to find out if lignin was reduced in the microgravity environment and if this caused plants to lose their strength.

Earlier Skylab and Salyut studies had already revealed that conditions in low-Earth orbit did indeed cause root and shoot growth to become somewhat disorientated, as well as increasing their mortality rates. However, very little was known about the *physical* changes. Understanding how plants behave and grow in the absence of gravity was – and still is – essential for long-duration missions to Mars, in which astronauts will be required to produce their own foodstuffs. On STS-3, Chinese mung bean, oat and slash pine seedlings were picked, because all three could grow in closed chambers and under relatively low lighting conditions. Additionally, pine is a 'gymnosperm', capable of synthesising large amounts of lignin, and it was thought that its growth was directly affected by gravity. Unlike the mung bean and oat seedlings, which were germinated only hours before Columbia's launch, the pine samples were germinated several days earlier.

The seedlings were used as part of three separate experiments. One looked at whether lignification was influenced by gravity or determined by genetic factors. Several of the mung beans did indeed experience orientation problems, although the oats appeared to suffer no ill effects, either on Earth or in space. The flight seedlings were all much shorter in stature than the ground control samples, but, overall, their levels of lignin production were only a few percent less than those grown on Earth. As such, although the results *did* point towards a reduction of lignin in space-grown plants, the difference was statistically insignificant. The second experiment used the mung beans and oats for chromosomal studies, revealing much fragmentation and breakage and confirming that their root cells *had* been affected by microgravity. A third study then investigated how the organisation of the gravity-sensing tissues of plants, including their root caps, were affected. Within hours of Columbia returning to Earth, the seedlings with removed from the orbiter, immersed and fixative, thin-sectioned and stained for microscopic analysis.

Returning to Earth, though, proved easier said than done. STS-3 was originally supposed to come home to Runway 23 at Edwards Air Force Base on 29 March, but unseasonal rain showers had left the Mojave Desert strip under several centimetres of water. Four days before Columbia was even *launched*, NASA called up White Sands as an additional option. Ironically, despite having 90-percent near-perfect weather, all year round, the Tularosa Valley site was battered by its worst wind and sand storm for 25 years ... on the *very day* that Lousma and Fullerton were due to land. One person supporting STS-3's landing at White Sands was rookie astronaut Charlie Bolden. "This dust storm was unlike *anything* I'd ever seen," he told the NASA oral historian. "It's gypsum and it's *very* fine, like talcum powder. *Everything* was covered with plastic; the windows were sealed, but it didn't make *any* difference. That was a hint that this was *not* a good place to land the Shuttle."

Blissfully unaware of the poor weather in New Mexico, on 29 March, the astronauts proceeded smartly through their preparations to fire Columbia's OMS engines for descent back through the atmosphere. Then, with less than half an hour to go before the burn, Mission Control advised them that conditions were too bad and they would have to try again the next day. The reason given was based upon higher-than-allowable surface gusts, but, in fact, high-altitude winds were also unacceptable. After seven busy days in space, Lousma and Fullerton were overjoyed at the chance of some extra time aloft.

"It was *terrific*," Fullerton said. "We got out of our suits and then we got something to eat and watched the world and I wouldn't have had it any other way. In fact, we flew *right over* White Sands, with the nose pointing straight down, and I could see this *monster* storm going on there. It looked like it was headed for Texas. It was *clearly* a good decision. It looked *really* bad down there." Chief astronaut John Young had already flown a number of weather reconnaissance sorties over the White Sands site and he reported that conditions were far from acceptable for Columbia to land. In fact, drifts of sand were blown into the public affairs areas of the site and gathered against the sides of the building, reaching depths of half a metre or more. "The runway got eroded by the wind," recalled Grady McCright, the White Sands facility manager, "so we had people driving a road grader that night to grade it, compact it and get it ready for landing the next morning. The wind didn't quit blowing until dark that night."

Fortunately, by dawn on 30 March, the sandstorm had subsided and the STS-3 crew repeated their preparations for re-entry. One of the Auxiliary Power Units was switched on just before the de-orbit burn and the other pair came online shortly after Columbia encountered the upper reaches of the atmosphere. Sixteen minutes into the fiery plunge back to Earth, the Kuiper Airborne Observatory successfully acquired its third set of infrared imagery, photographing the Shuttle's glowing belly and sides as she hurtled home at *fifteen times* the speed of sound. From his seat on the right-hand side of the cockpit, Fullerton was overwhelmed. Outside, the view resembled a blast furnace. "The entry was pretty cool," he said, using an entirely inappropriate choice of word, "because it was an early-morning landing, meaning that the main part of the entry is at night. We could see this *glow* from the ionisation really bright out there."

Columbia continued to fall, passing over Edwards Air Force Base and heading straight for the mountains of New Mexico. At an altitude of 3,000 m, Lousma tested the 'autoland' system, which NASA expected to use operationally on future missions, before taking manual control of his ship. After so many years preparing for this moment, there was *no way* that a seasoned pilot would allow a computer to land for him ... and Lousma, of course, was not a *test* pilot, having been selected by NASA before he had any opportunity to attend either the school at Edwards or Pax River. The other astronauts knew this, but they also knew that Lousma was, indeed, a highly experienced aviator. "Jack's a great guy," noted Fullerton. "He's *not* a test pilot, but a *very* capable guy and a great guy to work with. I couldn't have done better to have a partner to fly with."

Flying the Shuttle into White Sands and executing a manual landing was not the only worry. "The crews were very concerned that they had everything that they can at their control to make sure it goes well," explained Arnold Aldrich, "and what they worried about was not that the autoland system wouldn't fly the vehicle right, [but] if there was some glitch in the autoland system right at a critical point of approach and they had to take control back over. The [*transition*] of getting off the autoland and back onto manual control *might* be something they couldn't deal with." Charlie Bolden, who had followed the development of the Shuttle's early software during this period, was also unhappy about using autoland so close to touchdown, especially on a test flight. "We developed the procedures that we would use for autoland," he said, "how they would manually take over at the very last second and go ahead and land the vehicle. We recommended this was *not* a good thing to do. You're asking a person who's been *in space* to take over in this dynamic mode of flight and land the vehicle safely. Their *physical* gains, their *mental* gains, their *balance*; everything's *not there*. Not a smart thing to do, but the decision was made that we really need to demonstrate this, so we're only going to go to 500 feet anyway."

It was decided, before launch, to use *airspeed*, rather than *altitude*, as a cue to deploy Columbia's landing gear. The wheels began to lower about 30 m above the runway, but took longer than anticipated; they were only fully locked into position a mere *two seconds* before the Shuttle touched down. To observers, it was truly nail-biting to see the Shuttle streaking in to land at over 320 km/h with her gear still in the process of coming down. Thankfully, the landing was successful, although NASA would revert to using altitude, rather than airspeed, as a cue on future missions. On STS-3, the effect was that Columbia touched down 1.2 km past the runway threshold and required Lousma to apply differential braking to keep the vehicle close to the centreline. Although the vertical impact velocity of both the main and nose gears were within mission rules, it was still far harsher than expected and caused a gash-like scrape in one of her tyres, a cracked rotor in one of her brakes and extensive contamination by billowing gypsum dust.

So fine was the dust that it quite literally *saturated* the spacecraft and caused extensive damage which was not fully resolved in time for her next flight, STS-4 ... or, indeed, for the *rest* of Columbia's career. "I flew it several flights later," recounted Charlie Bolden, "on my first flight and when we got on orbit, there was *still* gypsum coming out of everything! They thought they had cleaned it ... but it was just *unreal* what it had done!"

STS-3's exact landing time was 9:04 am Mountain Standard Time (11:04 am in Florida and 10:04 am in Houston) on White Sands' Runway 17, setting a new Shuttle record of a little more than eight full days. As the gypsum-coated vehicle sped down the strip, with her forward gear in the process of coming down, the *nose* pitched suddenly, unexpectedly, back up into the air, again giving observers a moment or two with their hearts in their throats. Even the landing commentator's calm, professional voice seems laden with surprise as he counted down the number of feet to nose gear down and full weight on wheels: "Touchdown ... Nose Gears ... ten [feet] ... five ... four ... three ... ", when all at once the nose sprang back up. He paused for a moment, repeated himself – " ... Three ... " – and then, when the nose finally jolted its way harshly down and slapped onto the runway, "*Touchdown!*"

The effect, as Fullerton would relate, was "a kind of wheelie". The astronauts were trying to prevent what they thought might be a premature touchdown of the nose wheels. "It pointed out another flaw ... in the flight software. The gains between the stick and the elevons, that were good for flying up in the air, were *not* good when the wheels were on the ground. [Jack] kinda planted it down, but then came back on the stick and the nose came up. A lot of people thought this is a terrible thing, but we improved the software and so people don't do that anymore, but we discovered a susceptibility." In spite of concerns expressed by NASA managers at the time, STS-3 *was* still a *test flight*, as well as only the third voyage of the world's most advanced and complex spacecraft. The achievement was that the astronauts identified the problem before the Shuttle became operational and additional simulator runs by the STS-4 crew would use the 60 m altitude mark, rather than 500 km/h airspeed, as their cue to deploy the landing gear. The key point to be made with STS-3's landing is that it was *safe* and *successful*.

Charlie Bolden watched the landing attentively. "Everything seemed to be going well until just seconds before touchdown, when all of a sudden we saw the vehicle kinda *pitch up* and then kinda hard-nose touchdown. We found out that, just as Jack Lousma had trained to do, you need to move [the stick] an appreciable amount [to disengage the autoland]. We didn't realise that. The way he had trained was just to do a manual download with a stick. When he did that, he disengaged the roll axis on the Shuttle, but he *didn't* disengage the pitch axis, so the computer was still flying the pitch, although he was flying the roll. Gordon Fullerton just happened to look at the eyebrow lights and he noticed that he was still in auto in pitch. He told Jack and so Jack just kinda *pulled back* on the stick and it caused the vehicle to pitch up. Then he caught it and put it back down and he saved the vehicle."

As servicing vehicles encircled Columbia, the spacecraft sat motionless on the runway, in Fullerton's words, "surrounded by white gypsum". So severe was the damage that the flow rate from the purge units attached to the forward fuselage had to be increased and the aft compartment's vent doors were closed to prevent further contamination. On hand at White Sands to greet the astronauts, in addition to their families, were New Mexico Governor Bruce King and former Apollo astronaut Jack Schmitt, as well as the missile range's commanding officer, Major-General Alan Nord.

Within a few hours, the time-critical experiments, including the plants and samples from another unit, called the Electrophoresis Equipment Verification Test (EEVT), were removed from the middeck and returned to their respective research teams. The EEVT was a forerunner of later Shuttle experiments, in which 'electrophoresis' – a process in which electrical currents separate biological materials in fluids, without damaging the cells themselves – would permit scientists to study cell biology, aspects of immunology and conduct broader medical research. Electrophoresis on Earth is rendered difficult because heat produced by the electrical current induces buoyancy and the remixing of cells and fluids, thereby defeating the objective. During STS-3, the unit held red blood cells and live kidney cells, which Lousma and Fullerton placed into a series of glass columns for the separation process to take place. After an hour, the samples were removed and stored in a cryogenic freezer for the homeward journey. Unfortunately, at some point over the weekend of 3-4 April, as they underwent preparations to be flown back to Houston, the freezer failed and thawed the samples, totally ruining them. It was an intense disappointment for the EEVT scientists.

Columbia, meanwhile, was towed to a huge crane, known as the Stiffleg Derrick, which, with the assistance of a conventional crane, hoisted her atop the Boeing SCA for the return trip to Florida. On the afternoon of 6 April, following a refuelling stop at Barksdale Air Force Base in Louisiana, the Shuttle arrived back home on the East Coast. Had she returned to Edwards in California, as originally planned, her return to Florida was not anticipated before 9 April. This contributed in part to an earlier launch of her final test flight, STS-4, from July into the last week of June. Her early return to the Cape, though, was something of a double-edged sword for Shuttle engineers. On the one hand, it allowed NASA's management – who were, by now, becoming increasingly confident in the reusable spacecraft's abilities – to bring the STS-4 launch *forward*, but on the other hand it was alarmingly clear that she was *not* in good shape and would need extensive, and *expensive*, repairs before she could fly again.

As Charlie Bolden had remarked, Columbia was literally *saturated* with gypsum dust and, despite sterling efforts to remove it, the powdery stuff would remain in small quantities, hidden in nooks and crannies, for the rest of her spaceflying career. It was quite astonishing, therefore, that the 'flow' in the Orbiter Processing Facility to meet a late-June launch target was accomplished in just 42 days! This represented a significant reduction from the 610 days needed to prepare STS-1, the 104 days needed for STS-2 and the 68 days for STS-3. It seemed that NASA was making headway in its efforts to get the Shuttle ready to fly. Launches every two weeks, though, were still a long way off and, privately, many doubted that they would *ever* be achievable. The sheer technical challenges in preparing each payload for flight and tending to the refurbishment of the spacecraft and its thermal protection system were almost overwhelming. Even at its peak flight rate in 1985, the year before NASA lost its first orbiter and crew, only nine missions were accomplished in a single 12-month period. By 2002, the year before Columbia herself was lost in the second Shuttle disaster, that figure had *fallen* to just *six* flights.

With the benefit of hindsight, it was – and still is – naïve to suppose that a vehicle of such complexity could ever come close to becoming the spacegoing equivalent of a commercial airliner. Yet *that* was precisely the intention and it was a dream which would persist to 1986 and beyond. In fact, NASA's final Shuttle manifest, published a few weeks before the loss of Challenger, quoted plans for no fewer than 14 missions in 1986 and almost twice that number in 1987. Even without the loss of Challenger, insiders doubted that such flight rates were realistic. The plan to remove ejection seats from Columbia was also controversial. Bryan O'Connor, who would later command her in June 1991, remembered a conversation with fellow astronaut Ken Mattingly. "I told him I just didn't feel comfortable with how we could *possibly* get to a confidence level after such a short test programme," O'Connor recalled. Mattingly told him not to worry about the rhetoric from NASA Headquarters: "You and I both know that it will take a *hundred* flights before this thing will be operational!" In fact, it would take far more. Even in the summer of 2003, after 113 missions, as independent investigators and NASA itself mulled over the root cause of the second Shuttle disaster, a new realisation had dawned: the vehicle was, is and would remain, until the end of its days, purely *experimental* in nature. It would *never* attain the levels of reliability and frequency and cost-effectiveness that were demanded of it by its original designers, way back in the 1970s.

In the spring of 1982, it was a quite different story. On the outside, the Shuttle seemed to be prospering and it was with great anticipation that NASA set to work preparing Columbia for her fourth and final OFT mission. Thirty-six tiles and fragments of 14 others were found to have fallen from her nose and the aft body flap beneath her main engines, but none of those areas, thankfully, were subjected to exceptionally high temperatures during re-entry. The tiles had been closely inspected after each flight and a process of 'densification' had been ongoing since even before STS-1. This process involved the application of a silica solution to the tiles and was intended to improve their adhesion to a Nomex felt pad bonded onto Columbia's aluminium skin. Since the airframe expanded when heated, the tiles – which could not be permitted to open *any* gaps – were affixed to a 'dynamic' base. Most of the tiles in areas subjected to particularly high levels of heating, such as the belly, had been densified long before STS-1 and the remainder were completed between flights and during a year-long maintenance period which started after Columbia returned from her fifth mission.

Meanwhile, the payloads assigned to STS-4 were being brought up to speed and, by May 1982, most of them were aboard the Shuttle. The IECM unit, which Lousma and Fullerton had been unable to deploy on the previous mission, was returned to its manufacturer, inspected and then reinstalled in the payload bay. An important new commercial facility, called the Continuous Flow Electrophoresis System (CFES) – an 'operational' variant of the EEVT and destined to fly several Shuttle missions – was loaded into the middeck and the first 'real' Getaway Special (GAS) canister was mounted on Columbia's payload bay wall. These 180 kg, dustbin-sized canisters were to become frequent items on the Shuttle, with as many as 13 being flown at a time. They were part of a drive to encourage universities, government agencies, foreign nationals and even private individuals to develop their own scientific experiments for carriage into orbit. A 'practice' canister had been flown on STS-3 to demonstrate its viability; it was fitted with temperature sensors, accelerometers, acoustic detectors and pressure monitors to clarify the stresses which would be imposed on an experiment during launch, orbital flight and re-entry. The STS-3 test proved that the canisters could indeed support very sensitive experiments, including those carrying living creatures, such as insects. Gilbert Moore, a Morton Thiokol manager, paid several thousand dollars for a canister to be carried on STS-4 and donated it to Utah State University for student investigations. It was loaded with nine experiments, ranging from genetic studies of fruit flies and the growth of algae, duckweed and brine shrimp to a number of fluid physics and materials science investigations.

QUIET MISSION

One passenger aboard STS-4 which was *not* publicised as highly was the first classified Department of Defense payload. The US military, as we have seen, had long harboured a keen and active interest in the development and use of the Shuttle and an independent Air Force launch site had already been built at Vandenberg in California. In fact, as Columbia was being prepared for her fourth orbital voyage, the Air Force was midway through negotiations to buy nine missions from NASA to fly its top-secret reconnaissance and intelligence satellites and perform other experiments ... for the bargain-basement price of just $268 million. This remarkable deal of less than $30 million per flight, had been struck partly in recognition of the Air Force's support to NASA in gaining Congressional approval for the Shuttle, but also in anticipation of the plans to start flying military missions out of Vandenberg from 1986 onwards. One of the Shuttle fleet – most likely Discovery, then scheduled for completion in the summer of 1983 – was earmarked to be detailed to Vandenberg for either military payloads or polar-orbiting missions. It would represent the first time a manned spacecraft had flown from the West Coast.

Of course, history has shown that, in the wake of the Challenger disaster, it never happened. The loss of a vehicle and its crew imparted unwelcome public and political scrutiny on manned launches and eventually drew the Department of Defense back to using expendable rockets. Edward 'Pete' Aldridge was Undersecretary of the Air Force from 1981 until 1986 – and chair of the National Reconnaissance Office during the same period – and very quickly saw the problems coming. "I believe Jimmy Carter wrote a presidential directive [in 1978] that the Space Shuttle ... would meet *all* the demands of *all* the users," he told a NASA oral historian in May 2009. When Columbia finally launched on her first mission, it became abundantly clear that promises of weekly flights were a long way off, the turnaround times were far longer than anticipated and *two* of the four orbiters were too heavy for effective use by the Department of Defense. It did not bode well for the future.

"We only had two orbiters that could meet the DoD weight and size demands," Aldridge continued. "The cost was *not* one-third of the cost of an expendable [rocket]; it was more likely equal at best, and possibly much higher than that. When we first started to see this, we began to worry that we were not going to meet the demands of the Department of Defense. We had a requirement for 12 flights a year from the Shuttle. Our estimates of what we were seeing as turnaround time said 12 to 18 [flights per annum] was more likely the number. If it was going to be at the lower end, or even at 18 per year, we were going to take 12. We had a hard requirement to fly 12 flights. This meant the civil and commercial space business was not going to be as robust as we thought it was going to be. We – with national security priority – could *pre-empt* the launch of a commercial satellite in order to get a national security satellite up. It was highly uncertain whether or not any of the commercial or civil programs were going to have much viability if the orbiter flight rate was in the 12 to 18 per year. In 1983 I decided we ought to *not* terminate expendable launch vehicles. In 1978, we were to start phasing down the expendable launch vehicles, because we were no longer going to use them. All the production of the expendable launchers showed an end date that was going to be probably in the 1986 period. We were flying three different vehicles: a Delta, an Atlas and a Titan. The production lines were showing a tail-off of those. All the satellites that we had that were flying on the expendable launch vehicles … were so different that we had to redesign all the national security payloads to fit in the Shuttle bay and take the Shuttle environment. Since we were paying by the linear foot [in the payload bay] rather than the diameter, all the national security payloads got short and fat, because that's how they charged us."

That year, 1983, Aldridge approached Secretary of Defense Caspar Weinberger to advise *against* the termination of expendable rocket production. Weinberger agreed, and so too did President Ronald Reagan, and proposals were put in place to continue at least the Titan booster production for another five years. Furthermore, some payloads were reconfigured to fly aboard expendable vehicles. "NASA got *very* upset about it," added Aldridge. "Jim Beggs [who headed the agency from 1981 until 1985] saw that as a ploy of the Air Force to remove itself, ultimately, from the Shuttle." For his part, Beggs, in a March 2002 oral history, noted that he had no concerns about the Air Force opting for expendable launch vehicles – as a "backup" flight capability – but stressed that those rockets were equally as vulnerable as the Shuttle to failure. (In fact, *two* Titans would be lost in separate accidents, a few months either side of Challenger.) At length, after discussing the issue with Weinberger, a compromise would be reached between Beggs and Aldridge, in which the Department of Defense would buy up a third of each year's Shuttle missions for its purposes. The seeds of uncertainty had already been sown, however, and in the wake of Challenger, only a handful of flights would carry major national security payloads. The final dedicated mission took place in December 1992, by which time the military had long since reverted to expendable boosters.

Little of this could have been foreseen in the summer of 1982 and it was in anticipation of a flurry of classified missions that the first such payload was installed aboard Columbia for launching into orbit on STS-4. In the cryptic style of OSS-1 and OSTA-1, it was designated 'DoD-82-1', meaning that it was the first Department of Defense payload to be flown on the Shuttle in the financial year of 1982. Some details of this payload have slipped out over the years and the centrepiece would appear to have been a sensitive detector, known as the Cryogenic Infrared Radiance Instrument for Shuttle (CIRRIS), which would also fly another mission in 1991. Its objective was to test infrared sensors for an advanced 'staring-mosaic' surveillance satellite, called 'Teal Ruby', which, at the time of the Challenger loss, was scheduled to be aboard the first Vandenberg Shuttle mission in July 1986. In the wake of Challenger, and a three-year period on the ground, Teal Ruby was first shifted onto the STS-39 mission and finally cancelled. By the time STS-39 lifted off in April 1991, it carried *not* Teal Ruby … but an updated version of CIRRIS. Apparently, by the time it would have been ready to launch, the Teal Ruby technology – considered 'advanced' in the late 1970s – would be virtually obsolete, because sensor technology was advancing apace. Had it flown, the capabilities of the satellite would have included the passive detection and tracking of low-flying, air-breathing cruise missiles from space. The novelty of the sensor was that, previously, a scanning-spot had been used, which devoted only a fraction of its scan-time to any given point, whereas a 'staring' array would see every point in the sensor area on a *continuous* basis. This would enable it to see the faint signatures that a scanning-spot could not detect. In essense, Teal Ruby was an early application of then-super-secret CCD technology.

It was under an unusual shroud of secrecy that astronauts Ken Mattingly and Hank Hartsfield, commander and pilot of STS-4 and the last two-man Shuttle crew, rode the bus out to Pad 39A on 27 June 1982 for their flight into space. Mattingly had previously flown on Apollo 16 to lunar orbit, whilst Hartsfield would be embarking on his first mission. Originally, when NASA intended to fly six Orbital Flight Tests, before declaring the Shuttle fully operational, Mattingly and Hartsfield were named to fly the fifth (called the 'E') mission and when the agency reduced the number to four, they confidently expected to be reassigned to one of the operational flights instead. According to the initial plans, fellow astronauts Vance Brand and Bob Overmyer (the 'D' crew) were in line for STS-4, but their roles were switched with Mattingly and Hartsfield and it was Brand and Overmyer who wound up as the flight crew for the first operational voyage, STS-5. Many have considered the presence of a classified Department of Defense payload aboard STS-4, and the fact that Mattingly was the astronaut office's lead on DoD affairs, as the primary reason for the switch. "The idea of trying to get on an early test flight," Mattingly recalled many years later, "was what *every* pilot wants to do. Of course, *none* of us thought that it was going to take so many years before that first flight took place."

Having said this, by his own admission, Mattingly actually had little interest in the Shuttle when it was first conceived. He had, after all, flown to *the Moon* on his first flight and felt that a mission to Mars should be the *real* next step. "I believe that we needed to build a space station, first, so we could have hardware which would gather years of lifetime experience," he told the NASA oral historian in April 2002. He was drawn back to the Navy by promises from then-Secretary John Warner of prestigious roles – his own squadron, perhaps – but it soon became apparent that Vietnam was producing active-duty officers who had worked their way through the ranks on the *front line*. Mattingly knew that he could not simply walk into a squadron command and opted instead to remain with NASA. His career with the agency had taken several twists and turns of good and ill fortune and one of those episodes would later see him immortalised on the silver screen.

In August 1969, just a couple of weeks after the triumphant landing of Neil Armstrong and Buzz Aldrin on the Moon, the 33-year-old Mattingly was named as the command module pilot for Apollo 13, joining Jim Lovell and Fred Haise for what was to be the third lunar landing mission. With less than a week to go, an unfortunate exposure to German measles scrubbed Mattingly from the crew and he was replaced by his backup, Jack Swigert. He persevered on the ground, however, supporting a mission which narrowly averted disaster and in the spring of 1971 was assigned to Apollo 16.

Thomas Kenneth Mattingly II was born in Chicago, Illinois, on 17 March 1936 and received much of his schooling in Florida. Aviation, he would later recount, was in his blood: his father worked for Eastern Airlines and his earliest memories were of

Columbia begins the final Orbital Flight Test in fine fashion with a beautiful launch on 27 June 1982.

toy planes and model aircraft. "I built every model I could find," he explained, "ate every box of cereal that had a cut-out paper airplane on the back, all that sort of stuff." If the young Mattingly stayed out of trouble for long enough, his father would take him aboard an aircraft and fly to the end of a route and back. Weekends would be spent at the airport, watching planes take off and land. He studied aeronautical engineering at Auburn University, graduating in 1958 and securing membership of the Delta Tau Delta fraternity. Whilst there, as part of the Navy's Reserve Officer Training Corps, Mattingly had the chance to fly a propeller-driven attack aircraft and his choice of aeronautical engineering did little to hide an ultimate ambition to become a test pilot. Later in 1958, he enlisted in the Navy as an ensign and was vocal in his desire for flight training. When his gunnery officer, Lieutenant-Commander Glenwood Clark, asked if he *really* wanted to go through flight training, Mattingly, naturally, replied in the affirmative. "You're the dumbest ensign I've ever met," Clark announced. "Out!" More than a quarter of a century later, having flown to the Moon and commanded two Shuttle missions, Mattingly returned to active duty in the Naval Space and Warfare Systems Command ... and was introduced to his new commanding officer, one Vice-Admiral *Glenwood Clark*! Mattingly could hardly believe it. After briefing his new boss, the two senior officers sat down together.

"Admiral," Mattingly began, somewhat tentatively. "Do you remember me?"

Clark looked him straight in the eye.

"I sure do. You were the *dumbest* ensign I ever knew!" The two men laughed. They became firm friends after that.

So it was that the Navy's dumbest ensign completed initial flight instruction and received his wings in 1960. He went on to fly the A-1H Skyraider for three years and the A-3B Skywarrior for another two years, then his options included postgraduate school for a master's degree in aerospace engineering – which he was not too keen on – or test pilot school at Edwards Air Force Base. Mattingly picked the latter. In March 1965, witnessing the first manned Gemini launch planted the germ of an idea to someday become an astronaut. He knew that the Air Force was selecting candidates from the Edwards school for its MOL – so he reckoned he might get a shot at *that*. Another Navy pilot, Ed Mitchell, had the same idea and the pair applied for MOL ... and were rejected. Although they could not ordinarily apply to NASA, a sympathetic senior officer at Edwards, Lieutenant-Colonel John Prodan, gave them a chance and submitted their names to the civilian space agency. In April 1966, they were both selected. Mitchell, of course, had already completed test pilot school, but Mattingly would fly back to California in May for his graduation ceremony.

Three years later, when he was assigned to Apollo 13, Mattingly had gained a reputation as an expert in command module systems. Had he flown as planned, he would have been the first rookie command module pilot to fly solo in lunar orbit; quite an honour when one considers that he was just 34 at the time and *all* previous lunar CMPs had been picked from a pool of experienced astronauts. Eleven months would elapse after the unlucky voyage of Apollo 13 before Mattingly was finally named to a new lunar crew. However, it is interesting that he was one of only a few astronauts to have actually been given the *option* to decide which crew position he might take. He told the oral historian that in the autumn of 1970 Deke Slayton offered him two choices: he could either be the CMP of Apollo 16 or, if he wanted an actual landing, wait and become the LMP of Apollo 18.

"Well," Mattingly said, "I'd sure like to go down to the surface."

"I'll give you the choice," replied Slayton, "but I would always take a bird in the hand."

Then Mattingly was quiet for a moment. "You know, there won't be a chance to go back as a commander if I go on 16."

"It's your call. Just think about a bird in the hand."

Perhaps Slayton felt a pang of conscience for Mattingly's loss of Apollo 13 and his commitment to the subsequent rescue; certainly, such an offer was not common, particularly for a rookie astronaut. Years later, Mattingly would have mixed feelings about the choice he made. He *really* wanted to go down to the Moon – the journey would be incomplete, he felt, without actually touching its surface – but with the likelihood of impending mission cancellations it would be better to at least go *near it* than *not go at all*. Mattingly picked Apollo 16 and in March 1971 the formal NASA announcement was made. In the meantime, Apollo 18 had been formally cancelled in September 1970. During his training for Apollo 16, Mattingly worked closely with support crew member Hank Hartsfield, an Air Force test pilot chosen by NASA in September 1969. The two men developed an excellent, attentive-to-detail rapport and it is perhaps unsurprising that they wound up together as a Shuttle crew.

Henry Warren Hartsfield Jr came from Birmingham, Alabama, where he was born on 21 November 1933. "When I was a kid, all I thought about was *flying*," he told the NASA oral historian in June 2001. "The guy my dad worked for smoked Wings cigarettes; an old brand that has a little picture of an airplane on every pack. He used to bring them home to me [and] I collected those cards." Hartsfield graduated from local high school and was accepted into Auburn University, majoring in chemical engineering ... until he found out that he was "a natural disaster in the chemistry lab, always blowing things up and catching things afire". He switched to physics, which he thoroughly enjoyed, and received his degree in 1954. As a member of the Army's Reserve Officers' Training Corps of Engineers, he noticed a call from the Air Force for pilots and underwent tests at Maxwell Air Force Base in Montgomery, Alabama. Hartsfield entered the Air Force in 1955 and immediately began theoretical physics work, hoping to achieve a master's qualification. The service agreed to delay his entry into active duty until the following year, but, notwithstanding a strong letter of support from the head of his physics department, refused to grant him additional time to complete his degree. Thankfully, Hartsfield loved flying and decided to remain in the Air Force.

Primary training in Georgia and Texas was quickly followed by gunnery training at Williams Air Force Base in Arizona, at which time he flew F-86 Sabre jets, then the F-100 Super Sabre and eventually the F-105 Thunderchief. During this period, Hartsfield qualified as a fighter pilot and in 1959 decided to pursue a route into the space programme. He was selected to attend the Air Force Institute of Technology for a master's degree in the new field of 'astronautics', but left partway through his course to continue his flying career and build up enough hours to attend test pilot school. Allowed to resign from his degree without prejudice, Hartsfield found himself, within a month, flying an F-105 from the large air base at Bitburg in West Germany. The assignment lasted for three years and as soon as he passed the magic mark of 1,500 hours in high-performance jets, he applied for test pilot school..."and was selected, *right off the bat*. It was *first try*, so I was fortunate."

During his time at Edwards Air Force Base, Hartsfield met three other students, named Al Worden, Charlie Duke and Stu Roosa, all of whom had their sights firmly set on the space programme. At that time, there were two possible routes: either through NASA or the Air Force. If they applied for the latter, the Air Force would nominate them for either NASA or its own Manned Orbiting Laboratory project. Worden, Duke and Roosa applied only to NASA and were selected by the agency in the summer of 1966. Hartsfield applied to both ... and was picked for MOL. Shortly after the crushing disappointment of MOL's cancellation in June 1969, he opted for a master's degree in engineering science at the University of Tennessee and graduated in 1971. By that time, he had been selected as an astronaut candidate by NASA and throughout the next decade he worked on the development of the Shuttle, focusing specifically on its flight control systems.

By the time the first crew announcements for the OFT missions were made in March 1978, Hartsfield had retired from active duty in the Air Force and had been hired by NASA as a civilian astronaut ... yet a flight opportunity seemed vague. He knew that Young and Crippen were in line for the first mission (known as 'A') and that Engle and Truly would fly the second ('B'). By the end of 1979, Jack Lousma and Gordon Fullerton were training for the third ('C') mission, when, all at once, Mattingly and Hartsfield were called up to join them for the *same* training. "It was kinda funny," Hartsfield recalled, "because it *scared* them." Fearing that he and Fullerton were about to be replaced on the mission, Lousma made a panic call to Houston. In fact, Mattingly and Hartsfield would *back up* Lousma and Fullerton. "It was a little bit confusing as to the way the crews were announced, because *no one* really knew; it was a standard joke ... around the office, trying to figure out this crew structure and *how* it was going to work." Mattingly and Hartsfield need not have worried. On 1 March 1982, NASA formally announced their names to fly STS-4.

On the eve of launch, concerns were raised about their chances of launching safely, because on the night of 26 June a severe hailstorm damaged several of Columbia's thermal tiles and deposited water behind the covers of two reaction control thrusters. Despite the possibility that this water might freeze during ascent, it was eventually concluded that it would not pose a serious problem. Columbia lifted off precisely on time at 11:00 am Eastern Standard Time, right at the opening of a four-and-a-half-hour window for that day, and quickly arced her way into the clear Florida sky for a planned week-long mission. For Ken Mattingly, who had previously endured a bone-shaking Saturn V ascent, ten years earlier, the Shuttle, by complete contrast, offered the smoothest ride in the world. "It just *goes!*" he said later. Ascent, overall, was nominal, but a number of hydraulic sensors registered dramatic temperature drops in the nose landing gear wheel well – in one particular case, from 25°C down to -15°C – which required several hours to return to normal. Nothing like it had been seen in Columbia's three previous launches and it was attributed to rainwater penetration. More worryingly, the DFI sensors recorded evidence of moisture lurking behind a number of tiles. This reinforced a decision by NASA managers, before launch, to fly the Shuttle in a belly-to-Sun attitude in order to evaporate the water. This attitude was maintained for 12 hours, after which Mattingly and Hartsfield began preparing for the 'normal' attitudes planned for their mission. With the belly facing the Sun and the payload bay and overhead windows facing Earth, this offered them the opportunity for some magnificent views of orbital flight. "All of a sudden," Mattingly noted, "it was like you pulled the shades back on a bay window ... and the *Earth* appeared!"

The ascent, however, had gone somewhat awry for the twin Solid Rocket Boosters, both of whose parachutes failed during their descent. Both boosters sank after splashdown in the Atlantic Ocean and, although an underwater remote camera did succeed in photographing the wreckage, it was considered too expensive to recover them. The cause of the failure was subsequently traced to a new feature intended to separate the parachutes from the SRBs at the instant of splashdown to prevent them from being dragged through the water by their deflated canopies. This system had been 'active' on the first three Shuttle missions, but was partially disabled on STS-4 and frangible nuts holding one of two risers for each parachute were replaced by a pair of solid nuts which would not separate the riser. Preparations for Columbia's *next* mission, STS-5, would see the replacement of all frangible nuts with solid ones. As a result, when she lifted off in November 1982, *both* risers on *both* parachutes remained attached to the boosters until they could be removed by recovery forces. The fix was successful.

Meanwhile, despite the half-day 'soak' of Columbia in orbit, DFI data indicated that water *still* lingered in several nooks and crannies. It was feared that, during 'cold' periods, this might freeze and potentially crack the more sensitive tiles on areas which endured maximum atmospheric heating during re-entry. As a consequence, another 'solar inertial' run, this time lasting for no less than 23 hours, was added to the astronauts' timeline for 29 June. It resolved the problem. After the mission, several DFI-monitored tiles were removed and checked for traces of water; but none were found, thereby verifying this solar inertial 'conditioning' as a method for resolving future incidents.

In fact, manoeuvring the Shuttle into various attitudes to better understand its thermal behaviour had long been a key objective for STS-4. "We had been assigned to do a bunch of thermal tests," explained Mattingly, "where you put the orbiter in an attitude and get one side hot, and then one side cold, and then *spin it around*. They were collecting the data [because], after *this* flight, we wouldn't have the [DFI] to do that. It was something that had to be done, but was really not a glamorous kind of test that you can run."

Sitting in Mission Control throughout these exercises was Capcom Mike Coats, another of the new astronauts, picked by NASA in January 1978. He would later dub the thermal behaviour of the Shuttle as "the banana effect", for during orbital flight its skin flexed, just like the eponymous fruit, as one side 'expanded' and the other 'contracted'. During a practice opening and closure of Columbia's payload bay doors under extremely cold conditions, problems arose after one period in the belly-to-Sun attitude. A 'closed' indicator on one of the doors was not achieved, prompting Coats to advise the astronauts to reverse their attitude and instead warm up the orbiter's topside. This procedure, as well as ten hours of so-called 'barbecue roll' and another two hours in a tail-to-Sun orientation, thankfully rectified the difficulty and door opening and closure was satisfactorily demonstrated several times thereafter. A thorough understanding of the factors impacting the satisfactory closure of the payload bay doors was critical before the Shuttle could be declared fully operational; for if a crew was unable to close them properly, a safe return to Earth would be impossible. For the remainder of her mission, Columbia spent 67 hours in a tail-to-Sun attitude and on 3 July underwent a ten-hour barbecue roll to thermally stabilise herself for re-entry.

In effect, Mattingly and Hartsfield could insert the Shuttle into the required attitude, then leave her alone until it became necessary for another orientation change. In the meantime, they had a multitude of other tasks to undertake. Of pivotal importance among these were the final series of RMS tests and, to highlight the central importance of the Canadian-built arm on this and subsequent flights, a new console was set up in Mission Control: the 'RMS, Mechanical Systems and Upper Stages Systems Officer', known by the callsign of 'RMU'. Although the Engle and Lousma crews *had* been assisted by an RMS officer in Houston, the new RMU post-holder would also supervise the Shuttle's hydraulic systems, payload bay doors, Auxiliary Power Units and also the upper stages which would used to boost a pair of satellites into geosynchronous orbit on STS-5.

Mattingly and Hartsfield's primary work with the RMS was the deployment and manoeuvring of the Induced Environmental Contamination Monitor, which had been forestalled on STS-3, due to the failure of the arm's wrist camera. The IECM itself was somewhat different on STS-4 in that it had been fitted with an extra instrument to measure bursts from Columbia's forward RCS unit. In total, the astronauts completed *two* successful deployment and berthing exercises with the device, lasting around nine hours in total, and commented positively on the arm's "crisp and precise" handling characteristics. They also evaluated a new berthing device in the payload bay – a 'Retention Engagement Mechanism' (REM) – and acquired good images of the IECM on their monitors in Columbia's aft flight deck. Images of the bay itself, though, were restricted, to keep the Department of Defense's secret 82-1 payload carefully under wraps. It seems that, in addition to the CIRRIS sensor, there were six other instruments, all of them affixed to a cross-bay 'bridge', known as an Experiment Support Structure. These all bore their own elaborate titles and hinted at a range of atmospheric, navigational, plasma physics and attitude investigations: the Horizon Ultraviolet Program (HUP), the Autonomous Navigation and Attitude Reference Systems (ANARS), the Shuttle Effects on Plasma in Space (SEPS), the Sheath and Wake Charging (SWC), a set of passive cosmic-ray collectors and a pallet alignment model. Each apparently functioned without incident ... except, ironically, for CIRRIS itself, which was foiled by the failure of its lens cap to open. Mission controllers in Houston discussed the possibility of either knocking it free with the RMS or sending Mattingly outside on a spacewalk to manually open it, but it was eventually decided not to complicate what was, after all, an *experimental* mission. 'Payloads' on the OFT missions were a bonus and CIRRIS, despite the impatience of its sponsor, would have to await another chance to demonstrate its flair in space.

Not surprisingly, Ken Mattingly was keen to do such a spacewalk, having already performed one on his Apollo 16 mission, more than a decade earlier. Owing to the curtailed STS-2 flight, the new space suit for the Shuttle had yet to be tested and the procedures for putting it on and taking it off in weightlessness had to be fine-tuned. *That* honour fell to Mattingly and on STS-4 he *did* get the opportunity to suit himself up for a spacewalk. The *only* problem, he said later, was that "I *didn't* get to open the door!"

The military experiments had already drawn sufficient criticism to delete a planned reconnaissance camera from STS-4 and the dismal failure of CIRRIS only magnified the Air Force's embarrassment. In fact, the overly secret nature of the mission turned into something of a joke, especially for the crew. "A funny thing happened on that flight," recalled Hank Hartsfield. "On one experiment, they had a classified checklist [and] because we didn't have a secure comm link, we had the checklist divided up in sections that just had letter-names, like Bravo Charlie, Tab Charlie, Tab Bravo, that they could call out." Whenever the crew talked to Air Force controllers at the Satellite Control Facility in Sunnyvale, California, they would be told, for example, to 'Do Tab Charlie'. "We had a locker that we kept all the classified material," continued Hartsfield, "and it was *padlocked*, so once we got on orbit, we unlocked it and did what we had to do." As the end of the mission neared, Hartsfield packed away the remainder of the classified materials and secured the locker.

He told Mattingly. "I got all the classified stuff put away. It's all locked up."

"Great!" replied Mattingly.

Half an hour later, Mission Control in Houston called and told them that the military staff at Sunnyvale wanted to talk to them. The Air Force controller asked them, cryptically, to 'do Tab November'. The two astronauts looked at each other, bewildered. What the *hell* was Tab November? Neither of them could remember. The secretive nature of the military instruction and the lack of a secure communications link also meant they could *not* ask over the radio. The only option was to reopen the classified locker, dig through all the materials and find the checklist. Eventually, after much searching, Hartsfield finally found the glossary entry for Tab November.

It read: *Put everything away and secure it!*

DATES WITH DESTINY

Columbia's return home on Independence Day, 4 July 1982, was interesting for Mattingly and quite different again from his memories of coming home from his lunar mission. "Apollo had *aggressive* forces on launch and entry," he said, "whereas the Shuttle has just really soft forces and ... is just a piece of cake." Their arrival at Edwards Air Force Base was being watched closely by President Ronald Reagan and the astronauts had already been briefed by NASA Administrator Jim

Beggs and asked to think of some memorable words to mark the occasion. "We *knew* they had hyped-up the STS-4 mission, so that they wanted to make sure that we landed on the Fourth of July," Mattingly continued. "It was in no uncertain terms that we were *going* to land on the Fourth of July, *no matter what day* we took off. Even if it was *Fifth*, we were going to land on the *Fourth*! That meant, if you didn't do *any* of your test mission, that's okay, as long as you land on the Fourth ... because the President is going to be there! We thought *that* was kinda interesting."

Fortunately, Columbia's landing – the first on concrete Runway 22 – came precisely on time on Independence Day afternoon, completing a textbook seven-day flight. Unlike the experience of Lousma and Fullerton, the landing gear was deployed at an altitude of around 120 m, a full 20 seconds before touchdown, allowing all six wheels to be firmly locked into position with plenty of time to spare. The vehicle landed a couple of hundred metres past the runway threshold and Mattingly applied the brakes for 20 seconds to bring Columbia to a smooth halt. Now came his biggest challenge: *How* to welcome Reagan inside the Shuttle? He and Hartsfield had considered putting up a big notice, worded to the effect of: 'Welcome to Columbia: Thirty minutes ago, this was in space'. As circumstances transpired, Mattingly actually welcomed his commander in chief with a rather painful forehead ...

Immediately after wheelstop, he turned to Hartsfield and spoke. "I *am not* going to have somebody come up here and pull me outta this chair! I'm going to give every ounce of strength I've got and get up on my own!" Previous crews had come back to Earth, some feeling fine, others feeling nauseous and others requiring a gurney to carry them off for medical attention. *That* would not happen with the President Reagan in attendance. Mentally and physically set up to meet the chief, Mattingly pushed himself upwards out of his seat ... and smashed his head sharply on the overhead instrument panel! "Oh, did *I* have a *headache*," he recalled later. Nevertheless, the two returning space heroes composed themselves, Mattingly wiped away the few spots of blood, descended the steps and smartly saluted Reagan.

At the same time, atop a Boeing SCA, Challenger – the second space-rated orbiter – was ready to take off from Edwards on a cross-country journey to Florida, in anticipation of her maiden launch in the spring of 1983. Reagan paid tribute to both vehicles in his speech; Columbia as having cleared the Shuttle for operational service and Challenger for being a vision of the future. "Way out there at the end of the runway," he told his Edwards audience, "the Space Shuttle Challenger ... is about to start the first leg of a journey that will eventually put it into space. It's headed for Florida, *now*, and they're about to take off." And with co-ordinated precision, and careful timing, Reagan gave the ceremonial go-ahead: "Challenger, you are *free* to take off, *now*!" Without further ado, the 747 and its cumbersome rooftop passenger roared down the runway and into the clear California skies. Reagan loved it and even Mattingly's voice cracked with emotion as he spoke of his pride in the mission and in his partner, Hank Hartsfield, whom he labelled "the finest pilot". With a second Shuttle now complete, and *airborne*, the future seemed bright.

Indeed, that Fourth of July in 1982 represented a time when perhaps anything was possible. The calamity that would befall Challenger was more than three years away and in a span of less than 15 months, Columbia had flown no fewer than *four* times

Ken Mattingly (left) and Hank Hartsfield stand alongside President Ronald Reagan after their successful STS-4 mission. The president used the occasion to declare the Shuttle fully operational and to unveil America's new space policy for the 1980s.

into space, spent nearly *three weeks* off the planet and performed admirably, putting its critical systems through a multitude of tests and declaring itself operational in spectacular style. Later that same year, it would undertake its first commercial flight, STS-5, with pilots Vance Brand and Bob Overmyer joined by mission specialists Joe Allen and Bill Lenoir to launch two commercial satellites and begin the fulfilment of the promise NASA made to Congress a decade earlier. At the same time, the Soviet manned space programme was prospering and advancing in leaps and bounds. Near-permanent occupation of space was now a reality and a new space station, Salyut 7, was widely expected to surpass even the achievements of its predecessor. As Ken Mattingly and Hank Hartsfield orbited Earth, cosmonauts Anatoli Berezovoi and Valentin Lebedev were already two months into a record-breaking seven-month mission. A new round of international visitors were scheduled ... but *these* visitors would be different from Russia's Eastern Bloc cronies, flown aboard Salyut 6 as the prostrate and thankful slaves of the Soviet Union: *these* visitors would be representatives of Western-aligned nations, including France and India. By the *end* of the 1980s, other Soviets would have spent a full year in orbit and half a decade thereafter, one man – Valeri Polyakov – would push that experience to 14 months, a record which still stands to this day. More representatives from Western and other 'developed' nations, including Japan, Britain, Austria, the newly-unified 'Germany' and even the United States itself would eventually visit Salyut 7's successor, the Mir space station.

The achievements of both superpowers had advanced dramatically in a thousand ways during a pivotal decade which drew them from the remarkable co-operation of Apollo-Soyuz and once more polarised their attitudes towards one another into a far more belligerent, even warlike, stance. Only a year after the triumphant return of Columbia from STS-4, an obscure NATO exercise, codenamed 'Able Archer', brought them to what many observers have considered the closest point to nuclear conflict since the Cuban Missile Crisis. Feelings between the two superpowers had been simmering ever since the Helsinki Accords and, particularly, since the Soviet invasion of Afghanistan and America's renewal of diplomatic relations with China in 1979. On the Soviet side, fears of an imminent nuclear attack led them to implement the 'Operation Ryan' intelligence-gathering exercise, after which the United States executed a series of clandestine naval manoeuvres in the waters around Greenland, Iceland and the United Kingdom between 1981 and 1983. During this period, American bombers often flew directly towards Soviet airspace, only peeling off at the last minute, in order to test the vulnerability of their radar systems and the alertness of their air force. This did nothing to settle the paranoia of the aging Leonid Brezhnev and the head of the KGB, Yuri Andropov.

Matters escalated still further with the shooting-down of a South Korean airliner in September 1983, whilst in Soviet airspace, killing 269 passengers and crew. Ronald Reagan's Republican administration, which had entered office in January 1981, took a distinctly hawkish outlook, built military capabilities and spoke of the Soviet Union as an "evil empire" for the first time. NASA's leadership had also changed to reflect the growing importance of space commercialism and militarism: in 1981, Reagan appointed Jim Beggs, a former vice-president of General Dynamics and previously Undersecretary of Transportation, to head the agency and ex-Secretary of the Air Force Hans Mark was brought in as his deputy. At Reagan's directive, the Strategic Defense Initiative would be created, effectively taking plans for advanced weapons systems into space, and causing Andropov, who succeeded Brezhnev as General Secretary after the latter's death in November 1982, to accuse America of deliberately attempting to trigger nuclear war. At around the same time, the crew of the first classified Department of Defense Shuttle mission was announced by NASA, with Ken Mattingly in command. The deployment of Pershing II intermediate-range missiles to Western Europe, ostensibly to counter Soviet ballistic systems, greatly alarmed the new Andropov regime and hearts leapt into throats in September 1983 when an erroneous indication of a US missile launch almost triggered a retaliatory response. Six weeks later came Able Archer, a ten-day exercise involving the United States and other Western powers, including the United Kingdom and West Germany, which simulated 'conflict escalation', together with coded communication, radio silences, participation by various heads of state and a simulated 'DEFCON-1' nuclear alert. The Soviets mistakenly interpreted Able Archer as a genuine military build-up and placed their nuclear-capable air units in East Germany and Poland at full combat readiness. When the exercise ended on 11 November, Reagan remarked that he could not understand why the Soviets had perceived Able Archer as a *real* threat ...

At this tense period for both superpowers, it is quite astonishing to comprehend that, barely a decade later, the seeds of Apollo-Soyuz would steadily ripen and bear fruit and *real* co-operation in space, beyond a one-off mission, would begin ... and *endure* to this day. As late as 1988, in fact, efforts by NASA to interest their Washington masters in working with the Soviets in the manned spaceflight arena – docking a Shuttle to the Mir station, perhaps – fell on deaf ears and were met with stone-faced silence. Yet only a few years later, after the collapse of the Soviet Union, the first cosmonauts (no longer 'Soviet', but 'Russian', and bearing a distinctly different flag on the arms of their flight suits) would travel to the United States to train for Shuttle missions and the first astronauts would head for Star City to ready themselves for long-duration voyages which would eclipse those of even the Skylab record-holders. The casual observer would, of course, be perfectly correct in assuming that this rite of passage from steely-eyed enemies to spacegoing brothers was *not* fraught with problems and obstacles and many of these will be explored in the next volume of this history. Many military astronauts and cosmonauts would openly struggle to work with their former foes. One quite remarkable story, told by Mike Mullane, concerns Bill Shepherd, with whom he flew on a top-secret Department of Defense mission in December 1988. On *that* flight, Mullane and Shepherd were tasked with the deployment of a classified radar-imaging satellite to monitor the Soviet Union and even labelled their payload with the somewhat inflammatory remark: *Suck on this, you commie dogs*. Almost a dozen years after deploying that satellite, however, Shepherd would ride into orbit, shoulder to shoulder with a pair of Russian cosmonauts, to start the permanent occupation of the International Space Station. In a sense, this decade-and-a-half span, from 1982 until 1998, was a quite remarkable period, which, aside from its problems, *significantly* advanced the space dreams of both America and Russia and unified two disparate space programmes into one. In doing so, it not only laid the cornerstone for today's International Space Station, but it also served to sow the seeds for a real, and lasting, partnership in space.

Bibliography

'A Mission for Surveying and Mapping the Lunar Surface.' Brown Engineering Company, Inc., April 1966

'ATM Study Program: Final Report.' Ball Brothers Research Corporation, Boulder, Colorado, April 1966

'The Fall of the Valkyrie.' *Time*, 17 June 1966

'Manned Venus Flyby.' Bellcomm, Inc., Washington, DC, February 1967

'Summary Description of the AAP Apollo Telescope Mount.' Bellcomm, Inc., Washington, DC, April 1968

'Skylab Launch Site Changed.' NASA Manned Spacecraft Center News Release, Houston, Texas, 15 May 1970

'Wear Suits During LM Jettison.' NASA News Release, Manned Spacecraft Center, Houston, Texas, 19 July 1971.

'Skylab's ATM arriving in Houston for testing.' NASA Manned Spacecraft Center News Release, Houston, Texas, 8 September 1971

'International Rendezvous and Docking Mission.' Space Division, North American Rockwell Corporation, Downey, California, December 1971

'Skylab Flight Crews Named.' NASA Manned Spacecraft Center News Release, Houston, Texas, 18 January 1972

'Space Shuttle Decisions.' NASA Manned Spacecraft Center News Release, Houston, Texas, 15 March 1972

'NASA Negotiates Shuttle Engine Contract.' NASA Manned Spacecraft Center News Release, Houston, Texas, 4 April 1972

'Skylab Command and Service Module Systems Handbook (CSM 116-119).' Flight Control Division, Manned Spacecraft Center, Houston, Texas, 20 April 1972

'Four Companies Submit Proposals for Space Shuttle Program.' NASA Manned Spacecraft Center News Release, Houston, Texas, 12 May 1972

'News Conference on US-USSR Rendezvous and Docking Agreement.' NASA Headquarters, Washington, DC, 24 May 1972

'Skylab SMEAT Test.' NASA Manned Spacecraft Center News Release, Houston, Texas, 23 June 1972

'Skylab CSM 116 Delivered.' NASA Manned Spacecraft Center News Release, Houston, Texas, 17 July 1972

'ATM Shipment.' NASA Manned Spacecraft Center News Release, Houston, Texas, 21 September 1972

'Skylab Saturn IB Flight Manual.' George C. Marshall Space Flight Center, Huntsville, Alabama, 30 September 1972

'Contract for ASTP Hardware.' NASA Headquarters News Release, Washington, DC, 8 November 1972

'Astronaut Stafford Promoted.' NASA Manned Spacecraft Center News Release, Houston, Texas, 1 December 1972

'3-Day Skylab Simulation.' NASA Manned Spacecraft Center News Release, Houston, Texas, 15 January 1973

'ASTP Crew Named.' NASA Manned Spacecraft Center News Release, Houston, Texas, 30 January 1973

'Apollo-Soyuz Test Project Prime Crew Press Conference.' NASA Manned Spacecraft Center, Houston, Texas, 1 February 1973

'Apollo-Soyuz Test Project Press Conference.' NASA Johnson Space Center, Houston, Texas, 19 March 1973

'Paris Air Show Joint Exhibit.' NASA Manned Spacecraft Center News Release, Houston, Texas, 26 March 1973

'Skylab Launch Date Set.' NASA Johnson Space Center News Release, Houston, Texas, 4 April 1973

'Haise Moves to Orbiter Project Office.' NASA Johnson Space Center News Release, Houston, Texas, 18 April 1973

'Skylab 1 and 2 Preliminary Timeline.' NASA Johnson Space Center News Release, Houston, Texas, 19 April 1973

'Skylab Simulations Begin at MSFC.' NASA Johnson Space Center News Release, Houston, Texas, 15 May 1973

'Marshall Center Studies Skylab Sun Curtain.' NASA Johnson Space Center News Release, Houston, Texas, 16 May 1973

'SL-3 Launch Date.' NASA Johnson Space Center News Release, Houston, Texas, 1 June 1973

'Living It Up in Space.' *Time*, 25 June 1973

'Preliminary Time Line for Second Manned Skylab Mission.' NASA Johnson Space Center News Release, Houston, Texas, 12 July 1973

'NASA Board Reports on Skylab Meteoroid Shield Failure.' NASA Johnson Space Center News Release, Houston, Texas, 19 July 1973

'Second Manned Skylab Mission Planned For Up To 59 Days.' NASA Johnson Space Center News Release, Houston, Texas, 20 July 1973

'Skylab Crew to Carry Gyro Package.' NASA Johnson Space Center News Release, Houston, Texas, 23 July 1973

'Skylab's New Crisis: A Rescue Mission?' *Time*, 13 August 1973

'SL-4 Launch Readiness Date Reset.' NASA Johnson Space Center News Release, Houston, Texas, 14 August 1973

'Skylab Will Study Comet Kohoutek.' NASA Johnson Space Center News Release, Houston, Texas, 16 August 1973

'Mission Requirements: Skylab Rescue Mission.' Skylab Program Offices, NASA Johnson Space Center and Marshall Space Flight Center, 24 August 1973

'Skylab Observes Major Solar Flares.' NASA Johnson Space Center News Release, Houston, Texas, 13 September 1973

'Quiet Sun Not So Quiet.' NASA Johnson Space Center News Release, Houston, Texas, 17 September 1973

'Skylab Medical Experiments Altitude Test.' NASA Johnson Space Center, Houston, Texas, October 1973

'Operation Skylab/Barium.' NASA Johnson Space Center News Release, Houston, Texas, 20 November 1973

'Special Camera to Photograph Comet Kohoutek from Skylab.' NASA Johnson Space Center News Release, Houston, Texas, 20 November 1973

'Crew Cleared for Another Week in Space.' NASA Johnson Space Center News Release, Houston, Texas, 10 January 1974

'Skylab to be Left in Revisit Condition.' NASA Johnson Space Center News Release, Houston, Texas, 6 February 1974

'747 Selected for Space Shuttle Orbiter Ferry Flights.' NASA Johnson Space Center News Release, Houston, Texas, 17 June 1974

'Apollo-Soyuz Crews to Produce Own Solar Eclipse.' NASA Johnson Space Center News Release, Houston, Texas, 2 October 1974

'Apollo-Soyuz Experiments to Study Interstellar Helium.' NASA Johnson Space Center News Release, Houston, Texas, 15 October 1974

'ASTP Experiment Promotes Understanding of Energy Generation.' NASA Johnson Space Center News Release, Houston, Texas, 18 November 1974

'Soyuz 16 ASTP rehearsal.' *Flight International*, 12 December 1974

'Soyuz 16 clears the way for ASTP.' *Flight International*, 19 December 1974

'Rehearsal for 1975.' *Time*, 23 December 1974

'Plans for Apollo-Soyuz.' *Flight International*, 6 February 1975

'ASTP Crew Meeting with JSC Employees.' NASA Johnson Space Center News Release, Houston, Texas, 27 February 1975

'Crystals to be Grown in Space During ASTP Mission.' NASA Johnson Space Center News Release, Houston, Texas, 7 March 1975

'ASTP Experiment to Probe Ultraviolet Radiation Sources.' NASA Johnson Space Center News Release, Houston, Texas, 28 March 1975

'USA/USSR Discussions on Soyuz Launch Failure.' NASA Johnson Space Center News Release, Houston, Texas, 8 April 1975

'ASTP Experiment Planned on Ultraviolet Absorption.' NASA Johnson Space Center News Release, Houston, Texas, 24 April 1975

'Shuttle Orbiter Wing Delivery.' NASA Johnson Space Center News Release, Houston, Texas, 30 April 1975

'ASTP to Study Earth Mass Density Concentrations.' NASA Johnson Space Center News Release, Houston, Texas, 2 May 1975

'Notes from Visit to the Soviet Union, May 17-23 1975.' Office of the Administrator, NASA Headquarters, Washington, DC, 5 June 1975

'ASTP Press Kit.' NASA Headquarters, Washington, DC, 10 June 1975

'No Interference Between Soyuz 18/Salyut 4 and Apollo-Soyuz Flights, Technical Directors Agree.' NASA Johnson Space Center News Release, Houston, Texas, 29 June 1975

'ASTP: America and Russia together in space.' *Flight International*, 3 July 1975

'American/Soviet Experiments on ASTP Concern Fish.' NASA Johnson Space Center News Release, Houston, Texas, 10 July 1975

'Apollo ASTP Launched with Great Precision.' NASA Johnson Space Center News Release, Houston, Texas, 16 July 1975

'ASTP Science Briefing.' Apollo News Center, NASA Johnson Space Center News Release, Houston, Texas, 19 July 1975

'Apollo-Soyuz: Appointment in Space.' *Time*, 21 July 1975

'Successful ASTP Soyuz Recovery.' *Flight International*, 24 July 1975

'The Last Splashdown.' *Flight International*, 31 July 1975

'Apollo-Soyuz: A Dangerous Finale.' *Time*, 4 August 1975

'ASTP Crew Post-Launch Press Conference.' 9 August 1975

'Astronaut Deke Slayton to Undergo Surgery.' NASA Johnson Space Center News Release, Houston, Texas, 19 August 1975

'Space Shuttle Approach and Landing Test Crews Named.' NASA Johnson Space Center News Release, Houston, Texas, 24 February 1976

'NASA to Recruit Space Shuttle Astronauts.' NASA Johnson Space Center News Release, Houston, Texas, 8 July 1976

'Salyut 5 Awaits First Crew.' *Flight International*, 10 July 1976

'Salyut 5 Less Ambitious Than Expected?' *Flight International*, 31 July 1976

'Surprise Return by Salyut 5 Crew.' *Flight International*, 4 September 1976

'Soyuz 22 Recovered.' *Flight International*, 2 October 1976

'ALT 747 Maiden Flight.' NASA Johnson Space Center News Release, Houston, Texas, 3 December 1976

'Soyuz 24 Docks With Salyut 5.' *Flight International*, 19 February 1977

'Brezhnev's Rising Sun.' *Time*, 13 June 1977

'First Shuttle Payload to Investigate Earth Resources.' NASA Johnson Space Center News Release, Houston, Texas, 8 September 1977

'Soyuz 25 Mission Ends in Docking Failure.' *Flight International*, 22 October 1977

'Skylab Activation.' NASA Johnson Space Center News Release, Houston, Texas, 1 March 1978

'NASA Names Astronaut Crews for Early Shuttle Flights.' NASA Johnson Space Center News Release, Houston, Texas, 16 March 1978

'US and Russia Consider Shuttle/Salyut Flight.' *Flight International*, 1 April 1978

'NASA Plans Skylab Attitude Change Procedures.' NASA Johnson Space Center News Release, Houston, Texas, 18 May 1978

'Adrift in Orbit.' *Time*, 3 July 1978

'Records Galore for Salyut 6.' *Flight International*, 8 April 1978

'New Astronaut Candidates Arrive at JSC for Training.' NASA Johnson Space Center News Release, Houston, Texas, 29 June 1978

'Polish Cosmonaut Boards Salyut 6.' *Flight International*, 8 July 1978

'Skylab Reuse Study.' Martin Marietta Corporation, September 1978

'Progress 3 Leaves Salyut 6.' *Flight International*, 2 September 1978

'New Endurance Record for Salyut 6.' *Flight International*, 23 September 1978

'31 Into 29 Will Go.' *Flight International*, 23 September 1978

'Skylab Orbital Attitude to be Reversed.' NASA Johnson Space Center News Release, Houston, Texas, 31 October 1978

'Skylab Will Come Tumbling Down.' *Time*, 1 January 1979

'Skylab Flight Control Activities to be Reduced.' NASA Johnson Space Center News Release, Houston, Texas, 1 February 1979

'New Mexico Lakebed Airstrip Named as Shuttle Backup Landing Site.' NASA Johnson Space Center News Release, Houston, Texas, 1 March 1979

'Russians Plan New Soyuz-Salyut 6 Link-Up.' *Flight International*, 28 April 1979

'Skylab Maneuver Plan.' NASA Johnson Space Center News Release, Houston, Texas, 1 June 1979

'NASA Will Try to Adjust Skylab's Attitude.' NASA Johnson Space Center News Release, Houston, Texas, 15 June 1979

'Skylab's Fiery Fall.' *Time*, 16 July 1979

'Orbital Flight Test Program Extended.' NASA Johnson Space Center News Release, Houston, Texas, 27 May 1980

'Soviet Union Flies First Manned Soyuz-T Capsule.' *Flight International*, 21 June 1980

'Russia Launches First Vietnamese Cosmonaut.' *Flight International*, 2 August 1980

'STS Flight Assignment Baseline.' NASA Johnson Space Center, 23 September 1980

'Soyuz 38 Crew Safely Back on Earth.' *Flight International*, 11 October 1980

'New Crew Takes Over Salyut.' *New Scientist*, 4 December 1980

'Three-Man Soyuz Crew Return Safely to Earth.' *Flight International*, 20 December 1980

'BAe Approves Shuttle Flight of Spacelab Test-Pallets.' *Flight International*, 21 March 1981

'Soviet Union Launches 100th Astronaut.' *Flight International*, 21 March 1981

'STS-2 Crew Selected.' NASA Johnson Space Center News Release, Houston, Texas, 23 April 1981

'Soviet Crews Return Safely.' *Flight International*, 6 June 1981

'Columbia Prepares for Second Launch.' *Flight International*, 22 August 1981

'STS-3 Crew Selection.' NASA Johnson Space Center News Release, Houston, Texas, 30 November 1981

'NASA Puts Together a Third Shuttle.' *Flight International*, 23 January 1982

'NASA Names Crews for Three Missions.' NASA Johnson Space Center News Release, Houston, Texas, 1 March 1982

'Ground Freezer Failure Causes Loss of STS-3 Electrophoresis Test Samples.' NASA Johnson Space Center News Release, Houston, Texas, 6 April 1982

'Columbia Needs Only Minor Rework Before Final Orbital Test Flight.' NASA Johnson Space Center News Release, Houston, Texas, 19 April 1982

'Soviet Space Programs 1976-80: Supporting Vehicles and Launch Vehicles, Political Goals and Purposes, International Co-Operation in Space, Administration, Resource Burden, Future Outlook.' Prepared at the Request of Hon. Bob

Packwood, Chairman, Committee on Commerce, Science and Transportation, United States Senate, December 1982

'Anglo-Soviet Cosmonaut Faces Challenge.' *Flight International*, 28 June 1986

'Skylab Reactivation Mission Report.' NASA Marshall Space Flight Center Technical Memorandum, March 1980

August, Oliver, 'Genghis Khan is back in force as law names names.' *The Times*, 11 June 2004

Belew, Leland F. and Stuhlinger, Ernst (1973) *Skylab: A Guidebook*. NASA Marshall Space Flight Center, Huntsville, Alabama

Belew, Leland F. (ed.) (1977) *Skylab: Our First Space Station*. NASA Marshall Space Flight Center, Huntsville, Alabama

Brooks, Courtney G., Ertel, Ivan D. and Newkirk, Ronald W. (1977) *Skylab: A Chronology*. History Office, NASA Headquarters, Washington, DC

Burrough, Bryan (1998) *Dragonfly*. London: Fourth Estate

Clark, Phillip (1988) *The Soviet Manned Space Programme*. London: Salamander

Clark, Lenwood G., Kinar, William H., Carter, David J., Jr., and Jones, James L., Jr. (eds.) *The Long Duration Exposure Facility (LDEF) Mission 1 Experiments*. NASA Langley Research Center, Hampton, Virginia, 1984

Compton, W. David and Benson, Charles D. (1983) *Living and Working in Space: A History of Skylab*. Scientific and Technical Information Branch, NASA Headquarters, Washington, DC

Conrad, Nancy and Klausner, Howard A. (2005) *Rocketman*. New York: New American Library

Evans, Ben (2009) *Escaping the Bonds of Earth*. Chichester: Praxis

Evans, Ben (2010) *Foothold in the Heavens*. Chichester: Praxis

Hall, Rex and Shayler, David J. (2003) *Soyuz: A Universal Spacecraft*. Chichester: Praxis

Heppenheimer, T.A. (1999) *The Space Shuttle Decision*. NASA Office of Policy and Plans, NASA Headquarters, Washington, DC

Hitt, David, Garriott, Owen and Kerwin, Joe (2008) *Homesteading Space: The Skylab Story*. Lincoln & London: University of Nebraska Press

Jenkins, Dennis R. (2000) *Hypersonics Before the Shuttle*. NASA Office of Policy and Plans, NASA Headquarters, Washington, DC

Jenkins, Dennis R. (2001) *Space Shuttle: The History of the National Space Transportation System: The First 100 Missions*. Hinckley: Midland Publishing

Jenkins, Dennis R., Lander, Tony, and Miller, Jay (2003) *American X-Vehicles*. NASA Office of External Relations, NASA Headquarters, Washington, DC

Linenger, Jerry M. (2000) *Off the Planet*. New York: McGraw-Hill

Lundin, Bruce T. *et. al.* (1973) 'NASA Investigation-board Report on the Initial Flight Anomalies of Skylab 1.' NASA Headquarters, Washington, DC

Mullane, Mike (2006) *Riding Rockets*. New York: Scribner

Portree, David S.F. and Treviño, Robert C. (1997) *Walking To Olympus: An EVA Chronology*. Washington, DC: History Office, NASA Headquarters

Rosenblatt, Roger, 'Space Shuttle Columbia: Aiming High in '81.' *Time*, 12 January 1981

Scott, David and Leonov, Alexei (2004) *Two Sides of the Moon*. London: Simon & Schuster

Shayler, David J. (2002) *Apollo: The Lost and Forgotten Missions*. Chichester: Praxis

Shayler, David J. and Burgess, Colin (2007) *NASA's Scientist-Astronauts*. Chichester: Praxis

Shayler, David J. (2001) *Skylab: America's First Space Station*. Chichester: Praxis

Shepard, Alan and Slayton, Deke (1994) *Moon Shot*. London: Virgin Publishing Limited

Siemer, Hannah, 'Littering fine paid.' *The Esperance Express*, 17 April 2009

Slayton, Donald K. and Cassutt, Michael (1994) *Deke*. New York: Forge

Stafford, Thomas P. and Cassutt, Michael (2002) *We Have Capture*. Washington, DC: Smithsonian Books

Velupillai, David, 'Shuttle First Flight in Profile.' *Flight International*, 11 April 1981

Velupillai, David, 'Shuttle First Flight Debrief.' *Flight International*, 20 June 1981

White, Sarah, 'Salyut 6 cosmonauts break more space records.' *New Scientist*, 28 September 1978

Willis, Edward A., Jr., 'Manned Venus Orbiting Mission'. NASA Technical Memorandum, NASA Lewis Research Center, Cleveland, Ohio, 1967

Index